Silent Witness

SILENT WITNESS

Forensic DNA Analysis in Criminal Investigations and Humanitarian Disasters

Edited by Henry Erlich, Eric Stover, and Thomas J. White

with a Foreword by Scott Turow

UNIVERSITY PRESS

Oxford University Press is a department of the University of Oxford. It furthers
the University's objective of excellence in research, scholarship, and education
by publishing worldwide. Oxford is a registered trade mark of Oxford University
Press in the UK and certain other countries.

Published in the United States of America by Oxford University Press
198 Madison Avenue, New York, NY 10016, United States of America.

© Oxford University Press 2020

All rights reserved. No part of this publication may be reproduced, stored in
a retrieval system, or transmitted, in any form or by any means, without the
prior permission in writing of Oxford University Press, or as expressly permitted
by law, by license, or under terms agreed with the appropriate reproduction
rights organization. Inquiries concerning reproduction outside the scope of the
above should be sent to the Rights Department, Oxford University Press, at the
address above.

You must not circulate this work in any other form
and you must impose this same condition on any acquirer.

Library of Congress Cataloging-in-Publication Data
Names: Erlich, Henry A., 1943– editor. | Stover, Eric, editor. | White, Thomas J. editor.
Title: Silent witness : forensic DNA analysis in criminal investigations
 and humanitarian disasters / Henry Erlich, Eric Stover, and Thomas J.
 White, editors.
Description: New York : Oxford University Press, 2020. |
Includes bibliographical references and index. |
Identifiers: LCCN 2020022220 (print) | LCCN 2020022221 (ebook) |
ISBN 9780190909444 (hardback) | ISBN 9780190909451 (paperback) |
ISBN 9780190909475 (epub)
Subjects: LCSH: DNA fingerprinting. | Forensic genetics. | DNA—Analysis.
Classification: LCC RA1057.55 .S55 2020 (print) | LCC RA1057.55 (ebook) |
DDC 614/.1—dc23
LC record available at https://lccn.loc.gov/2020022220
LC ebook record available at https://lccn.loc.gov/2020022221

This book is dedicated to the memory of
Dr. Cristián Orrego Benavente in honor
of his many contributions to advancing science
in the service of human rights.

(Credit: Andrea Lampros)

CONTENTS

Foreword by Scott Turow ix
Acknowledgments xiii
List of Contributors xv

Introduction: Genetics for Justice 1

PART I: DNA Technology and Individual Identification
1. In the Beginning: Forensic Applications of DNA Technologies 15
   Henry Erlich
2. Exonerating the Wrongfully Convicted 34
   Justin Brooks and Desiree Moshayedi
3. Analysis of Forensic Mixtures 49
   Michael Coble, Bruce Budowle, and Henry Erlich
4. Forensic DNA Data Banks and Data Mining: Balancing Societal Interests and Public Safety 67
   Frederick R. Bieber
5. Recent Developments in Forensic DNA Technology 105
   Henry Erlich, Cassandra Calloway, and Steven B. Lee
6. Microbial Forensics: From Epidemiology to Crime Investigations 128
   Antti Sajantila and Bruce Budowle

PART II: Human Rights and Humanitarian Disasters
7. The Living Disappeared: Forensic DNA Typing and the Search for Argentina's Stolen Children 149
   Mariana Herrera Piñero, Eric Stover, Melina Tupa, and Víctor B. Penchaszadeh
8. Disappeared, Not Lost: Finding El Salvador's Missing Children 173
   Andrea Lampros, Montserrat Martínez Gómez, Cristián Orrego Benavente, and Patricia Vásquez Marías
9. Large Scale Identification of the Missing: Experiences and Perspectives of the International Commission on Missing Persons 193
   Andreas Kleiser and Thomas J. Parsons
10. Tracing Windblown Seeds: Genetic Information as a Biometric for Tracking Migrants 208
    Sara H. Katsanis

11. Preventing a Third Death: Identification of Missing Migrants at the US-Mexico Border   238
    Sara H. Katsanis and Katherine M. Spradley
12. Taking Stock: DNA Testing and Its Complex Truths   268
    Dawnie Steadman and Sarah Wagner

PART III: Challenges and Debates
13. Admissibility of DNA Evidence in Court   291
    Andrea Roth
14. Immediacy and Authority: Identification Efforts in Bosnia and Herzegovina and the World Trade Center   311
    Amy Mundorff and Sarah Wagner
15. Forensic Genetics, Ethics, Privacy, and Public Policy   329
    Thomas J. White and Steven B. Lee

Conclusion: The Future of Forensic DNA Analysis   363

Index   373

FOREWORD

SCOTT TUROW

In contrast to the many experts whose fine essays make up this book, I am a DNA foot soldier. I have practiced criminal law for decades, first as a federal prosecutor and now, for more than 30 years, as a defense lawyer. Forensic DNA has proved critical in a couple of murder cases I have handled. And because I have also been blessed to make a career as a writer of novels about crime, DNA analysis has also been pivotal to the plots of several books I've published in the last decade.

Standing back, I see my experience as illustrative of the massive impact the forensic use of DNA has had on criminal law and the substantial ripples it has caused for everyday Americans in their understanding of the world. But despite my exposure to the subject, I profess no technical expertise. In each instance I have learned what I have had to in order to do my job, but on every occasion I have found that the underlying science has advanced so quickly that what I knew before is largely irrelevant.

Like most people in the world, I first heard about the forensic application of DNA from reading about it in the newspapers, when it was being proposed as a way to exonerate the wrongfully convicted. I was nearing the end of my time as a prosecutor, and my gut reaction to the whole notion was pretty much like that of other prosecutors: it sounded like bunk. Prisoners are full of inmates who, as a result of character flaws or canny strategy, committed themselves long ago to never admitting their guilt. From a prosecutor's perspective, performing tests aimed at freeing someone already convicted is nothing more than an unwarranted do-over, a Hail Mary from the Big House for somebody who has nothing to lose by demanding to have his guilt verified scientifically. Proving someone guilty beyond a reasonable doubt is always a huge task, as it should be, much akin to rolling a boulder up a steep hill. All this talk of some witch-doctor-ish scientific voodoo that could proof test convictions sounded like it was going to turn the job of the prosecutor into the tale of Sisyphus.

But in 1990, after I had changed sides and was practicing as a criminal defense lawyer at a firm now known as Dentons, I joined the pro bono appellate team representing two men who had been convicted and sentenced to death for the kidnapping, rape, and murder of a 10-year-old girl in the Chicago suburbs several years before. After our clients were first convicted, a third man was apprehended for kidnapping, raping, and murdering another little girl in the area. Under interrogation, he'd admitted not only to *that* murder, but also to the crimes for which our clients, at the time, were on death row. Once our clients had been granted a retrial due to prosecutorial misconduct, defense lawyers, after a long struggle, won the right to do DNA testing on the sperm fragment recovered from the victim. Analysis of the alleles included the third man as the potential perpetrator of the sexual assault, excluded my client, but could not exclude my client's codefendant.

By this point in time, the case had become the object of fierce and bitter controversy. A couple of prosecutors had built their careers on the case, and the parents of the dead child continued to assert that the right men had been found guilty. But the defense insisted that our clients were convicted due only to a shocking pattern of prosecutorial misconduct in the investigation and trial of each man. (The defense allegations were far from fanciful. Four cops and three prosecutors were ultimately indicted for obstruction of justice by a special prosecutor, albeit ultimately acquitted at trial.)

As a result of all of this, the prosecutors maintained for a decade that they had made no mistakes. Their response to the initial DNA test was to theorize that my client must have abetted his codefendant in raping the victim. But the arc of the case, to paraphrase Dr. King, finally began bending toward justice when Dr. Edward Blake performed newly devised polymerase chain reaction testing on the sperm specimen. Blake was able to exclude not only my client but also his codefendant. More pertinently, the tests provided a DNA "match' to the third man, who had confessed repeatedly by then, meaning that he was among the 3 out of 10,000 white males who could have contributed the genetic evidence recovered from the victim.

The DNA testing in that case not only ultimately helped free both of the wrongfully convicted men, but it also altered much of my thinking about the way the criminal process works. As several essays in this book demonstrate, the forensic use of DNA has exposed countless errors in cases seen as dead-bang certainties by juries, judges, prosecutors, and police.

Although it took some time, prosecutors began to recognize that DNA analysis was, as the lawyers like to say, a sword as well as a shield. Carefully employed by highly qualified scientists, DNA would not only free the innocent but also convict the guilty. In fact, DNA analysis has changed the prosecution of whole classes of cases. When I began my career, sexual assault cases were the bad dream of a prosecutor's world. No matter how compelling the victim was in the office, the cases were presented in court as the word of the defendant against the word of the victim, meaning there was reasonable doubt almost by definition. The emergence of DNA analysis finally brought justice for many sexual assault victims, whose attackers could now be identified with scientific certainty.

It also needs to be said that although mistakes in the criminal process are highlighted in the press, forensic use of DNA does not always lead to freedom. In the early 2000s, I took on another postconviction murder case, after Illinois's then governor, George Ryan, had commuted the sentence of every person on death row to life imprisonment without parole. In that atmosphere, the managing partner at Dentons, Errol Stone, was concerned that the defense attorneys would be less zealous about exploring viable claims of innocence by those convicted of murder now that their lives were no longer at stake. Looking for the right case to potentially exonerate someone who'd been sentenced to life, I contacted a number of public defenders I knew, asking them to identify murder convictions that haunted them as grave miscarriages of justice.

The case we chose to take up involved a young man twice convicted of murdering his grandmother, even though the evidence against him seemed to me, as it had to his public defender, completely improbable. For example, in an effort to account for the lack of our client's footprints at the murder scene, the prosecutors had argued to the jury that he might have hopped barefoot through the snow. There had already been DNA testing of some crime scene evidence, but contamination issues had arisen. After a couple years of motions in court and behind the scenes searches, we located the original trial evidence in storage at the state police crime lab. Included was a bloody towel that had been found at the murder scene but hadn't been tested the first time. The prosecutors reluctantly agreed to new DNA tests. To my amazement and chagrin, the results were not favorable and offered no basis to question the original verdicts.

While I was dealing with DNA in the courtroom, I also found myself writing about it in my novels. The book that put me on the map as a novelist was *Presumed Innocent*. As those who have read the book or seen Alan J. Pakula's movie based on it, starring Harrison Ford, may remember, a sperm specimen recovered from the victim was pivotal to the plot. Despite the landmark success of the book, I waited 20 years before approaching a sequel, *Innocent*. When I did, I knew that the forensic use of DNA that had become common in the intervening period had created a fundamental problem. Whatever doubts there might have been about the perpetrator of the murder in *Presumed Innocent*, forensic DNA would now leave no question about who had contributed the sexual fluids recovered from the victim.

The next novel I wrote, *Identical*, concerned identical twins, one of whom had been convicted of murdering his girlfriend decades before. Again DNA was central to the plot. This time I learned, notwithstanding what I thought I knew about genetics, that experts could now distinguish the DNA of identical twins. The novel also demonstrated another issue I'd confronted in my pro bono work: when DNA proves that the men presumed to be the fathers of various characters are not. The burden of having to tell a client that his biological father is not who he thinks it is has shown me that DNA testing can provide unexpected emotional blows, even to the convicts who supposedly have nothing to lose by seeking the tests.

My most recent novel, *Testimony*, shows another pioneering use of forensic DNA analysis: identifying the victims of war crimes. It was through research for *Testimony* that I came to know Eric Stover, one of the editors of this volume. *Testimony* is a fictional account about an investigation by the International Criminal Court of the alleged massacre of 400 Romas in the course of the Bosnian War. That conflict ripped open century-old grievances and religious antagonisms between Orthodox Serbs, Catholic Croats, and Muslim Bosniaks. One of the most poignant things Eric told me was that when examining mass graves, he and other experts were well aware that genotyping would be a powerful tool for identifying the skeletal remains of victims—a process that involved extracting and comparing DNA from the teeth and bones of the victims to genetic samples provided by family members. But, he added, genotyping would not provide any insight into the ethnic origin of the deceased. From a DNA perspective, the Serbs and Croats and Bosniaks who'd been murdered came from a common gene pool and, no matter how deeply they were divided by religion, the truth was that if you went back far enough, all three ethnic groups probably had common ancestors. In effect, those who carried out the killings had literally been slaughtering their own extended families.

So that is the largest point. Forensic DNA analysis has not just illumined individual cases. It has offered, in myriad ways, new understandings of the human condition.

ACKNOWLEDGMENTS

The editors would like to express our profound thanks to the following people:

Jonathan Cobb, science editor, for his review of the manuscript and helpful criticism and advice.

Thomas Callaghan, a senior biometric scientist at the US Federal Bureau of Investigations Laboratory, for his criticism and suggestions for additional citations for several chapters.

Students at the University of California, Berkeley, who served as researchers for the book: Kavya Nambiar, Alexa Barrett, Lili Spira, Gia Park, Isabella Luong, Ava Wu, and Laure Barthelemy.

CONTRIBUTORS

EDITORS

Henry Erlich is a senior scientist at Children's Hospital Oakland Research Institute and a full adjunct professor at the University of California San Francisco Benioff Children's Hospital.

Eric Stover is faculty director and adjunct professor of law and public health at the Human Rights Center, School of Law, University of California, Berkeley.

Thomas J. White is an adviser to the Human Rights Center, School of Law, University of California, Berkeley, and was the 2012–2013 regents lecturer at the University of California, Berkeley.

AUTHORS

Cristián Orrego Benavente was a research fellow in forensic genetics at the Human Rights Center, School of Law, University of California, Berkeley. He died at his home in San Salvador on December 12, 2018.

Frederick R. Bieber is a medical geneticist at Brigham and Women's Hospital and a member of the Faculty of Medicine at Harvard University.

Justin Brooks is a professor of law at California Western School of Law and director of the California Innocence Project.

Bruce Budowle is a professor at the University of North Texas Health Science Center and director of the Center for Human Identification.

Cassandra Calloway is an assistant scientist at Children's Hospital Oakland Research Institute, where she has led a research program in forensic science and mitochondrial genetics over the last 17 years.

Michael Coble is associate professor of microbiology, immunology and genetics at the University of North Texas Health Science Center.

Montserrat Martínez Gómez is an investigator with the Asociación Pro-Búsqueda de Niñas y Niños Desaparecidos in El Salvador.

Sara H. Katsanis is a research assistant professor, Lurie Children's Hospital of Chicago, Feinberg School of Medicine, Northwestern University, Chicago.

Andreas Kleiser is director for policy and cooperation at the International Commission of Missing Persons.

Andrea Lampros is associate director of the Human Rights Center, School of Law, University of California, Berkeley.

Steven B. Lee is a professor and former director of the Forensic Science Programs in the Department of Justice Studies at San Jose State University and a professor in the International Forensic Research Institute at Florida International University.

Patricia Vásquez Marías is a geneticist at the Asociación Pro-Búsqueda de Niñas y Niños Desaparecidos and professor of molecular biology at the Universidad Dr. José Matías Delgado in El Salvador.

Desiree Moshayedi is a member of the Human Rights Investigation Lab of the Human Rights Center, School of Law, University of California, Berkeley.

Amy Mundorff is a biological anthropologist at the University of Tennessee, Knoxville, who specializes in forensic anthropology and disaster identification management.

Thomas J. Parsons is director of science and technology at the International Commission of Missing Persons.

Víctor B. Penchaszadeh is a professor of genetics and human rights at the Universidad Nacional de Tres de Febrero, Buenos Aires.

Mariana Herrera Piñero has a PhD in biology with a specialization in forensic genetics. She is the director of the Banco Nacional de Datos Genéticos in the Ministry of Science, Technology and Innovation in Argentina.

Andrea Roth is a professor of law and a faculty director of the Berkeley Center for Law and Technology at the School of Law, University of California, Berkeley.

Antti Sajantila is professor of forensic medicine at the University of Helsinki and a senior medical officer in the Finnish Institute of Health and Welfare.

Katherine M. Spradley is a professor of anthropology at Texas State University.

Dawnie Steadman is a professor of anthropology and director of the Forensic Anthropology Center at the University of Tennessee, Knoxville.

Melina Tupa is a freelance filmmaker and producer.

Scott Turow is a lawyer and novelist who has written three nonfiction and twelve fiction books, including *Testimony, Presumed Innocent*, and *The Last Trial*. He has won numerous literary awards, most notably the Silver Dagger Award of the British Crime Writers' Association.

Sarah Wagner is an associate professor of anthropology at George Washington University.

Introduction

Genetics for Justice

Since its introduction in the late 1980s, DNA analysis has revolutionized the forensic sciences. It has helped to convict the guilty, exonerate the wrongfully convicted, identify victims of mass atrocities, prosecute human traffickers, and reunite families whose members have been separated by war and repressive regimes. Yet many of the scientific, legal, societal, and ethical concepts that underpin forensic DNA analysis remain poorly understood, and their application is sometimes controversial.

Over the past 20 years, numerous accounts have been published on the use of forensic DNA analysis in the criminal justice system.[1] Our book builds on these earlier

---

1. These publications include Allan Jamieson and Scott Bader, eds., *A Guide to Forensic DNA Profiling* (Chichester, UK: John Wiley and Sons, 2016); John S. Buckleton, Jo-Anne Bright, and Duncan Taylor, eds., *Forensic DNA Evidence Interpretation* (Boca Raton, FL: CRC Press, 2016); Justine Burley and John Harris, eds., *A Companion to Genetics* (Hoboken, NJ: Wiley-Blackwell, 2004); John M. Butler, *Advanced Topics in Forensic DNA Typing: Methodology* (San Diego, CA: Elsevier Academic Press, 2012); John M. Butler, *Advanced Topics in Forensic DNA Typing: Interpretation* (San Diego, CA: Elsevier Academic Press, 2015); Peter Gill, *Misleading DNA Evidence: Reasons for Miscarriages of Justice* (London: Elsevier, 2014); Sheldon Krimsky and Tania Simoncelli, *Genetic Justice: DNA Data Banks, Criminal Investigations, and Civil Liberties* (New York: Columbia University Press, 2011); David Lazar, ed., *DNA and the Criminal Justice System: The Technology of Justice* (Cambridge, MA: MIT Press, 2004); Erin Murphy, *Inside the Cell: The Dark Side of Forensic DNA* (New York: Nation Books, 2015); National Research Council, Committee on DNA Technology in Forensic Science, *DNA Technology in Forensic Science* (Washington, DC: National Academies Press, 1992); National Research Council, Committee on DNA Technology in Forensic Science, *The Evaluation of Forensic DNA Evidence* (Washington, DC: National Academies Press, 1996); National Research Council, *Strengthening Forensic Science in the United States: A Path Forward* (Washington, DC: National Academies Press, 2009); National Academies of Sciences, Engineering, and Medicine, Committee on Strengthening Forensic Science at the National Institute of Justice, *Support for Forensic Science Research* (Washington,

volumes by examining the application of new and emerging DNA technologies in criminal investigations and efforts to identify missing persons in the aftermath of mass violence and humanitarian disasters. We also explore the scientific, legal, and psychosocial challenges that have arisen as forensic DNA analysis has spread around the globe.

Popular television programs like *CSI* and *Law and Order* have made viewers everywhere aware of forensic DNA analysis and how it can be applied in crime scene investigations. But this increased awareness, as John M. Butler suggests in *Fundamentals of Forensic DNA Typing*, tends to evoke a sense that every case can be "solved in a simple and straightforward manner" (Butler, 2010). Sometimes this is true, but often it is not. A forensic specimen collected at a crime scene—whether it is blood, saliva, skin, or semen—may contain the DNA of several different contributors, only one of whom might be a perpetrator. It might also be compromised by exposure to factors such as light, heat, moisture, or chemicals, making proper analysis difficult or impossible.

The initial cases in the United States and United Kingdom in which genetic analysis was introduced in court used different DNA technologies to analyze genetic variation and to compare the DNA pattern obtained from the crime scene evidence with that of a reference sample from a suspect or victim. The UK method was based on variation in the length of specific DNA fragments (a method known as restriction fragment length polymorphism, or RFLP), while the US method was based on what is termed the polymerase chain reaction (PCR) amplification of specific segments of DNA, which vary among individuals and are known as genetic markers. When DNA evidence generated by these two methods was introduced into US courts in the late 1980s, it was hailed by many as foolproof. The technique had been used in at least 80 cases of murder and rape in 27 states, and many scientists and lawyers had assumed the tests were infallible.

Yet some of the methods used in these early cases were later found to be unreliable or to have been applied improperly. And prosecutors and defense attorneys often misrepresented the probative value of such evidence. In August 1989, in a pretrial hearing on the admissibility of RFLP evidence in a murder case (*People*

DC: National Academies Press, 2015); R. J. Parker and Peter Vronsky, *Forensic Analysis and DNA in Criminal Investigations: Including Cold Cases Solved* (New York: RJ Parker Publishing, 2015); Jaiprakash G. Shewale and Ray H. Liu, eds., *Forensic DNA Analysis: Current Practices and Emerging Technologies* (Boca Raton, FL: CRC Press, 2014); Jay D. Aronson, *Genetic Witness: Science, Law, and Controversy in the Making of DNA Profiling* (Piscataway, NJ: Rutgers University Press, 2007); Norah N. Rudin and Keith Inman, *An Introduction to Forensic DNA Analysis* (Boca Raton, FL: CRC Press, 2001); Richard Hindmarsh and Barbara Prainsack, *Genetic Suspects: Global Governance of Forensic DNA Profiling and Databasing* (Cambridge, UK: Cambridge University Press, 2010); Dorothy C. Wert and John C. Fletcher, *Genetics and Ethics in Global Perspective* (Edinburgh: International Library of Ethics, Law, and the New Medicine; and Royal Society and Royal Society of Edinburgh, 2004); and The Royal Society, *Forensic DNA Analysis: A Primer for Courts* (London: Creative Commons, 2018).

v. Castro) before the New York Supreme Court, Justice Gerald Sheindlin declared that DNA evidence that had been presented by the prosecution could be used to *exclude* the accused person but was inadmissible to *include*, or positively identify, a person, on the grounds that demonstrating a match was not scientifically reliable given the method used. Establishing that two DNA profiles are different is straightforward, but a "match" requires both accepted criteria for determining that the profiles are the same and statistical data to estimate the probability of an adventitious match. As Justice Sheindlin put it: "The testing laboratory failed in several major respects to use the generally accepted scientific techniques and experiments for obtaining reliable results, within a reasonable degree of scientific certainty" (McFadden, 1989).

Months before the court's ruling, the Federal Bureau of Investigation's (FBI's) newly formed Scientific Working Group on DNA Analysis Methods (SWGDAM) brought together 31 scientists representing 16 forensic laboratories in the United States and Canada to develop reliable principles and methods for analyzing single-source DNA samples for criminal proceedings. As a group, the scientists developed recommendations and minimum requirements for providing both prosecutors and defense counsel with copies of all lab results, statistical probability calculations, and chain-of-custody documents. They also developed guidelines for laboratories conducting RFLP DNA analysis and later for the newer DNA technology of PCR amplification (SWGDAM, 2018; President's Council of Advisors on Science and Technology, 2016; National Research Council, 1992)

In the early 1990s, crime labs—first in the United States and the United Kingdom—advocated for the creation of DNA databases to aid in criminal investigations, particularly in those without a suspect. To be effective, a forensic DNA database requires several components. First, local and state law enforcement agencies have to provide specimens suitable for DNA analysis and store the genetic profiles in the database. Second, a common set of DNA markers must be established so that results can be compared among all samples entered into the database and the evidence profile. Finally, a database's software and computer formats have to be compatible with other databases so data can be transferred between laboratories in a secure and consistent manner (Butler, 2010). Searching the databases consisting of the genetic profiles of convicted felons (or in some states, of arrestees) for a match with the profiles of crime scene evidence has proved highly effective in identifying persons of interest in suspect-less crimes. The search parameters have recently been extended to detect "partial matches," indicating that the evidence sample might have been contributed by a relative of the individual in the database. This strategy, known as "familial searching," has achieved some dramatic successes, as in the identification and conviction of the Grim Sleeper in 2016, but it remains controversial (Piquado et al., 2019).

In recent years the application of DNA analysis to criminal investigation has taken on a new twist as law enforcement has begun searching genealogical databases to identify relatives whose profiles might match the profiles of evidence samples. Many of these databases have been created by direct-to-consumer genetic testing services for people seeking their biogeographic ancestry, long-lost relatives, or clues to hereditary diseases. More than 15 million people in the United States have purchased direct-to-consumer genetic testing kits from a range of public and private companies, including GEDmatch, Family Tree DNA, MyHeritage, Ancestry.com, and 23andMe (Hernandez, 2019). The collection of DNA profiles these companies have amassed as a result is already large enough to allow second or third cousins of 60% of Americans of European ancestry—the primary consumers of commercial genealogy company services—to be potentially identified, even if they have not submitted their own DNA (Mervosh, 2019).

Law enforcement has begun to mine these genealogical data to crack cold cases—mostly murder and rape—that have lingered unsolved in police files, sometimes for 20 years or more. This strategy, used in suspect-less crimes, drew widespread attention in 2018 after state and federal authorities used an online genealogy database to arrest a suspect in the case of the Golden State Killer. After detectives found a well-preserved DNA specimen from one of the crime scenes, they uploaded the profile generated from the evidence (that of the presumed perpetrator) to GEDmatch's public database, which linked it to several relatives of the individual in the database. By a lengthy process of elimination, the detectives were able to link the crime scene sample profile to a former police officer named Joseph James DeAngelo, who was connected to more than 50 rapes and 12 murders committed throughout California from 1976 to 1986. Since DeAngelo's arrest, genealogical sleuthing techniques applied to other cases have led to arrests in Washington State, Pennsylvania, North Carolina, and Minnesota.

These cases have raised ethical concerns about law enforcement using genetic information from people who might have allowed their DNA profiles to be searched to find information about their heritage, without knowing the results could be used to help detectives track down close or distantly related relatives as potential criminal suspects (Mervosh, 2019). While commercial genealogy databases have the potential to help law enforcement agencies generate an extraordinary number of leads, ethicists and human rights activists are calling for the creation of formal guidelines on the use of such data in criminal investigations to ensure proper privacy controls and the informed consent of those who have contributed their DNA (Kolata & Murphy, 2018). The US Department of Justice's interim policy on forensic genetic genealogical analysis and searches requires that "an investigative agency must seek informed consent from third parties before collecting reference samples that will be used for searches" (US Department of Justice, 2019).

Searching DNA databases has also contributed invaluable information in human rights and humanitarian cases. The creation of databases of the DNA extracted from remains and from relatives of the missing and presumed dead can, by matching genetic profiles, help identify persons who were killed in a police massacre; a terrorist attack; or a mass disaster, such as a plane crash, large fire, or tornado (Krimsky & Simoncelli, 2011). Such databases can also be used to identify the offspring of those missing and presumed dead. In missing person investigations, the genetic profiles of remains recovered from an individual or mass grave or a mass disaster site are compared to (1) the profiles of biological relatives in an effort to determine if there is a familial relationship (partial match) or (2) the DNA from personal items, such as a toothbrush, a comb, a razor, or even dirty laundry, belonging to the presumed victim, to make an identification (complete match). One of the first missing persons databases was established in 1989 in Argentina to offer state-of-the-art services to relatives of children who had been kidnapped by the military and given up for illegal adoption to childless military and police couples. Since then, similar missing persons databases have been established in a number of other countries, including El Salvador, Guatemala, Colombia, Peru, and the former Yugoslavia (Joyce & Stover, 1991; Wagner, 1998; Rosenblatt, 2015; Wertz & Fletcher, 2004).

A recent groundbreaking advance in DNA technology has been the development of Rapid DNA instruments, which are fully automated devices the size of a microwave oven that can produce a profile of 20 specific genetic markers from a cheek swab sample. Unlike traditional DNA testing, which can take at least two days and requires processing at a laboratory, portable Rapid DNA devices can give results within 90 minutes. Yet unlike DNA labs, Rapid DNA machines do not have rigorous protocols governing the handling of samples and at present are intended for analyzing reference samples collected from an individual rather than crime scene evidence. Local and state law enforcement agencies may nevertheless be tempted to use Rapid DNA devices to analyze potential evidence, such as a discarded soda can or cigarette butt found at a crime scene. However, such objects could have a mixture of DNA from different sources or be contaminated by their surroundings, complications that these devices are not equipped to resolve (Kofman, 2016).

In December 2015, then FBI director James Comey testified before the US Senate in favor of a bill introduced by then senator Orrin Hatch (R-UT) called the Rapid DNA Act. Rapid DNA devices, Comey claimed,

> would help us change the world in a very, very exciting way. [They will] allow us, in booking stations around the country, if someone's arrested, to know instantly—or near instantly—whether that person is the rapist who's been on the loose in a particular community before they're released on bail and get away or to clear somebody, to show they are not the person. (Kofman, 2016)

Adopted in 2017, the law allows profiles collected by Rapid DNA to be entered into the FBI's national database, known as the Combined DNA Index System, or CODIS. By early 2018 the FBI was working with five states—Arizona, California, Florida, Louisiana, and Texas—to use Rapid DNA devices to upload genetic profiles to CODIS. While the FBI will not allow the submission of unknown crime scene DNA from Rapid DNA machines to the CODIS database, concerns remain that some local law enforcement agencies will use the devices beyond their intended purposes, which in turn could result in false leads and potentially wrongful convictions (Jackman, 2018).

Properly using DNA information in the criminal justice system and in the aftermath of mass violence or humanitarian disasters thus remains a regulatory and governance challenge that will involve close monitoring by the scientific and legal communities, civil society, the press, and congressional bodies.

ORGANIZATION OF THE BOOK

Silent Witness is divided into three parts. Part I examines the development of forensic DNA technologies since the mid-1980s and how they have been applied in criminal investigations. Part II describes the ways in which DNA analysis has been used in a range of human rights and humanitarian settings, from El Salvador to the western Balkans to the US-Mexico border, and part III explores social, ethical, and legal issues surrounding DNA collection and analysis that have emerged since the 1990s.

> We define these general and frequently used terms as follows: *DNA analysis* refers to the step-by-step process used to generate a genetic profile from crime scene evidence. *DNA fingerprinting* was first introduced by Alec Jeffreys in 1985 to refer to the complex DNA patterns that reflect variation in the variable number tandem repeat (VNTR) loci (regions of the genome that have a variable number of a repeated DNA sequence) and their potential use for individual identification (Jobling, 2013). This method has not been used for the past 25 years; all current DNA analysis is based on PCR amplification. *Genetic profile* is a general term referring to a pattern of genetic variation, or genotype, derived from the evidence. The genetic markers used could be based on either length or sequence polymorphisms. The current standard is the analysis of the short tandem repeat (STR) loci, whose length variation is determined by capillary electrophoresis, a method of separating molecules based on their size, which produces an "electropherogram" pattern. *Evidence* is used in this book in two different ways. A crime scene specimen can be thought of as evidence. At the same time, the DNA profile obtained through analysis of that sample can also be thought of as evidence, introduced in court proceedings. Similarly, we use the term *match* in two different ways: a "complete match" and a "familial match." In comparing the genetic profile of an evidence sample and the profile for a reference sample (e.g., the suspect), a match means the two profiles are indistinguishable. In the search to identify relatives of a missing person, a partial or familial match means the profile of the remains of that missing person and that of the putative family member are sufficiently similar that they indicate a familial relationship.

PART I: DNA TECHNOLOGY AND INDIVIDUAL IDENTIFICATION

Chapter 1 provides a historical overview of the use of DNA analysis in determining individual identification in criminal cases. Authored by Henry Erlich, the chapter discusses the methods of DNA analysis used in the mid-1980s in the first court cases in which DNA evidence was introduced in the United States and United Kingdom. It also explains the concepts of inclusionary and exclusionary results and discusses how an inclusionary result can be interpreted in statistical terms for presentation to the jury, as well as the contentious history of DNA admissibility in the courtroom.

The chapter closes by reviewing the evolution of PCR-based DNA assays testing both sequence (e.g., mitochondrial DNA) and length polymorphism, while emphasizing the value of objective and scientifically validated data that can help convict the guilty, protect the innocent, and exonerate the wrongfully convicted.

Chapter 2, coauthored by Justin Brooks and Desiree Moshayedi, examines the critical role DNA analysis has played in exonerating the wrongfully convicted. Since the first DNA exoneration in 1988, of Gary Dotson, who was falsely convicted of rape in Illinois, hundreds of people have been exonerated through DNA analysis, including many who were on death row. Brooks and Moshayedi tell the story of the establishment in 1992 of the Innocence Project at the Cardozo School of Law at Yeshiva University in New York and its efforts to exonerate wrongly convicted people through the use of DNA analysis and to reform the criminal justice system to prevent future injustice.

Chapter 3, authored by Michael Coble, Bruce Budowle, and Henry Erlich, discusses the problem posed by forensic DNA specimens with multiple contributors. These mixed samples from two or more individuals are common in forensic cases, and appropriately comparing the profile of a suspect with the complex profile of the mixture is one of the most difficult problems in forensic DNA analysis. Mixtures with limited amounts of DNA create particular challenges for interpretation because of the possibility of missing data. The authors review specific cases involving the interpretation of mixtures and current approaches to teasing apart contributions to the mixtures, including use of the recently developed software for probability-based statistical analysis of mixed samples and the potential for next generation sequencing (NGS). They also examine several statistical approaches to presenting the results of mixture analysis, such as the combined probability of inclusion and the likelihood ratio.

In chapter 4, Frederick R. Bieber examines the use of DNA databases for the investigation of crimes for which there is no suspect, but DNA evidence is available from the crime scene. A database is essentially a collection of computer files containing entries of DNA profiles that can be searched to look for potential matches. Initially these searches involved databases of convicted offenders, but more recently some states have been collecting and searching databases of people recently arrested for a crime. The "cold hit" matches that result from these searches have led to scores of convictions and, in some cases, exoneration of wrongfully convicted persons by identifying the true perpetrator. Bieber discusses 'familial searching" of databases, an investigative tool used to identify "persons of interest" who are relatives of individuals in the database. He also examines the ethical issues involved when law enforcement searches consumer genealogy databases like Family Tree DNA, GEDmatch, and MyHeritage.

Chapter 5, by Henry Erlich, Cassandra Calloway, and Steven Lee, examines recent developments in forensic DNA technology, including NGS and Rapid DNA technology. NGS, also known as massively parallel sequencing (MPS), is particularly well-suited to rapid quantitative analysis of mixtures such as those discussed in chapter 3. NGS also offers the opportunity to analyze different categories of genetic markers (e.g., STRs, SNPs or single nucleotide polymorphisms, mitochondrial DNA) on the same platform. Chapter 5 also describes the development and implementation of Rapid DNA technology and the commercial instruments that automate the process of obtaining an STR profile from a reference buccal swab in a few hours, as well as analysis of genetic markers for physical appearance and their potential use in developing investigational leads in criminal cases.

Part I closes with an examination of the genetic analysis of microbial DNA, derived from the nonhuman organisms that inhabit our bodies and constitute our microbiomes. Coauthored by Antti Sajantilla and Bruce Budowle, chapter 6 reviews the recent research on DNA sequencing of microorganisms for forensic purposes. Geneticists have found that humans are born with approximately 20,000 protein-coding genes and quickly gain over a million additional genes via their human microbiomes. Because the mix of species and strains in a person's microbiome is highly individual and varies by gender, age, and lifestyle, such molecular signatures could be used to place someone at a crime scene. However, as Sajantilla and Budowle caution, the field is still in its infancy, and further research is necessary before the technique can be applied in criminal investigations.

PART II: HUMAN RIGHTS AND HUMANITARIAN DISASTERS

Part II takes us out of the laboratory and into the field by describing how forensic DNA analysis has been—or could potentially be—applied to investigate terrorist attacks, violations of international human rights and humanitarian law, and humanitarian disasters. Chapter 7, authored by Mariana Herrera Piñero, Eric Stover, Melina Tupa, and Víctor B. Penchaszadeh, follows the Abuelas de Plaza de Mayo (Grandmothers of the Plaza de Mayo) in their search for as many as 500 children who were kidnapped by the government or born in custody during the military dictatorship that ruled Argentina from 1976 to 1983. In most cases, the security forces killed the parents and gave the newborns and young children—referred to by their abductors as *botín de guerra*, or war booty—to childless military and police couples and others favored by the regime. Using falsified documents, the "adoptive families" would then register the children as their own biological sons and daughters. Over the past twenty-five years, Argentine geneticists have used DNA analysis to identify 114 children (now young adults), most of whom have been reunited with their biological families. The authors describe the barriers these scientists confronted in explaining this new form of scientific evidence to the courts, as well the complex and often painful process these young people have faced as they learned decades later who their true biological parents were and the role their "adoptive parents" may have played in the deaths of their original parents.

In chapter 8, Andrea Lampros, Cristián Orrego Benavente, Monserrat Martínez Gómez, and Patricia Vásquez Marías tell the story of La Asociación Pro-Búsqueda de Niñas y Niños Desaparecidos de El Salvador, or Pro-Búsqueda (the Pro-Search Association of Missing Children of El Salvador), an organization established in 1994 by the Catholic priest Jon de Cortina and families of children who went missing during El Salvador's civil war (1980–1992). Many, if not most, of the children were given up for illegal adoption in Central America, the United States, and Europe. Since the early 1990s, Pro-Búsqueda, which unlike the Grandmothers of the Plaza de Mayo has received no government support, has used DNA analysis to reunite more than 400 Salvadorian children with their biological families. In its early years, Pro-Búsqueda had to rely on geneticists in California, who volunteered their time to perform the analysis of family reference samples. Later the organization hired Salvadorian geneticist Patricia Vásquez, who established an in-house database of forensic genetic profiles of family references and the missing children, who are now young adults. In addition to confirming biological kinship, the database has also provided investigatory leads where no other clues were available.

In chapter 9, Andreas Kleiser and Thomas J. Parsons describe how the International Commission of Missing Persons (ICMP) developed a "DNA-led" approach to identify thousands of missing persons in the aftermath of the wars in the Western Balkans of the early 1990s. Working with local and foreign geneticists, the ICMP developed a database of STR profiles from recovered human remains and from families of missing persons. The ICMP later applied this same approach to large-scale victim identification efforts after humanitarian disasters in Asia, the Americas, and Eastern Europe. In order to obtain family reference DNA profiles, the ICMP and local family organizations launched a massive public outreach campaign in the region and in Europe and North America, calling on relatives of missing persons to provide reference samples. Although the ICMP's efforts have been largely humanitarian, geneticists at the organization have provided expert testimony before the International Criminal Tribunal for the former Yugoslavia. At the time of writing, the ICMP database holds more than 101,189 family reference DNA profiles and has issued DNA match reports on 20,034 individuals.

Chapter 10, authored by Sara H. Katsanis, examines the inherent privacy and societal issues of using genetic information as a biometric for human identification. Policymakers, especially in the United States, are turning to genetic information as a biometric for identifying migrants, tracking refugee claims, and screening for human trafficking. Because many migrants travel without documentary proof of identity, genetic information can be useful for establishing individual identity and verifying family relationships. Since 2009 the United States has collected DNA from immigrant detainees for inclusion in its criminal database, in large part to detect repeat border crossers and immigrants who commit crimes in the United States, including cross-border labor and sex trafficking. In the future, US immigration courts might even consider DNA testing for ancestral origin to verify refugees' ethnicity claims. Establishing broad DNA databases (as proposed by at least one country) could potentially benefit both efforts to identify disaster victims and terrorism investigations. But such expanded uses of genetic information beyond traditional criminal investigations could also be employed to stigmatize individuals or entire populations. Moreover, the "geneticization" of families and individuals undermines the social constructs that underlie human relations and self-identity and could lead to discrimination against nontraditional families or the unintentional revelation of family secrets.

In chapter 11, Sara H. Katsanis and M. Katherine M. Spradley examine the logistical challenges of identifying—and the ethical considerations of failing to identify—the remains of migrants found near the US-Mexico border. Each year hundreds of human remains are discovered along the southern US border. Many of these remains are simply left in situ, while others are taken to be cremated or buried in local cemeteries without a grave marker or collection of a DNA sample to enable future identification. Historically, US authorities have shown little interest in identification of remains believed to be those of illegal immigrants and have failed to launch criminal investigations into the manner and cause of death. Moreover, the legal infrastructure in place for DNA identification of US citizens who have gone missing is inadequate for identifying missing migrants, as their family members may be residing in another country or living as undocumented migrants in the United States. Fear of US immigration police and other law enforcement authorities further complicates the normal processes for reporting missing migrants and thereby limits DNA collection from family members.

The final chapter in part II, chapter 12, authored by Dawnie Steadman and Sarah Wagner, explores the evolving role of forensic genetics in the postmortem identification of missing persons who have died as a result of extrajudicial executions or as casualties of war. How has DNA's increasingly privileged place as a line of evidence affected the fields of forensic science and transitional justice and increased expectations of identifying loved ones among surviving kin and their communities? Drawing on interviews with leading figures in these fields and buttressed by ethnographic analysis of exhumation and identification efforts in Bosnia and Herzegovina and northern Uganda, the authors provide an overview of and commentary on the technology's complicated place in unearthing truths and affecting efforts at social repair in the aftermath of mass violence.

PART III: CHALLENGES AND DEBATES

The first chapter in part III provides an overview of legal issues related to the admissibility of forensic DNA evidence in court cases. Authored by Andrea Roth, chapter 13 describes the admissibility standards that jurisdictions around the world use for allowing expert testimony based on DNA evidence, with a particular focus on the two primary legal standards applied in US courts: the so-called Frye general acceptance and Daubert reliability tests. While courts throughout the United States have accepted PCR-STR results for robust single-source nuclear DNA typing, lay jurors still struggle to avoid statistical fallacies in interpreting DNA match statistics. Roth closes the chapter with a review of emerging admissibility issues, including privacy challenges related to DNA databases, trade secrets litigation with respect to discovery of DNA software source code, mitochondrial DNA typing, "phenotyping," and Rapid DNA.

In chapter 14, Amy Mundorff and Sarah Wagner compare and contrast the efforts to identify remains from the genocide in Bosnia and Herzegovina in the 1990s with those in the humanitarian disaster resulting from the attack on the World Trade Center in 2001. Focusing on the temporal relationship between death and the decisions taken in response to it in the aftermath of mass fatalities, the authors examine the identification processes that the International Commission on Missing

Persons and the Office of the Chief Medical Examiner in New York City used and their impact on the families of the missing or deceased.

The final chapter in part III, chapter 15, coauthored by Thomas J. White and Steven B. Lee, covers ethical, privacy, and public policy issues that have arisen from applying DNA analysis in a range of criminal investigations and humanitarian disasters. These include informed consent, securing storage of samples and genetic profiles, data security, privacy, searches of genealogical or clinical trial databases for matches to forensic evidence, and government misuse of genetic profiles (Lazer, 2004; Murphy, 2015; Phillips, 2016; Hazel & Slobogin, 2018). In particular, genetic privacy has "few, if any, applicable legal doctrines or enactments [that] provide adequate protection or meaningful control to individuals over disclosures that may affect them" (Clayton et al., 2019). Today, DNA analysis is extolled as the "gold standard" for forensic evidence (Saks & Koehler, 2005). Indeed, there is little question that DNA analysis is a significant improvement over older forensic methods. Yet as White and Lee remind us, it, too, has its limits, and without proper peer review, regulation, and oversight, it can be abused and fall prey to error. These concerns notwithstanding, the introduction of DNA analysis in criminal investigations and humanitarian disasters has contributed enormously to the pursuit of justice.

REFERENCES

Buckleton, J. S., Bright, J., & Taylor, D., eds. (2016). *Forensic DNA evidence interpretation*. 2nd ed. Boca Raton, FL: CRC Press.

Burley, J., & Harris, J. (2002). *A companion to genethics*. Hoboken, NJ: Wiley-Blackwell.

Butler, J. M. (2010). *Fundamentals of forensic DNA testing*. Burlington, MA: Academic Press.

Butler, J. M. (2012). *Advanced topics in forensic DNA typing: Methodology*. San Diego, CA: Academic Press.

Butler, J. M. (2015). *Advanced topics in forensic DNA typing: Interpretation*. San Diego, CA: Elsevier Academic Press.

Clayton, E. W., et al. (2019). The law of genetic privacy: Applications, implications and limitations. *Journal of Law & Biosciences, 96*, 1–36.

Hazel, J. W., & Slobogin, C. (2018). Who knows what, and when? A survey of the privacy policies proffered by U.S. direct-to-consumer genetic testing companies. *Cornell Journal of Law & Public Policy, 28*, article 3.

Hernandez, S. (2019, February 7). Using DNA databases to find your distant relatives? So is the FBI. *BuzzFeedNews*. Retrieved from https://www.buzzfeednews.com/article/salvadorhernandez/using-dna-databases-to-find-your-distant-relatives-so-is.

Jackman, T. (2018, December 13). FBI plans "Rapid DNA" network for quick database on arrestees. *The Washington Post*. Retrieved from https://www.washingtonpost.com/crime-law/2018/12/13/fbi-plans-rapid-dna-network-quick-database-checks-arrestees/.

Jamieson, A. & Bader, S., (2016). *A guide to forensic DNA profiling*. 1st ed. Chichester, UK: John Wiley and Sons.

Jobling, M. A. (2013, November 18). Curiosity in the genes: The DNA fingerprinting story. *Investigative Genetics*. Retrieved from https://www.ncbi.nlm.nih.gov/pmc/articles/PMC3831598/.

Joyce, C., & Stover, E. (1991). *Witnesses from the grave: The stories bones tell*. Boston, MA: Little, Brown and Company. Kofman, A. (2016, February 24). The troubling rise of Rapid DNA testing. *The New Republic*. Retrieved from https://newrepublic.com/article/130443/troubling-rise-rapid-dna-testing.

Kolata, G., & Murphy, H. (2018, April 7). The Golden State Killer is tracked through a thicket of DNA, and experts shudder. *The New York Times*. Retrieved from https://www.nytimes.com/2018/04/27/health/dna-privacy-golden-state-killer-genealogy.html.

Krimsky, S., & Simoncelli, T. (2011). *Genetic justice: DNA data banks, criminal investigations, and civil liberties*. New York, NY: Columbia University Press.

McFadden, R.D. (1989, August 15). Reliability of DNA testing challenged by judge's ruling. *The New York Times*. Retrieved from https://www.nytimes.com/1989/08/15/nyregion/reliability-of-dna-testing-challenged-by-judge-s-ruling.html.Mervosh, S. (2019, February 17). Jerry Westrom threw away a napkin last month: It was used to charge him in a 1993 murder. *The New York Times*. Retrieved from https://www.nytimes.com/2019/02/17/us/jerry-westrom-isanti-mn.html.

Murphy, E. (2015). *Inside the cell: The dark side of forensic DNA*. New York, NY: Nation Books.

National Research Council, Committee on DNA Technology in Forensic Science. (1992). *DNA technology in forensic science*. Washington, DC: National Academies Press.

Phillips, A. M. (2016). Only a click away—DTC genetics for ancestry, health, love . . . and more: A view of the business and regulatory landscape. *Applied & Translational Genomics, 8*, 16–22.

Piquado, T., et al. (2019). *Forensic familial and moderate stringency DNA searches: Policies and practices in the United States, England, and Wales*. Santa Monica, CA: RAND Corporation. Retrieved from https://www.rand.org/pubs/research_reports/RR3209.html.

President's Council of Advisors on Science and Technology. (2016, September). *Forensic science in criminal courts: Ensuring scientific validity of feature-comparison methods*. Retrieved from https://obamawhitehouse.archives.gov/blog/2016/09/20/pcast-releases-report-forensic-science-criminal-courts.

Rosenblatt, A. (2015). *Digging for the disappeared: Forensic science after atrocity*. Stanford, CA: Stanford University Press.

Saks, M. J., & Koehler, J. J. (2005, August 5). The coming paradigm shift in forensic identification science. *Science, 309*, 892.

SWGDAM. (2018). Scientific Working Group on DNA Analysis Methods (SWGDAM). *About Us*. Retrieved from https://www.swgdam.org/about-us.

US Department of Justice. (2019, September 2). *Interim policy: Forensic genetic genealogical DNA analysis and searching*. Retrieved from https://www.justice.gov/olp/page/file/1204386/download.

Wagner, S. (1998). *To know where he lies: DNA technology and the search for Srebrenica's missing*. Berkeley, CA: University of California Press.

Wertz, D., & Fletcher, J. C. (2004). *Genetics and ethics in global perspective*. Geneva, Switzerland: Springer.

PART I
DNA Technology and Individual Identification

Anthony Wright (pictured here with his granddaughter) endured two trials and 25 years in prison before a jury found him not guilty of the 1991 rape and murder of an elderly woman in Philadelphia. On August 23, 2016, he became the 344th DNA exoneree in the United States.
Photo: Kevin Monko/Innocence Project

CHAPTER 1
In the Beginning

Forensic Applications of DNA Technologies

HENRY ERLICH

The use of DNA evidence in the courtroom has become so integral to the criminal justice system that it is easy to forget how recently it was introduced. The first criminal cases using DNA technology for individual identification took place in 1986 (*Pennsylvania v. Pestinikas US*, 421 Pa. Super. Ct. 371 [1992]) and 1987 (the *Colin Pitchfork* case, UK; Wambaugh, 1989). These two cases ushered in the dawn of the DNA era in forensics at about the same time, but the DNA technologies used were very different. What *Pestinikas*, a negligent homicide case requiring DNA analysis of autopsy samples, and *Pitchfork*, a rape/murder case involving DNA analysis of semen samples, did have in common was a comparison of the DNA profile generated from the evidence with a reference sample (e.g., from a suspect or victim).

Prior to these two cases, comparing patterns obtained from evidence samples to a reference sample was performed by blood group typing or analysis of the physical properties of proteins found in the blood. The critical element in these comparisons was the genetically determined variation in the human population of these patterns, either blood group types (A, B, AB, O) or the size and electric charge of specific proteins. The basic logic of this approach is simple: Does the evidence pattern match the reference pattern? If not, the evidence could not have come from the suspect. If the patterns are indistinguishable (i.e., they "match"), the evidence specimen *could* have come from the suspect, but it could also have come from someone else with, say, the same blood type. Consider a blood stain at the crime scene that is blood type AB and a suspect sample that is also AB. The issue in the courtroom, of course, is how *likely* it is that the evidence came from the suspect. In other words, what is the probability of a coincidental match, the scenario that someone else who happened to have the same pattern (AB blood type) actually left the biological specimen at the crime scene? The frequency of the AB blood group in the general US population is about 3.4%, so the

Henry Erlich, *In the Beginning* In: *Silent Witness*. Edited by: Henry Erlich, Eric Stover and Thomas J. White, Oxford University Press (2020). © Oxford University Press. DOI: 10.1093/oso/9780190909444.003.0002

probability of a coincidental match can be estimated as 1/29. Thus, the probative significance of this particular match (in which both the evidence and reference samples are AB), while valuable in the context of other evidence, is relatively modest.

Thanks to the introduction of automated DNA sequencing technology in the 1980s, many of the parts of the human genome that vary among individuals (less than 1% of the genome) have been identified. With the sequencing of many different human genomes in recent years, this variation has been exhaustively documented. DNA sequence variation is termed genetic *polymorphism*, and the polymorphic sites used in DNA testing are often referred to as genetic *markers*. Since the human genome is 3×10^9 base pairs long, there are over one million polymorphic sites[1] that could, in principle, serve as genetic markers for individual identification. In practice, only a small number of DNA markers are required. The revolutionary impact of DNA-based genetic testing for individual identification is that the probabilities of a coincidental match of less than 1 in 100 million, in contrast to the 1 in 29 estimate for the AB blood group type, can be achieved by analyzing multiple genetic markers. Equally important in the transformative power of DNA testing in forensics is that *all* biological specimens contain DNA, so *any* crime scene evidence (hair, skin, saliva, semen, etc.) can potentially be used to identify its source.

The DNA technologies used in the *Pestinikas* and *Pitchfork* cases had been introduced to the scientific community just a year before their application in these two cases. In 1985, two scientific papers that transformed the field of molecular genetics as well as judicial systems around the world appeared in the journal *Science*. Alec Jeffreys, a distinguished molecular geneticist, and his colleagues at Leicester University described a method of analyzing complex patterns of DNA that varied in the human population. This variation (genetic polymorphism) could be, and in this seminal paper was, used for individual identification (Jeffreys et al., 1985a; Jeffreys et al., 1985b). The method analyzed DNA fragments, generated by cutting the DNA at specific sites, that varied in length at many different regions of the genome. These regions, known as the variable number tandem repeat (VNTR) loci, could be analyzed using a technique known as restriction fragment length polymorphism (RFLP) genotyping.[2] The polymorphism among individuals in the length of these

---

1. A polymorphism is defined as a variant that is present in the population at > 1%. There are many rarer variants present in human populations, but these would not be termed "polymorphic" and would not be generally useful in forensic individual identification.

2. A restriction fragment results from cutting the double-stranded DNA molecule with an enzyme, known as "restriction endonuclease," at specific sites in the DNA. The length of these "restriction" fragments can vary from individual to individual. The size of these restriction fragments can be measured by a common biochemical technique, electrophoresis, that separates molecules based on their size. The pattern of the DNA fragments generated by separation by electrophoresis in a gel could be detected by a method known as "Southern blotting," a technique for transferring the DNA fragments from the gel to a membrane and visualizing the pattern with a radioactive probe and X-ray film. In Jeffreys' DNA fingerprinting method, the variation in the length of the DNA fragments did not result from the presence or absence of specific restriction enzyme sites, as in some RFLP applications, but by variation in the number of tandem repeat copies. His method could best be described by the combination acronym RFLP-VNTR.

DNA fragments results from variation in the number of copies of a repeated specific DNA sequence. These copies are adjacent to each other, hence *tandem repeat* in VNTR. Jeffreys called this strategy *DNA fingerprinting*, an apt term since it involves comparing very complex patterns derived from the evidence and reference samples.

A few months later, my colleagues and I at Cetus Corporation, a biotech company based in the San Francisco Bay area, reported the development of a method of multiplying a specific segment of human genomic DNA using an enzyme, DNA polymerase, to create millions of copies in a test tube (Saiki et al., 1985). Kary Mullis, who came up with the idea, named the method the polymerase chain reaction (PCR) amplification because of the exponential increase in the amount of DNA amplified at every cycle of the reaction.[3] The initial demonstration amplified part of a gene for hemoglobin, the blood protein that carries oxygen, and therefore could be used for the diagnosis of sickle cell anemia, a disease caused by a mutation in this gene (Saiki et al., 1985). Subsequently, a system was developed for amplifying and analyzing the sequence variation (genotype) in a gene involved in the immune system (Saiki et al., 1986). Because this gene, HLA-DQA1, is highly variable (polymorphic) in humans, this PCR-based test could be used for individual identification and was, in fact, used in *Pestinikas* (Blake et al., 1991).

The procedures involved in the two transformative DNA technologies described in the two *Science* papers differ from each other in several respects. RFLP analysis requires large amounts of intact DNA, that is, DNA that is present as large fragments many thousands of base pairs long. In contrast, PCR can be used to amplify trace amounts of DNA and has allowed the genetic profiling of individual hairs, minute amounts of saliva, blood, bone fragments, and semen stains—essentially any biological specimen from which DNA can be extracted. Moreover, PCR analysis does not require intact DNA. Most forensic specimens, whether they are blood, semen, skin, saliva, or bone found at the crime scene, contain DNA that is present in short fragments and in small amounts. Consequently, for the past 25 years, all forensic DNA analyses have been based on PCR amplification.

In Jeffreys' initial 1985 RFLP analysis, the complex pattern derived from many different parts of the human genome was, like fingerprint patterns, difficult to interpret in statistical terms. With these complex DNA patterns, as with any genetic marker used for individual identification, the critical question about any observed *match* between the evidence and reference sample is: What is the probability of a coincidental match—that is, what is the probability that someone else randomly selected from the population would also match?

This probability is known as the random match probability (RMP), and geneticists use it to interpret the significance of a match, the observation of two indistinguishable

---

3. PCR is a test tube system that targets a specific region of DNA with short synthetic pieces of DNA known as primers, which flank the targeted region. Using DNA polymerase, an enzyme that copies DNA within the cells, the targeted DNA in the test tube is copied, approximately doubling at every cycle. Multiple cycles of PCR are carried out so that in 20 cycles, the amount of DNA that is synthesized (amplified) in this exponential reaction is around 1 million times the initial amount.

patterns. A match is also known as an *inclusion*, as opposed to an *exclusion*, the situation in which the genetic profile of the evidence is different from that of the suspect. For the complex patterns seen in the early DNA fingerprinting profiles, the statistical meaning of a match between the complex DNA pattern observed in the evidence sample and the pattern of a suspect (i.e., the RMP) could not be easily calculated. Because many different genetic regions, or loci, were being analyzed together, it was not always possible to tell whether two particular DNA fragments were variants (alleles) of the same locus or were derived from different loci. Nonetheless, the transformative idea and experimental demonstration that these complex DNA patterns could be used, like conventional fingerprinting, for individual identification was launched.

Geneticists were subsequently able to simplify these complex DNA patterns by focusing on single genomic regions, and the probabilities of a random match at each *individual* locus could be estimated based on population genetics data. By multiplying the probabilities estimated at each individual locus, statistical estimates of RMPs could be calculated and presented to the jury to help interpret the significance of matching profiles. This "product rule" is based on the assumption of statistical independence between the allele and genotype frequencies for the individual genetic loci. The same assumption is used when estimating the probability of obtaining three successive heads in a heads-tails coin toss by multiplying $1/2 \times 1/2 \times 1/2$ to yield a probability of $1/8$. The probability of a random match for a particular genetic profile at a single genetic locus is based on the frequency of that profile in a random population database. The use of this product rule is why RMPs of less than 1/100 million can be calculated from population databases with only several hundred individual samples.

While the RMP is a statistic that is often presented to juries, an alternative metric is the likelihood ratio (LR), based on the DNA evidence. The LR is the ratio of the probabilities of the evidence, given the prosecution's hypothesis (the evidence came from the suspect) and the defense's hypothesis (the evidence came from someone else). In the analysis of a *single source* sample, the LR is, typically, the reciprocal of the RMP because the probability of the evidence given the prosecution's hypothesis is 1 and the probability of the evidence coming from a coincidental match (the defense's hypothesis) is the RMP. (This is not the case if the sample is a mixture or if the Balding-Nichols formula, which addresses potential population substructure, is used.) Some of the earliest critiques of DNA forensic typing challenged the assumption of statistical independence among genetic markers and the use of the product rule in generating estimates of the RMP and/or the LR. (The statistical issues involved in interpreting mixtures—wherein DNA from more than one person is present on a specimen—are more complicated and are discussed in chapter 3.)

COLIN PITCHFORK CASE

Jeffrey's system of DNA fingerprinting had been used in a few immigration cases in the UK (Jeffreys et al., 1985a) before its first application in a criminal case, in 1987 in

Leicestershire, England.[4] The case involved the rape and murder of two young girls. The first incident took place in 1983 on the grounds of a psychiatric hospital and the second in 1986 near the town of Narbourough. Jeffreys compared RFLP profiles from semen samples from the two victims and concluded that they had been killed by the same man but *not* by the prime suspect at the time, Richard Buckland, a local 17-year-old who had admitted to the second murder under questioning. Jeffreys noted later that he had no doubt Buckland "would have been found guilty had it not been for DNA evidence."

In an attempt to identify the true assailant, the Jeffreys lab and the UK Forensic Sciences Service carried out an unprecedented mass DNA screening. They conducted a six-month-long investigation in which 5,000 local men aged 17 to 24 were asked to provide either saliva or blood samples. The turnout was 98%, but no matches were found. Then, in August 1987, a woman overheard a colleague, Ian Kelly, boasting in a local pub that he had given a sample posing as a friend of his, Colin Pitchfork. This conversation was reported to the police, and a month later, Pitchfork was arrested. His RFLP profile matched that of the two semen stains, and he pled guilty to the two rapes/murders.

PENNSYLVANIA V. PESTINIKAS

Pestinikas was in many respects an unlikely case for the introduction of DNA evidence into the US judicial system. The case involved a suspected homicide in which the victim, Joseph Kly, a 92-year-old retired coal miner, was found dead in the Stagecoach Inn, a rooming house where he had been living. The Stagecoach Inn was operated by Walter and Helen Pestinikas, who had agreed to superintend Kly's welfare and to see to his eventual burial because Walter Pestinikas was also a funeral home director. The results of the autopsy examination indicated that Kly was extremely malnourished and dehydrated; the body weighed only 65 pounds. In 1985, Walter and Helen Pestinikas were arrested on a homicide charge, and the prosecutor had the body exhumed and re-autopsied. A second autopsy found results for many of the tissue specimens that were incompatible with those of the first autopsy. Based on the observed inconsistencies, the prosecution developed a hypothesis that some of the tissue specimens had been replaced or tampered with between the two autopsies. To test this hypothesis, formaldehyde-preserved tissue samples from both autopsies were sent out by the prosecutor to determine whether the specimens came from the same or different persons.

The prosecution hired Ed Blake, a young forensic scientist at Forensic Sciences Associates, a small forensics lab housed in the same building as Cetus in Emeryville, California, to analyze the autopsy samples that had been stored in formaldehyde. Blake's analysis of proteins by the standard forensic assays failed. Blake then

---

4. The case, with its many melodramatic twists and turns, has been described in great detail in *The Blooding* by Joseph Wambaugh (New York: Bantam, 1989).

contacted my colleagues and me at Cetus. We extracted DNA from the autopsy samples and, initially, tried a variant of the RFLP analysis. Like Blake's attempt with proteins, this too failed because the DNA was present only in small fragments due to the formaldehyde treatment of the specimen. Our initial attempt at PCR amplification with the degraded DNA from the autopsy samples using the DNA polymerase from *E. coli* also failed. However, when a newly isolated and much more efficient DNA polymerase from *Thermus aquaticus*, a bacterium growing in the Yellowstone Hot Springs, was used in the PCR (Saiki et al., 1988), we were eventually able to amplify and analyze variation in the HLA-DQA1 gene. (The current official name for this gene is HLA-DQA1; the initial test referred to the gene as HLA-DQ alpha.) The justification for using this recently developed technology in a criminal case was that everything else that had been tried had failed.

The PCR analysis gave the same results for tissue specimens from the first and second autopsy, and Blake's report on the *Pestinikas* case concluded: "These findings fail to reveal a genetic difference between any of the tissues from the first and second autopsy of Joseph Kly." A year or so later, a second-generation version of the HLA-DQ alpha forensic test with increased discrimination power also strongly supported the interpretation that the two samples came from the same individual. Since the results of the test requested by the prosecution failed to support their hypothesis, the prosecutor tried to have the DNA typing evidence excluded. The defense objected and, given the understandable reluctance of the prosecution to attack the reliability of tests that they had requested, the judge allowed the introduction of the DNA evidence to the jury. The lack of obstacles in the *Pestinikas* case to the admissibility of this novel DNA evidence turned out to be highly atypical, and fierce battles attended the admissibility hearings for DNA evidence for many years. (The technical and legal issues associated with admissibility hearings for DNA evidence are discussed in more detail in chapter 13.)

As it turned out, the tissue-switching hypothesis was not really central to the prosecution's case, and after a lengthy trial, the Pestinikases were found guilty of murder in the third degree and of reckless endangerment. But the prosecution's request for genetic testing and the defense's insistence that the results be presented were momentous and initiated the long and contentious introduction of DNA evidence into the courtroom, a process that has transformed the US judicial system.

While the first use of DNA evidence in a US court was the PCR-based genotyping of a single gene in *Pennsylvania v. Pestinikas*, the first instance in which DNA evidence was used to obtain a conviction was the Tommy Lee Andrews case (Andrews v. State, 533, U.S., 841 (1988), based on the RFLP analysis of a semen sample. Andrews, a serial rapist, had been arrested in several cases in the 1980s. While already serving a 22-year sentence for rape, Andrews was accused of raping and stabbing a 27-year-old Orlando woman in her home on May 9, 1986. The victim identified Andrews as her assailant. RFLP analysis of DNA samples of a semen stain from the crime scene matched the profile of a blood sample that had been taken from Andrews after his arrest. Lifecodes, Inc., a small, commercial forensics lab in Stamford, Connecticut, carried out the tests. The prosecution claimed that the match between the RFLP profile of the semen and the suspect's blood meant that there was only a 1 in 10 billion chance someone else would have the same profile and presented this astronomical number to the jury. Andrews was convicted of rape, aggravated battery, and burglary on November 6, 1987.

PEOPLE V. CASTRO

In the first few years of forensic DNA testing by RFLP, neither the criteria for establishing a match nor the calculations of RMPs were subject to rigorous challenge. The first real challenge to the introduction of DNA evidence came in 1989 in *People v. Castro* (144 Misc. 2d 956, 545 N.Y.S.2d 985 (Sup. Ct. 1989)). in New York. Once again, the RFLP analysis was carried out by Lifecodes. Eric Lander, a distinguished mathematician and geneticist and currently director of MIT's prestigious Broad Institute, was the expert witness for the defense. Richard Roberts, a Nobel laureate who discovered and studied the bacterial enzymes (restriction endonucleases) that cut DNA at specific sites and were used to carry out part of RFLP analysis, was the expert witness for the prosecution.

Jose Castro was accused of murdering his neighbor and her daughter. The prosecution maintained that RFLP analysis of a bloodstain on his watch revealed a match with the victim. Lander was highly critical of the quality of the RFLP data as well as the lack of standards in declaring a match. In an unusual development in admissibility challenges, Roberts, the prosecution expert witness, and Landers, the defense expert witness, met, discussed the experimental data, and issued a scathing joint report to the court.

After an extensive admissibility hearing, the court ruled that the DNA evidence could be used to demonstrate that the bloodstain did not come from Castro but could *not* be used to show that it came from the victims. An exclusion, the apparent difference in DNA profiles between the evidence and a reference sample, is relatively straightforward to detect and to interpret. A match or inclusion, on the other hand, is more complicated. The criteria necessary to determine that DNA profiles from two different samples are actually identical must be established, as must the appropriate procedures for calculating the probability of a coincidental match (the RMP). The joint report by Landers and Roberts was critical of the Lifecodes data and procedures on both scores.

The court's ruling in this case proved to be critical in all subsequent US admissibility hearings on DNA evidence. While finding that DNA forensic techniques were "generally accepted" by the scientific community, it stated that pretrial hearings should be required to evaluate whether the testing laboratory's results were consistent with scientific standards. The court recommended extensive requirements for discovery of laboratory data and protocols for future admissibility hearings. The "general acceptance" by the relevant scientific community is part of the *Frye* standard for admissibility of evidence in the courtroom. (Chapter 13 presents a detailed discussion of the issues controversies, and the history associated with the admissibility of DNA evidence.)

Castro was convicted of murder, but the challenge to the presentation of the RFLP evidence in this case marked the beginning of a concerted effort by the government and the forensic community to establish standards and guidelines for all DNA forensic applications.

THE "DNA WARS"

The *Castro* case and the associated challenges raised in the admissibility hearing formed a pivotal moment in the history of DNA evidence in the courtroom and led to a period of dueling scientific publications known as the "DNA Wars." The battle started with a 1989 editorial in the highly respected journal *Nature*, "DNA Fingerprinting on Trial," by Eric Lander (1989), based on his experience as a defense expert witness in *Castro*. He focused on the lack of standards in the application of RFLP analysis for forensic specimens. He also challenged the assumption of statistical independence of the various probability estimates for the individual genetic markers.

Two years later, in 1991, two well-known and respected population geneticists, Dan Hartl and his Harvard colleague Richard Lewontin, wrote a blistering critique in *Nature* of the forensics use of DNA analysis (Lewontin & Hartl, 1991). While they were highly critical of many aspects of forensic DNA analysis, their main criticism focused on the assumptions made in the statistical estimates of RPMs, in particular the use of the product rule. Like Lander, they argued that available population genetics data for RFLP profiles did not support the simplifying assumption that the frequency of a given DNA profile at one genetic locus was independent of the frequency at other loci. Two other distinguished geneticists, Ranajit Chakraborty and Kenneth Kidd, responded in the pages of *Nature* (Chakaborty & Kidd, 1991), vigorously defending the interpretation of the population genetics data and the methods used to estimate RMPs. The following year another well-known statistical geneticist, Bruce Weir, entered the fray (Weir, 1992), arguing that the assumption of statistical independence was, in fact, consistent with the genotype frequency data in population databases.

A truce in the battle over the appropriate interpretation of population genetics data and the statistical estimates of RMPs was established with the publication of "DNA Fingerprinting Dispute Laid to Rest" in *Nature* (Lander & Budowle, 1994), coauthored by Bruce Budowle of the FBI and Eric Lander, who had fired the first shot in the DNA Wars in 1989. The article presented a "new consensus" on DNA evidence in 1994 and represented a ceasefire leading to a negotiated peace, as did the guidelines set forth in the 1992 National Research Council Report, *DNA Technology in Forensic Science*. One of the developments that helped end the early phase of the DNA Wars was the establishment, by the FBI and other members of the forensic community (the Scientific Working Group on DNA Analysis Methods, or SWGDAM), of the validation guidelines, specific experiments, and procedures to be carried out for any new forensic genetic technique to be accepted in the courtroom.

With the PCR-based genetic markers in current use, the genetic profiles are better defined and, with the expanded population databases for all the genetic markers, the calculation of RMPs for a single-source sample is no longer contentious, though some controversy still persists in the analysis of mixed samples (see chapter 3) and in the interpretation of the matches that result from searching criminal databases.[5] The one remaining area of continuing admissibility challenges for single-source DNA is the "low copy DNA" or "trace DNA", sometimes referred to as "touch DNA." The concern in the case of trace DNA analysis is that the amount of starting DNA in the evidence sample is so low that some of the alleles present in the sample may not be detected, leading to a potentially erroneous genotyping result.

While the debate about appropriate statistical interpretation of RFLP data and calculation of RMPs raged in journals and the courtroom, the RFLP technology itself was gradually being replaced by PCR-based assays. The introduction of PCR gave rise to a new set of challenges in admissibility hearings, namely the possibility of contamination. Opposing arguments often cited the ability of PCR to create millions of DNA copies from a few initial copies present in the evidence sample as a potential problem. In response to this concern, laboratory guidelines were established about how PCR amplification was performed, to minimize the possibility of contamination (Kwok & Higuchi, 1989). Control experiments, such as the analysis of samples with no DNA (a "negative control"), were also performed to detect contamination, should it occur. Over time, these objections to the admissibility of PCR-based data based on potential contamination diminished.

One often cited example of contamination or secondary DNA transfer is the 2013 case of Lukis Anderson, who was charged with murdering Raveesh Kumra, based on the finding of trace amounts of Anderson's DNA on the victim's body (Chinn, 2013). Anderson had been drunk and unconscious at a local hospital at the time of the murder, however. As it turned out, the same two paramedics who treated Anderson for intoxication were at the Kumra murder scene a few hours later and evidently unwittingly transferred Anderson's DNA to the victim's fingernails. Anderson was released from jail after five months. Although this case is often cited as an example in the "DNA is not infallible" narrative, the DNA testing itself was correct. It illustrates instead the potential pitfalls of trying to interpret trace amounts of DNA and is a valuable reminder that identifying the source of a crime scene DNA sample can be unrelated to the crime and the issue of guilt.

In 1996, a National Research Council report, *The Evaluation of DNA Evidence*, concluded: "The state of the profiling technology and the methods for estimating frequencies and related statistics have progressed to the point where the admissibility of properly collected and analyzed DNA data should not be in doubt." Some

---

5. A database search is an investigative tool to identify a suspect. If the profiles of the evidence and the suspect match, the RMP is, arguably, still the relevant statistic to be presented to the jury. A contrary view is that the size of the database should be multiplied times the RMP. For example, if the RMP is 1/100 million and the database that was searched is 1 million, this perspective argues that the number 1/100 is the one that should be presented to the jury. This calculation, however represents the probability of getting a "hit" (match) with the database and not the probability of a coincidental match between the evidence and the suspect, which is the more relevant metric for interpreting the probative significance of a match.

controversies persist, however. The interpretation of mixtures and of matches obtained by searching databases is discussed in chapters 3 and 4, respectively. As the new technologies discussed in chapter 5 are introduced into the courtroom, we can anticipate that they, too, will be challenged. These, however, are mere skirmishes; the DNA Wars are over—and DNA won.

EXCLUSIONS, EXONERATIONS, AND THE *DOTSON* CASE

The interpretation of an exclusionary result is straightforward. The source of the evidence sample is *not* the individual, usually the suspect, who provided the reference sample. In the initial casework analyses using PCR and a single gene, HLA-DQA1, about a third of the cases involved an exclusionary result (Blake et al., 1991). The evidence sample genotype did not match that of the suspect, and usually the state did not continue its prosecution of the excluded individual. The ability of PCR to amplify old forensic evidence suggested it might be possible to re-examine some past cases involving potential wrongful conviction.

The first such exoneration by DNA analysis was the Gary Dotson case (*People v. Dotson, 424N.E.2d1319 (Ill.App.Ct.1981)*). He had been convicted in Illinois of aggravated kidnapping and rape in 1979, based on the victim's testimony and analysis of pubic hair. In 1985, the victim recanted her testimony, explaining that her fabrication was intended to hide having had sex with her boyfriend. The prosecution, however, was unwilling to overturn Dotson's conviction, and in 1987, the Appellate Court of Illinois affirmed the conviction. In 1988, Dotson's new attorney sent the 11-year-old semen stain obtained from the (alleged) victim to Alec Jeffreys, who tried to apply RFLP analysis to these samples but failed to obtain a result. Shortly thereafter, the forensic samples were sent to Cetus and Ed Blake (Forensic Sciences Associates), where the PCR-based HLA-DQ alpha testing gave a clear result. The HLA DQ alpha profile did *not* match that of Gary Dotson but *did* match the profile of the alleged victim's boyfriend. Still, it took well over a year for the State of Illinois to overturn Dotson's conviction. Since the initial DNA-based exoneration of Gary Dotson in 1989, more than 350 convictions have been overturned based on DNA evidence as of 2018 (Innocence Project), 20 of them in death penalty cases. (Chapter 2 provides a detailed discussion of DNA-based exonerations of the wrongfully convicted.)

THE EARL *WASHINGTON JR* CASE AND THE PROBLEM OF MIXTURES

In most of the cases involving DNA exonerations, it was later analysis of the DNA evidence that revealed the wrongful conviction. In some cases, however, the original conviction was based on a misinterpretation of DNA evidence. Mixtures—specimens with more than one contributor—have been notably difficult to interpret. From the beginning of PCR-based forensic DNA typing with a single gene, the HLA-DQA1 locus, to the current technology with multiple STR markers, the analysis and interpretation of mixed forensic samples in casework have been challenging and contentious. In 1994, I submitted a report based on my interpretation of the HLA-DQ alpha results observed in a mixed sample that, ultimately, led to overturning the wrongful conviction for rape and murder of Earl Washington Jr. in Virginia (Washington v. Com. 323 S.E.2d 577 (1984)) Washington was scheduled to be executed within days of the submission of my report, which argued for an exclusion.

In 1982, Rebecca Williams, a 19-year-old girl, was raped and stabbed to death in her apartment in Culpeper, Virginia. Before she died, she identified her attacker as a black man. A year later, Earl Washington Jr., a mentally challenged 22-year-old African American, was arrested and charged with burglary and assault in a nearby county. The police accused Washington of Williams's rape and murder as well and, following interrogation, he confessed. At his trial, however, Washington maintained his innocence. Nevertheless, he was convicted and sentenced to death by a Virginia court in 1984. His execution was scheduled for January 1994. Washington's attorney, Barry Weinstein, was convinced of his innocence, and a scant ten days before his client's scheduled execution, asked me to review the results of the PCR HLA-DQ alpha testing of the semen stain and vaginal swab sample carried out by the State of Virginia's lab.

The DNA analyst in the Virginia lab, Jeff Ban, performed the test on DNA from a vaginal swab taken during the autopsy performed on Williams as well as on a semen stain found on a blanket at the scene of the crime. Washington was excluded when the semen stain was analyzed, but the vaginal swab specimen was a mixture, with DNA from more than one contributor, a notoriously difficult kind of forensic sample to analyze and interpret. Ban reported that although the test identified a genetic variant (allele) absent in Washington, the victim, and the victim's husband, he could not eliminate Washington as a potential source of the semen. The prosecution argued, on the basis of the lab report, that the DNA results indicated an inclusion, consistent with the guilty verdict, because his alleles were present in the mixture. My interpretation of this complex mixture, however, was that Earl Washington Jr.'s genotype was not present in the mixture (an exclusion) and, therefore, that he was likely innocent of the charges for which he'd been convicted. The report was submitted on January 13, and on January 14, Governor Douglas Wilder commuted Washington's sentence to life in prison but declined to pardon him.

As is typically the case with a mixed sample, the DNA test results on the vaginal swab taken from the victim were not at all straightforward. There was clear evidence that at least two different individuals had contributed to the sample; this mixture was inferred from the presence of three different HLA-DQ alpha alleles. The test distinguished six different alleles: 1.1, 1.2, 1.3, 2, 3, and 4. The Virginia lab report stated that three different HLA-DQ alpha alleles were identified in the vaginal swab: 1.1, 1.2, and 4. The reference sample for Mr. Washington was a 1.2/4 genotype and for the victim, a 4/4 genotype. The conclusion of the Virginia lab report was that since the mixture contained both a 1.2 and a 4 allele, the analysis of the swab was compatible with Mr. Washington's being one of the contributors to the sample, and that therefore, the DNA evidence provided no basis on which to reconsider his conviction and imminent execution.

But alleles do not exist independently in nature; they occur in pairs, as genotypes, in diploid somatic cells. In the case of sperm (haploid) from a 1.2/4 individual, one would expect an equal amount of 1.2-bearing and 4-bearing sperm. Consider a mixture of these three alleles in which the alleles, based on the test results, were present in equivalent amounts. This situation would pose major difficulties in interpreting the data because this result is consistent with a mixture of many possible genotypes.

In the HLA-DQ alpha forensic test, the different alleles were identified by a blue color corresponding to a specific DNA probe. The intensity of the blue color reflected the amount of amplified DNA that had bound to the probe and therefore the relative amounts of the alleles in the mixed sample. In the case of the HLA-DQ alpha results in the Washington report, the intensity of the blue dots identifying the 1.1 and 1.2 alleles was strong, while the intensity of the blue dot for the 4 allele was much weaker.

In interpreting this mixture, I made two critical assumptions. First, based on the intensity of the blue dots, the two DQ alpha alleles 1.1 and 1.2 constituted the majority genotype (1.1/1.2), and there was a trace amount of a third allele, DQ alpha 4. Second, since the separation of sperm from a semen stain or vaginal swab often results in residual trace amounts of the victim's epithelial cells, I assumed that the minority component in this mixture was a 4/4 contribution from the victim. Alternatively, but less likely, the minor DQ-alpha 4 allele could have come from the husband. This analysis implicated an individual with the DQ-alpha 1.1/1.2 genotype as the source of the sperm, and it excluded Earl Washington Jr.

Though saved from imminent execution, Washington remained in jail in spite of the HLA-DQ alpha evidence in his favor. Then, in 2000, following significant advances in forensic DNA technology and continuing doubts on other grounds about Washington's guilt, then governor Jim Gilmore ordered the Virginia lab to perform additional testing with the current STR genotyping systems. A comparison of the genetic profile obtained with STR technology with the profiles in the state's DNA database yielded a match with Kenneth Tinsley, a convicted rapist. In 2007 Tinsley, previously convicted of rape in Illinois and again in Virginia, where he had been sentenced to life in prison, confessed and pled guilty to the rape and murder of Rebecca Williams. Later that year, Washington was formally exonerated; Governor Tim Kaine issued a full pardon, acknowledging that Washington had been wrongfully convicted of rape and murder 23 years earlier.

EVOLUTION OF FORENSIC GENETIC TYPING

The initial DNA fingerprinting analysis used a radioactive DNA probe that recognized many different parts (the VNTR loci) of the genome, resulting in a very complex pattern; consequently, matches were difficult to interpret statistically. Subsequent development of the technology, pioneered in the United States and later applied to casework by two companies, Lifecodes and Cellmark, yielded multiple "single-locus" probes, which generated a much simpler pattern for each individual VNTR locus. This single locus pattern could be interpreted statistically. Large numbers of random individuals in a given population could be analyzed with these single-locus probes, and the frequencies of the different genetic profiles at a single locus could be estimated. The probability of a random match at one locus for a given genetic profile is assumed to be equal to the frequency with which that genetic profile occurs in the population. Consider a bowl with black and red balls. The probability of randomly picking a red ball is equal to the frequency of the red balls in the bowl. (In a bowl with 20 red balls and 80 black ones, the probability of picking a red ball would be estimated at 20%.) Assuming statistical independence of the various individual VNTR genetic profiles, the probability of a multilocus profile match could be estimated by multiplying the probabilities of each individual single-locus profile match (the product rule).

Defining a match/inclusion was still complicated using the multiple single-locus probe RFLP system due to difficulties determining whether the DNA fragment lengths derived from two different samples were truly identical. But the significance (RPM) of a match that was estimated by multiplying the probabilities calculated for each individual locus was very impressive, on the order of 1 in hundreds of millions.

During the late 1980s, when PCR genetic typing was being introduced into forensics, a common view was that it was an impressively powerful tool, allowing the analysis of forensic specimens with little and/or degraded DNA, but that it was less discriminating (i.e., less informative for inclusions) than RFLP analysis. This perspective was based on the availability at the time of only one PCR genetic marker, the HLA-DQA1 locus, compared to multiple genetic markers for RFLP analysis. The value of multiple genetic markers is that if they are statistically independent, the probability of a match at one location can be multiplied by the probabilities at the other sites, generating an impressively low RPM.[6]

PCR genetic typing was much more robust, faster and easier to perform, less expensive, and required much less DNA. Unlike RFLP, it did not require radioactive labels. It could also analyze many forensic samples, such as those with degraded DNA, as in the *Pestinikas* and *Dotson* cases, in which attempts at RFLP analysis had failed. The solution to this dilemma of versatility versus discrimination was clear: develop more PCR-based genetic markers and have the best of both worlds.

The second PCR-based forensics test was based on five different genes, which had sequence variants that could be distinguished by DNA probes. The Polymarker test, launched in 1994, was designed to be used in conjunction with the HLA-DQ alpha forensics kit (see Figure 1.1). The informativeness of a given test for individual identification can be measured by a metric known as the power of discrimination (Pd), which represents the probability that a given genetic profile generated by a test will

---

6. The statistical independence of genetic markers, the random association of alleles at one locus with the alleles at another locus, reflects their physical independence. If they are located on different chromosomes, they segregate independently at cell division, as shown by Gregor Mendel. If they are located on the same chromosome, genetic recombination between the loci can create random association. In addition, analysis of population genetic data supporting statistical independence is a critical part of the validation studies required for forensic markers.

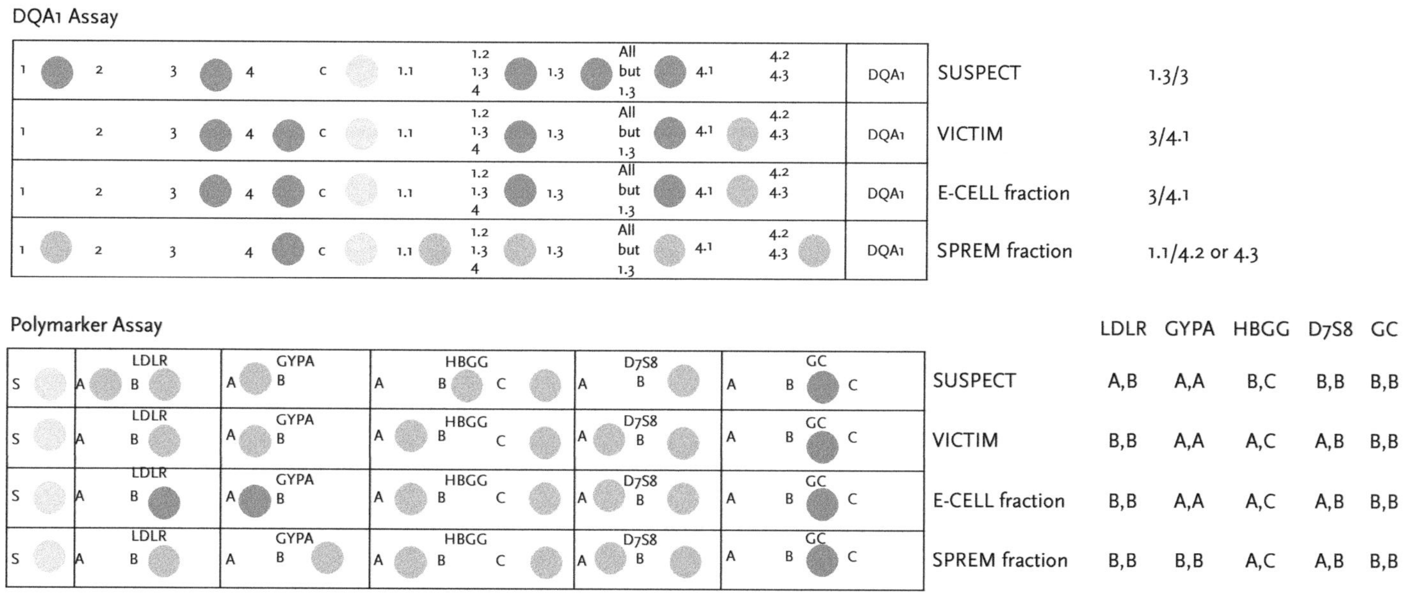

DQA1 Assay

1	2	3	4	C	1.1	1.2, 1.3, 4	1.3	All but 1.3	4.1	4.2, 4.3	DQA1	SUSPECT	1.3/3
1	2	3	4	C	1.1	1.2, 1.3, 4	1.3	All but 1.3	4.1	4.2, 4.3	DQA1	VICTIM	3/4.1
1	2	3	4	C	1.1	1.2, 1.3, 4	1.3	All but 1.3	4.1	4.2, 4.3	DQA1	E-CELL fraction	3/4.1
1	2	3	4	C	1.1	1.2, 1.3, 4	1.3	All but 1.3	4.1	4.2, 4.3	DQA1	SPREM fraction	1.1/4.2 or 4.3

Polymarker Assay

	LDLR		GYPA		HBGG			D7S8		GC			LDLR	GYPA	HBGG	D7S8	GC
S	A, B		A, B		A, B, C			A, B		A, B, C		SUSPECT	A,B	A,A	B,C	B,B	B,B
S	A, B		A, B		A, B, C			A, B		A, B, C		VICTIM	B,B	A,A	A,C	A,B	B,B
S	A, B		A, B		A, B, C			A, B		A, B, C		E-CELL fraction	B,B	A,A	A,C	A,B	B,B
S	A, B		A, B		A, B, C			A, B		A, B, C		SPREM fraction	B,B	B,B	A,C	A,B	B,B

Figure 1.1 HLA-DQ alpha and Polymarker immobilized probe dot-blot assays. The circles indicate the position of oligonucleotide probes that bind specifically to sequences in the amplified DNA for the HLA-DQ alpha locus and the five genetic loci included in the Polymarker assays. The genotype results from the suspect, from the victim, and for the epithelial cell fraction and the sperm fraction of the semen stain are shown to the right. The genotypes of the sperm fraction and suspect are different, so the suspect is *excluded* as a source of the sperm DNA.

have a match in the general population. This metric is averaged over the distribution of all possible genotypes in the relevant population. These six genes, the five genes in the Polymarker test plus the HLA-DQ alpha, were statistically independent of one another, so the probabilities of a match at each of the individual loci could be multiplied. The combination of the Polymarker and HLA-DQ alpha tests greatly increased the Pd for PCR-based tests, but it was still less than that of multigene RFLP analysis.

PCR MEETS VARIABLE NUMBER TANDEM REPEATS (VNTRS)

The HLA-DQ alpha and Polymarker tests were based on detecting variation in the *sequence* of the amplified DNA, but PCR could also be used to distinguish *length* variation, that is, variation in the number of tandem repeat copies for a given genetic segment. PCR amplification is based on positioning short synthetic DNA fragments, known as PCR primers, on either side of the targeted DNA sequence. If the PCR primers were designed to flank the tandem repeat copies, the length of the PCR product generated by amplification would vary according to the number of tandem repeat copies (see Figure 1.2). In this PCR-based system, the number of allelic variants was less (typically 5–20 variants) than a typical VNTR locus, and each discrete variant could be identified and "named" as the number of tandemly repeated copies. For example, at a given locus, a heterozygote genotype might be that one allele on the paternal chromosome had 10 repeat copies and another on the maternal chromosome had 8 repeat copies. In contrast, RFLP analysis of the VNTR loci involved distinguishing among hundreds of less well-defined allelic variants. Thus, calling a match was more straightforward with the discrete PCR-based length variants. In keeping with the long and distinguished tradition of acronyms in molecular biology, these PCR-based tests for length variation were initially called "AmpFLPs."

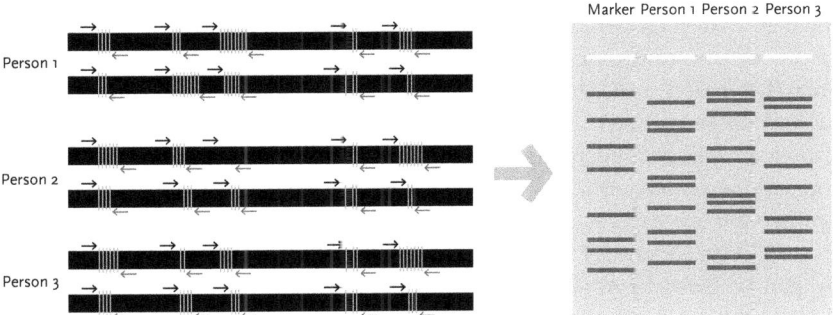

Figure 1.2 PCR amplification of VNTR loci. The arrows in the left-hand diagram represent the position of PCR primers flanking the VNTR loci. The bands in the gel electrophoresis analysis (right-hand diagram) represent the variable length PCR products (amplified DNA fragments) from different individuals. The bands labeled "Marker" represent DNA fragments of known length.

A subset of VNTR loci that had short repeats of four bases (e.g., variable numbers of, say, CTAG or any other tetranucleotide) proved to be well-suited to PCR amplification and analysis of length variation by capillary electrophoresis. These so-called short tandem repeat (STR) markers were developed commercially by Applied BioSystems (now ThermoFisher) and Promega. The primers for amplifying these four base pair repeat regions (figure 1.2) scattered around the genome could be labeled with one of four fluorescent dyes, and the size-based separation of the variable length PCR products was automated by capillary electrophoresis instruments made by ABI. This automated system became the standard for forensic analysis and, with minor modification, remains how the vast majority of forensic specimens and virtually all database and reference samples are currently analyzed. However, the automated analysis of STRs was not without technical issues. "Stutter bands,' artifactual minor PCR products that are slightly longer or shorter than the true DNA variant fragment, can complicate interpretation of the results. Nonetheless, automated genotyping analysis using STR panels with capillary electrophoresis continues to be the standard method for forensic DNA analysis. One of the rationales for moving to the STR markers was that unlike the HLA or blood group genes used in the initial forensic tests, the STR loci did not encode proteins and were presumed to convey no medical information,[7] a potential advantage in terms of protecting personal medical information.

The first commercial panel in the United States involved analyzing 13 different STR segments amplified together in a single PCR test. This 13-marker panel was used for many years as a standard (Combined DNA Index System, or CODIS) genotyping method for forensic specimens and for criminal and missing person databases as well as for the population databases that are used for estimating RPMs. In Europe, another set of STR markers was used.

Currently, the commercial DNA kits include 24 different STR markers that include the initial core sets used in the United States and Europe. All STR panels include STR loci on the X and Y chromosomes. In particular, the amelogenin gene, which exists on both X and Y chromosomes but can be distinguished based on the size of the amplified DNA fragment, is included in these panels and can be used to determine the sex of the contributor to an evidence sample. Unlike the STR markers on autosomal chromosomes (the non-sex chromosomes), the product rule cannot be used to estimate RMPs on the STR markers on the Y chromosome because they are not statistically independent. Unlike the autosomes, the Y chromosome does not participate in genetic recombination, and the allelic variants of the Y chromosome are linked together and inherited as a "haplotype." Currently, the commercial PCR-STR-based kits used most commonly in forensic analysis in criminal cases in the United States are manufactured by Applied Biosystems (such as ProFiler/CoFiler, which test 13

---

7. Recent studies demonstrating that variants at the STR loci may be associated with variants at other genomic regions by a phenomenon known as "linkage disequilibrium" (Edge et al., 2018) call this assumption into question.

STR loci, including Amelogenin[8]; IdentiFiler (testing 16 loci, including Amelogenin); and, most recently, GlobalFiler (testing 24 loci) and Promega's PowerPlex kits.

THE PROBLEM OF DEGRADED DNA

The DNA extracted from most forensic specimens is degraded; that is, it is present only in short fragments. Degraded DNA can pose a significant challenge for STR genotyping. If a STR allele has, for example, 25 copies of a four-base tandem repeat, DNA fragments in the evidence sample that are less than 100 base pairs long[9] will not be amplified and genotyped.

One approach to the problem of degraded DNA in evidence samples has been to genotype single nucleotide polymorphisms (SNPs) instead of or in addition to STRs. These are individual sites in the genome that can have one of two alternative bases. Since the vast majority of SNPs are bi-allelic (some rare SNPs have three alleles), and the STRs typically have 10–20 alleles, the potential of one SNP to identify an individual is much less than that of one STR locus. SNP markers are also less well suited to the detection and analysis of mixtures than are STRs; the signature observation indicating a mixture is the detection of more than two alleles. However, there are millions of SNPs in the genome and, with the recent development and application to forensics of Next Generation Sequencing (NGS), hundreds of thousands of SNPs can now be sequenced in parallel. In general, SNPs have both distinct advantages and disadvantages relative to STR markers (Butler, Coble, and Vallone; STRs vs SNPs, 2007). SNPs, for example, have a lower mutation rate than STR markers, making them better suited for kinship analysis and paternity testing, and do not have the problem of stutter PCR products.

Another strategy for addressing the problem of degraded DNA is to analyze sequence polymorphism in PCR-amplified mitochondrial (mt) DNA, pioneered by Mark Stoneking while he was a visiting scientist at Cetus (Stoneking et al. 1991). Mitochondrial (mt) DNA is one of the most valuable genetic markers for analysis of forensic samples, due to the high copy number per cell. Mitochondria are the cytoplasmic cell organelles responsible for energy generation and are transmitted from mother to child in the egg (matrilineal inheritance). Because each cell contains hundreds to thousands of copies of the mtDNA genome, in contrast to the two copies of chromosomal genetic markers, many evidence samples with very limited and/or degraded DNA can be analyzed only with mtDNA genetic markers. Until very recently, most forensic analyses focused on the most polymorphic segment of the mtDNA genome, the "hypervariable regions." Although very powerful for the analysis of evidence samples with limited and degraded DNA as well as for mixed samples, analysis

---

8. Amelogenin is not an STR but a gene that is present on both the X and Y chromosomes but has a different length on the X and Y and is used to determine the sex of the contributor to the DNA evidence.

9. In fact, most DNA fragments that are less than 150 base pairs will also not be amplified, because both PCR primers must bind to unique DNA sequences flanking the STR region.

of mtDNA for individual identification has a few limitations. The matrilineal pattern of inheritance means that siblings and maternal relatives will have the same mtDNA sequence. Moreover, like Y-STRs, the polymorphic sequences in the mtDNA genome do not recombine and are linked and inherited together as a haplotype. Since these polymorphic sequences are not statistically independent, the product rule is not applicable. As with Y-STR haplotypes, the statistical significance of a match between mtDNA sequences is estimated by the "counting method," building a confidence interval around the number of matching haplotypes found in a relevant mtDNA population database.

These STR tests and mtDNA sequencing assays represent the current standard of forensic DNA testing. Their use in interpreting forensic mixtures is discussed in chapter 3, and recent technological developments, like NGS, which offer some significant advantages over the existing STR systems, are discussed in chapter 5.

Since the initial work applying PCR to the analysis of a few autopsy samples from the *Pestinikas* case in the summer of 1986, millions of samples have been analyzed and many thousands of cases have been resolved using what are basically the same DNA techniques. Within the United States, DNA testing has raised challenging and difficult questions about the death penalty as well as about the reliability of eyewitness testimony, other forensic techniques such as hair morphology and tooth bite analysis, and even confessions. Identifying a perpetrator in crimes without a suspect (cold cases) by checking an evidentiary genetic profile against a databank of genetic profiles of convicts (or in some cases, arrestees) has already become standard practice, although exactly how this should be done remains somewhat controversial, as described in chapter 15. Now, just over 30 years since *Pennsylvania v. Pestinikas*, it is hard to imagine our justice system without access to the DNA testing of biological evidence.

REFERENCES

Blake, E., et al. (1991). Polymerase chain reaction (PCR) amplification and human leukocyte antigen (HLA)-DQa oligonucleotide typing on biological evidence samples: Casework experience. *Journal of Forensic Sciences*, 37(3), 700–726. PMID: 1629670. C p. 4, 17.

Butler, J. T., Coble, M. D., & Vallone, P. M. (2007). STRs versus SNPs. Thoughts on the future of DNA testing. *Forensic Sci Med Pathol*, 3, 200–205.

Chakraborty, R., & Kidd, K. K. (1991). The utility of DNA typing in forensic work. *Science*, 254(5039), 1735–1739. PMID: 1763323. C p. 14.

Chinn, J. (2013) How an innocent man's DNA was found at a crime scene. Californiainnocence project.org

Edge, M. D., et al. (2017, May 15). Linkage disequilibrium matches forensic genetic records to disjoint genomic marker sets. *Proceedings of the National Academy of Sciences USA*, 114, 5671–5676.

The Innocence Project. (n.d.). https://www.innocenceproject.org/. C p. 18.

Jeffreys, A. J., Brookfield, J. F. Y., & Semeonoff, R. (1985a, October 31). Positive identification of an immigration test-case using human DNA fingerprints. *Nature*, 317, 818–819. PMID: 4058586. C p. 3, p. 7.

Jeffreys, A. J., Wilson, V., & Thein, S. L. (1985b, July 4). Individual-specific "fingerprints" of human DNA. *Nature, 316,* 76–79. PMID: 2989708. C p. 3.

Kwok, S., & Higuchi, R. (1989, May 18). Avoiding false positives with PCR. *Nature, 339,* 237–238. PMID: 2713852. C p. 16.

Lander, E. S. (1989, June 15). DNA fingerprinting on trial. *Nature, 339,* 501–505. PMID: 2567496. C p. 13.

Lander, E. S., & Budowle, B. (1994, October 27). DNA fingerprinting dispute laid to rest. *Nature, 371,* 735–738. PMID: 7818670. C p. 14.

Lewontin, R. C., & Hartl, D. L. (1991, December 20). Population genetics in forensic DNA typing. *Science, 254,* 1745–1750. PMID: 1845040. C p. 14.

National Research Council. (1992). *DNA technology in forensic science.* Washington, DC: The National Academies Press. PMID: 25121318. C p. 13, 15.

National Research Council. (1996). *The evaluation of forensic DNA evidence: An update.* Washington, DC: The National Academies Press. PMID: 25121324. C p. 17.

Saiki, R. K., et al. (1985, December 20). Enzymatic amplification of beta-globin genomic sequences and restriction site analysis for diagnosis of sickle cell anemia. *Science, 230,* 1350–1354. PMID: 2999980. C p. 4.

Saiki, R. K., et al. (1986, November 13–19). Analysis of enzymatically amplified beta-globin and HLA-DQa DNA with allele-specific oligonucleotide probes. *Nature, 324,* 163–166. PMID: 3785382. C p. 4.

Saiki, R. K., et al. (1988, January 29). Primer-directed enzymatic amplification of DNA with a thermostable DNA polymerase. *Science, 239,* 487–491. C p. 9.

Stoneking, M., et al. (1991, February). Population variation of human mtDNA control region sequences detected by enzymatic amplification and sequence-specific oligonucleotide probes. *American Journal of Human Genetics, 48,* 370–382. PMID: 1990843. C p. 24.

ThermoFisher Scientific. (n.d.). *PCR amplification for forensic DNA profiling.* Retrieved from https://www.thermofisher.com/us/en/home/industrial/forensics/human-identification/forensic-dna-analysis/pcr-amplification-forensic-dna-profiling.html.

Wambaugh, J. (1989). *The Blooding.* New York, NY: Bantam.

Weir, B. S. (1992, December 15). Population genetics in the forensic DNA debate. *Proceedings of the National Academy of Sciences of the USA, 89,* 11654–11659. PMID: 1465380. C p. 14.

CHAPTER 2

Exonerating the Wrongfully Convicted

JUSTIN BROOKS AND DESIREE MOSHAYEDI

Forensic DNA analysis has played a role in roughly a quarter of the 2,000 exonerations that have taken place in the United States since it was first used in 1988, to clear Gary Dotson, who had been falsely convicted of rape in Illinois (as discussed in chapter 1). In the last decade, the number of DNA exonerations has held steady at about 20 cases per year. Most of these cases have involved individuals who were wrongfully convicted of murder and/or sexual assault, with minority groups disproportionately represented (approximately 70%). Moreover, in tens of thousands of cases the prime suspects have been identified and pursued until DNA testing prior to conviction proved they were wrongly accused.

This chapter examines the modern innocence movement, beginning with Centurion Ministries' efforts to free innocent people from prison, continuing with the founding of the Innocence Project, and concluding with the current state of the work, involving more than 100 innocence organizations worldwide. DNA evidence has been the key to hundreds of exonerations, leading to the review of other cases without DNA evidence and more than 2,000 documented cases of wrongful conviction in the United States alone.

EARLY INNOCENCE CASES

"It is better 100 guilty Persons should escape than that one innocent Person should suffer," Benjamin Franklin once said (Franklin, 1785). Though the founders of the United States understood the importance of protecting against "cruel and unusual punishment" such as putting an innocent person behind bars, our criminal justice system has flaws that have for centuries sometimes enabled the incarceration of the innocent. Many casually date the innocence movement to when DNA was first used to exonerate Gary Dotson. There is no doubt that DNA was, and is, the most powerful forensic evidence ever used to prove innocence and guilt; however, there have

Justin Brooks and Desiree Moshayedi, *Exonerating the Wrongfully Convicted* In: *Silent Witness*. Edited by: Henry Erlich, Eric Stover and Thomas J. White, Oxford University Press (2020). © Oxford University Press.
DOI: 10.1093/oso/9780190909444.003.0003

been cases with equally powerful evidence of innocence before that time, such as the famous case of William Jackson Marion, in which the alleged victim showed up alive after Marion was executed in 1887 for murdering him.

The modern innocence movement began in 1979, when Jim McCloskey, a student chaplain at Princeton Theological Seminary, met a prisoner named Jorge De Los Santos, who had been sentenced to life for the murder of a car dealer (Centurion Ministries, n.d.-c). De Los Santos persuaded McCloskey that he was innocent, and McCloskey then recruited a lawyer named Paul Casteliero to work on the case. De Los Santos was ultimately freed when it was shown that he had been convicted based on a lying jailhouse informant (*De Los Santos v. O'Lone*, 1983).

After graduating from Princeton, McCloskey decided to make freeing innocent people his life's work. In 1984 he founded Centurion Ministries, which to date has helped to free fifty-four innocent men and women, bringing more media attention to the problem of wrongful conviction and serving as a model for later innocence organizations.

DUE PROCESS AND CRIME

During the 1960s a "due process" revolution of sorts took the form of three important cases that expanded the rights of criminal suspects. In 1961, *Mapp v. Ohio* placed restrictions on the admissibility of evidence found through illegal searches and seizures (*Mapp v. Ohio*, 1961). Two years later, *Gideon v. Wainwright* created a right to counsel that has been expanded to most criminal cases (*Gideon v. Wainwright*, 1963). Then, in 1966, *Miranda v. Arizona* established for the first time that defendants must be informed of their right to remain silent and to the presence of an attorney to represent them during interrogation (*Miranda v. Arizona*, 1966). These due process rights set the stage for recognition of the rights needed to reopen criminal cases after conviction.

The US Supreme Court also decided a number of cases in the late 1960s and 1970s that resulted in protecting prisoners' basic constitutional rights to religious freedom and correspondence (Death Penalty Information Center, n.d.-a). The late 1960s and 1970s also gave birth to the modern anti-death penalty movement. In *Furman v. Georgia* (1972), the Supreme Court ruled that since "capital cases resulted in arbitrary and capricious sentencing" and punishment had become increasingly racialized, the death penalty would be suspended (Death Penalty Information Center, n.d.-a). As "race and crime became inextricably intertwined during this era," the anti-death penalty movement became a part of the broader civil rights movement (Norris, 2017, p. 117).

Progressive reforms experienced a setback in the late 1970s and 1980s when a crime wave swept the United States. Racial tensions had risen, Nixon's war on drugs had created a culture of incarceration, and "tough on crime" rhetoric became a ticket to elected office. In *Gregg v. Georgia*, the Supreme Court upheld the constitutionality of the death penalty, and the practice was reinstated (*Gregg v. Georgia*, 1972). Later Supreme Court cases then upheld the constitutionality of executing minors aged 16 or 17 or "persons with 'mental retardation,'" making it easier for people to be sentenced to capital punishment (Death Penalty Information Center, n.d.-a). The number of inmates on death row jumped from fewer than 700 in 1980 to more than 3,000 in 1995 (Death Penalty Information Center, n.d.-b). More generally, "The federal sentencing guidelines took effect in 1987 and by the mid-1990s, the number of federal inmates nearly doubled" (Norris, 2017, p. 118). Huge increases in the numbers of people being prosecuted, incarcerated, and on death row strained the resources of the criminal justice system; more mistakes were made, and the number of people wrongfully convicted also increased. Some experts have estimated that "1–5% of all felonies may have been wrongful convictions" (Norris, 2017, p. 3). Those percentages carry great weight considering that the number of people incarcerated in the United States rose from 329,821 in 1980 to 771,243 in 1990 (Cohen, 1991). With the anti-crime policies of the 1970s and 1980s filling the prisons, the development and forensic application of DNA typing in the mid-1980s could not have come at a better time, challenging the notion that all people convicted of crimes were guilty and highlighting problems in the criminal justice system that lead to the conviction of the innocent.

Approximately 99.9% of human DNA is the same in all people.[1] It is the difference in the remaining 0.1% of DNA that separates us from one another (Chial, 2008). Within that 0.1%, if there is adequate, trustworthy DNA evidence at the scene of a crime, a defendant can be proven definitively innocent, and another defendant can be proven guilty[2] to a near certainty because "the probability of two individuals having the same DNA fingerprint [is] . . . less than one in 33 billion" (Norris, 2017, p. 31). Using the standard set of Combined DNA Index System (CODIS) short tandem repeat (STR) markers, the estimated probability could be less than one in several billion. Forensic DNA evidence had the potential to revolutionize the legal system, giving confidence to a court's decision that only science could bring when the evidence was available.

ADMISSIBILITY OF DNA EVIDENCE

With the development of DNA forensic technology (described in Chapter 1), a key issue in wrongful convictions was the extent to which judges would allow DNA evidence in the courtroom. *Frye v. United States* first defined what was acceptable science-based evidence in court in 1923 (*Frye v. United States*, 1923). To be admissible in court, the *Frye* standard required a new form of scientific evidence to be

---

1. The human genome has 3×10^{-9} base pairs. Around 0.1% of these are "polymorphic"; that is, they vary in the human population, and the variants are present at > 1% frequency. There are many more rare variant sites, but they are not considered "polymorphic" and are not generally useful in forensic individual identification.
2. DNA evidence can indicate the source of a biological specimen at the crime scene; it does not per se establish guilt or innocence.

generally accepted in the relevant scientific community. The admissibility of DNA evidence was decided in 1988 in two cases, *People v. Wesley* and *People v. Bailey*, which led to what is commonly referred to as the Wesley-Bailey hearing. Both cases involved rapes, and in both the prosecution wanted to use DNA evidence from the blood of the defendants to prove their guilt. The question that needed to be answered was whether the DNA tests done by Lifecodes met the *Frye* standard. Judge Joseph Harris ruled in favor of admitting the DNA test results, and between 1987 and 1989 alone, similar results were admitted in more than 100 cases in the United States (Norris, 2017, p. 37).

The reliability of DNA evidence was briefly called into question in 1987 when Barry Scheck and Peter Neufeld collaborated on the defense of Joseph Castro and challenged the methods under which unregulated private companies performed DNA tests (*People v. Castro*, 1989). This case, which was discussed in some detail in chapter 1, brought together prosecutors, geneticists, and mathematicians, who found that many laboratory technicians would sometimes call a match when the tests only showed some similarities. As a result of Lifecode's unreliable tests in this case, national standards for the practice of forensic DNA testing began to emerge.

In 1993 the Supreme Court changed the principles upon which scientific evidence could be admitted into federal court, finding that under Rule 702 of the Federal Rules of Evidence, the judge should be the gatekeeper for admitting scientific evidence. This decision allowed evidence to be admitted without general acceptance of the scientific community, as previously required, if the judge was convinced the evidence was reliable (*Daubert v. Merrell Dow Pharmaceuticals*, 1993). *Daubert* opened up the possibility of getting new DNA testing techniques and results admitted into federal courts and in state courts that followed the *Daubert* standard. (The issue of the admissibility of DNA evidence is discussed in more detail in chapter 13.)

EXPANSION OF THE INNOCENCE MOVEMENT

The work Jim McCloskey began with Centurion Ministries exploded in the 1990s with the widespread introduction of DNA evidence in the criminal justice system. Seeing the value of DNA technology in righting potentially wrongful convictions, Barry Scheck and Peter Neufeld founded the Innocence Project in 1992 as a law school clinic at Cardozo Law School (Innocence Project, n.d.-a). Throughout the 1990s similar organizations were formed in Arizona, Illinois, Washington, Wisconsin, and California. International efforts to free the innocent spread to Canada, where legendary exoneree Rubin "Hurricane" Carter was the first executive director of the Association in Defense of the Wrongly Convicted (American Program Bureau, n.d.).

The initial organizations were mostly housed in law schools, and the education of law students was a fundamental part of their mission. The projects fit perfectly into existing law school clinical programs, which were expanding in the 1990s based on a growing perception that law schools should provide more practical experiential legal education to prepare students for the practice of law. A 1992 report of the American Bar Association Section on Legal Education and Admission to the Bar Chaired by Robert MacCrate (commonly referred to as the MacCrate Report) explored the gap between traditional legal education and law practice and laid the foundation for this expansion (American Bar Association, 1992). The rise in popularity of experiential education, combined with the growing stories about wrongful convictions, provided fuel for the innocence movement.

In 1993, when Kirk Bloodsworth, who had been convicted for the rape and murder of a nine-year old girl, became the first person on death row to be exonerated by DNA evidence, the Innocence Movement also began to power the anti-death penalty movement (Innocence Project, n.d.-g). In June 2000, researchers at Columbia University Law School wrote a comprehensive account of American death penalty cases and their pitfalls, "A Broken System: Error Rates in Capital Cases, 1973–1995" (Death Penalty Information Center, 2000). Among other findings, they concluded that there had been reversible error in 7 out of 10 capital cases. Many of these errors had led to innocent people being convicted, and this revelation changed the way people talked about the death penalty. Through cases such as Kirk Bloodsworth's, the innocence movement brought to light the high risk of executing an innocent person. This became a key component of the death-penalty abolition movement's position.

The new millennium rang in with two key events that shaped the innocence movement. The first was the introduction of the Innocence Protection Act in 2000 by Democratic senator Patrick Leahy, cosponsored by Republican senator Gordon Smith as a bipartisan bill (Innocence Protection Act, 2004). The goals of the act were to make it easier to gain access to postconviction DNA testing, increase capital defendants' access to adequate legal representation, and provide compensation for the wrongfully convicted. This was the beginning of legislative action on death-penalty cases around the country; all 50 states now give defendants statutory access to postconviction DNA testing, 32 states have wrongful conviction compensation laws, 24 states require interrogations to be recorded, 23 states have laws that require evidence be preserved for future DNA testing, and 19 states have introduced eyewitness identification reforms that improve the reliability of identifications (Innocence Project, n.d.-h).

The other major event of 2000 was the first unified meeting of the innocence organizations, which took place in Chicago. Project directors shared information about casework and the structure of their programs. By 2004 a loosely affiliated network was created, and in 2005, 15 organizations became the founding members of that network (Innocence Network, n.d.).

Since 2005, more than 100 innocence organizations have developed in the United States and around the world, including groups active in Armenia, Australia, Canada, the Czech Republic, Germany, Holland, Ireland, Israel, Italy, Japan, New Zealand, Poland, South Africa, Taiwan, and the United Kingdom. There is also an independent network of innocence organizations in Latin America called Red Inocente doing work in Argentina, Bolivia, Chile, Colombia, Costa Rica, Ecuador, Mexico, Panama, Peru, and Puerto Rico (Red Inocente, n.d.).

WHAT HAVE WE LEARNED FROM THE INNOCENCE MOVEMENT?

As Peter Neufeld has put it, "when you hold DNA up to the mirror and that mirror is the criminal justice system, the DNA points out all the cracks in that mirror in a way that we never saw it before" ("What Jennifer Saw," 1997). Tracking DNA analyses sheds light on why wrongful convictions occur and what we can do to prevent them.

Sometimes People Lie

According to a report by the National Registry of Exonerations, perjury or false accusations account for the largest percent (55%) of the first 1,600 exonerations in the United States. Perjury can be the result of informant testimony, in which the informant is seeking a better deal for his or her own criminal acts or tells lies for other reasons.

It was a lie told by a 12-year-old boy that put Ricky Jackson in prison for 39 years, the third longest time any exonerated prisoner has been behind bars (National Registry of Exonerations, n.d.-c). In 1975, when Harold Franks, a money order salesman, was murdered at a grocery store, 12-year-old Edward Vernon said he saw Jackson and two others commit the murder from his school bus. When Vernon tried to back out of the lie at the time of the lineup, he was told by the police that it was too late and that his parents would be arrested for perjury if he recanted. Though there was no physical evidence presented, and none of the defendants had a criminal record, the word of a 12-year-old boy was enough to sentence Jackson to death. Finally, in 2013, Vernon confessed to having lied, and Mark Godsey and Brian Howe from the Ohio Innocence Project picked up Jackson's case. In November 2014, Jackson was released. Upon his release, Jackson stated, "I spent my 20s, my 30s, my 40s and nearly all of my 50s in prison. . . . Time is just something that you can't get back, so I'm not going to really cry about it. But I had plans for my life" (Sutton, 2017, Chapter 1).

Sometimes the Crime Never Happened

The first person to be exonerated through DNA evidence, Gary Dotson, was put in jail for a crime that did not even happen, due to lies told combined with misleading forensic evidence, as discussed briefly in chapter 1 (Innocence Project, n.d.-f). In fact, misleading forensic evidence was involved in 23% of the first 1,600 exonerations, according to the National Registry of Exonerations. In 1977, 16-year-old Kathleen Crowell fabricated a story of rape to cover up a possible pregnancy through consensual sex with her boyfriend. Dotson matched the description she gave and was arrested (National Registry of Exonerations, n.d.-b).

A key piece of evidence in the trial was that the pubic hair found in Crowell's rape kit was "similar" to Dotson's. A 352-page report by the National Academy of

Sciences, "Strengthening Forensic Science in the US: A Path Forward," highlighted forensic issues in microscopic hair analysis, shoe print comparisons, and bite mark comparisons. The report identified significant weaknesses in certain common forensic tests and recommended ways to improve the field (National Academy of Sciences, 2009). The largest postconviction DNA study conducted by the FBI was released in 2015. Focused on cases handled before 2000, the review found that 26 of 28 FBI analysts either "provided . . . testimony with erroneous statements or submitted laboratory reports with erroneous statements" (Norris, 2017, p. 108).

In Dotson's case, the hair analysis was based on the assumption of the validity of vague terms like *consistent* and *similar* and the idea that every person's hair is unique. The forensic review should not have placed as much weight on the microscopic hair analysis, especially because the forensic scientist failed to exclude Dotson as the person who committed the rape. Both type A and type B blood were found in the rape kit, and both the victim and Dotson have type B blood. While Dotson could have been the source of the type B blood, he could not have been the source of the type A blood.

Crowell recanted her testimony in 1985, and in 1989, after spending 10 years in prison, Dotson was exonerated through the DNA analysis of the semen found in Crowell's rape kit, as noted in chapter 1 (National Registry of Exonerations, n.d.-b). "As is the case of many of these exonerations, her decision to come forward was initially met with great skepticism by both prosecutors, police, and the general public" (Sutton, 2017, Chapter 2). Nevertheless, Crowell wrote a book called *Forgive Me* in 1985 and gave Dotson $17,000 from the proceeds of its sales. As Dotson noted upon his release, "It's been 12 long, long grueling years and I'm relieved it's over. . . . The stigma remains. It's something I have to deal with. I've been referred to as a 'convicted rapist.' Now, at least, I'm no longer 'convicted'" (Bluhm Legal Clinic, n.d.).

Sometimes the Government Fabricates Evidence

There are countless stories of corrupt and incompetent police officers fabricating evidence to close cases or for other reasons. However, such fabrication has also occurred within the scientific community. One of the most egregious examples is that of Fred Zain, a West Virginia forensic laboratory expert who testified in hundreds of criminal cases and was notorious for falsifying DNA profiles to obtain convictions (Giannelli, 2010):

> The acts of misconduct on the part of Zain included: (1) overstating the strength of results; (2) overstating the frequency of genetic matches on individual pieces of evidence; (3) misreporting the frequency of genetic matches on multiple pieces of evidence; (4) reporting that multiple items had been tested when only a single item had been tested; (5) reporting inconclusive results as conclusive; (6) repeatedly altering laboratory records; (7) grouping results to create the erroneous impression that genetic markers had been obtained from all samples tested; (8) failing to report conflicting results; (9) failing to conduct or to report conducting additional testing

to resolve conflicting results; (10) implying a match with a suspect when testing supported only a match with the victim; and (11) reporting scientifically impossible or improbable results. (Matter of Investigation of West Virginia State Police Crime Laboratory, Serology Div., 190 W.Va. 321, 438 S.E.2d 501, 503 [1993])

Unfortunately, Fred Zain's misconduct was not an isolated set of incidents. There have been cases of scientific fraud throughout the United States that have led to scandals in several major city crime labs, including fraudulent DNA analysis (Marshall Project, 2017).

Sometimes People Identify the Wrong Person

Eyewitness misidentifications often cause wrongful convictions. They have played "a role in more than 70% of convictions overturned through DNA testing" (Innocence Project, n.d.-d). Many factors trigger misidentifications, including the pressure to make an identification when the wrong person has been detained, misremembering elements of a traumatic event, and misleading police procedures designed to obtain identifications. These factors are particularly relevant in rape cases, in which the trauma of surviving a rape affects the "ability to accurately identify a person" (Garrett, 2011, p. 51). In fact, according to the National Registry of Exonerations, 72% of wrongful sexual assault convictions were the result of mistaken witness identification (National Registry of Exonerations, 2012).

Ronald Cotton spent 10 years in prison for a crime he did not commit after Jennifer Thompson, then a 22-year-old college student, identified him as the man who broke into her apartment and raped her. As she later described on the news show *60 Minutes*, "[I was] trying to pay attention to a detail, so that if I survived . . . I'd be able to help the police catch him" (CBS, 2009). Even with that attention to detail, Thompson identified the wrong man, and Cotton was sentenced to life in prison plus 54 years. Later DNA testing led police to the actual rapist, and Cotton was exonerated in 1995. Cotton and Thompson are now friends and travel the country as advocates for eyewitness identification reform (Innocence Project, n.d.-i).

Sometimes People Confess to Crimes They Didn't Commit

On July 10, 1994, a woman was kidnapped and assaulted by two men in Waukegan, Illinois. She told the police that the kidnappers' car was a dark, four-door sedan with tinted windows. Angel Gonzalez was later seen leaving the victim's apartment building driving a dark sedan with tinted windows. He was arrested and identified by the victim (Innocence Project, n.d.-b).

The police read Gonzalez his Miranda rights in English, and he was initially "interrogated in English before a Spanish speaking detective took over. Gonzalez was asked to write a confession in Spanish, which, his lawyers would say later, did not match the crime he was being accused of; the detective then 'translated' the confession into English, with details now matching the rape allegations, and had Gonzalez sign it" (Sutton, 2017, Chapter 3).

The police had no physical evidence linking Gonzalez to the crime and did not investigate his alibi. The signed confession and the victim's misidentification led to a sentence of 40 years in prison for kidnapping and sexual assault. He served 20 of those years before he was exonerated in 2015, when his DNA was excluded from biological material in the rape kit.

There were false confessions in 13% of the first 1,600 documented innocence cases in the United States (National Registry of Exonerations, 2012). As Supreme Court Justice Byron White once put it, "the defendant's own confession is probably the most probative and damaging evidence that can be admitted against him" (*Bruton v. United States*, 1968). Why would an innocent person say he or she was guilty of a crime he or she didn't commit?

Brandon L. Garrett, the author of *Convicting the Innocent*, reported that for those who confessed among the first 250 cases of exoneration, "almost all of their interrogations were prolonged affairs, lasting many hours or even days" (Garrett, 2011, p. 21). In wrongful convictions, many of the confessions are either coerced-compliant confessions—in which innocent people confess in order to end interrogations by telling the police what they "want to hear"—or coerced-internalized confessions—in which innocent people come to believe their own guilt as a result of aggressively suggestive interrogations (Nesterak, 2014).

In a case of false confession, mentally impaired David Vasquez was led to believe by police that he would receive the death penalty if he didn't confess to the murder of a woman in Arlington County, Virginia. The "confession," which was taped and transcribed, revealed the official entrapment:

DETECTIVE: Did she tell you to tie her hands behind her back.
VASQUEZ: Ah, if she did, I did.
DETECTIVE: Whatcha use?
VASQUEZ: The ropes?
DETECTIVE: No, not the ropes. Whatcha use?
VASQUEZ: Only my belt.
DETECTIVE: No, not your belt. . . . Remember being out in the sun room, the room that sits out to the back of the house? . . . [A]nd what did you cut down? To use?
VASQUEZ: That, uh, clothesline?
DETECTIVE: No, it wasn't a clothesline, it was something like a clothesline. What was it? By the window?. . . Think about the venetian blinds, David. Remember cutting the venetian blind cords?
VASQUEZ: Ah, it's the same thing as rope.
DETECTIVE: Yeah.

The policemen then guided Vasquez through the details of the actual killing.

DETECTIVE: Okay, now tell us how it went, David . . . tell us how you did it.
VASQUEZ: [S]he told me to grab the knife and, and, stab her, that's all.

DETECTIVE (RAISING HIS VOICE): David, no, David.
VASQUEZ: If it did happen, and I did it, and my fingerprints were on it . . .
DETECTIVE (SLAMMING HIS HAND ON THE TABLE AND YELLING): You hung her!
VASQUEZ: What?
DETECTIVE: You hung her!
VASQUEZ: Okay, so I hung her. (Priest, 1989)

Although the "confession" was riddled with coercion and contamination, the trial judge did not suppress it, and Vasquez was convicted and sentenced to 35 years in prison. Vasquez was exonerated when another man was convicted of three identical crimes. (Innocence Project, n.d.-c). Because false confessions are such a major factor of wrongful conviction, there has been a national movement to record all confessions. As of the end of 2016, 23 states have some form of recording requirement for confessions (Sutton, 2017, Introduction).

Sometimes Lawyers Do a Bad Job

The Supreme Court set the constitutional standard for competent criminal defense representation in *Strickland v. Washington*, finding: "Counsel is unconstitutionally ineffective if his performance is both deficient, meaning his errors are 'so serious' that he no longer functions as 'counsel,' and prejudicial, meaning his errors deprive the defendant of a fair trial" (Garrett, 2011). In *Convicting the Innocent*, Garrett describes the decision as the "foggy mirror test," in which, if you put a mirror under a defense lawyer's nose and the mirror fogs, then the counsel is deemed effective (Garrett, 2011, p. 205).

In his dissent in *Strickland*, Justice Thurgood Marshall wrote, "To tell lawyers and the lower courts that counsel for a criminal defendant must behave 'reasonably' and must act like 'a reasonably competent attorney' is to tell them almost nothing." This prophetic statement has played out over the many years since that statement was made. There has been an uneven application of the *Strickland* standard, and many innocent people have gone to prison based on ineffective representation by counsel.

Rafael Madrigal was working on the line in a factory in July 2000 when a shooting occurred 35 miles away. Madrigal was identified by a single witness from a picture taken when he was a teenager. There was no physical evidence attaching him to the crime, no motive, and no confession. Nonetheless, he was convicted because his defense attorney failed to investigate his alibi. There was definitive proof of his innocence in the form of time cards, witnesses, and a statement from his supervisor that Madrigal was the only person who knew how to use a piece of manufacturing equipment, and without him on the assembly line at the time of the crime 35 miles away, production would have stopped. Furthermore, there was proof of his innocence in the form of a jailhouse recorded statement from another suspect. In 2009, Madrigal was exonerated after nine years of wrongful incarceration for a crime he did not commit, with the help of the California Innocence Project (California Innocence Project, n.d.). This is but one example of the hundreds, possibly thousands, of wrongful convictions that have resulted from incompetent representation.

Sometimes Wrongful Convictions Are Race Based

In her 2010 book *The New Jim Crow*, Michelle Alexander compares mass incarceration to the segregationist "Jim Crow" laws:

> Arguably the most important parallel between mass incarceration and Jim Crow is that both have served to define the meaning and significance of race in America. Indeed, a primary function of any racial caste system is to define the meaning of race in its time. Slavery defined what it meant to be black (a slave), and Jim Crow defined what it meant to be black (a second-class citizen). Today mass incarceration defines the meaning of blackness in America: black people, especially black men, are criminals. (Alexander, 2010, p. 197).

The reasons for the disproportionate number of racial minorities in prison in the United States range from selective policing in neighborhoods with a high proportion of racial minorities, to improper jury selection, to high rates of poverty among racial minorities, to blatant prejudice. According to the study *Race and Wrongful Convictions in the United States*:

> African Americans are only 13% of the American population but a majority of innocent defendants wrongfully convicted of crimes and later exonerated. They constitute 47% of the 1,900 exonerations listed in the National Registry of Exonerations (as of October 2016), and the great majority of more than 1,800 additional innocent defendants who were framed and convicted of crimes in 15 large-scale police scandals and later cleared in "group exonerations" (National Registry of Exonerations, 2017).

CHALLENGES FACING THE INNOCENCE MOVEMENT

The innocence movement faces substantial challenges. The most difficult is obtaining cooperation from police, prosecutors, and judges in obtaining evidence and reopening old cases. Politics and a distaste for revisiting old convictions often make cooperation difficult to obtain. However, a wave of conviction integrity units is opening in prosecution offices around the country, providing hope that prosecutors will reopen and remedy their own cases (Garrett, 2011, p. 242).

Enacting policies that decrease the number of wrongful convictions and remedy the ones that have occurred continues to be a challenge. Interrogations must be recorded, and we must reform practices such as the "Reid technique," which focuses on getting a suspect to agree with a particular theory of the police whether or not it is the truth. This technique has been linked to many false confessions and yet is still a training staple for hundreds of thousands of police officers around the country and the world.

States must also adopt eyewitness reform throughout their policing agencies. For example, the Eyewitness Identification Reform Act (2007), enacted by the North Carolina Actual Innocence Commission, requires investigators to say, "the perpetrator might or might not be present in the lineup" and "it is as important to exclude innocent persons as it is to identify the perpetrator" (Garrett, 2011). In addition, the act requires that eyewitnesses express the degree of their confidence in the identification they have made on camera in order to avoid false confidence in trial. New Jersey was the first state to adopt the double-blind lineup, in which neither the police nor the witness knows who the suspect is in a lineup. This is critical, as police have a tendency to suggest who the suspect is through feedback or body language. Reforms in eyewitness identification include using sequential photo arrays, in which photos are shown one at a time, as opposed to six-pack photo arrays, which have been shown to be ineffective and lead to a high number of misidentifications.

We must review how we handle jailhouse informants. Inmates may be a valuable asset in prosecution; however, with the incentive of lowering their sentences or gaining parole, they can often be unreliable witnesses. Before allowing jailhouse informants to testify, judges should at least assess their reliability. In addition, "since contaminated informant testimony may appear uncannily reliable, all conversations with jailhouse informants could be electronically recorded to ensure that case information is not disclosed and that any deals are documented" (Garrett, 2011, p. 256).

We must continue to review and challenge unreliable forensic science reports and overstated findings. As Peter Neufeld has explained, "With hair, a biological kind of evidence, you can go back and do DNA testing [to check its accuracy]. But if inappropriate exaggerated testimony was used to explain a bullet or to compare tire prints found at the scene to the tires on a defendant's car. . . . the FBI wants to have published standards of what's considered appropriate testimony by their forensic scientists" (quoted in Sutton, 2017 Chapter 7). In addition, there should be nationwide, even global, standards for testing evidence. As Neufeld has also said, "We're talking about science—there should be no disparity. The results of a forensic test should be the same in Buffalo as they are in Dallas, the same way the results of, say, a clinical laboratory's test for strep throat provides the same results whether you are in Buffalo or Dallas" (quoted in Sutton, 2017, Chapter 7).

These scientific and procedural remedies to problems in the justice system are easier to envision and even effect than the more deep-seated challenges, such as those that lead to a disproportionately punitive system for minorities and poor people in our communities. Similarly, we face huge challenges in reintegrating those who have been wrongfully convicted back into society and into good lives. Exonerees struggle to find jobs, rebuild relationships, deal with PTSD, and face a host of other challenges that come from losing a significant portion of their lives to a crime they never committed.

CONCLUSION

DNA technology has changed lives. What began as a scientific discovery in a lab has led to a revolutionary movement to reform our criminal justice system. The collective work of scientists, lawyers, law students, and activists around the world is reflected in the hundreds of innocent men and women who have walked out of prison.

Beyond the exonerations, the work of innocence organizations around the world has resulted in the training of thousands of law students who have gone on to be excellent attorneys. Furthermore, there have been some profound changes in the criminal justice system over the past two decades that are directly attributable to the innocence movement. We have seen the adoption of evidence-preservation laws around the world so that later scientific testing is possible long after trial. There have been reforms in the way identifications and confessions are obtained. And there have been reforms in forensic science, with junk science being debunked and other science following tighter protocols. All of this makes for a better justice system, one that can never be perfected but can always be improved.

REFERENCES

Alexander, M. (2010, January 5). *The New Jim Crow Mass Incarceration in the Age of Colorblindness*. New York, NY: The New Press, 197.

American Bar Association. (1992, July). *Legal education and professional development*. Retrieved from https://www.americanbar.org/content/dam/aba/publications/misc/legal_education/2013_legal_education_and_professional_development_maccrate_report).authcheckdam.pdf.

American Program Bureau. (n.d.). *Rubin "Hurricane" Carter*. Retrieved from http://www.apbspeakers.com/ speaker/rubin-hurricane-carter.

Bluhm Legal Clinic. (n.d.). *First DNA exoneration: Gary Dotson*. Retrieved from Center on Wrongful Convictions, Northwestern Pritzker School of Law website: http://www.law.northwestern.edu/legalclinic/wrongfulconvictions/exonerations/il/gary-dotson.html.

Bruton v. United States, 391 U.S. 123 (1968).

California Innocence Project. (n.d.). *Rafael Madrigal*. Retrieved from https://californiainnocenceproject.org/read-their-stories/rafael-madrigal/.

CBS. (2009, March 11). Eyewitness: Anatomy of a story. *60 Minutes*. https://www.cbsnews.com/news/eyewitness-anatomy-of-a-story/

Centurion Ministries. (n.d.-a). *About us*. Retrieved from http://centurion.org/about-us/.

Centurion Ministries. (n.d.-b). *How and why it was created*. Retrieved from http://centurion.org/about-us/at-a-glance/.

Centurion Ministries. (n.d.-c). *Jorge De Los Santos*. Retrieved from http://centurion.org/cases/jorge-de-los-santos/.

Chial, H. (2008). DNA sequencing technologies key to the Human Genome Project. *Nature Education*, 1(1), 219. Retrieved from https://www.nature.com/scitable/topicpage/dna-sequencing-technologies-key-to-the-human-828.

Cohen, R. (1991). Prisoners in 1990. *Bureau of Justice Statistics Bulletin*, https://www.bjs.gov/content/pub/pdf/p90.pdf

Daubert v. Merrell Dow Pharmaceuticals, 509 U.S. 579 (1993).

De Los Santos v. O'Lone, No. 82-1717, *slip op*. (D.N.J. July 6, 1983).

Death Penalty Information Center. (n.d.-a). *Part 1: History of the death Penalty*. Retrieved from https://deathpenaltyinfo.org/part-i-history-death-penalty.

Death Penalty Information Center. (n.d.-b). *Searchable execution database*. Retrieved from https://deathpenaltyinfo.org/views-executions.

Death Penalty Information Center. (2000, December). *2000 year end report: A watershed year of change*. Retrieved from https://deathpenaltyinfo.org/2000-year-end-report-watershed-year-change.

Devers, L. (2011, January 24). Background. *Plea and Charge Bargaining*. Retrieved from Bureau of Justice Assistance website: https://www.bja.gov/Publications/PleaBargainingResearchSummary.pdf.

DNA Forensics. (n.d.). *DNA fingerprinting: The discovery of DNA fingerprinting*. Retrieved from http://www.dnaforensics.com/DNAFingerprinting.aspx.

Federal Bureau of Investigation (FBI). (n.d.). *Combined DNA Index System (CODIS): Overview*. Retrieved from https://www.fbi.gov/services/laboratory/biometric-analysis/codis.

Franklin, B. (1785, May 14) Letter to Benjamin Vaughan. In Albert H. Smyth, ed., *The Writings of Benjamin Franklin (1906)*, vol . 9, p. 293.

Frye v. United States, 293 F. 1013 (D.C. Cir. 1923).

Furman v. Georgia 408 U.S. 238, 92 S. Ct. 2726 (1972)

Garrett, B. (2011, April 4). *Convicting the Innocent*. Cambridge MA: Harvard University Press.

Giannelli, P. (2010). Scientific Fraud. *Criminal Law Bulletin*, 46(6). Retrieved from https://scholarlycommons.law.case.edu/cgi/viewcontent.cgi?article=1097&context=faculty_publications.

Gideon v. Wainwright, 372 U.S. 335 (1963).

Gregg v. Georgia, 428 U.S. 153 (1972).

Innocence Network. (n.d.). *About*. Retrieved from http://innocencenetwork.org/about/.

Innocence Project. (n.d.-a). *About the organization*. Retrieved from https://www.innocenceproject.org/about/.

Innocence Project. (n.d.-b). *Angel Gonzalez*. Retrieved from https://www.innocenceproject.org/cases/angel-gonzalez/.

Innocence Project. (n.d.-c). *David Vasquez*. Retrieved from https://www.innocenceproject.org/cases/david-vasquez/.

Innocence Project. (n.d.-d). *Eyewitness misidentification*. Retrieved from https://www.innocenceproject.org/causes/eyewitness-misidentification/.

Innocence Project. (n.d.-e). *Exonerate the innocent*. Retrieved from https://www.innocenceproject.org/exonerate/.

Innocence Project. (n.d.-f). *Gary Dotson*. Retrieved from https://www.innocenceproject.org/cases/gary-dotson/.

Innocence Project. (n.d.-g). *Kirk Bloodsworth*. Retrieved from https://www.innocenceproject.org/cases/kirk-bloodsworth/.

Innocence Project. (n.d.-h). *Policy reform*. Retrieved from https://www.innocenceproject.org/policy/.

Innocence Project. (n.d.-i). *Ronald Cotton*. Retrieved from https://www.innocenceproject.org/cases/ronald-cotton/.

Innocence Protection Act of 2004, P.L. No. 108-405, 118 Stat. 2260.

Mapp v. Ohio, 367 U.S. 643 (1961).

Matter of Investigation of West Virginia State Police Crime Laboratory, Serology Div., 190 W.Va. 321, 438 S.E.2d 501 (1993).

Miranda v. Arizona, 384 U.S 436 (1966).

Myers, G. (2016, June 6). Former football star Brian Banks, who served five years in prison for a rape he didn't commit, disgusted by Brock Turner ruling. *New York Daily News*. https://www.nydailynews.com/sports/football/wrongfully-convicted-brian-banks-disgusted-brock-turner-ruling-article-1.2663595

National Academy of Sciences. (2009). *Strengthening forensic science in the US: A path forward*. https://www.ncjrs.gov/pdffiles1/nij/grants/228091.pdf

National Registry of Exonerations. (n.d.-a). *About*. Retrieved from www.law.umich.edu/special/exoneration/Pages/about.aspx.

National Registry of Exonerations. (n.d.-b). *Gary Dotson*. Retrieved from https://www.law.umich.edu/special/exoneration/Pages/casedetail.aspx?caseid=3186.

National Registry of Exonerations. (n.d.-c). *Longest incarcerations*. Retrieved from https://www.law.umich.edu/special/exoneration/Pages/longestincarceration.aspx.

National Registry of Exonerations. (2012). *The first 1,600 exonerations*. Retrieved from https://www.law.umich.edu/special/exoneration/documents/1600_exonerations.pdf.

National Registry of Exonerations. (2017). *Race and wrongful convictions in the United States*. (Gross, Possley, and Stephens). http://www.law.umich.edu/special/exoneration/Documents/Race_and_Wrongful_Convictions.pdf

Nesterak, E. (2014, October 21). Coerced to confess: The psychology of false confessions. *The Psych Report*. Retrieved from http://thepsychreport.com/conversations/coerced-to-confess-the-psychology-of-false-confessions/.

Norris, R. (2017, May 16). *Exonerated: A History of the Innocence Movement*. New York: NYU Press.

100-to-1 Rule. (2007, November 15). *The New York Times*. Retrieved from http://www.nytimes.com/2007/11/15/opinion/15thu3.html

People v. Bailey, 65 Ill. App. 2d 261, 205 N.E.2d 756 (1965).

People v. Castro, 143 Misc.2d 276 (1989).

People v. Wesley, 83 N.Y.2d 417 (1994),

Priest, D. (1989, July 16). At each step, justice faltered for Va. man. *The Washington Post*. Retrieved from http://www.law.virginia.edu/pdf/faculty/garrett/falsconfess/vasquez_david_wash_post.pdf.

Red Inocente. (n.d.). *Exoneraciones Mexico*. Retrieved from http://redinocente.org/exoneraciones-mexico/.

Strickland v Washington Citation: 466 U.S. 668 (1984).

Sutton, B. (2017, February 17). *TIME Innocent: The Fight Against Wrongful Conviction*. New York: Time Inc.

US Const. amend VII.

US National Library of Medicine (n.d.). The Francis Crick papers: The discovery of the double helix, 1951–1953. Retrieved from https://profiles.nlm.nih.gov/SC/Views/Exhibit/narrative/doublehelix.html.

What Jennifer saw: Interview with Peter Neufeld. (1997, February 25). PBS *Frontline*. https://www.pbs.org/wgbh/pages/frontline/shows/dna/interviews/neufeld.html

Wilke, A., & Mata, R. (2012). Cognitive bias. In *Encyclopedia of human behavior*. 2nd ed. Retrieved from https://adweb.clarkson.edu/~awilke/Research_files/EoHB_Wilke_12.pdf.

CHAPTER 3
Analysis of Forensic Mixtures

MICHAEL COBLE, BRUCE BUDOWLE, AND HENRY ERLICH

After many years of peer-reviewed research publications, admissibility hearings, quality assurance standards, and validation studies, the DNA analysis of single-source forensic evidence samples and the calculation of random match probabilities (RMPs) are well established and well accepted in the criminal justice system. The analysis and interpretation of mixture samples—samples containing DNA from multiple contributors—remains often technically difficult (depending on the nature of the mixture) and sometimes subject to legal challenge (see chapter 13). The analysis of mixtures with two contributors can in some cases be relatively straightforward, and multilocus profiles for each individual can be determined, or at least the number of explanations to describe the potential contributors will be relatively small. Consequently, for these relatively straightforward two-person mixtures, a statistic can be calculated when an individual (because of his or her reference profile) cannot be excluded as a contributor of the mixture. The statistics traditionally applied to these simple mixtures typically are in the form of an RMP, the combined probability of inclusion (CPI), or a likelihood ratio (LR).

The current standard of forensic genotyping is the analysis of the short tandem repeat (STR) loci. As described in chapter 1 and elsewhere (Flores et al. 2014; Ensenberger et al. 2016; Kraemer et al., 2017), STR genotyping systems are based on the use of fluorescently labeled PCR primers that flank specific STR loci, resulting in the amplification of PCR products of varying length. The PCR products are separated based on DNA fragment length by an automated capillary electrophoresis system and visualized as a pattern of fluorescent peaks, known as an *electropherogram* (see figure 3.2, later in the chapter). The relative heights of the peaks, representing alleles, reflect the relative amounts of fluorescent signal (called relative fluorescent units, or RFUs), which are directly related to the effective amount of input DNA being tested. The interpretation of DNA evidence from mixed samples is based on the observed patterns of STR peaks.

The process of determining the individual components of the mixture is termed *deconvolution*. In some two-person mixtures, the profile of one contributor may be known (e.g., one may assume that the DNA from the female complainant of a sexual assault will be on her vaginal swab, obtained as evidence of that assault), so the genotype of this individual can be considered in the composite set of genotypes in the mixture. In such cases, this assumed known contributor's alleles can be "subtracted" from the mixture alleles, with the remaining alleles being attributed to the unknown contributor. The unknown contributor's profile can be treated in a similar manner to that of a single-source profile, and thus the weight of the evidence can be calculated using the RMP statistic.

When this is not possible, many laboratories use the probability of inclusion (PI) for each genetic marker and combine the PIs of the eligible markers of the profile to generate a CPI. Basically, this calculation conveys the portion of the population that could contribute to the mixture that includes the genetic profile of the person of interest. While these methods of interpretation and accompanying statistical approaches are valid for two-person mixtures, the interpretation of complex mixtures (i.e., three or more contributors and/or low-quantity or low-quality samples) is more complicated. Clarification of the application of the traditional methods of interpretation has helped reduce misinterpretation; however, substantial data often were deemed inconclusive and simply not used. Recent technical developments in the form of probabilistic genotyping (PG) now enhance interpretation of complex mixtures and make possible analysis of low-quantity samples that previously were not susceptible to study. Probabilistic genotyping exploits the power of biological models and computer algorithms to enable interpretation of complex mixtures in a reliable, robust, and more objective manner than previous methods. These developments are discussed later in the chapter.

The analysis of samples with multiple contributors is not a special or rare case. A common category of mixed samples is semen stains from sexual assault/rape cases in which sperm from the assailant and epithelial cells from the victim are both present, for example. Mixed blood stains may also be found at the crime scene. With the increased sensitivity of current PCR-based tests, combined with increased sensitivity of the capillary electrophoresis instruments used to detect DNA profiles, evidence that has been touched by several different individuals, "touch DNA," has become another common category of forensic mixtures encountered in casework. Touching a cell phone or a handgun can transfer skin cells from the handler to the item. For low-copy, or trace, DNA samples, the proportion of mixtures is likely to be substantially greater than 50% (Gill et al., 2015). In the analysis of such trace DNA mixtures, one of the major challenges is the increased likelihood of missing data due to failure to sample, amplify, and/or detect some of the alleles of some or all of the contributors, a phenomenon known as "allele drop-out."

Although the *interpretation* of mixed samples can be challenging, the *detection* of a mixture is relatively straightforward. In a single-source sample, except for rare genetic reasons, a marker should display no more than two alleles (two alleles if the individual is heterozygous at the marker and one allele if the individual is homozygous at the marker). If three or more alleles are observed, then it is most likely that more than one individual has contributed DNA to the sample. The logic is simple. Somatic cells, all cells other than gametes—sperm and eggs—are all diploid; that is, they contain two copies of each autosomal chromosome, (i.e., chromosomes other than X or Y) and therefore two copies of each and every marker on those chromosomes. (Red blood cells are an exception because they lose their nuclei as they mature.) For any given marker, a single individual has two alleles, one inherited from the father and one from the mother. If the alleles from each parent have a different number of repeats (e.g., 12 copies of the tandem repeat sequence from the father and 17 repeats from the mother), the marker will show two peaks in the DNA profile (often displayed in an electropherogram, the pattern generated by capillary electrophoresis of the PCR products), and the genotype is called *heterozygous*. If the alleles from each parent have the same number of repeats (e.g., 12 repeats from the father and 12 repeats from the mother), the marker will only show one peak on the electropherogram (the 12s will stack upon each other), and the genotype is called *homozygous*.

Markers with many different possible alleles are better at detecting mixtures than are markers with just a few alleles. The forensic community has opted for a specific class of genetic markers known as short tandem repeats (STRs), which are highly polymorphic and well suited for mixture deconvolution. Current STR panels, or kits, contain 20–30 markers and thus have sufficient capacity, in theory, to individualize a sample. However, the quantity and quality of the DNA derived from an evidence sample will affect the overall evidentiary power of an STR panel. Many labs require the presence of more than two alleles at two markers to confidently claim the presence of a mixture.

Even a single genetic marker with two alleles, however, can reveal the presence of multiple contributors. In a single-source sample, the ratio of the two alleles should be around 50:50 for a heterozygote, and a homozygote with two copies of the same allele should be represented by a single allele peak. If the balance between the two alleles differs substantially from the 50:50 ratio expected for a heterozygous single-source sample (e.g., 75:25), the presence of a mixture, with a major homozygous contributor, can be inferred.

As the number of DNA contributors increases from the relatively straightforward two-person mixture, the data become more difficult to deconvolute. Primarily, it is more challenging to unequivocally determine the true number of contributors to a complex mixture. Consider a mixture with three alleles, A, B, and C, detected at a single genetic marker (see figure 3.1). If the data indicate more of the A and B alleles and a lesser amount of the C allele, then the simplest interpretation, assuming two contributors, is that the majority component of the mixture is an AB genotype and

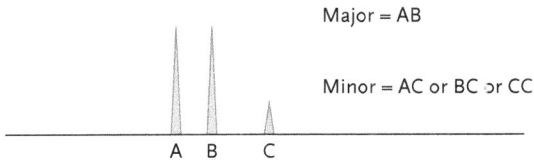

Figure 3.1 A hypothetical mixture with three alleles (A, B, and C). The genotype of the inferred major is AB. For the minor contributor, the inferred genotypes could be AC, BC, or CC.

a minor component could be CC, AC, or BC (or a C partnered with any allele possible for that marker due to the possibility of allele drop-out). With data from multiple genetic markers, the complexity of the interpretation increases. Alleles stacking on one another, combined with the possibility of missing alleles due to drop-out, can now influence the estimation of the number of contributors in the mixture. In addition, the uncertainty in the inferred genotype assignment at one genetic locus must now be combined with the inferred genotypes at the other loci to determine the genetic profiles of the individual contributors. We discuss in the following section the current strategies employed to analyze these very common and challenging forensic samples.

ANALYSIS OF MIXTURES BY STR MARKERS AND CAPILLARY ELECTROPHORESIS

Currently, mixed and single-source samples are routinely analyzed using commercial STR kits, manufactured and distributed by commercial entities such as ThermoFisher Scientific, Promega Corporation, and Qiagen. The most recent kits include greater than 20 different STR markers as well as markers on the X and Y chromosomes to allow for determination of the biological sex of the sample's donor. As noted previously, the relative heights of the peaks, representing alleles, reflect the relative amounts of florescent signal (of RFUs), which are related to the amount of input DNA. In cases where there is abundant DNA and the relative amounts of the two contributors to a simple mixture are clearly different, the interpretation of the electropherogram pattern is straightforward. The "major" and "minor" components are clearly resolved, and there are no missing data. In other cases, the interpretation can be difficult because the contributors are not readily resolved and/or there are missing data, requires making assumptions and subjective judgments.

There are a number of technical challenges to the interpretation of STR profiles in mixed samples. The height of the peaks measured in the electropherograms does not always precisely and quantitatively reflect the amounts of the DNA fragments, particularly for low-quantity samples. For example, the amplification of low quantities of DNA with very few copies of the target DNA region is subject to *stochastic* (randomly generated) variation. Stochastic variation reflects sampling fluctuation due to low numbers and the operation of chance. As a result, the PCR-generated signal of the alleles (peak height) may not represent the amount of DNA that was in the original sample; the resulting profile may show significant imbalance in the height of allele peaks. The largest effect is allele drop-out (no detected peak) or missing data. In addition, very low-level peaks may be due to "background noise" inherent in the technology; if scored as true alleles, they would create false positive data.

Another challenge to the interpretation of STR profiles is *stutter*, a PCR artifact that results from the amplification of STR regions. Stutter (see figure 3.2) typically is an amplification byproduct (showing up as a peak) that is one or two repeat copies shorter or longer than the true allelic repeat length. For single-source samples, stutter is dealt with readily, as the stutter peak height is far less than the true allele

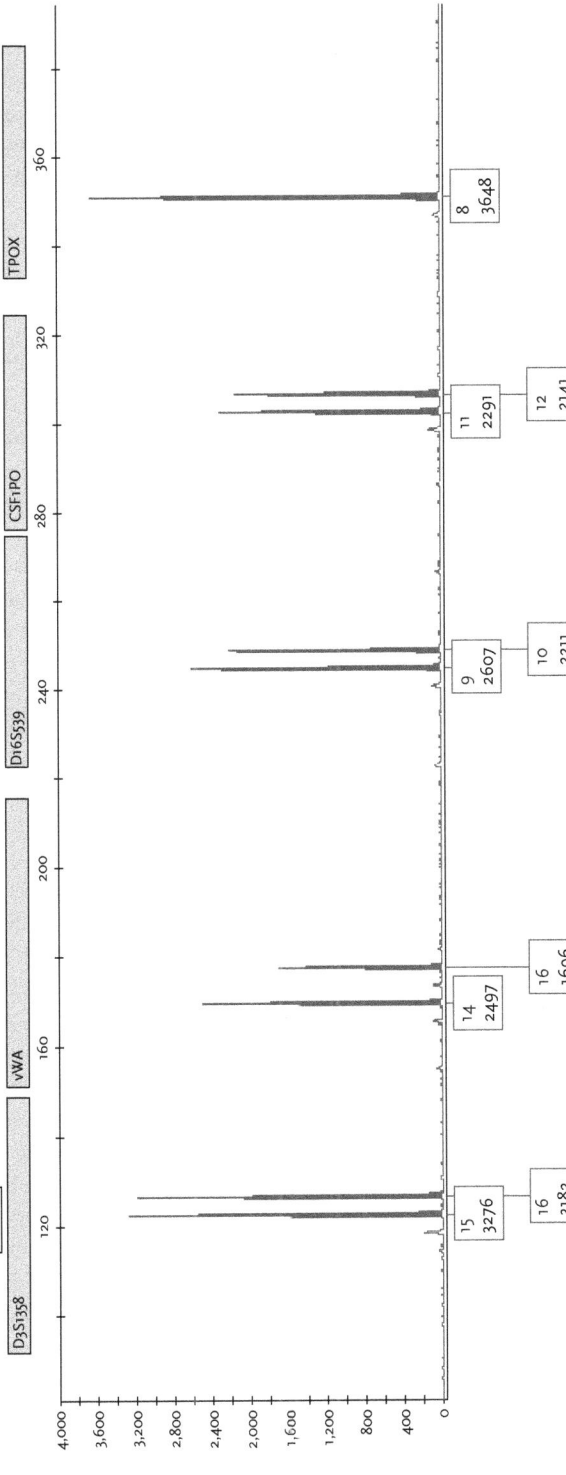

Figure 3.2 An example of an electropherogram from a single-source individual. Five genetic markers are shown in this "blue" fluorescent dye channel: D3S1358, vWA, D16S539, CSF1PO, and TPOX. The values along the Y-axis represent the relative fluorescent units (RFUs) that indicate how much fluorescence was detected by the capillary electrophoresis instrument. Along the X-axis, in the box, are two numbers that represent the number of repeats (top number) and the RFUs of the peak (bottom number). The D3S1358 marker shows that the individual is heterozygous (15, 16) at this locus, while the individual is homozygous (8,8) at the TPOX locus. The small, unlabeled peak to the left of the 15 allele at D3S1358 is a representation of a "stutter" artifact from PCR. The vWA locus in this electropherogram is an example of peak imbalance but is well within the acceptable peak height ratio for an expected single-source profile.

(e.g., around 10%) and easily recognized as an artifact. However, in mixtures with a minor contributor(s), the combination of true minor peaks and the artifactual stutter peaks could confound the interpretation of the STR pattern generated from the mixture. Finally, in samples with low input DNA, the potential of allele drop-out, the failure to detect a true allele, complicates the interpretation of mixtures and the calculation of CPI, RMP, and LR to convey the weight of the evidence to fact finders, judges, and juries.

THE USE OF A STOCHASTIC THRESHOLD FOR MIXTURE INTERPRETATION

Generally, if sufficient amounts of DNA are placed into the PCR reaction (e.g., 1.0 nanograms [ng] of DNA), then one would expect the amplification of a heterozygous locus to have nearly equal (balanced) peak heights. For example, if an individual has the 6, 8 genotype at a specific locus, then the height of the 6 allele peak would be approximately the same as the 8 allele peak. For example, a perfect balance of peak heights for both alleles would give a peak height ratio (PHR) of 1.0. Amplification of low quantities of DNA (e.g., 0.1 ng) may generate a locus profile with significant allele imbalance, for example, a PHR of 0.2 (20%). Even more problematic is the extreme imbalance when one allele of the genotype is sampled or amplified and the other allele fails to be amplified to a sufficient level to be detected in the electropherogram (e.g., the 6 allele is observed, but the 8 allele is missing in the profile). This "missing" allele is said to have dropped-out of the profile, as previously noted. The analyst may indicate in the case notes the possibility of allele drop-out by noting the genotype as 6, F—where "F" means there was the possibility that the sister allele "failed" to amplify.

Missing data can confound interpretation, since only one allele from a heterozygous genotype can be incorrectly interpreted as a homozygous genotype. Research publications (Moretti et al., 2001; Budowle et al., 2009) and forensic DNA-focused organizations such as the DNA Commission of the International Society for Forensic Genetics (ISFG) (Gill et al., 2006) and the Scientific Working Group on DNA Analysis Methods (SWGDAM, 2010, 2017) have recommended laboratories adopt a stochastic threshold (ST) to interpret DNA profiles. This threshold provides a level of confidence based on peak height that an allele has not dropped out of the profile.

A *stochastic threshold* is a quantitative line drawn across the profile in which a partner allele with a peak having a height above this threshold will most likely be observable (i.e., an indication that drop-out is unlikely). Thus, heterozygote and homozygote types are assigned with greater confidence. If a peak falls below this stochastic line, the profile is interpreted as potentially having missing data (dropped-out alleles). Using figure 3.1 as an example, if allele "C" falls below the stochastic threshold, then there are now four possible genotypes or genotype categories for the minor contributor: AC, BC, CC, and CQ—where the "Q" allele represents a missing allele in the profile that is any other allele besides A, B, or C (e.g., D, E, F). It is also possible to have a profile in which both of the minor alleles have dropped out of the mixture, resulting in a "QQ" genotype as an explanation for the missing alleles. The consequence of having the "Q" allele in the list of possible genotypes from the mixture is that the locus can no longer be used for a statistical calculation like CPI, because this statistic does not have a mechanism to accommodate the possibility of allele drop-out. Essentially *any* individual in the population could now be a contributor to the mixture. Thus, the data at this marker are uninformative regarding the CPI.

FURTHER CHALLENGES IN INTERPRETING MIXTURES

There are a number of other challenges with interpreting mixtures, especially mixtures with low-level DNA profiles (also called low-template DNA profiles). These issues include (a) uncertainty in determining the number of contributors, (b) allele stacking, (c) allele drop-in, and (d) stutter peaks.

Determining Number of Contributors

When interpreting a mixture, the analyst must determine the number of contributors to that mixture. The true number of contributors to a mixture may never be known in a casework sample, so a sensible estimation is required. Analysts typically use peak heights and the maximum number of observed alleles at a locus to estimate the number of contributors. For example, in a single-source sample (one individual), one expects to observe one (homozygous) allele or two alleles (heterozygous) at any particular marker. (It is possible to observe three alleles at a locus but, as previously noted, this tends to be an extremely rare event.) Generally, if the analyst observes three or four alleles at a particular marker, then a mixture of at least two contributors is indicated. Observing five or six alleles at a marker would suggest a mixture of at least three individuals.

In addition to the maximum number of alleles observed, peak height information can be used to estimate a likely number of contributors. It is necessary to estimate the number of contributors for (1) assessing the potential for allele drop-out, as the potential increases with an increase in the number of contributors; and (2) making statistical calculations, especially to assign LRs where the number of contributors is used to propose two mutually exclusive hypotheses to explain the mixture and account for the observed data.

Allele Stacking

Allele stacking, or allele sharing, can often increase the uncertainty in estimating the number of contributors to a mixture. Suppose there are three individuals in a mixture with the genotypes of AB, BC, and BD at a particular locus. Suppose they have also contributed equal amounts of DNA. The profile would show three "minor" peaks (A, C, and D) that are about equal in peak height, along with a "major" peak, that for the B allele, about three times the height of the minor alleles. Note that relying simply on the number (four) of observed allele peaks at this locus would indicate that this profile could be a two-person mixture. However, observing the relatively large peak height at the B allele suggests that there are more than two contributors in the mixture. In addition, the relatively large B allele peak could be mistakenly interpreted as the presence in the mixture of a homozygous BB genotype. Although the hypothetical mixture scenario does not include a BB individual, this possible explanation is consistent with the data, and the presence of the BB type cannot be excluded. This uncertainty must be included when estimating the possible contributors to the mixture. Data from the other markers of the profile, however, may provide insight into the proportions of each contributor in this hypothetical mixture, which may allow elimination of a major BB contributor from the interpretation.

Allele Drop-in

Allele drop-in, the presence of spurious additional peaks, is another confounding phenomenon in STR profile interpretation. Such peaks, a product of background noise, are sometimes called *contaminating alleles*, but they tend to be random as opposed to a true contamination from a secondary source. In a true contamination event, one may observe additional allele peaks at multiple loci from the secondary foreign profile present in the mixture or in a negative control. Drop-in events differ in that there is often only one, or, maybe two, additional peaks. For example, suppose a low-level 10 allele peak is observed in a profile that displays a 7, 8 genotype, suggesting possible mixture in what appears to be a single-source sample based on all the other STR loci. If a reamplification and genotyping of the DNA reveals only the 7, 8 genotype without detecting the 10 allele, the previous detection of the 10 allele can be attributed to background noise.

Stutter Peaks

Finally, *stutter peaks*, artifacts created by the PCR process,[1] can create challenges to mixture interpretation since these artifacts can be confused with authentic peaks from a minor contributor's alleles that are about the same height and repeat copy number as stutter peaks. For example, a stutter peak (n-1) for a true 10 copy allele peak might look like a 9-copy peak corresponding to a potential 9-copy allele.

---

1. Typically stutter PCR products are around 10% or less of the true allele product (depending on the locus and the size of the allele).

Additional factors such as DNA degradation and inhibition of PCR amplification, which can both lead to missing data (drop-out) of alleles in the profile, can also confound mixture interpretation. All these factors make mixture interpretation one of the greatest challenges for the DNA analyst. Uncertainty in mixture interpretation, compounded by lack of a vigorous validation study or interpretation protocols developed by the testing laboratory, can contribute to misinterpretations.

How Mixtures Can Be Misinterpreted

While the procedures for interpreting simple mixtures are well defined and robust if followed correctly, there are still some situations in which misinterpretations can occur. It is important that analysts be cognizant of such situations so they can avoid misinterpreting the data and overstating the statistical weight of the evidence. Figure 3.3 provides an example of how a mixed profile can lead to misinterpretation. The example is a summary from an actual case, and for simplicity, only three loci are shown. Based on the totality of the profile, this profile presents as a mixture from two contributors: one major contributor and one minor contributor. Viewing from left to right across the profile, the major contributor shows two alleles (AB) at the first locus, two alleles (DE) at the second locus, and two alleles (FG) at the third locus. Also present in the mixture are minor alleles attributed to the minor contributor at both the first (C allele) and third (H allele) loci. Since these alleles are not in stutter positions, they are authentic alleles and not artifacts. The analytical threshold (solid line) is set at 50 RFU, and the stochastic threshold (dashed line) () is set at 150 RFU.

The alleles of the major contributor can confidently be interpreted as having no drop-out (they're above the stochastic threshold), and thus all data are present. The minor contributor does have a risk of drop-out, since the "C" allele at locus 1 and the "H" allele at locus 3 are below the stochastic threshold. At the interpretation stage, so far, the two minor alleles can be confidently assigned to the minor contributor, and the minor alleles can be compared with those of a reference

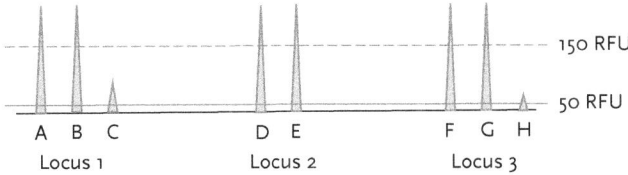

Figure 3.3 A hypothetical mixture at three loci based on an actual case. The genotype of the inferred major is AB at locus 1, DE at locus 2, and FG at locus 3. For the minor contributor, the inferred genotypes must include the C allele at locus 1 and the H allele at locus 3. The analytical threshold is set at 50 RFUs (solid line), and the stochastic threshold is set at 150 RFUs (dashed line).

profile (a person of interest, such as a suspect). If the suspect does not possess the alleles observed in the minor contributor, then the suspect may be excluded as the minor contributor to the mixture. In contrast, if the suspect carries alleles that are concordant with the minor alleles of the minor contributor, then the suspect cannot be excluded as a contributor to the mixture. In the statistical analysis of the strength of the evidence (using CPI) for inclusion of the suspect, both locus 1 and 3 *cannot* be used in the calculation, since the minor alleles at these two loci are below the stochastic threshold, and there is a possibility that their partner (sister) alleles have dropped out.

Locus 2 shows no hint of any minor alleles. Therefore, the possibilities are that (1) the minor alleles are masked by the major contributor's alleles, (2) the minor alleles have dropped out, or (3) a combination of the events of 1 and 2 has occurred. In this particular case, the testing laboratory decided that since *no alleles were observed below the stochastic threshold*, then this locus *could* be used for including the person of interest (POI) as a potential contributor and thus used this locus in the CPI calculation. A more precise or defensible interpretation would be to call results at this locus inconclusive and not include the locus for statistics (Bieber et al., 2016). In other words, if drop-out is possible at locus 1 and locus 3, then it is also possible that drop-out may be happening at locus 2. The main issue with the way this laboratory was interpreting mixtures is that the stochastic threshold was used as a "binary" method of interpretation: if alleles were below the stochastic threshold, then the locus was excluded for statistics, and if all alleles were above the stochastic threshold, then the locus was used in the statistical calculation. This logic is inappropriate and will inflate the strength of the CPI (Bieber et al., 2016). In this case, the laboratory should have deemed the profile too complex to interpret using the CPI, approach since drop-out is possible across the entire profile.

While the analysis of the STR electropherograms from mixed samples can in many cases be interpreted in a relatively straightforward manner, the interpretation often involves assumptions and subjective judgments, such as that just noted or, for example, where to set the threshold for peak detection—that is, how to distinguish a low true STR peak from background noise and PCR artifacts. The potential for subjective judgment to give rise to inconsistent interpretations has also been demonstrated in a number of multi-lab studies. Interlaboratory studies can be useful exercises to help laboratories monitor consistency of analysis following their mixture interpretation within the laboratory and among different laboratories. Studies from European laboratories (Rand et al., 2002; Crespillo et al., 2014; Toscanini et al., 2016) have typically shown wide variation in the results within and between the participants' laboratories.

The National Institute of Standards and Technology (NIST) conducted two interlaboratory studies in 2005 and 2013, MIX05 and MIX13 (Butler et al., 2018), involving laboratories performing STR genotyping in the United States and Canada using sets of contrived mixtures. The overall results were concordant with the studies from Europe: a wide range of interpretations and thus results was observed both within and between laboratories participating in the studies.[2] Four examples in MIX13 (ranging from relatively simple two-person mixtures to more complex three-person mixtures with stochastic issues), for instance, revealed the need for more training and education for mixture interpretation. Several responses from the study in fact showed the same issues with interpretation highlighted by figure 3.3.

TECHNICAL INNOVATIONS THAT AID ANALYSIS OF MIXTURES
Probabilistic Genotyping

The inclusion of a stochastic threshold has been necessary for more accurate interpretations of low-level mixtures where drop-out is possible. The limitation with using a stochastic threshold, however, is that it creates an "all or none" situation in which alleles are considered "binary": they are present or their potential for absence is accepted according to the peak heights of the observed alleles. This approach is necessary to reduce the chance of falsely including an individual as a contributor to a mixture, but it throws away potentially useful data. An example of how this traditional approach ignores useful data is displayed in figure 3.4, which shows two electropherograms with essentially the same data for a single marker. Each has three alleles displayed (for the sake of simplicity, stutter is not shown in this example), which are designated 14, 15, and 17, with peak height RFUs below the allele number (1625, 1555, and 300 or 299).

Based on the assumption that this profile is of a two-person mixture, alleles 14 and 15 would be consistent with being from a major contributor, and the major contributor would be typed as a heterozygote 14,15. Allele 17 would be consistent with originating from a minor contributor. The question then is what the possible types of the minor contributor in this electropherogram are. In this example, a stochastic threshold has been set at 300 RFUs. The minor contributor could be one of three possibilities: first, 17,17 homozygote; second, a heterozygote 14,17 or 15,17 (if one of the minor contributor's alleles is in the same position as either allele 14 or 15 and thus is masked by the major contributor); or third, a heterozygote designated, for example, 17,F (in which F represents any possible allele) if the observed peak height for allele 17 is sufficiently low such that allele drop-out is possible. These considerations are based on the fundamental molecular biology of DNA typing, and all methods of interpretation need to consider the possible explanations for the minor contributor. To determine which of the possible explanations for the minor contributor is better supported, the peak heights are assessed. Traditionally, a stochastic threshold was

---

2. One example (sample 05) from the MIX13 study was over-engineered to be problematic for the participants; this sample was atypical of casework samples, and the results should be viewed with caution.

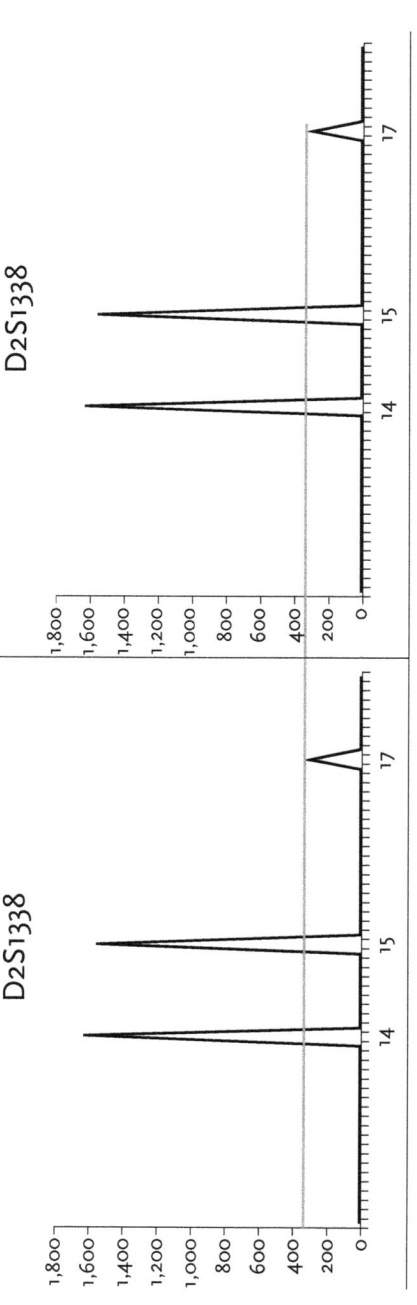

Figure 3.4 An example of a hypothetical profile of the same locus (D2S1338), where the minor allele differs by 1 RFU. On the left, the minor allele is exactly the same RFU as the stochastic threshold (300 RFU), providing confidence that no other sister alleles have "dropped out" of the profile. On the right, the minor allele is now 1 RFU below the stochastic threshold (299 RFU). There is now a risk that a sister peak may be missing from the profile.

invoked. Generally, high peak heights have little or no possibility of allele drop-out, and as the heights decrease (usually below the stochastic threshold), the chance of allele drop-out increases.

In this example, the stochastic threshold line is drawn, only for demonstration purposes, at 300 RFUs across the electropherograms (the actual stochastic threshold is determined based on internal validation studies performed by the laboratory). In the panel to the left, the height of allele 17 is 300 RFUs. Since the peak height of allele 17 is at the stochastic threshold, using the traditional method, it would be deemed that allele drop-out of a partner allele is highly unlikely; thus, the possible types for the minor contributor would be 17,17; 14,17; and 15,17. Under the traditional manual method, all three genotypes were equally possible, and all would be addressed equally statistically. However, given the relatively equal heights of the major peaks and that a smaller allele tends to have a higher height than a larger allele of a heterozygous person, as well as the possible additive effects that a minor contributor might display, the more probable explanation is that the minor is a 17,17, while the 14,17 and 15,17 types, though still possible, are less likely to explain the mixture. For the panel on the right, the DNA data are essentially the same as the electropherogram in the left panel, except allele 17 has a peak height of 299 RFUs (only 1 RFU difference compared with the left panel). Because the peak height for allele 17 is below the stochastic threshold, albeit by only 1 RFU, the possibility of allele drop-out must be considered. Therefore the genotype is assigned unequivocally as a 17,F, even though peak heights of 299 and 300 are not essentially different for practical purposes. This traditional approach to interpretation is valid but does not make the best use of the data; it moves from the assumption of no allele drop-out to considering the possibility of allele drop-out with a change in peak height of only 1 RFU. A better approach would be to consider all possible allele combination scenarios, with weighting based on the observed peak heights. Until recently, use of a stochastic threshold was necessary because there was no way to compute the probabilities of the myriad genotype scenarios.

With the advent of advanced computational methods, however, it is now possible to consider the possible genotypes giving rise to the observed peaks probabilistically. *Probabilistic genotyping*, according to SWGDAM (2015), refers to the use of biological modeling, statistical theory, computer algorithms, and probability distributions to calculate LRs and/or infer genotypes for the DNA typing results of forensic samples ("forensic DNA typing results"). Human evaluation and review are still required for the interpretation of forensic DNA typing results in accordance with the FBI director's Quality Assurance Standards for Forensic DNA Testing Laboratories. Probabilistic genotyping can be seen, then, as a tool to assist the DNA analyst in the interpretation of forensic DNA typing results. Probabilistic genotyping is not an "expert system" and is not intended to replace the human evaluation of the forensic DNA typing results or the human review of the output prior to reporting.

Probabilistic genotyping is particularly valuable in the analysis of mixed samples with very limited amounts of DNA, in which potential allele drop-out is likely. Probabilistic genotyping software can model (or depending on the software, will incorporate) a probability of alleles "dropping out" as well as "dropping in." Unlike the stochastic threshold method, in which all considered genotype combinations were equally probable, probabilistic software provides a statistical weighting to the different genotype combinations. The probabilistic approach "enhances the ability to distinguish true contributors and non-contributors" (SWGDAM, 2015), as opposed to traditional methods that would have deemed the data inconclusive.

All probabilistic genotyping software uses an LR as the reporting statistic. In general, the LR for a true contributor to the mixture tends to be much greater than 1, while the LR obtained for a noncontributor in the mixture tends to be less than 1. Several probabilistic software programs for mixture interpretation are currently available (see Coble & Bright, 2018), the two most widely used commercial packages in the United States being STRmix (Taylor et al., 2013) and TrueAllele (Perlin & Sinelnikov, 2009). The SWGDAM guidelines recommend extensive validation studies in forensic labs for all of these software. Other groups, such as the International Society for Forensic Genetics (Coble et al., 2016) and the United Kingdom Forensic Science Regulator (Forensic Science Regulator, 2018), have also provided guidance and recommendations for validation.

It is important to note that not all of the probabilistic software programs are built on the same models; it is likely (expected!) that two different programs will provide different LRs on the same profile. Some of these differences can arise from the human interpretation of the mixture. For example, some programs require the analyst to make decisions on stutter products and to remove these peaks before inputting the information into the program. Other programs will model stutter in the profile, and removal of these peaks prior to analysis is not necessary. An interlaboratory study using a "semicontinuous" program (the peak heights of the alleles in the mixture were ignored; only the alleles present were used by the software) found a wide range of variation in the results despite all of the labs using the same software, the same allele frequencies, and the same hypotheses to build the LR. Much of the variation was a result of different analysts making decisions about which stutter peaks were included (or not) in the software program (Barrio et al., 2018). Researchers have also reanalyzed the MIX13 study with several probabilistic genotyping programs and found that they outperformed present methods such as CPI (Buckleton et al., 2018). Finally, a recent interlaboratory study using STRmix (Bright et al., 2019) from 174 participants found a high degree of repeatability and reproducibility among the resulting LRs, with most differences almost always minor or conservative.

Both STRmix and TrueAllele use fully continuous methods of interpretation, in which peak heights are used and a variety of parameters are modeled through a process of simulations to determine which combinations of genotypes best explain the mixture. Although these and other software programs use similar approaches, the models within the software can be and are often different. This difference was highlighted in a case in New York in which a young boy was killed and a minor contributor profile was found in a mixture recovered from his fingernails (Augenstein, 2016). An analysis of the mixture using TrueAllele gave an inconclusive result for the POI, while the analysis by STRmix gave an inclusionary LR for the POI. Ultimately the judge, who found STRmix to be reliable, excluded the results (ruling them inadmissible) generated by STRmix in this case because there was a lack of internal validation studies performed by the laboratory in determining the parameters used by the software. It is important that one understand the models being used by the software program and determine the properties and limits of the software through a rigorous validation study.

With probabilistic genotyping, such as STRmix, the data in figure 3.4 can be evaluated more effectively than using a strict stochastic threshold approach. Probabilistic genotyping would consider all possible scenarios (or at least probable scenarios) and weight them accordingly. With probabilistic genotyping, the 17,17 genotype for the minor contributor would be weighted the most probable for both panels. The other scenario genotypes would be given lower weights (lesser probabilities), as they are still possible given the known intrinsic variation of DNA typing results prior to any interpretation of the profile.

Subsequently, the probability of the DNA profile would be assessed given particular genotype(s) under alternate propositions. This simplified example illustrates that probabilistic genotyping is more logical and hence more powerful than the traditional methods. Under this approach, the statistical significance of inclusion in the mixture will be expressed as an LR given the genotypes of a suspect versus that of an unknown, unrelated individual in the greater population. It is worth noting that with the CPI approach, the panel on the right in figure 3.4 would be deemed inconclusive and no data would be used—neither the major nor the minor peaks in the profile. Clearly the development and implementation of probabilistic genotyping represents a huge advance for forensic genetics over the CPI approach.

Haploid Lineage Markers and Next Generation Sequencing

One of the challenges in interpreting mixtures with STR markers, or any autosomal genetic marker, is combining detected alleles into inferred genotypes. The use of haploid lineage genetic markers, polymorphic sequences that reside on the Y chromosome or in the mitochondrial genome, can simplify the analysis of some mixtures because each contributor donates only one allele per marker, compared with the two alleles for autosomal markers. For example, a mixture of three different individuals would be expected to display at most three different mitochondrial DNA (mtDNA) sequences, each corresponding to each single contributor; for a single autosomal genetic marker, however, there could be three to six different alleles present, and many possible genotypes would have to be considered to explain the data.

The recent development of DNA sequencing technologies, referred to collectively as next generation sequencing (NGS) or massively parallel sequencing (MPS), offers the capability to quantify the contributors of a mixture and thus may allow for expansion of the use of mtDNA for typing mixtures.

Next generation sequencing is discussed in chapter 5 and thus is not described in detail here. Its value lies in performing sequencing in a "massively parallel" fashion; that is, millions of individual sequencing reactions are performed simultaneously in parallel, which provides a high throughput system for DNA analyses. Second, these many parallel reactions are each "clonal"; that is, each is initiated with a single DNA molecule. Because each clone is sequenced separately, each contributor's mitochondrial genome is sequenced individually. The advantage is that the contributions to the mixture can be quantified by counting individual sequence reads. These features provide the information to make it possible to deconvolve mtDNA mixtures.

CONCLUSION

The interpretation of complex mixtures is arguably the greatest challenge in the application of DNA analysis to forensic specimens. This difficulty can lead, and has led, to misinterpretation of results and in some cases (e.g., the Earl Washington Jr. case described in chapter 1) wrongful convictions. Fortunately, some recent developments have made the deconvolution of mixtures much more robust and reliable. The development of probabilistic genotyping has made the interpretation of mixed STR profiles more objective and more statistically powerful. Given that the different software uses slightly different models, it might be valuable to use more than one PG program in the analysis of mixed profiles whenever possible. The use of haploid lineage markers, such as Y chromosome markers and mtDNA polymorphisms, can also significantly simplify the deconvolution of mixtures, although these provide less discrimination power than autosomal STRs. Finally, given the clonal and massively parallel nature of NGS/MPS technology, quantitative analysis of mixtures can now be achieved by counting sequence reads.

Complex mixtures, in sum, can now be interpreted with more confidence, but they still raise technical and statistical issues; careful and rigorous education and training will be necessary to address the remaining challenges.

REFERENCES

Augenstein, S. (2016, August 26). Prosecutor's DNA evidence tossed from upstate N.Y. murder trial. *Forensic Magazine*. Retrieved from https://www.forensicmag.com/news/2016/08/prosecutors-dna-evidence-tossed-upstate-ny-murder-trial?cmpid=horizontalcontent.

Barrio, P. A., et al. (2018). GHEP-ISFG collaborative exercise on mixture profiles (GHEP-MIX06) reporting conclusions: Results and evaluation. *Forensic Science International: Genetics, 35*, 156–163.

Bieber, F. R., Buckleton, J. S., Budowle, B., Butler, J. M., & Coble, M.,D. (2016). Evaluation of forensic DNA mixture evidence: Protocol for evaluation, interpretation, and statistical calculations using the combined probability of inclusion. *BMC Genetics*, *17*, 125.

Bright, J. A., et al. (2019). STRmix™ collaborative exercise on DNA mixture interpretation. *Forensic Science International: Genetics*, *40*, 1–8.

Buckleton, J. S., et al. (2018). NIST interlaboratory studies involving DNA mixtures (MIX13): A modern analysis. *Forensic Science International: Genetics*, *37*, 172–179.

Budowle, B., et al. (2009). Mixture interpretation: Defining the relevant features for guidelines for the assessment of mixed DNA profiles in forensic casework. *Journal of Forensic Sciences*, *54*, 810–821.

Butler, J. M., Kline, M.C., & Coble, M. D. (2018). NIST interlaboratory studies involving DNA mixtures (MIX05 and MIX13): Variation observed and lessons learned. *Forensic Science International: Genetics*, *37*, 81–94.

Coble, M. D., & Bright, J. A. (2019). Probabilistic genotyping software: An overview. *Forensic Science International: Genetics*, *38*, 219–224.

Coble, M. D., et al. (2016). DNA Commission of the International Society for Forensic Genetics: Recommendations on the validation of software programs performing biostatistical calculations for forensic genetics applications. *Forensic Science International: Genetics*, *25*, 191–197.

Crespillo, M., et al. (2014). GHEP-ISFG collaborative exercise on mixture profiles of autosomal STRs (GHEP-MIX01, GHEP-MIX02 and GHEP-MIX03): Results and evaluation. *Forensic Science International: Genetics*, *10*, 64–72.

Ensenberger, M. G., et al. (2016). Developmental validation of the PowerPlex(®) Fusion 6C System. *Forensic Science International: Genetics*, *21*, 134–144.

Flores, S., Sun, J., King, J., & Budowle, B. (2014). Internal validation of the GlobalFiler™ Express PCR Amplification Kit for the direct amplification of reference DNA samples on a high-throughput automated workflow. *Forensic Science International: Genetics*, *10*, 33–39.

Forensic Science Regulator. (2018). Software validation for DNA mixture interpretation. *FSR-G-223* (1). Retrieved from https://assets.publishing.service.gov.uk/government/uploads/system/uploads/attachment_data/file/730994/G223_Mixture_software_validation_Issue1.pdf.

Gill, P., et al. (2006). DNA commission of the International Society of Forensic Genetics: Recommendations on the interpretation of mixtures. *Forensic Science International*, *160*(2–3), 90–101.

Gill, P., et al. (2015). Genotyping and interpretation of STR-DNA: Low-template, mixtures and database matches-Twenty years of research and development. *Forensic Science International: Genetics*, *18*, 100–117.

Kraemer, M., et al. (2017). Developmental validation of QIAGEN Investigator® 24plex QS Kit and Investigator® 24plex GO! Kit: Two 6-dye multiplex assays for the extended CODIS core loci. *Forensic Science International: Genetics*, *29*, 9–20.

Moretti, T., et al. (2001). Validation of short tandem repeats (STRs) for forensic usage: Performance testing of fluorescent multiplex STR systems and analysis of authentic and simulated forensic samples. *Journal of Forensic Sciences*, *46*, 647–660.

Perlin, M. W., & Sinelnikov, A. (2009). An information gap in DNA evidence interpretation. *PLoS One*, *4*(12), e8327.

Rand, S., Schürenkamp, M. & Brinkmann, B. (2002). The GEDNAP (German DNA profiling group) blind trial concept. *International Journal of Legal Medicine*, *116*, 199–206.

Scientific Working Group on DNA Analysis Methods (SWGDAM). (2010). *SWGDAM interpretation guidelines for autosomal STR typing by forensic DNA testing laboratories*. Retrieved from http://www.forensicdna.com/assets/swgdam_2010.pdf.

Scientific Working Group on DNA Analysis Methods (SWGDAM). (2015). *Guidelines for the validation of probabilistic genotyping systems*. Retrieved from http://media.wix.com/ugd/4344b0_22776006b67c4a32a5ffc04fe3b56515.pdf.

Scientific Working Group on DNA Analysis Methods (SWGDAM). (2017). *Interpretation guidelines for autosomal STR typing by forensic DNA testing laboratories.* Retrieved from https://docs.wixstatic.com/ugd/4344b0_50e2749756a242528e6285a5bb478f4c.pdf.

Taylor, D., Bright, J. A., & Buckleton, J. (2013). The interpretation of single source and mixed DNA profiles. *Forensic Science International: Genetics, 7*(5), 516–528.

Toscanini, U., et al. (2016). Analysis of uni and bi-parental markers in mixture samples: Lessons from the 22nd GHEP-ISFG Intercomparison Exercise. *Forensic Science International: Genetics, 25,* 63–72.

CHAPTER 4

Forensic DNA Data Banks and Data Mining

Balancing Societal Interests and Public Safety

FREDERICK R. BIEBER

This chapter provides an overview of forensic and other DNA data banks and a discussion of metrics used for evaluating their effectiveness. It includes an introduction to novel data mining methods for proxy searching of DNA databases, including using open-access public data sets for genealogy searches for missing persons and crime suspects through their relatives. A look at the future of data mining, based on scientific, legal, and societal privacy considerations, addresses the need for structured policy development at the national and international levels to help balance privacy concerns with security interests.

Forensic data banks contain actual biological samples and DNA extracts or the computerized databases of coded DNA profiles of convicted offenders, arrestees, and crime scene samples. When used for investigative and law enforcement purposes, forensic DNA data banks have provided key investigative leads in thousands of criminal investigations. Many of these crimes would never have been resolved without use of such data banks. In addition, in many forensic investigations the exclusion of suspects whose DNA profiles can be found in an offender or arrestee database can save valuable investigative time. These data banks have expanded to allow searching for the identity of missing persons and recovered human remains.

Proxy searching, known as *familial searching*, is conducted in more than a dozen US states, and in many other countries, to investigate high-profile crimes when no direct match is found between crime scene DNA evidence and someone in the offender/arrestee database. This search process involves rank ordering DNA profiles in the offender/arrestee database based on the likelihood of a close biological relationship to DNA found associated with the crime in question.

Recently, searching crime scene DNA profiles against those held in public, open-access, ancestry DNA databases has enabled identification of profiles contributed by even distant relatives of those who have left DNA evidence at a crime scene. In scores of instances this has led investigators, using metadata and genealogy tools, to locate viable suspects. After direct confirmation of a DNA match to crime scene evidence, suspects have been arrested and charged with very serious crimes that had been unsolved for years or even decades (Bieber, 2018).

In addition to the government-controlled forensic DNA data banks just described, human tissues and DNA genotyping data are collected from millions of individuals with no criminal history. These include the millions of blood spot cards held by the US military for identification purposes as well as DNA and other stored human tissues held by universities, hospitals, medical centers, and many other private organizations. These institutions collect tissue and blood samples, then extract DNA for genotyping or full DNA sequencing for research projects related to disease diagnostics or therapeutic treatments.

Along with these tissue and DNA collections for humanitarian and health purposes, a growing interest in genealogy, ancestry, and phenotype prediction has led to the growth of a new direct-to-consumer (DTC) industry that offers such testing for a fee and the collection of millions of biological samples for DNA typing. In return, individuals receive a report that summarizes basic information about purported biogeographic ancestry, admixture, and in some instances, health information, such as risk for specific diseases.

Innovative methods have recently dramatically extended the reach of data mining and proxy searching capabilities and led to introduction of the concept of *genealogics* (aka forensic genetic genealogy) to describe the search for more distant relatives of those whose DNA is collected for forensic, medical, or research purposes. Since April 2018, investigative approaches have employed searches of some of these public databases, which contain DNA profiles based on testing of hundreds of thousands of single nucleotide polymorphisms (SNPs). This is done to identify individuals who willingly deposited their DNA and whose profiles might reveal that they share sufficient numbers of SNP haploblocks to indicate they might be close or distant relatives of an owner of crime scene DNA. Such DNA data, stored in private and open-access public ancestry databases, has enabled identification of distant relatives of those leaving crime scene DNA evidence. After hours of genealogy searching efforts using a panoply of public records, many of these investigations have led to the identification and arrest of suspects in high-profile cases involving murder and sexual assault (Kennett, 2019). These developments create a new dynamic tension between the sometimes-competing interests of privacy and public safety.

Despite the public safety and, in the case of disasters, humanitarian benefits of DNA data collection, in thousands of criminal investigations relevant DNA samples are not collected or administrative and laboratory logjams delay prompt database searches. Furthermore, hundreds of DNA database matches (hits) with the evidence languish without any follow-up by law enforcement or prosecutors due to workload, case backlogs, unavailable or deceased witnesses, or simple inadvertence. This prevents or delays potential confirmatory matches with DNA from the implicated suspect and therefore can leave the public at risk of potential harm from recidivistic offenders who otherwise might have been apprehended. Even now, after several decades of use, data compilations on meaningful metrics of success, which might identify further shortcomings in criminal investigations, are critically lacking. This leaves legislators and policy analysts with inadequate data on which to judge the overall effectiveness of offender/arrestee DNA data banking programs. Data collection and research are needed to improve the effectiveness of forensic DNA data banks in meeting the stated goal of enhancing public safety. Systematic tracking and follow-up of database hits and prioritizing case management must become a high priority (Bieber, 2006).

GENERAL BACKGROUND

DNA testing has become the gold standard for forensic identification since its introduction in the late 1980s. Such testing involves determining both DNA length and sequence variation found in animal and plant genomes, including genomes of microbial organisms (Bieber, 2004; Bieber et al., 2016; Mapes et al., 2015). Having successfully withstood continuous legal admissibility challenges, particularly in the United States, DNA analysis's usefulness for individualization (i.e., its potential inculpating power and therefore its probative value) is now beyond reasonable doubt, making DNA analysis a central part of routine criminal investigations around the world (Buckleton et al., 2016; Lazer, 2004). Moreover, the discriminating power of DNA technology permits rapid exclusion of suspects when DNA crime scene specimens are available and has led to DNA-based exonerations of hundreds of individuals who had been wrongfully convicted before the modern era of forensic DNA testing (Gross et al., 2005). In addition to its use for criminal investigations, DNA testing is also widely used in civil paternity cases, immigration and probate disputes, and missing persons recovery efforts, and for identification of human remains in the aftermath of mass disaster, political repression, or war (Biesecker et al., 2005; Budowle et al., 2005). Most of the techniques and technologies utilized in DNA analysis are shared by basic researchers and medical professionals, as well as by scientists working outside the field of forensics (Bieber, 2004; Butler, 2014; Ge et al., 2012; Gill et al., 2006; Primorac & Schanfield, 2014). Despite the successes of forensic DNA technology, problems and challenges do persist in some aspects of collection, preservation, analysis, interpretation, and jury understanding of forensic evidence (Kloosterman et al., 2014; Thompson, 1995).

Most industrialized nations now collect biological samples from crime scenes and from those convicted of serious crimes for entry into government DNA data banks. These data banks consist of a variety of separate computerized databases that hold the DNA profiles of convicted offenders, crime scene samples, missing persons, mass disaster samples, or family members who have come forward in hopes of identifying loved ones lost in a mass disaster (Biesecker et al., 2005; Budowle et al., 2005; Ge et al., 2011a).

The rationale for establishing forensic data banks and computerized databases is primarily its value in resolving unsolved past, present, and future crimes when biological evidence is available for analysis. The reality of criminal recidivism (likelihood to repeat offend) was clearly recognized when DNA technologies became relevant to forensics in the 1980s, as well as the potential of using DNA typing to help link crime scene DNA evidence to a known prior offender in cases where multiple crimes were committed by a single perpetrator. The ability of DNA typing to identify recovered human remains in order to reunite individuals with family members for humanitarian purposes was also anticipated.

The United States initiated the rapid development of forensic DNA data banks in the early 1990s, operated by state and local crime labs but centralized by the Federal Bureau of Investigation (FBI) as part of the US Department of Justice. The US program is known as the Combined DNA Index System (CODIS), which began as a pilot project in 1990, first serving fourteen state and local laboratories. The FBI's authority to establish a national DNA index for law enforcement purposes was formalized by the DNA Identification Act of 1994 (Pub. L. No. 103-322). CODIS is organized as three separate hierarchical systems at the local, state, and national levels. DNA profiles are generated at the local level (LDIS) and can then flow to the state (SDIS) and national (National DNA Index System, NDIS) levels. This tiered approach enables local or state agencies to operate their own databases in accordance with specific statutory or regulatory mandates. The DNA Analysis Backlog Elimination Act of 2000 (H.R. 4640, 42 U.S.C. 14135 *et seq.*) formally allows US states to both collect and perform DNA analyses and to enter and then search the profiles in CODIS. Table 4.1 summarizes current CODIS statistics for the United States.

In 1995, the first European DNA database for offender identification was established in the United Kingdom as the National DNA Database (NDNAD). The number of DNA profiles in the NDNAD is shown in table 4.2.

A few years later, in 2000, the National DNA Data Bank of Canada was established by the DNA Identification Act (1998, c. 37). The number of DNA samples stored in the Canadian data bank is summarized in table 4.3.

DNA databases have been introduced in much of Europe. In October 1988, a group of forensic scientists from various European countries joined together to form the European DNA Profiling Group (EDNAP) for organizing and sharing use of DNA technology for crime investigations. The Interpol DNA Unit was established in Lyon, France, following passage of Resolution Number 8 of the 67th General Assembly of

Table 4.1 DNA SAMPLES STORED IN THE US CODIS (MAY 2020)

Sample Type	Number of Samples
Offender profiles	14,240,876
Arrestee profiles	3,998,467
Forensic profiles	1,026,054

Table 4.2 DNA PROFILES KEPT IN THE UK NATIONAL DNA DATABASE (NDNAD) (MARCH 31, 2020)

Sample Type	Number of Samples
England and Wales estimated total number of individuals retained on NDNAD	5,604,185
Total number of subject sample profiles retained on NDNAD from volunteers	4,339
Total number of crime scene sample profiles retained on NDNAD	647,378

Table 4.3 DNA PROFILES CONTAINED IN THE NATIONAL DNA DATA BANK OF CANADA (MARCH 31, 2020)

Sample Type	Number of Samples
Currently in Convicted Offender Index (CODIS)	401,546
Currently in the Crime Scene Index (CODIS)	173,292

Interpol (in Cairo in 1998) to advance international cooperation on the use of DNA in criminal investigations, assist member states, and encourage DNA profile comparison across international borders (Martha, 2010).[1] Police in participating member countries can submit a DNA profile from offenders, crime scenes, missing persons, or unidentified bodies to Interpol's automated DNA database, the DNA Gateway, which by December 2017 contained more than 173,000 DNA profiles contributed by more than 84 member countries.

Every two years, Interpol conducts a global DNA profiling survey to monitor DNA profiling and DNA databases in various countries. In January 2017, Interpol's 190 National Central Bureaus were asked to provide their national DNA statistics as of the end of 2016. Ninety-five (50% of) member countries responded to the 2016 survey, of which 84 reported using DNA profiling in police investigations, 69 reported having a national DNA database (a centralized searchable repository), 73 reported using Y-STR analysis, and 31 reported using mitochondrial DNA (mtDNA) analysis. The estimated global total of DNA profiles reported by member countries and member countries' overseas sub-bureaus is 35,413,155 DNA profiles (see Interpol Global DNA Profiling Survey Results, 2016).[2]

Regardless of the jurisdiction, DNA laboratory and programmatic operations have some standard features. Basically, computerized searching of archived DNA

---

1. https://www.interpol.int.
2. https://www.interpol.int/en/How-we-work/Forensics/DNA

profiles allows comparison of those profiles from biological samples collected from different crime scenes with those of known offenders or arrestees. The legally mandated collection of a DNA sample occurs in some cases after conviction and in others upon arrest. Blood samples or oral swabs are typically used for DNA collection. The DNA markers used for identification are various noncoding regions of the human genome showing DNA sequence or repeat-length variation among individuals in the population. Between 13 and 25 of these genomic regions are typically tested using standardized commercial kits to determine the genotype with regard to the variable number of four or five DNA base-pair short tandem repeats (STRs) (Bieber, 2004; Buckleton et al., 2016; Butler, 2014; Ge et al., 2012; Primorac & Schanfield, 2014). These STR markers are inherited as Mendelian traits and are not known to be associated with predilection to human disease; the forensic DNA typing results thus do not predict the present or future health status of the individuals from whom the samples are collected. This is important, as medical and health data are typically protected under patient safety and privacy guidelines. Nevertheless, at least in the United States, the biological evidence and the extracted DNA are usually retained by the law enforcement laboratories, making additional genetic testing possible as new loci and methods are identified and new DNA testing kits are developed.

DNA DATABASE EXPANSION

Initially, statutory legislation allowed collection of DNA only from those convicted of murder or sexual crimes, but criteria have steadily expanded to include all felonies, including nonviolent property crimes, in most US states, following the UK's lead. As criminologists have long known, recidivism rates for property crimes and many violent felonies are at least as high as for sexual offenses, and probably higher than for murders (McCleary, 1992; Schlesinger, 2001). Furthermore, many violent sex offenders have previous arrests and convictions for burglary and other property crimes. Realization of these facts has led to passage of revised statutes expanding the criteria for inclusion in offender DNA databases.

Enacted legislation now allows for DNA collection from most felons convicted in federal and military courts as well as in criminal courts in US states, the District of Columbia, and Puerto Rico. Moreover, 31 US states have now passed laws to mandate DNA collection from arrestees for certain crimes, with similar pending legislation in several other states. In addition to legislation at the state level, on January 5, 2006, then president George W. Bush signed into law Public Law No. 109-162, allowing DNA collection from those arrested in the District of Columbia, Puerto Rico, and the military for certain federal crimes.[3] Federal and state monies have been provided for outsourcing the backlog of DNA samples from convicted offenders

---

3. Codified at 42 U.S.C. § 14135a (2005).

to high-throughput commercial laboratories, leading to an increase in the overall number of offender and crime scene profiles stored in CODIS.

CODIS uses the metric known as "Investigations Aided" to track the number of criminal investigations in which CODIS has allegedly added value to the investigative process. As of May 2020, CODIS has reported over 514,982 "hits" assisting in more than 503,968 investigations (FBI, 2019).[4] CODIS contains DNA profiles from about 4% of the US population. By contrast, in Canada the National DNA Databank holds DNA samples from ~1% of the Canadian population (Public Safety Canada, 2019).[5]

Court challenges have been raised to maintaining offender/arrestee DNA data bank programs and to expansion of existing criteria for DNA collection. In the United States, none of the challenges to such collection based on Fourth Amendment considerations have yet to prevail on appeal to the highest courts. In *U.S. v. Kincade* (9th Cir. (2004) 379 F.3d 813), the Ninth Circuit Court of Appeals reheard its prior three-judge panel decision that the compelled production of a DNA sample from a parolee for inclusion in a nationwide DNA database is an unlawful search. In a close 6–5 ruling, the Court, sitting *en banc*, decided in 2004 that a parolee can be forced to provide a DNA sample for deposit into the FBI's national DNA database.

TAKING DNA UPON ARREST

At the time of this writing, 31 states authorize some form of DNA collection upon arrest for certain categories of crime (National Conference on State Legislatures, 2018).[6] Provisions for expunging a profile, if a qualifying charge is dismissed or results in acquittal, are required if profiles are uploaded to the national DNA index. While some US states initiate this expungement process, in many the burden is on the individual (Kaye, 2001b, 2006)

In 1997, Louisiana became the first state to pass legislation authorizing the collection of DNA samples from arrestees, specifically from "a person arrested for a felony sex offense or other specified offense on or after September 1, 1999." Other states soon followed with similar legislation. In California, for example, Proposition 69, approved by 62% of the voters in 2004, allowed the collection of DNA from anyone *arrested* for a felony. The DNA samples are forwarded to CODIS and are accessible to the FBI as well as to state and local police. In April 2018, in a 4–3 vote, the California Supreme Court overturned a prior appeals court ruling, refusing to throw out that part of Proposition 69 that led to the storing of DNA profiles of tens of thousands of people arrested but never charged or convicted (see *People v. Buza*,

---

4. https://www.fbi.gov/services/laboratory/biometric-analysis/codis/ndis-statistics.
5. Royal Canadian Mounted Police. Statistics for National DNA Data Bank, March 31, 2020 http://www.rcmp-grc.gc.ca/nddb-bndg/stats-eng.htm.
6. National Conference on State Legislatures http://www.ncsl.org/Documents/cj/ArresteeDNA Laws.pdf.

People v. Buza (2018) 413 P.3d 1132. In this matter, Mark Buza had been arrested for arson and related felonies and transported to jail. At booking he refused to provide a DNA sample as then required by law. At trial, a jury convicted him of both the arson-related felonies and the misdemeanor offense of refusing to provide a specimen required by the DNA Act.

The practice of taking DNA upon arrest was also challenged unsuccessfully in Maryland in a case subsequently heard before the US Supreme Court in February 2013 (*Maryland v. King*, (2013), 569 U.S. 435. Alonzo Jay King Jr. was arrested for first- and second-degree assault. In accord with the Maryland DNA Collection Act, DNA was taken from King at the time of his arrest and the profile was then entered into Maryland's database, where it was searched and the profile was found to match that from evidence in an unsolved rape committed in 2003. When this was presented to a grand jury, an indictment was issued, and a warrant was obtained for a second reference buccal DNA sample. King, through his counsel, tried to suppress the DNA evidence, arguing that this collection infringed his Fourth Amendment rights. His motion was denied, and King pleaded not guilty to the charge of rape and appealed the ruling. The Maryland Court of Appeals reversed the lower court decision, agreeing that the DNA sampling was a violation of the Fourth Amendment and could not be used as evidence. The State of Maryland then appealed this ruling to the Supreme Court, which reversed the Maryland court judgment in a 5–4 vote, holding that "when officers make an arrest supported by probable cause to hold for a serious offense and bring the suspect to the station to be detained in custody, taking and analyzing a cheek swab of the arrestee's DNA is, like fingerprinting and photographing, a legitimate police booking procedure that is reasonable under the Fourth Amendment."[7] In their dissenting opinion, the late associate justice Antonin Scalia, joined by Associate Justices Ruth Bader Ginsburg, Sonia Sotomayor, and Elena Kagan, argued that the Fourth Amendment forbids searching a person for evidence of a crime when there is no basis for believing the person is guilty of the crime or is not in possession of incriminating evidence (citing 133 S. Ct. at 1980) (see chapter 13 in this volume).

Even before the King decision, various agencies had initiated clandestine collection of "abandoned DNA" from those not (yet) arrested or convicted for any crime, performed usually by local law enforcement agencies operating outside of CODIS. These DNA samples can be collected from items discarded by known individuals who, while never (yet) convicted of or arrested for any crime, may be under suspicion for some reason(s), and their DNA can be compared in the database to other items (e.g., firearms) (Murphy, 2015).[8] Nothing in the King decision would seem to preclude such collections and comparisons.

---

7. U. S. Const. amend. IV.

8. The Trace, 2017, Sept 25. In New York City, Gun Cases Fuel Growing, Unregulated DNA Database https://www.thetrace.org/2017/09/new-york-city-gun-crime-dna-database/.

DNA SAMPLE COLLECTION FROM IMMIGRATION DETAINEES: A NEW RULE BY THE US JUSTICE DEPARTMENT

Effective April 8, 2020, the US Department of Justice amended regulations regarding DNA collections from individuals detained under the authority of the United States (see Docket Number OAG-164 AG Order No. 4646-2020). This amendment removes a provision authorizing the secretary of homeland security to exempt from the sample-collection requirement certain aliens from whom collection of DNA samples is not feasible because of operational exigencies or resource limitations. This restored the attorney general's plenary legal authority to authorize and direct all relevant federal agencies, including the Department of Homeland Security, to collect DNA samples from individuals who are arrested, facing charges, or convicted, and from non-United States persons who are detained under the authority of the United States (see 85 Fed. Reg. 13483).

COLLECTION OF BLOOD (FOR DNA TYPING) FROM MILITARY MEMBERS

After the First Gulf War in 1991, the US Department of Defense (DOD) developed a program to use DNA samples to identify the remains of service members killed in combat or in training accidents. To do so, the DOD developed a plan for mandatory collection and storage of blood spot or cheek swab reference cards from all members of the active duty and reserve forces, and also from some State Department and other personnel stationed overseas. This was called the "DOD DNA Registry," developed within the Armed Forces DNA Identification Laboratory (AFDIL), part of the Armed Forces Institute of Pathology (AFIP), and was established pursuant to a December 16, 1991, memorandum from the deputy secretary of defense. Blood spot cards were obtained from active duty and reserve military personnel, typically upon initial enlistment, reenlistment, or preparation for operational activation and deployment (Deputy Secretary of Defense, 1991).[9]

The actual specimens are housed in secure facilities located at Dover Air Force Base in Delaware. Any DNA typing information is kept on separate, secure servers, and DNA profiles are not to be released to any other agencies for any other purposes, nor are they uploaded to any other federal forensic databases. As of January 2019, the AFDIL Repository held blood spot or buccal swab specimens from almost eight million current or former service members. While the repository was initially conceived solely to identify the remains of service members, the Culberson amendment to the 2003 National Defense Authorization Act, signed by President George W. Bush on December 2, 2002, overrides former Pentagon policy that the DNA samples be used

---

9. Defense POW/MIA Accounting Agency. Armed Forces Medical Examiner System, DNA Identification Laboratory; June 25, 2018, http://www.dpaa.mil/Resources/Fact-Sheets/Article-View/Article/590581/armed-forces-medical-examiner-system-dna-identification-laboratory/.

almost solely to identity troops killed in combat, as it now allows limited access to the repository for law enforcement purposes under certain circumstances when authorized by a military or federal judge. It has not often been used in that way as of this writing. The Culberson Amendment followed the January 2002 rape of a soldier at Fort Hood, Texas, by a fellow soldier. The 2002 Culberson provision reads:

§ 1565a. DNA samples maintained for identification of human remains: use for law enforcement purposes

(a) Compliance with a court order.

> (1) Subject to paragraph (2), if a valid order of a Federal court (or military judge) so requires, an element of the Department of Defense that maintains a repository of DNA samples for the purpose of identification of human remains shall make available, for the purpose specified in subsection (b), such DNA samples on such terms and conditions as such court (or military judge) directs.
> (2) A DNA sample with respect to an individual shall be provided under paragraph (1) in a manner that does not compromise the ability of the Department of Defense to maintain a sample with respect to that individual for the purpose of identification of human remains.

(b) Covered purpose. The purpose referred to in subsection (a) is the purpose of an investigation or prosecution of a felony, or any sexual offense, for which no other source of DNA information is reasonably available.

(c) Definition. In this section, the term "DNA sample" has the meaning given such term in section 1565(c) of this title.

The military DNA collection program was opposed by two members of the United States Marine Corps based in Hawaii, Cpl. Joseph Vlacovsky and Lance Cpl. John C. Mayfield III, who refused to provide samples before being deployed overseas in 1995. The men were charged under Uniform Code of Military Justice rules with violation of a direct order from a superior commissioned officer, and a military court sentenced the two Marines to a seven-day restriction to their base in 2016. Lieutenant Cmdr. Peter Straub, who presided over the two-day court-martial, ordered that a letter of reprimand go in each man's file. While they each could have faced six months in prison and dismissal from the service, the ruling allowed both Marines to complete their military service with honorable discharges.

RESEARCH- AND HEALTH-RELATED DATA BANKS

Besides collecting DNA for forensic purposes, many nations and organizations (public and private) now solicit and collect personal data (and often DNA) from private citizens, intended to enhance the health of populations and individuals. Since the late 1990s, DNA collection in Iceland has resulted in about a third of the Icelandic population of 320,000 having contributed their DNA to a private company (deCODE

Genetics, purchased by Amgen in 2012 for $415M), and full or partial DNA sequence data have been tabulated (Jonsson, 2018).[10] While this research effort has led to many useful results (Jonsson et al., 2017), this program has recently led to controversy as the nation struggles with whether citizens should be informed about their cancer risk based on results of DNA testing (Castellanos, et al., 2020).[11]

More recently, the United Kingdom established the UK Biobank with support from the Wellcome Trust, the Medical Research Council, and the Scottish government, with additional funding from the Welsh government, British Heart Foundation, Cancer Research UK, and Diabetes UK. This UK Biobank is open to bona fide researchers anywhere in the world, including those funded by academia as well as industry. Biobank participants provide baseline measurements and saliva samples, and some have agreed to wear 24-hour activity monitors.[12] In addition to collection of electronic health records from the participants, blood chemistry is being analyzed, and genotyping has been undertaken on all 500,000 participants; these data are being used in health research. British GlaxoSmithKline & Regeneron have applied to exome sequence the samples.

In July 2018, GlaxoSmithKline also announced investment of $300 million in the consumer genetics company 23andMe, in a planned collaboration to discover new pharmaceuticals of clinical value using genetic data. 23andMe is but one of more than a dozen direct-to-consumer (DTC) companies that solicit customers to submit biological specimens (e.g., saliva, buccal swabs) for DNA testing for ancestry and health purposes. In the United States, the National Institutes of Health (NIH) has funded a consortium of Boston hospitals in the All of Us Research Program to help speed up health research and medical breakthroughs by creation of a research program to collect health data from one million or more people who agree to share their health information with researchers.[13]

GOALS OF FORENSIC DNA DATA BANKING AND RESULTS TO DATE

Before discussing evaluation of the effectiveness of offender DNA data banks and the offender and crime scene databases, it is important to consider the stated mission of such programs. The goals and values inherent in DNA legislation vary somewhat but generally focus on efforts to apprehend criminals and reduce crime. The commission of prior nonviolent property crimes by some who later commit serious

---

10. Kirby, E. J. (June 19, 2014) Iceland's DNA: The world's most precious genes? https://www.bbc.com/news/magazine-27903831
11. Leavenworth, S. (June 15, 2018). Iceland faces DNA dilemma: Whether to notify people carrying cancer genes https://www.seattletimes.com/nation-world/iceland-faces-a-dna-dilemma-whether-to-notify-people-carrying-cancer-genes//
12. BioBank UK https://www.ukbiobank.ac.uk/.
13. Partners Healthcare. The All of Us Research Program https://www.partners.org/Medical-Research/All-of-Us-Program.aspx

violent offences has been used as a justification for the expansion of database inclusion criteria. The extent to which these legislative and programmatic goals are being met and can be improved is considered here.

DNA DATA BANKING: INDIVIDUAL SUCCESSES IN ARREST AND CONVICTION

DNA comparisons using local, state, and national database searches have led to the arrest and conviction of thousands of criminals who probably would not have been apprehended—or even identified as suspects—without such DNA collection. Also, from a law enforcement perspective, if an individual is already in an offender DNA database, failure to find a match after comparison of a crime scene DNA profile against all those in the offender index can effectively exclude that individual without the need to locate that person or collect another sample, thus allowing investigators to focus their attention on more promising investigative leads.

The many successes based on forensic DNA data bank programs include the following examples. In the United States, one of the more dramatic successes occurred in Houston in November 2003 when CODIS identified a match of DNA evidence recovered from one victim, which was used to help apprehend a bike-riding predator who had sexually assaulted young boys at knifepoint. Police closed in on a registered sex offender, Andrew Wayne Hawthorne, who later pleaded guilty to assaulting three boys and was given a prison sentence of 60 years. This occurred while another man, Ricardo Rachell, had been wrongfully convicted and had served over six years in prison for similar attacks on young boys in his Houston neighborhood, which continued after his arrest. A judge ordered Rachell, 51, released because new DNA evidence cleared him of any involvement in the assaults.

In another case, Massachusetts's highest court affirmed the 2006 conviction of a pizza delivery man, Martin Guy, for the savage murder of a Foxboro woman on December 1, 1998. Another man, Edmund Burke, had been arrested for the crime in part based on bite mark comparisons but was later released when he was excluded by DNA testing. Four years after the charges against Burke were ordered dropped, the CODIS DNA database linked the evidence in the murder to Guy, whose DNA had been entered into the DNA database after he was convicted of stabbing his neighbor, Christopher Payne, to death in September 1999.

In April 2003, the National DNA Data Bank of Canada and Interpol identified a convicted offender in an Alberta jail whose DNA profile matched that of a person wanted for sexual assault and homicide charges in Ohio. A key tip came from a viewer in western Canada of the television show *America's Most Wanted* who thought the suspect in the Ohio murder looked familiar. The Royal Canadian Mounted Police found that the suspect was one Thomas McCray, then serving time under an alias in an Alberta prison for a secondary offense. The judge in the case had ordered a blood sample sent to the National DNA Data Bank. Interpol Ottawa then coordinated the comparison of DNA from the Ohio crime scene with that collected from McCray in Canada. The DNA profiles matched, linking McCray to the Ohio murder scene. He was deported to the United States for trial after his Canadian prison term had been fully served.

Another Interpol success involved the capture of Afghan national Ali Achekzai, who was wanted for sexually assaulting three women in Orange County and San Diego, California, in 2002 and 2004. He fled to Canada and later changed his name several times, obtaining identification under the identities Ali Achekzai, Ali Achekza, Ali Achelczia, Waleed Nawabi, Walid Nawabi, David Azizi, and Wali Ahmed Shoja. Later, in Austria, he was accused of rape in Salzburg and was arrested under the name Wali Ahmed Shoja on April 24, 2009. Local authorities worked with Interpol to make his DNA profile available worldwide. He was arrested again on January 26, 2010, after Austrian authorities discovered DNA from that case matched DNA samples from the California rapes. This led to Achekzai being extradited to California in August 2010 to stand trial. After his conviction he was sentenced to 61 years in prison in Orange County, California, in January 2012.

These examples are but a few of the thousands of successful investigations concluded, in part, based on positive identification of persons of interest or suspects based on searching offender and arrestee DNA databases.

NOVEL DNA DATABASE SEARCHING METHODS: FAMILIAL SEARCHING

Even more remarkable are cases solved after searches of a DNA database lead to the identification of a suspect who is not actually in the database but is closely related to someone who is. Such search successes can occur unexpectedly when close, but not quite identical, DNA matches are observed between crime scene samples and the DNA profiles of convicted offenders. For example, in a case in North Carolina, retrospective DNA testing of evidence excluded Darryl Hunt, who had been wrongfully convicted of the 1984 murder of Deborah Sykes and had served 18 years in prison. After Hunt's conviction was overturned in 2003, laboratory scientists compared the crime scene DNA profile to the 40,000 offender profiles in the North Carolina state DNA database. While no perfect DNA matches were found, an almost-perfect match was noted in one of the offender profiles derived from a man named Anthony Brown. Police discovered that Brown had a brother, Willard, and while surveilling him, police secured a discarded cigarette butt for DNA comparison. A "perfect DNA match" was found, and Willard Brown subsequently confessed to the crime.

In the United Kingdom, deliberate searches of forensic database profiles have been performed in some high-profile cases to identify suspects when no perfect DNA matches have been found between crime scene samples and known offenders (Maguire et al., 2014). These searches were performed simply by direct allele count comparisons or by searching for rare alleles at individual loci. One remarkable case, the Valentine's Day 1988 stabbing murder of 16-year-old Lynette White in Cardiff, Wales, was finally solved in 2003. Three men had had their, as it turned out, wrongful convictions quashed due to allegations of police misconduct. After their convictions were set aside, investigators went back to the crime scene, collected additional trace evidence, and through more modern STR analysis identified a single rare allele at one locus in a blood spot that had previously gone unnoticed. Then a search was made of the UK National DNA Database to produce a list of all the individuals in the database who had this rare allele at that individual locus. The search identified over 600 such individuals, but one, a 14-year-old boy, stood out, with a very similar overall DNA profile to that of the crime scene evidence. This led police to his paternal uncle, Jeffrey Gafoor, who had the same DNA profile as the evidence and subsequently confessed to the crime. Gafoor was given a life sentence in 2003 and ordered to serve a minimum of 13 years for the 1988 murder. After twice refusing him parole, in March 2020 the Parole Board said Gafoor was "suitable" for a move to an "open prison," which would prepare him for possible release.

More recently, formal genetic kinship analysis of an entire offender database has also proven to be highly successful in aiding investigations. In 2006 this author, with colleagues Charles Brenner and David Lazer, demonstrated that if no direct DNA match is found after searching an offender database, all is not lost. We demonstrated, through use of simulations, how kinship analysis of forensic DNA databases would likely identify in the database close biological relatives of the unidentified suspects. That is, formal kinship analysis of the type routinely used in DNA-based humanitarian family reunification efforts could be used to efficiently evaluate a crime scene DNA sample to determine whether it might be derived from a close (i.e., first-degree: a parent, child, or sibling) biological relative of someone already in an offender DNA database (Bieber et al., 2006). The strategy is similar in concept to that used in humanitarian DNA-based reunifications when family members voluntarily come forward to donate their own DNA in hopes of identifying remains of their lost family members (Biesecker et al., 2005).

Others have confirmed the value of such an approach and have shown that the kinship analyses using likelihood ratio estimates (parent-child index, sib index) are more efficient than simple tabulation of the extent of allele sharing (Balding et al., 2013; Chung et al., 2010; Chung & Fung, 2013; Curran & Buckleton, 2008; Ge et al., 2011a; Ge et al., 2011b; Ge & Budowle, 2012; Hicks et al., 2010; Slooten & Meester, 2011). We noted that, taking these methods to the logical extreme, typing large numbers of SNPs should greatly enhance the ability to identify even more distant relatives, if they were present in a database. This potential is quite obvious to geneticists and has been demonstrated dramatically by those who have searched genealogy sites operated by groups focused on finding relatives and determining ancestry (Bieber, 2018; Greytak et al., 2019, Kennett, 2019).

The theoretical basis for the immediate and striking successes of proxy searching (familial searching) of DNA databases is consonant with the observations some have made that multiple members of the same kindred are not uncommonly involved in criminal activities that involve leaving biological evidence (i.e., DNA), irrespective of their so-called race or ethnicity (Butterfield, 2018; Farrington, 1995; Farrington et al., 1996; Farrington et al., 2001; Farrington & Crago, 2016). For example, in studies of crime, Farrington and colleagues have demonstrated a probabilistic relationship between conviction rates of children and that of their parents, while US Justice Department surveys document that a high proportion of prison inmates have relatives with criminal histories.[14] If this were not the case, familial searching would not be successful. Further, not only do parent/child and sib-pairs heavily populate prison populations; hundreds of pairs of identical twins populate offender DNA databases. For example, contained within the Convicted Offender Index of the National DNA Databank of Canada as of September 30, 2018, were DNA profiles from 341 pairs of identical twins, both of whom were convicted of qualifying offenses, among the total of 375,282 offender samples (R. Fourney, personal communication, October 15, 2018), while in Massachusetts in 2018, 93 pairs of identical twins were identified among ~144,000 convicted offenders (S. Collins, personal communication, December 17, 2018). The troubling relationship between criminal activity and family ties is beyond the scope of this chapter, but there is an extensive literature on this subject (Butterfield, 2018; Farrington, 1995; Farrington et al., 1996; Farrington et al., 2001; Farrington & Crago, 2016; Putkonen et al., 2002; Smith & Farrington, 2004).

The first use in the United States, of the formalized kinship methods for familial searching was in 2008 in California, after then attorney general Jerry Brown sanctioned this method of proxy searching (aka familial searching). This first use of kinship methods in California to search an offender DNA database led to the identification and subsequent arrest and conviction of Lonnie Franklin Jr. (aka the Grim Sleeper) for the murder of nine women and a teenage girl in Los Angeles between 1985 and 2007. This dramatic success of familial searching on the very first attempt resolved an investigation that had lasted over 20 years, costing millions of dollars and thousands of hours of investigative time. In this case, a close overall DNA profile match was found by comparing the crime scene DNA profile to profiles in the California convicted offender index and making the subsequent identification of a Y-STR match, indicative of a possible patrilineal relationship. The offender DNA sample in the CODIS database turned out to come from a male whose DNA was collected after his conviction on a firearms possession charge. This then led police to surveillance of his biological father, Lonnie Franklin Jr., and to collection of his discarded DNA, which was found to completely match the STR profile of the crime scene evidence. After his arrest in 2010, a known reference DNA sample confirmed that result. Franklin was convicted in 2016 of the 10 murders and soon after received a death sentence for each of them. He now resides on California's death row.

After the robbery and sexual assault of a coffeehouse worker in Santa Cruz, California, on March 19, 2008, authorities used the same kinship searching method to identify a suspect. Investigators searched the California DNA database in late 2010, finding a close partial match of the rape kit DNA to a man whose DNA profile was in the California offender index for an unrelated crime. His biological son, Elvis Lorenzo Garcia, was then surveilled, and after an STR DNA match was found he was arrested, tried, and convicted, in 2011, of sodomy; he

---

14. Survey of State Prison Inmates, Bureau of Justice Statistics, US Department of Justice, *Correctional Populations in the United States* 1996, p. 62, table 4.18.

was later sentenced to 65 years in prison. Since these two initial successes of familial searching of offender databases in California, dozens more have taken place in the United States and other countries (Maguire et al., 2014; Myers et al., 2011; Slooten & Meester, 2011).

Apart from such familial searching using DNA databases, simple kinship analysis in the form of paternity testing has been used to compare crime scene samples to a single possible relative/suspect. In the "BTK" serial killer investigation in Wichita, Kansas, once police had identified Dennis Rader as a suspect in 2005, they obtained a warrant to secure his daughter's medical biopsy sample, without her specific knowledge or consent, for DNA analysis. DNA analysis was then performed on the biopsy and the crime scene evidence left by the perpetrator. This was followed by paternity analysis of the DNA results, confirming that the likely perpetrator of the murders could indeed be Rader. This led, in part, to Rader's arrest and to his subsequent confession. He was sentenced in August 2005 to 10 consecutive life sentences, with a minimum of 175 years.

Despite concerns expressed by some about equity and fairness (Epstein, 2009; Kaye, 2001a; Krimsky & Simoncelli, 2010; Rohlfs et al., 2013), after so many successes, familial searching is winning favor in the law enforcement community.[15] Hundreds of applications of familial searching of offender databases in criminal investigations have been reported around the world, with successful identification of suspects in dozens of these searches. There have been constructive critiques of the concept of familial searching (Greely et al., 2006; Haimes, 2006; Kim et al., 2011; Murphy, 2010; Nelkin & Andrews, 1999; Ram, 2011; Rohlfs et al., 2013; Rosen, 2009), two US jurisdictions (Maryland and D.C.) disallow it,[16] and it remains uncertain whether current Canadian legislation would permit it. There have, however, been no successful court challenges, in the US, to such methods, based on Fourth, Sixth, or Fourteenth Amendment claims, in either arrests or convictions based on using such DNA database searching tools (Kaye, 2013). To date, proactive family searching methods are underutilized in most countries, in part due to lack of trained personnel and standardized software tools. This is surprising given the potential power of such analysis to identify suspects indirectly and the fact that legislation, aside from that previously noted, does not bar use of the databases in this manner. This should change, as the most recent version of the CODIS software (version 8.0, released in 2018) has upgraded capability to perform such searches at the state level (SDIS) (Debus-Sherrill & Field, 2019).[17]

---

15. http://www.latimes.com/local/lanow/la-me-familial-dna-20161023-snap-story.html
16. DC Stat 22-4151, https://code.dccouncil.us/dc/council/code/titles/22/chapters/41B/; and MD Code § 2-506, 2010, http://law.justia.com/codes/maryland/2010/public-safety/title-2/subtitle-5/2-502/.
17. https://www.americanbar.org/content/dam/aba/events/criminal_justice/2017/DNA_Familial_Searches.pdf.

SEARCHES OF PUBLIC AND PRIVATE GENEALOGY DATABASES FOR SUSPECTS (GENEALOGICS)

In developed countries like the United States, most of the population can be found in some sort of database, from birth, marriage, or death records to voter registration rolls, real estate records, credit bureaus, and census data. Millions voluntarily and freely offer their own DNA to ancestry databases; accordingly, the odds are quite high that one has at least one fourth cousin or closer relative in such a database. Most adults freely offer other personal data through online shopping and volunteer information about themselves to open-access sources, chat rooms, forums, and dating sites. With current crowdsourced efforts to facilitate genealogical research, such tools now increase the chance to identify the individual biological source of evidence collected at many unidentified crimes, especially those such as murders and sexual assaults, with DNA that has been left behind.

While searches of offender and arrestee forensic DNA databases have been the most widely used by law enforcement agencies to investigate DNA-related crimes, other databases have not been ignored. There is nothing to stop local authorities from collecting discarded DNA (e.g., dropped cigarette butts, drinking vessels) for storage/typing and comparison in private databases. Indeed, many other DNA collections are common and include medical, research, and DTC testing.

The personal DTC DNA testing market has become enormously popular and successful for identification of lost or missing relatives, identification of recovered human remains, and discovery of biogeographic ancestry and genetic admixture. More than a dozen commercial companies offer testing of SNPs for ancestry determination and for finding or locating missing relatives, and more than 30 million individuals have purchased such kits to submit their own saliva or cheek swab samples. Beyond the DTC arena, university researchers as well as forensic scientists have now clearly recognized the value of DNA for phenotype prediction, which has applications for humanitarian identification of human remains and for creation of suspect profiles in criminal investigations. The power of typing large numbers (hundreds of thousands) of targeted SNPs and thereby the potential of finding relatives for forensic investigation purposes has not gone unnoticed by law enforcement agencies. SNPs can be chosen to assist with prediction of ethnicity/ancestry and for phenotype prediction (e.g., eye, hair, and skin color with a significant genetic basis) (Chaitanya et al., 2018; Hysi et al., 2018; Ingold et al., 2018).

Searching an SNP-based DNA profile against a publicly or privately held genealogical database will predictably yield a very large number of possible biological relatives, who may be widely dispersed across the globe. This is because these genealogic and ancestry databases contain profiles from millions of volunteers and test for the types of DNA variation known as SNPs. Maternally inherited mtDNA, along with Y-chromosome-specific DNA in males, can be used to document matrilineal and patrilineal relationships, respectively. Many individuals, who may be third-, fourth-, or even fifth-generation distant relatives, can thereby be identified. This is used only to develop investigative leads; once a person of interest is identified as a possible suspect, a subsequent DNA test would be needed to confirm that the DNA profile from the evidence matches that from the known DNA sample obtained from the suspect.

The arrest in April 2018 of Joseph James DeAngelo Jr., a former police officer, as the suspect in the California Golden State Killer investigation, is remarkable for the striking success of finding a suspect by searching the SNP profile of the DNA left at one of the crime scenes against the DNA deposited by volunteers in an open access public genealogic database, GEDmatch (Bieber 2018; Greytak et al., 2019). GEDmatch was founded in 2010 by Curtis Rogers, a retired businessman, and John Olson, a transportation engineer, in Lake Worth, Florida, with the goal to help "amateur and professional researchers and genealogists," including adoptees searching for their birth parents. It has since been sold to Verogen, Inc.

Searching a genealogic database was successful in identifying DeAngelo, the suspect in the Golden State Killer investigation, because a distant relative of his had reportedly freely uploaded his or her own DNA SNP profile to a public online website to try to find biological relatives. By doing the same thing with the DNA profile of biological evidence from one of the Golden State Killer crimes, California authorities, with the assistance of genealogist Barbara Rae-Venter, a retired patent attorney, eventually located this relative and thereby were able to focus their investigation on a specific suspect (DeAngelo). They then surreptitiously collected his DNA for direct comparison to the STR profile from old crime scene evidence from one of the cases that had been stored for decades. DeAngelo has now been charged with 13 counts of murder and 13 counts of kidnapping. He is reportedly suspected of committing over 50 rapes and more than 100 residential burglaries during a 10-year period (1976–1986). At an April 2019 court proceeding, prosecutors announced that they would seek the death penalty. On March 4, 2020, DeAngelo offered to plead guilty if the death penalty were taken off the table. A trial date has not yet been scheduled. Charges cannot be laid for the suspected rapes or burglaries, as the statute of limitations has expired for those offenses.

Less than one month after the arrest of DeAngelo as the alleged Golden State Killer in California, Canadian investigators, working closely with Washington State authorities, announced that the same strategy had been used to identify and later arrest an alleged killer, William Earl Talbott II. Using evidence from the scene of the murders of Jay Cook and Tanya Van Cuylenborg (of British Columbia, Canada) that had occurred 30 years earlier, police took advantage of available genealogy websites to identify possible relatives of the killer using biological evidence on a blanket on Cook's body, which was found in an abandoned vehicle in Washington State. While Talbot's name was reportedly not one of those 300+ persons given to police by tipsters, using DNA data and a genealogy website, a half cousin once removed (and two second cousins) of Talbott were found, leading police directly to him. He was convicted in Washington State at trial by jury in June 2019 and received two consecutive life sentences.

The growth in ancestry data collection has been enormous. While at the time of DeAngelo's arrest in April 2018, use of public ancestry/genealogy databases for forensic purposes was not widespread, it quickly became so, with one to two new identifications/arrests reported almost weekly in the United States during the 12 months after DeAngelo's arrest. This merger of genomics and genealogy searching has been explosive, with several companies and forensic laboratories (e.g., Parabon NanoLabs, Bode Technology) establishing formal relationships with genealogists to assist in the search for suspects in cold cases. Both GEDmatch (now Verogen, Inc.) and FamilyTreeDNA have mechanisms that can permit law enforcement agencies to upload and search DNA profiles from crime scene evidence against their database of profiles of individuals who have voluntarily submitted their own data. (On December 9, 2019, Verogen, Inc., a California-based forensic genomics company, acquired GEDmatch.) I have referred to this new wave of DNA-based ancestry investigations as *genealogics*. It has been referred to by others as *forensic genetic genealogy*.

Erlich and colleagues recently reported their studies of genomic data of over one million persons tested by consumer genetics (Erlich et al., 2018). These authors project, for example, that more than 60% of searches of consumer databases for those of European descent will result in identification of at least one individual as close as a third-degree relative (e.g., first cousins, who are "third-degree" relatives, share on average 1/8 of their genes based on common shared ancestry).

Several websites[18] allow anyone to perform database searches on their own DNA or that of those for whom they have guardianship. The terms of service displayed on the GEDmatch (now Verogen, Inc.) website make it abundantly clear that the data may be used by law enforcement agencies (i.e., by third parties). Once a list of possible relatives is obtained, law enforcement could then collaborate with genealogists using metadata, such as geography and age, to focus attention on specific individuals most likely living near the crime scenes who have certain relevant characteristics. For example, in the California Golden State Killer case, because the crimes occurred in the 1970s and 1980s, the investigators reasoned that by 2018 the perpetrator must have been at least 65 years old (DeAngelo, charged with these crimes, was born in 1945). DNA technology is now rapidly evolving to predict many physical characteristics about someone simply from analysis of a DNA sample (Chaitanya et al., 2018; Hysi et al., 2018; Ingold et al., 2018).

As more data are collected from all manner of public and private DNA collection, assembly of large, multigenerational pedigrees has become possible (Kaplanis et al., 2018). Recently, Kim and colleagues (2018) have considered how relatives may be identified from searches of databases containing research or medical samples. As such new tools and technologies are increasingly being used for both humanitarian purposes and for solving crimes, public officials in the United States and Canada are now looking more closely at how they are used and by whom. Use of current genotyping platforms, which interrogate almost a million SNPs, raises many compelling questions about medical privacy, as meaningful kinship analysis does not require testing so many SNPs. This creates possible conundrums for forensic laboratories, if their genotyping results show evidence of disease risk, in samples collected covertly from possible distant relatives.

18. Zhang, 2018; Zhang, 2019.

While concerns may loom large for some about new approaches to data mining methods (Berkman et al., 2018; Bieber & Lazer, 2004; Murphy, 2010), it is worth keeping in mind that cases like that of the Golden State Killer and William Talbott II notwithstanding, most criminal cases involving DNA are likely to be solved by direct comparison of crime scene DNA to a known suspect or through a direct search of an offender/arrestee database. The DNA kinship searching methods used in familial searches of offender/arrestee databases also generate close matches, leading to identification of suspects in a substantial number of cases (Bieber et al., 2006). Such kinship searching methods, known popularly as familial searching, involve a multistep search of local or state databases containing only DNA profiles of convicted offenders, collected under state law. In this approach the offender DNA profiles are ranked in decreasing likelihood of being derived from a close relative of the crime scene DNA. The DNA from a small number of closely matching profiles, along with the crime scene evidence, is then subjected to further testing, known as Y-STR testing, to help winnow out nonrelatives (only if the suspect is male). When successful, a familial search yields a single DNA profile of an individual likely related to the perpetrator. The name of that convicted offender whose DNA profile is a close, but not perfect, match to the crime scene DNA profile can then be released to authorities. Further investigation is then needed to determine whether a close relative could be a viable suspect in the investigation. A known DNA sample is then needed for direct comparison before proceeding with arrest or charges.

Because it is quite common for criminals to have a relative whose DNA profile is in the offender database, such familial searching methods have now led to arrests and convictions in several US states and other countries. Although the California Department of Justice attempted this familial searching method several times in the Golden State Killer investigation, without success, overall these more focused formal kinship searching methods are an effective tool to help resolve crimes, with minimal intrusion on privacy.

More recent investigations involving searches of public DNA genealogy data sets, which include SNP profiles from millions of individuals, may take hundreds or even thousands of hours of additional investigative time by seasoned genealogists in attempts to locate potential persons of interest based on distant biological relationships (e.g., third or fourth cousins). As individuals continue to freely give up their private information in so many ways, government organizations and big-data companies may choose to leverage the data that are part of public data sets. These choices require the public to weigh the delicate balance between competing interests of privacy and public safety (Greytak et al., 2018; Kim et al., 2011; Lazer & Bieber, 2010; Murphy, 2018; Rosen, 2009).

In September 2019, the U.S. Department of Justice proposed its Interim Policy on Forensic Genetic Genealogical DNA Analysis and Searching, to become effective November 2019 (see https://www.justice.gov/olp/page/file/1204386/download), calling for comments and discussion about the recommended protocols and procedures to limit intrusion on privacy (Callaghan, 2019). While this interim policy provides some useful guidance to law enforcement and to crime labs, it has also generated considerable discussion and debate. One reason for this is that the policy proposes to allow covert collection of discarded DNA from individuals who, though not implicated in any crime themselves, are possibly distantly related to possible suspects, simply to help narrow the search for possible perpetrators. The policy does recommend obtaining a warrant for the actual DNA/SNP analysis of the discarded DNA, but only after it has been surreptitiously collected without consent. This sequence of steps seems illogical to many and may open doors for court challenges to admissibility.

GENEALOGICS AND THE FOURTH AMENDMENT OF THE US CONSTITUTION

In the aftermath of a flurry of dozens of identifications/arrests in 2018–2019 using the genealogics methods just described (i.e., searching open-access public DNA SNP databases for relatives of those who leave DNA at a crime scene), some will question whether such searches violate the Fourth Amendment to the US Constitution (which protects persons against some types of government search and seizure without a warrant) or otherwise disturb core American values. In the United States, neither existing Fourth Amendment case law nor any extant statutes appear to directly protect genetic data in public or private databases from scrutiny by law enforcement. That is, law enforcement may claim to have the same right as any member of the public to search the databases (D. Kaye, personal communication). Two of the ancestry sites (Verogen, formerly GEDmatch, and FamilyTreeDNA) have openly declared that law enforcement may be using the data for investigations, though with some limitations based on user consent.

In May 2019, new terms of service were announced by GEDmatch, which held one of the largest ancestry DNA databases. GEDmatch revised its terms of service and privacy policy, which describe four classes of DNA data on the site:

1) "Private" DNA data are not available for comparisons with others;
2) "Public + opt-in" DNA data are available for comparison to any Raw Data in the GEDmatch database;
3) "Public + opt-out" DNA data are available for comparison to any Raw Data in the GEDmatch database, except DNA kits identified as being uploaded for Law Enforcement purposes; and
4) "Research" DNA data are available for one-to-one comparison to other Public or Research DNA.[19]

---

19. https://www.forensicmag.com/559058-Verogen-CEO-GEDmatch-Will-Be-Improved-Not-Changed/.

These terms of service have been largely continued by the new owner, Verogen, Inc. Besides having their IP addresses captured, those who wish to use Verogen's (formerly GEDmatch) services to search for relatives will be required to "opt in" to allow law enforcement officials access to their DNA data. This could make it much more difficult for law enforcement to gain access to the website's data, as a large proportion of the DNA profiles contained could be beyond reach for investigators (at the time of this writing, June 2020, Verogen had almost 250,000 profiles in their database from individuals who had elected to "opt-in"). There is, however, no formal mechanism in place to positively assure that those searching the database to find relatives are not in fact law enforcement personnel (i.e., they may choose not to disclose their real purpose when uploading a DNA profile or conducting a search). Consequently, some have speculated that the search process itself may be "discoverable" in court proceedings, such that any lack of candor by police could jeopardize the investigation. Thus there is likely to be a trend for police to consider requesting formal search warrants in attempts to gain access to the millions of DNA profiles now stored by the many genealogy/ancestry companies.

It is likely that the government need not demonstrate probable cause for a subpoena to force the database owner to give up the information, though some state courts have construed their state constitutions as providing greater protection for medical records. Some constitutional pundits have predicted that the situation could change with a ruling in favor of greater privacy protection, pointing out that searches of houses require individualized suspicion (indeed, they require probable cause), but it is left to our imagination to ponder how the search of a database is akin to searching a house (Greytak et al., 2018; Kaye, 2013).

Some might argue that suspects identified through familial searches cannot be said to have voluntarily shared their genetic profiles in a database of known individuals, even if a genetic relative has done so. Left unexplained is why a voluntary act (leaving DNA at a crime scene) by the suspect matters when his relatives willingly and voluntarily placed their DNA data in public view. After all, if two siblings (e.g., brother and sister) are suspected of committing a crime, and surveillance of the sister produces evidence against the brother, the sister would seem to have no standing and cannot argue that she did not consent to scrutiny of her brother (Greytak et al., 2018; Kaye, 2013; Murphy, 2018).

Others might suggest that the exclusionary rule[20] may apply to results of DNA/genealogy searches conducted without a warrant by law enforcement officials, resulting in excluding such evidence at trial as "fruit of the poisonous tree." It seems likely, however, that one or more of the standard exceptions to the exclusionary rule would apply (e.g., independent source, good faith, inevitable discovery).[21]

20. Mapp v. Ohio, 367 U.S. 643 (1961).
21. Nardone v. United States, 308 U.S. 338 (1939); and Silverthorne Lumber Co. v. United States, 251 U.S. 385 (1920).

THIRD-PARTY DOCTRINE AND DNA COLLECTION/ DATABASE SEARCHES

In a recent decision (*Carpenter vs. United States*, 585 U.S. ___ [2018]), the US Supreme Court ruled on the nature of what legal authorization a government agency requires to compel a third-party wireless service provider to produce past records revealing the actual physical locations of cell phones during use. Gaining access to historical cell site records, the government was able to determine that Carpenter's cell phone communicated within a two-mile radius of four armed robberies, and he was charged and arrested. A jury later convicted Carpenter on several counts of aiding and abetting a robbery that affected interstate commerce and another count of aiding and abetting the use or carriage of a firearm during a federal crime of violence. Carpenter was sentenced to more than 100 years in federal prison.

Carpenter, through his counsel, argued that a search warrant should have been obtained to track the location of his cell phone for so long. The government had reasoned that the authorization should be by a simple court order, the requirements for which are typically less stringent than for a formal search warrant. However the Supreme Court held, in a 5–4 decision, that the government had violated the Fourth Amendment to the Constitution by accessing historical records containing the physical locations of cell phones without a search warrant.

The Supreme Court has consistently rejected the claim that subpoenas of financial records or inspection of discarded rubbish placed outside the property for pickup are "searches" within the meaning of the Fourth Amendment.[22] Prior to the *Carpenter* decision, the Supreme Court had consistently held that persons have no legitimate expectation of privacy regarding information they have voluntarily turned over to a third party (e.g., a bank), and therefore no search warrant is needed to obtain such information. This concept, known widely in US jurisprudence as the third-party doctrine, was firmly established by several Supreme Court cases, including *United States v. Miller* (1976), and *Smith v. Maryland* (1979), which, respectively, determined the rights of individuals concerning bank and phone communications.

The ruling in *Carpenter*, however, restricts the third-party doctrine exposure, at least involving cell phone location data. The Court held that, with some exceptions, a search warrant is needed to search cell phone records. The majority opinion in *Carpenter* may not predict how the courts would view government uses of ancestry and genealogy software tools (like Verogen, formerly GEDmatch). It is likely, though not certain, that a defendant's raw DNA data are not available in a typical ancestry database, and the government is simply searching for his or her relatives' ancestors' data. If so, the defendant would not therefore appear to have legal standing to assert a Fourth Amendment claim on his own behalf. However, this could change if, by some chance, the defendant had submitted his own data earlier to a company like GEDmatch. Even so, the majority *Carpenter* analysis might still predict that he has

---

22. Smith v. Maryland, 442 U.S. 735 (1979); and United States v. Miller, 425 U.S. 435 (1976).

no reasonable expectation of privacy for the raw data, since the specific purpose of anyone submitting data to an ancestry database like GEDmatch (now Verogen) is to compare it and share it with others. Only time will tell whether this Supreme Court ruling in *Carpenter* has any implications for the genealogy database searches.

It is perhaps noteworthy that in *Carpenter*, Associate Justice Neal Gorsuch wrote a dissent in which he seemed to agree with the majority's decision but disagreed with their reasoning. He did not seem to agree that the Fourth Amendment provides the right to a "reasonable expectation of privacy." The Supreme Court first noted in *Katz v. United States*[23] in 1967 that the Fourth Amendment provides the right to a "reasonable expectation of privacy." Gorsuch seems to believe that *Katz* may have been incorrectly decided because the original meaning of the Fourth Amendment does not provide for a "reasonable expectation of privacy." This point notwithstanding, Gorsuch argued that cell phone location records are indeed the property of cell phone owners and under the Fourth Amendment, law enforcement agencies cannot search a person's property without a warrant.

An interesting issue in *Carpenter*, suggested only indirectly by Gorsuch in his dissent, is whether the seizure of discarded DNA (e.g., from a drinking cup, cigarette butt, rubbish), which the police then use to compare with the scene specimen, might constitute a Fourth Amendment violation (in Gorsuch's dissent, he specifically mentions 23andMe). Current case law suggests that seizure of discarded/abandoned DNA is not a violation, as there is no precedent for a "reasonable expectation of privacy" in discarded or abandoned items. Some have pondered whether under both the majority opinion and Gorsuch's dissent in *Carpenter*, a search warrant could be required by future courts for a law enforcement agency to search a publicly accessible DNA database; this matter has yet to be tested in any case (Ohm, 2019).

DNA AND PRIVACY INTERESTS

The extent to which expanded DNA collections, whether for research, medical, or forensic purposes, become intrusive on the privacy interest of individuals or "categories" of people in society has been raised and evaluated (Clayton et al., 2019; Greely et al., 2006; Haimes, 2006; Kim et al., 2011; Krimsky & Simoncelli, 2010; Williams & Johnson, 2006). The sense of urgency to confront these concerns is greater now that DNA collection is so widespread and data sharing is so common. Moreover, identifying the source of an individual sample, even if it has been "anonymized" (e.g., for research or medical study), is now possible given the power of genotyping using SNP profiling or direct DNA sequencing. It seems that few individuals are aware that by donating or depositing their DNA, they are also displaying the profiles of their relatives (e.g., all that of an identical twin; half that of their parents, children, and siblings; one-eighth that of their first cousins). Even so, in the context of forensic investigations, DNA analysis often results in less intrusion upon at least some

---

23. Katz v. United States, 389 U.S. 347 (1967).

citizens than some traditional police investigative methods (e.g., stop and frisk, racial profiling), as individuals who might otherwise be under suspicion are easily excluded as sources of crime scene evidence.

Medical institutions, research laboratories, and police agencies have the responsibility to perform their work in a timely manner and to avoid intrusions on uninvolved parties. That said, existing privacy laws do little to meet the common public expectations of privacy they assume they enjoy (Clayton, 2019). Even health information that includes genetic information is not beyond the reach of non-health providers, as many entities that may have health information are not subject to the HIPAA privacy rule, including employers, most state and local police or other law enforcement agencies, state agencies like child protective services, and most schools and school districts. Furthermore, in certain situations a HIPAA-covered entity may disclose protected health information, even turning over biological samples, to law enforcement without the individual's signed consent or HIPAA authorization. In 2018, two data protection acts were created that in some ways could restrict access to certain DNA collections without authorization or consent by those providing DNA samples. In June 2018, California governor Jerry Brown signed into law the California Consumer Privacy Act (CCPA), intended to give consumers an effective way to control their personal information and thereby provide better protection of their own privacy and autonomy.[24] The CCPA, which took effect on January 1, 2020, with implementation regulations scheduled for July 2020, defines personal information broadly to include information that can identify, relate to, describe, be associated with, or be reasonably linked directly or indirectly to a consumer or household. It may not apply to HIPAA-covered entities or to information collected by a HIPAA-covered entity or business associate that is part of a clinical trial.

While legislative debate on this privacy bill was taking place, global events raised heightened concerns about privacy of personal information and data. First was the news about unauthorized use of over 86 million Facebook clients' personal information by the British political consulting firm Cambridge Analytica to target political advertising prior to the 2016 US presidential election.[25] Then the General Data Protection Regulation (GDPR) took effect, on May 24, 2018, affecting the European Union's 28 member states. The GDPR makes revisions to the EU laws protecting data privacy.

The identification and arrest of a suspect using genealogics, while certainly stunning, should not be surprising. Ancestry and genealogy sites have potential indirect access to millions of DNA samples, so the odds of having a distant cousin in its database are reasonably high, thus providing access to most family trees. This highlights the networked nature of privacy in the twenty-first century. Technology connects us, and in so doing, also makes our privacy interdependent. The Cambridge Analytica case highlighted this for Facebook, at which decisions

---

24. TITLE 1.81.5. California Consumer Privacy Act of 2018 [1798.100–1798.199] (*Title 1.81.5 added by Stats. 2018, Ch. 55, Sec 3*).
25. Meyer, 2018.

by a few hundred thousand individuals disclosed sensitive information from tens of millions. In the presence of interdependent privacy, the decision of how data can be used needs to be considered both at the societal level and by individuals (Kennett, 2019).

THE OTHER SIDE OF DNA EVIDENCE: AN INNOCENT MAN IS FREED

Besides the forceful and often probative inculpating power of DNA evidence, its role in proving actual innocence cannot be overemphasized (Berger, 2006; Gross, 2005). Retrospective analysis of old evidence using modern DNA methods has led to reversals of convictions, based on DNA exclusions, for hundreds of persons whose convictions occurred in the era prior to DNA testing (see chapter 2).

Some of these exonerations occurred secondary to identification of the true perpetrator after searching offender DNA databases. Perhaps the most dramatic example occurred in California, when Kevin Green was freed after serving 16 years in prison for the rape and beating of his pregnant wife, causing her to miscarry their near-full-term fetus. His fortune changed in 1996 when Orange County California forensic experts matched crime scene profiles from a string of unsolved rape-murders in the 1970s with that of a former Marine, Gerald Parker, who was in the database because of convictions for sex crimes in the 1980s. Parker then confessed to the 1970s murders and admitted beating and raping Green's wife. The court not only freed Green but also found that he was completely innocent. Parker was sentenced to death in 1999.

TALLYING OUTPUT (HITS AND INVESTIGATIONS AIDED)

Despite the documented utility of forensic DNA data banks in helping solve hundreds of individual violent crimes, the actual outcomes of thousands of "cold hits" (i.e., DNA matches of crime scene evidence to known offenders in the database) to date are mostly unknown. The resolution of bona fide DNA matches is uncertain, as hits are not necessarily prioritized or followed up on by police or prosecutors—efficiently or at all.

The overall success of such DNA matching programs has not been carefully evaluated in a systematic way by the justice system. This is lamentable, as without such monitoring, it is impossible to identify new ways to improve the effectiveness of these data banks. Ultimately such evaluation must occur, as these collections constitute a costly government program, and, with limited resources, law enforcement agencies must balance competing demands on budgets and personnel. Furthermore, these data banks have expanded greatly in scope since the initial implementation, and now DNA collection is mandated for relatively minor offenses. This expansion often overwhelms the already overburdened public safety agencies due to staffing and budget constraints.

It has often been assumed that DNA data banks are effective in increasing criminal convictions and enhancing public safety. But in fact we know little about the outcomes of most "hits" or about how most of these investigations would have proceeded—or whether they would eventually have been resolved—if DNA database searches were not performed. Sadly, despite years of operation of these programs around the world, no peer-reviewed, hypothesis-driven research has been published to measure outcomes (case resolutions); only output (e.g., number of DNA matches) has been measured.

Metrics of success for forensic DNA data banks have heretofore been limited to tallies of two outputs, "hits" and "investigations aided." These tallies in turn may be used by enthusiastic legislators as a justification for further expansion of criteria for inclusion. Several types of hits are tallied, including case-to-offender hits (DNA match between crime scene evidence and known offender in the database) and case-to-case hits (linking various crime scenes together). In addition to such hit counting, if an agency determines that the DNA comparison aids an investigation, it is included in the second tally. As of May 2020, CODIS statistics report over 514,982 hits, which are claimed to have assisted in more than 503,968 investigations, ranging from just 541 in Puerto Rico to over 88,000 in California (which also collects DNA upon arrest).[26]

While these two tallies of output are certainly interesting, reliance on them as indicators of overall success seems misguided, as they may provide only an incomplete picture of overall DNA database performance. For several reasons, both hit counting and enumerations of investigations aided are, alone, rather poorly defined concepts with limited use in evaluating the overall effectiveness of these programs:

First, not all database "cold hits" are truly unexpected, in the sense of providing the first identification of an individual criminal suspect. In fact, often investigators have already identified a key suspect and, if he or she is already known to be in the offender database, they expect a database match; they simply wait for such a hit report before making further steps toward an arrest, search warrant, or other investigative step. There is no standardized mechanism in place to verify the true nature of purported cold hits.

Second, DNA database matches are not necessarily probative, and it can certainly be argued that any investigative act whatsoever (including any DNA comparison) aids an investigation (whether or not a DNA match is identified from the database search). Moreover, it is well known that a single DNA hit could lead to several arrests or aid multiple investigations. We have little, if any, information on which hits were used and which were not, or why.

Third, and most important, counting hits or investigations aided does not help assess case resolution, the most meaningful and relevant outcome. Case resolution is largely a downstream activity that is the responsibility of police, prosecutors, and

---

26. FBI Codis Statistics https://www.fbi.gov/services/laboratory/biometric-analysis/codis/ndis-statistics.

the courts and is too little studied. Only a few agencies appear to have expended any time or energy on collecting and analyzing such important data.[27]

In Canada, for example, the number of hits could be much higher if judges issued bench orders for collection of DNA after conviction. They do so only about 50% of the time, even for primary offenses. Consequently, only 1% of the Canadian population is included in the convicted offender index, seriously limiting the potential effectiveness of the DNA data bank in that country, as most of the criminology literature suggests that 5–7% of the population is responsible for the vast majority of violent crime. The National DNA Data Bank of Canada reports 62,568 offender hits (crime scene to offender) and 6,855 forensic hits (crime scene to crime scene) as of March 31, 2020.[28]

The number of indirect hits could also be increased if familial searching tools were used more often on selected cases. Regarding investigations aided, there are no clearly accepted criteria to define what precisely aids an investigation (e.g., if two crime scenes without known suspects are linked by DNA comparison, is this one investigation aided or two?).

MEASURING DESIRABLE OUTCOMES: CASE RESOLUTION, CRIME PREVENTION, AND SOCIETAL INTERESTS

Aside from some admittedly dramatic successes, for the most part little, if any, follow-up of hits and their role in case resolution is documented or even attempted. The few US states (New York and Virginia) that have followed up on a small number of hits have reported a low rate of convictions following their DNA database hits.[29]

The connection between DNA hits generated by forensic database searches and the subsequent investigative follow-up of such hits to case resolution deserves immediate and careful study to document the full value of DNA collections, to examine weaknesses in the system, and more important, to seek ways to improve such programs to make them even more effective in accomplishing their stated goals.

Resolving crimes is an important goal, but not necessarily the only one. Besides solving or resolving crimes, possible outcomes from DNA database searches include eliminating some suspects as potential perpetrators, reduction of future crimes, and societal and personal interests in public safety and collective security.

Case Resolution

A presumed benefit of DNA database hits is an increase in convictions, with subsequent sentencing of the offender. This may not actually be the outcome in a large

---

27. Collins, 2013.
28. Royal Canadian Mounted Police https://www.rcmp-grc.gc.ca/en/the-national-dna-data-bank-canada-annual-report-20182019#a3.
29. Samuels et al., 2013.

proportion of cases. There are many possible explanations for limited conviction rates after DNA database hits. First is the assumption, often incorrect, that finding a DNA match between a crime scene DNA profile and that from a known individual will be considered useful or probative to the finders of fact—the jury. This is certainly recognized by seasoned investigators and prosecutors, who may eventually decide to ignore a DNA database hit if the match is unlikely to be a key piece of evidence. Finding someone's DNA at a crime scene does not necessarily reveal when or how it was deposited (obvious exceptions occur, as in sexual assaults). Similarly, failure to find someone's DNA at a crime scene does not necessarily eliminate him or her as a viable or likely suspect. Second, the degree and timeliness of law enforcement follow-up after the crime laboratory reports a DNA database match are highly variable. Laboratories may be so backlogged that samples sit waiting to be analyzed and entered into the database; consequently, the perpetrator remains unknown and at large. Breakdowns in reporting and overloaded detectives and prosecutors may provide little incentive to follow up on old cases from years ago, considering the pressure to solve current cases. Other problems that contribute to failure to resolve cases (by arrest, indictment, and conviction) include difficulty locating key witnesses, deceased victims or suspects, memory lapse or waning interest in testifying, witness intimidation, and missing evidence, even after bona fide DNA matches link crime scene evidence to known offender profiles in the database.

In the end, convictions alone are probably not the most reliable indicator of success in the stated goal of solving crimes. Case resolution may be a more suitable and reasonable metric, given the many reasons for failure to observe or to document high conviction rates in the aftermath of DNA database hits or investigations aided.

Reduction in or Prevention of Future Crimes

Some proponents have argued that DNA databases will reduce or prevent crime. Whether systematic DNA collection for inclusion in offender databases can, by itself, reduce crime rates is probably impossible to verify with confidence. Predictors of crime rates include community structure and social and demographic factors, as well as variables in policing practices.[30] Interpreting crime rate statistics presents many challenges, as definitions vary widely and reporting accuracy is uncertain. Many crimes go undetected or unreported, and definitions of crime may differ in different jurisdictions. Plea bargaining is so common in the United States that pleas are frequently entered (and accepted) for offenses far less than the actual crime. Furthermore, secular trends in crime reporting can lead to mistaken interpretations. If cultural conditions allow victims to report date rape more readily, for example, the statistics will indicate, perhaps falsely, that sexual assaults are increasing. Conversely, not reporting date rape may lower inappropriately the number of recorded sexual assaults.[31]

30. Taylor et al., 2016.
31. Noonan & Vavra, 2007; and Truman & Rand, 2009.

If DNA database programs were evaluated with regard to effects on crime rate, it is unclear how we could appropriately normalize the available data to assess the impact of DNA databases on such rates. If DNA databases produced the desired outcome at the micro level (i.e., increased suspect identifications, arrests, and recorded convictions), then observed recidivism rates would appear to increase. Conversely, measured crime rates could fall because of DNA collections, perhaps due to a deterrent effect or to incarceration of offenders.

Certainly, when DNA hits lead to arrest and conviction of offenders, lengthy prison sentences might theoretically reduce crime rates by preventing crimes by these offenders (at least during the time they remain incarcerated). Nevertheless, prison crowding, light sentencing, conditional sentencing (or no sentencing), and early release programs probably account, in part, for increases in crime rates due to recidivism, quite independent of any database effects. In Canada, the practice of "restorative justice" has resulted in one of the lowest worldwide incarceration rates after guilty verdicts, ranging from only 24% in Saskatchewan to 58% in Prince Edward Island (the national average is 35%). These offenders, when released and if recidivistic, are then free to reoffend again and again. Evaluation of proper database performance metrics could provide insight into this issue.

A very high level of repeat offenses is well documented for most categories of felony crime, including property crimes. Once they are released, offense rates by parolees and probationers are strikingly common for similar and even more serious crimes. Estimates suggest that as many as one-third of all violent crimes are committed by those on probation and parole. It seems clear that many factors contribute to crime rates and changes in them.[32] Thus, proof that any single new program in the justice system, such as the development of DNA databases, directly reduces crime rates would be difficult to convincingly demonstrate statistically. Nevertheless, we know that rapid apprehension and conviction of true perpetrators identified by DNA database matches will, at the very least, prevent some very serious crimes by particularly dangerous serial offenders.

Societal Interests

In addition to the admirable goals of DNA data banks to resolve unsolved crimes and reduce crime rates, there are several important societal interests potentially served by such programs. These include real and perceived public safety and security in homes and communities, though these interests must be balanced against often competing privacy interests.

A good example of problems that can crop up with errors and delays in the process of DNA analysis occurred on Cape Cod, Massachusetts, in the investigation of the January 2002 murder of Truro resident Christa Worthington. While a "person of interest," rubbish hauler Christopher M. McCowen, agreed to give a DNA sample within

---

32. Taylor et al., 2016.

weeks of the murder, the sample was not collected until March 2004, and it then sat in the Cape Cod police barracks for more than five months because samples were held at the barracks for transport to the crime laboratory until 10 had accumulated, so they could be transported to the crime lab and analyzed as a batch. In 2005, as the investigation grew colder, police and the district attorney decided to initiate what became a very controversial (and completely unproductive) DNA dragnet (Wagner, 2009), asking all 790 adult male residents in the area to "volunteer" a sample for comparison to crime scene evidence. Only after collecting several hundred samples and testing more than 40 did the State Police Crime Laboratory receive and—after a further delay of several months in the laboratory—finally test the key sample from McCowen. Finding a DNA match, police finally arrested McCowen, and he was indicted for the crimes. He was convicted of first-degree murder in 2007, and his request for a retrial was denied in 2010. The "voluntary" exclusion samples collected from over 200 uninvolved Truro residents have been retained by the Massachusetts State Police.

Another example of the seriousness of delays in DNA data bank programs relates to the substantial DNA backlogs that exist in many labs. DNA and evidence backlogs result in delays in processing old casework and in processing offender samples and quickly adding the profiles to the DNA databases. One such example gained widespread attention in June 2004 when police in Columbus, Ohio, arrested Robert N. Patton Jr. in connection with dozens of rapes in one neighborhood. Patton was arrested based on DNA taken more than two years before, but which was delayed in being entered into the state DNA database. After this was finally done, police linked him to the crimes within hours. Altogether, Patton was indicted in the rapes of 37 women—shockingly, 13 of whom were attacked while his DNA sample sat waiting to be processed. Patton's DNA was part of 11,000 convicted offender samples that accumulated after a federal grant for Ohio expired in mid-2001. Clearly, funding issues are crucial determinants of success. Careful research is needed to determine the scope and nature of these types of systemic failure and how program improvements through performance management practices can prevent recurrences.

RECOMMENDATIONS FOR IMPROVEMENT

To date, the public has had little input into standards of performance for DNA-related public safety programs, despite the expenditure of large amounts of public funds. Commitments for assessment and evaluation of such DNA collections have had little discussion in public forums, and little oversight by those outside the agencies responsible for their implementation.

Crime laboratories perform a Herculean task by responding to legislative mandates to collect, analyze, and report results of DNA profiling on millions of offender/arrestee and crime scene samples. This is costly, difficult, and sometimes overwhelming work.

Nevertheless, a thorough audit and review of police and laboratory practices is in order to identify ways to speed throughput and timely identification and reporting of hits. Time lags between the collection of evidence, laboratory analysis, confirmation of a DNA hit, and subsequent investigative follow-up vary considerably and are very worrisome. In some states such lags are measured in months for the laboratory component alone. As one way to address such problems, in the United States the FBI announced the formation of the Rapid DNA Program Office in 2010 to facilitate the development and integration of Rapid DNA technology for use by law enforcement. The office works with the Department of Defense, the National Institute of Standards and Technology, the National Institute of Justice, and other federal agencies to ensure the coordinated development of this new technology among federal agencies. The office also works with state and local law enforcement agencies and state bureaus of identification through the FBI's Criminal Justice Information Services Division Advisory Policy Board to facilitate the effective and efficient integration of Rapid DNA in the booking environment.[33] Under The Rapid DNA Act of 2017 (Pub. L. No. 115-50), signed into law by President Trump on August 18, 2017, the FBI director is authorized to "issue standards and procedures for the use of Rapid DNA instruments and resulting DNA analyses." Accordingly, the FBI is testing new technology that may allow processing a Rapid DNA upload and search in the CODIS software. Such a practice could identify some unknown perpetrators by finding DNA database matches and apprehending suspects within hours of law enforcement's arriving on the crime scene.

Several government agencies and organizations oversee and regulate the laboratory methods used, inspection of the labs themselves, and rules regarding use of government DNA databases. However, both common law and policy development is lagging in this area, and the government agencies are challenged to keep up with the rapid pace of new technologies and data mining strategies.[34] In addition to the interests of local law enforcement authorities and of victims and their advocates, along with public expectations for safety at home and abroad, come other, sometimes competing legitimate privacy concerns, with justifiable fear of an overgoverned society. Accordingly, a national debate over data mining of genetic and genealogy data is likely in many countries, as the internet age gives us ease of access to all manner of personal data held in public repositories (Berkman et al., 2018; Bieber & Lazer, 2004; Greely et al., 2006; Greytak et al., 2018; Kennett, 2019; Kim et al., 2011; Murphy, 2018). In early 2019, a lone Maryland legislator, Delegate Charles E. Sydnor III (now a state senator), proposed a bill (H.B. 30) that would prohibit searches of consumer genealogical databases for the purpose of identifying an offender in connection with a crime through a biological relative's DNA samples.[35] After a hearing in 2019, that bill never made it out of committee.

---

33. H.R.510—Rapid DNA Act of 2017 https://www.fbi.gov/services/laboratory/biometric-analysis/codis/rapid-dna.

34. The late US federal judge Learned Hand once described the justice system as a three-horse chariot, with the science and technology horse racing faster ahead than the horses representing the law and ethics/policy, making for a very rocky ride.

35. https://legiscan.com/MD/text/HB30/2019.

As part of the debate, there have even been calls for a universal DNA database (Kaye & Smith, 2004), and including more recently (Hazel et al., 2018), sparking new dissent (Joly et al., 2019) and response (Hazel et al., 2019). Such proposals, while interesting to debate, seem terribly costly and inefficient, as less than 10% of the population is implicated in the majority of reported serious crime, and the likelihood of planting false and misleading evidence would seem great. Given that heel-stick blood spot cards are collected at birth in all US states for mandated genetic testing for serious but treatable genetic metabolic disorders, such universal collection is theoretically possible, but would predictably be unacceptable to the public.

CONCLUSION

This chapter summarizes the current state of public and private DNA collection aimed at public safety efforts to identify suspects, identify missing persons, and aid in reunification of human remains in the aftermath of war or natural disaster.

A lack of integration between the DNA laboratories and the other components of the justice system responsible for following up on results of database searches is perhaps the biggest weakness in evaluating the effectiveness of forensic DNA data bank programs, in that desirable outcomes have not been clearly defined or carefully researched. Systems of performance management are needed to assess specific DNA collection program activities in relation to specified outcomes. An integrated approach would facilitate real hypothesis testing, rather than relying on anecdotal stories of success in individual cases, which may not be representative of most outcomes.

DNA data banks and database expansion now include arrestees in many jurisdictions; military, health, and DTC DNA collections now number in the tens of millions of samples worldwide. This large number of samples also provides access, by proxy, to close and even distant relatives. This has been accomplished by use of data mining methods such as familial searching of offender databases and by searching open-access public or private DNA databases to compare the number and location of large blocks of closely linked SNPs, followed up by ancestry searching using the tools of the genealogist.

Access to most persons' DNA, either by DNA collections or by proxy searching methods, raises questions of privacy and policy that now confront both scholars and the public. While not yet addressed by the courts or the court of public opinion, legal concepts like the third-party doctrine, standing, and search and seizure may apply. Federal and state agencies, commercial DNA laboratories, DTC DNA companies, and public open-access ancestry data collections will need to work together to address questions of data security, access, and use of DNA collections.

Enhanced familial searching (i.e., genealogics) involves trade-offs that highlight some of the paradoxes of our networked age (Neblo et al., 2017 Rahwan et al., 2019; Ram et al., 2018). With investment for technology and new investigatory processes involving genealogical data, powerful new investigative leads are being generated daily for many hundreds of cold cases, perhaps increasing public security and providing closure to victims. However, such enhanced familial searching poses many more potential privacy intrusions than does traditional familial searching, as each genealogical search might easily involve scanning thousands of individuals. This is likely to identify a painful disjuncture between genealogical records and genetic reality (Anderson, 2006; Jonsson et al., 2017). In the end, we are all connected by our DNA, and if we pull strongly enough at these strands, we will all come under scrutiny.

ACKNOWLEDGMENTS

The author thanks attorney Rockne Harmon; Honorable David Deakin; Professors David Lazer, David Kaye, and Donald Shelton; and Dr. Carll Ladd for useful conversations. This chapter includes adaptations and some portions of an earlier work, by this same sole author, published previously (see Bieber, 2006).

REFERENCES

Anderson, K. G. (2006). How well does paternity confidence match actual paternity? Evidence from worldwide nonpaternity rates. *Current Anthropology*, 47(3), 513–520.

Balding, D. J., et al. (2013). Decision-making in familial database searching: KI alone or not alone? *Forensic Science International: Genetics*, 7(1), 52–54.

Berger, M. A. (2006). The impact of DNA exonerations on the criminal justice system. *Journal of Law, Medicine & Ethics*, 34(2), 320–7.

Berkman, B. E., Miller, W. K., & Grady, C. (2018). Is it ethical to use genealogy data to solve crimes? *Annals of Internal Medicine*, 169(5), 333–34.

Bieber, F. R. (2004). Science and technology in forensic DNA Profiling. In D. Lazer, ed., *DNA and the criminal justice system: The technology of justice* (pp. 23–62). Cambridge, MA: MIT Press.

Bieber, F. R. (2006). Turning base hits into earned runs: improving the effectiveness of forensic DNA data bank programs. *Journal of Law, Medicine & Ethics*, 34(2), 222–233.

Bieber, F. R. (May 2, 2018). All in the family: Finding criminals through the DNA of their relatives. *The Harvard Crimson*. https://www.thecrimson.com/article/2018/5/2/bieber-all-in-the-family/

Bieber, F. R., Brenner, C. H., & Lazer, D. (2006). Finding criminals through DNA of their relatives. *Science*, 312(5778), 1315–1316.

Bieber, F. R., & Lazer, D. (2004). Guilt by association: Should the law be able to use one person's DNA to carry out surveillance on their family? Not without a public debate. *New Science*, 184(2470), 20.

Bieber, F. R., et al. (2016). Evaluation of forensic DNA mixture evidence: Protocol for evaluation, interpretation, and statistical calculations using the combined probability of inclusion. *BMC Genetics*, 17(1), 125.

Biesecker, L. G., et al. (2005). DNA identifications after the 9/11 World Trade Center attack. *Science*, 310(5751), 1122–1123.

Buckleton, J. S., Bright, J., & Taylor, D. (2016). *Forensic DNA evidence interpretation*. 2nd ed. CRC Press.

Budowle, B., Bieber, F. R., & Eisenberg, A. J. (2005). Forensic aspects of mass disasters: Strategic considerations for DNA-based human identification. *Legal Medicine* (Tokyo), 7(4), 230–243.

Butler, J. M. (2014). *Advanced topics in forensic DNA typing: Interpretation*. Cambridge, MA: Academic Press.

Butterfield, F. (2018). *In my father's House: A new view of how crime runs in the family*. New York: Knopf.

Callaghan, T. F. (2019). Responsible genetic genealogy. *Science, 366*(6462), 155.

Castellanos, A. et al. (2020). Disclosure of genetic risk revealed in a research study. *New England Journal of Medicine, 382*(8), 763–765.

Chaitanya, L., et al. (2018). The HIrisPlex-S system for eye, hair and skin colour prediction from DNA: Introduction and forensic developmental validation. *Forensic Science International: Genetics, 35*, 123–135.

Chung, Y. K., & Fung, W. K. (2013). Identifying contributors of two-person DNA mixtures by familial database search. *International Journal of Legal Medicine, 127*(1), 25–33.

Chung, Y. K., Fung, W. K., & Hu, Y. (2010). Familial database search on two-person mixture. *Computational Statistics & Data Analysis, 54*(8), 2046–2051.

Clayton, E. W., et al. (2019). The law of genetic privacy: Applications, implications, and limitations. *Journal of Law and the Biosciences 6*, 1–36.

Collins, D. (2013). The Massachusetts forensic DNA database: the utility of offender hits and the promise of familial searching. Thesis. Harvard University, Cambridge, MA.

Curran, J. M., & Buckleton, J. S. (2008). Effectiveness of familial searches. *Science & Justice, 48*(4), 164–167.

Debus-Sherrill, S., & Field, M. B. (2019). Familial DNA searching—an emerging forensic investigative tool. *Science & Justice, 59*(1), 20–28.

Deputy Secretary of Defense. (1991). Memorandum # 47803, to Secretaries of the Military Departments, subject: Establishment of a Repository of Specimen Samples to Aid in Remains Identification Using Genetic Deoxyribonucleic Acid (DNA) Analysis, 16 Dec. 1991.

Epstein, J. (2009). "Genetic surveillance"—The bogeyman response to familial DNA investigations. *Journal of Law, Technology and Policy, 2009*(1) 141.

Erlich, Y., et al. (2018). Identity inference of genomic data using long-range familial searches. *Science, 362*(6415), 690–94.

Farrington, D. P. (1995). The Twelfth Jack Tizard Memorial Lecture: The development of offending and antisocial behaviour from childhood: Key findings from the Cambridge Study in Delinquent Development. *Journal of Child Psychology and Psychiatry, 36*(6), 929–964.

Farrington, D. P., Barnes, G. C., & Lambert, S. (1996). The concentration of offending in families. *Legal and Criminological Psychology, 1*(1), 47–63.

Farrington, D. P., & Crago, R. V. (2016). The concentration of convictions in two generations of families. In A. Kapardis & D. P. Farrington, eds., *The Psychology of Crime, Policing and Courts* (pp. 21-37). Abingdon, UK: Routledge.

Farrington, D. P., et al. (2001). The concentration of offenders in families, and family criminality in the prediction of boys' delinquency. *Journal of Adolescence, 24*(5), 579–596.

Federal Bureau of Investigation. (2019). *National DNA Index System Operational Procedures Manual*. Version 8, May 2019. https://www.fbi.gov/file-repository/ndis-operational-procedures-manual.pdf/view

Ge, J., & Budowle, B. (2012). Kinship index variations among populations and thresholds for familial searching. *PLoS One, 7*(5), e37474.

Ge, J., Budowle, B., & Chakraborty, R. (2011a). Choosing relatives for DNA identification of missing persons. *Journal of Forensic Science, 56*(Supp. 1), S23–28.

Ge, J., et al. (2011b). Comparisons of familial DNA database searching strategies. *Journal of Forensic Science, 56*(6), 1448–1456.

Ge, J., Eisenberg, A., & Budowle, B. (2012). Developing criteria and data to determine best options for expanding the core CODIS loci. *Investigative Genetics, 3*, 1.

Gill, P., et al. (2006). The evolution of DNA databases--recommendations for new European STR loci. *Forensic Science International, 156*(2–3), 242–244.

Greely, H. T., et al. (2006). Family ties: The use of DNA offender databases to catch offenders' kin. *Journal of Law, Medicine & Ethics, 34*(2), 248–262.

Greytak, E. M., Moore, C., & Armentrout, S. L. (2019). Genetic genealogy for cold case and active investigations. *Forensic Science International, 299*, 103–113.

Greytak, E. M., et al. (2018). Privacy and genetic genealogy data. *Science, 361*(6405), 857.

Gross, S. R., Jacoby, K., Matheson, D. J., & Montgomery, N. (2005). Exonerations in the United States, 1989 through 2003. *Journal of Criminal Law and Criminology, 95*(2), 523–560.

Haimes, E. (2006). Social and ethical issues in the use of familial searching in forensic investigations: insights from family and kinship studies. *Journal of Law, Medicine & Ethics, 34*(2), 263–276.

Hazel, J. W., et al. (2018). Is it time for a universal genetic forensic database? *Science, 362*(6417), 898–900.

Hazel, J. W., et al. (2019). Risks of compulsory genetic databases-Response. *Science, 363*(6430), 940.

Hicks, T., et al. (2010). Use of DNA profiles for investigation using a simulated national DNA database: Part II, statistical and ethical considerations on familial searching. *Forensic Science International: Genetics, 4*(5), 316–322.

Hysi, P. G., et al. (2018). Genome-wide association meta-analysis of individuals of European ancestry identifies new loci explaining a substantial fraction of hair color variation and heritability. *Nature Genetics, 50*(5), 652–656.

Ingold, S., et al. (2018). Body fluid identification using a targeted mRNA massively parallel sequencing approach—results of a EUROFORGEN/EDNAP collaborative exercise. *Forensic Science International: Genetics, 34*, 105–115.

Interpol. (2016). *Global DNA Profiling Survey Results 2016*. www.interpol.int

Joly, Y., Marrocco, G., & Dupras, C. (2019). Risks of compulsory genetic databases. *Science, 363*(6430), 938–940.

Jonsson, H., et al. (2017). Parental influence on human germline de novo mutations in 1,548 trios from Iceland. *Nature, 549*(7673), 519–522.

Jonsson, H., et al. (2018). Multiple transmissions of *de novo* mutations in families. *Nature Genetics, 50*(12), 1674–1680.

Kaplanis, J., et al. (2018). Quantitative analysis of population-scale family trees with millions of relatives. *Science, 360*(6385), 171–175.

Kaye, D. H. (2001a). Bioethical objections to DNA databases for law enforcement: questions and answers. *Seton Hall Law Review, 31*(4), 936–948.

Kaye, D. H. (2001b). The constitutionality of DNA sampling on arrest. *Cornell Journal of Law and Public Policy, 10*(3), 455–509.

Kaye, D. H. (2006). Who needs special needs? On the constitutionality of collecting DNA and other biometric data from arrestees. *Journal of Law, Medicine & Ethics, 34*(2), 188–198.

Kaye, D. H. (2013). The genealogy detectives: A constitutional analysis of familial searching. *American Criminal Law Review, 50*, 109.

Kaye, D. H., & Smith, M. S. (2004). DNA databases for law enforcement: The coverage question and the case for a population-wide database. In *DNA and the Criminal Justice System: The Technology of Justice* (pp. 247–284). Cambridge, MA: MIT Press.

Kennett, D. (2019). Using genetic genealogy databases in missing persons cases and to develop suspect leads in violent crimes. *Forensic Science International, 301*, 107–117.

Kim, J., et al. (2011). Policy implications for familial searching. *Investigative Genetics, 2*, 22.

Kim, J., et al. (2018). Statistical detection of relatives typed with disjoint forensic and biomedical loci. *Cell, 175*(3), 848–858.

Kloosterman, A., Sjerps, M., & Quak, A. (2014). Error rates in forensic DNA analysis: definition, numbers, impact and communication. *Forensic Science International: Genetics, 12*, 77–85.

Krimsky, S., & Simoncelli, T. (2010). *Genetic justice: DNA data banks, criminal investigations, and civil liberties*. New York: Columbia University Press.

Lazer, D., & Bieber, F. R. (2010). "Familial searching": Its promise and perils. *Los Angeles Times*. July 10, 2010 https://www.latimes.com/archives/la-xpm-2010-jul-10-la-oe-lazer-grim-sleeper-dna-20100710-story.html

Lazer, David, ed. (2004). *DNA and the criminal justice system: The technology of justice*. Cambridge, MA: MIT Press.

Maguire, C. N., et al. (2014). Familial searching: A specialist forensic DNA profiling service utilising the National DNA Database® to identify unknown offenders via their relatives—The UK experience. *Forensic Science International: Genetics*, 8(1), 1–9.

Mapes, A. A., Kloosterman, A. D., & de Poot, C. J. (2015). DNA in the criminal justice system: The DNA success story in perspective. *Journal of Forensic Science*, 60(4), 851–856.

Martha, R. S. J. (2010). *The Legal Foundation of Interpol*. Portland, OR: Hart.

Meyer, R. (2018). The Cambridge Analytica scandal, in three paragraphs. *The Atlantic*, March 20.

McCleary, R. (1992). *Dangerous Men: The Sociology of Parole*. Albany, NY: Harrow and Heston.

Murphy, E. (2010). Relative doubt: Familial searches of DNA databases. *Michigan Law Review*, 109.

Murphy, E. (2015). *Inside the Cell*. New York, NY: Nation.

Murphy, E. (2018). Law and policy oversight of familial searches in recreational genealogy databases. *Forensic Science International*, 292, e5–e9.

Myers, S. P., et al. (2011). Searching for first-degree familial relationships in California's offender DNA database: Validation of a likelihood ratio-based approach. *Forensic Science International: Genetics*, 5(5), 493–500.

National Conference on State Legislatures. (2018). DNA Arrestee Laws. https://www.fbi.gov/file-repository/ndis-operational-procedures-manual.pdf/view

Neblo, M. A., et al. (2017). The need for a translational science of democracy. *Science*, 355(6328), 914–915.

Nelkin, D., & Andrews, L. (1999). DNA identification and surveillance creep. *Sociology of Health & Illness*, 21(5), 689–706.

Noonan J. H., & Vavra, M. C. (2007). Crime in schools and colleges: A study of offenders and arrestees reported via national incident-based reporting system data. *Federal Bureau of Investigation, U.S. Dept Justice*. https://ucr.fbi.gov/nibrs/crime-in-schools-and-colleges-pdf

Ohm, P. The many revolutions of Carpenter. (2019). 32 Harvard J. Law & Tech. 357. Spring, 2019. Copyright © 2019 by the President and Fellows of Harvard College, Cambridge, MA.

Primorac, D., & Schanfield, M. (2014). *Forensic DNA Applications: An Interdisciplinary Perspective*. Boca Raton, FL: CRC Press.

Public Safety Canada. (2019). *Evaluation of Public Safety Canada's Roles in Support of DNA Analysis*. © Her Majesty the Queen in Right of Canada, 2019 Cat. No.: PS4-253/2019E-PDF ISBN:978-0-660-30984-2

Putkonen, A., et al. (2002). The quantitative risk of violent crime and criminal offending: a case-control study among the offspring of recidivistic Finnish homicide offenders. *Acta Psychiatrica Scandinavica Suppl* 106(412), 54–57.

Rahwan, I., et al. (2019). Machine behaviour. *Nature*, 568(7753), 477–486.

Ram, N. (2011). Fortuity and forensic familial identification. *Stanford Law Review*, 63(4), 751–812.

Ram, N., Guerrini, C. J., & McGuire, A. L. (2018). Genealogy databases and the future of criminal investigation. *Science*, 360(6393), 1078–1079.

Rohlfs, R. V., et al. (2013). The influence of relatives on the efficiency and error rate of familial searching. *PLoS One*, 2013;8(8):e70495. Published 2013 Aug 14. doi:10.1371/journal.pone.0070495

Rosen, J. (2009). Genetic surveillance for all. *Slate*.

Samuels, J. E., Davies, E. H., & Pope, D. B. (2013). Collecting DNA at arrest: policies, practices, and implications, final technical report document No.: 242812, The Urban Institute, Justice Policy Center, Washington, D.C.

Schlesinger, L. B. (2001). *Serial offenders*. New York, NY: CRC Press.

Slooten, K., & Meester, R. W. J. (2011). Statistical aspects of familial searching. *Forensic Science International: Genetics Supplement Series*, 3(1), E167–E69.

Smith, C. A., & Farrington, D. P. (2004). Continuities in antisocial behavior and parenting across three generations. *Journal of Child Psychology and Psychiatry*, 45(2), 230–247.

Taylor, R. B., Groff, E. R., Elesh, D., & Johnson, L. (2016). Intra-metropolitan crime patterns predictors, and predictions. Document No.: 249740, U.S. Dept. of Justice.

Thompson, W. C. (1995). Subjective interpretation, laboratory error and the value of forensic DNA evidence: Three case studies. *Genetica*, 96(1–2), 153–168.

Truman, J. L., & Rand, M. R. (2010). National Crime Victimization Survey. Criminal Victimization, 2009, U.S. Department of Justice, Office of Justice Programs, Bureau of Justice Statistics. https://www.bjs.gov/content/pub/pdf/cv09.pdf

Wagner, J. K. (2009). Just the facts, ma'am: Removing drama from DNA dragnets. *N.C. Journal of Law & Technology*, 11(1), 51.

Williams, R., & Johnson, P. (2006). Inclusiveness, effectiveness and intrusiveness: Issues in the developing uses of DNA profiling in support of criminal investigations. *Journal of Law, Medicine & Ethics*, 34(2), 234–247.

Zhang, S. (2018). How a tiny website became the police's go-to genealogy database. The Atlantic. June 1.

Zhang, S. (2019). A DNA company wants you to help catch criminals. The Atlantic. March 29.

CHAPTER 5

Recent Developments in Forensic DNA Technology

HENRY ERLICH, CASSANDRA CALLOWAY, AND STEVEN B. LEE

The history of DNA in the criminal justice system has been one of continual technological development since the first cases in the mid-1980s. In the beginning of forensic DNA analysis, technologies for analyzing both sequence and length polymorphisms were developed (as described in chapter 1). Following the application of restriction fragment length polymorphism (RFLP) analysis of variable number tandem repeat (VNTR) loci in the first few years (see chapter 1), all subsequent forensic DNA tests were performed using polymerase chain reaction (PCR) amplification (Butler, 2015). As noted in the introduction and chapter 1, routine forensic DNA analysis is currently performed using DNA extraction and quantification, followed by PCR amplification of autosomal and Y chromosome short tandem repeat (STR) loci. In addition to the STR genetic markers, analyses of single nucleotide polymorphisms (SNPs) and/or mitochondrial DNA (mtDNA) have been used in some cases, particularly with degraded DNA samples (Jobling & Gill, 2004.) The current standard (as of 2020) for forensic genotyping, however, remains the analysis of STR polymorphisms by capillary electrophoresis, performed in an accredited forensic lab.

The trend of dramatic technological developments has continued over the last few years, with the most consequential recent innovations being (1) the development of next generation sequencing (NGS), also known as massively parallel sequencing (MPS); and (2) the implementation of commercially available Rapid DNA chemistries and instruments that automate DNA typing of all Combined DNA Index System (CODIS) core STR loci from sample to profile in 90 minutes (Grover et al., 2017). Although not a DNA technology innovation per se, the recent development and implementation of probabilistic genotyping software using biological modeling, statistical theory, and computer algorithms to interpret mixed DNA samples has also had a significant impact on forensic DNA analyses of STR genotyping data (see chapter 3; Liu & Harbison, 2018; Coble & Bright, 2019).

In addition to the implementation of MPS and Rapid DNA technologies, significant advances have been achieved in the development of new forensic genetic marker sets for physical appearance and biogeographical ancestry, with great potential for developing investigative leads (Kayser, 2019; McCord & Lee, 2019). This chapter discusses these recent technical developments as well as their impact on the criminal justice system.

MASSIVELY PARALLEL SEQUENCING (MPS): ADVANTAGES, TECHNOLOGIES, AND APPLICATIONS

Virtually all felon databases, including the federal CODIS and the many state and local databases, are collections of STR profiles; to enable searches of these databases, analysis of criminal evidence will continue to use STR genotyping well into the foreseeable future, even as new DNA technologies and new genetic markers are introduced into the criminal justice system. In fact, analysis of STR length polymorphisms by MPS has increased the discrimination power of these standard STR loci by revealing some sequence variation among STR alleles of the same size. For example, two alleles that both have five copies of a CCGG repeat might differ in a base substitution in one of the repeats (e.g., CC**A**G) known as isoalleles, or isometric heterozygotes (Silva et al., 2018); this sequence polymorphism within the STR would not be detected by the standard capillary electrophoresis method based on distinguishing the length of the PCR product (Gettings et al., 2018; Phillips et al., 2018).

Various commercial MPS systems are available today, all with two properties in common: (1) they are massively parallel, with millions of simultaneous sequencing reactions; and (2) each reaction is initiated with a single DNA molecule or from a clonal population generated from a single DNA molecule. The two most commonly used platforms (Illumina and Ion Torrent, now ThermoFisher) achieve clonal sequencing by limiting dilution so that the amplification and sequencing starts from a single DNA molecule. This clonal sequencing property of MPS facilitates the analysis of mixtures (reviewed in chapter 3) because the individual components of a mixture can be sequenced separately. MPS has been applied to autosomal STR loci (Alonso et al., 2018; Budowle et al., 2017; Bruijns et al., 2018; de Knijff, 2019); Y and X STR loci, ancestry, identity and phenotype-informative SNPs (Phillips et al., 2018; Moreno, 2018; Bulbul & Finoglu, 2018); and mtDNA (Just et al., 2015; Holland et al., 2018; Kim et al., 2015). One important benefit of the MPS technology is that it offers the opportunity to analyze different categories of genetic markers (e.g., STRs, SNPs, mtDNA polymorphisms) in parallel on the same instrument for multiple samples. Relevant details of MPS technology, additional advantages, and applications are discussed further in the rest of this chapter.

Next generation sequencing or massively parallel sequencing (Le Gallo et al., 2017) provides low-cost, high-throughput, rapid, accurate sequencing capabilities that have been applied to both fundamental and applied research questions in a variety of scientific fields.[1]

ADVANTAGES OF MPS IN FORENSIC DNA

There are several advantages to applying MPS to forensic analysis. These systems enable simultaneous analysis of a large number of forensically relevant genetic markers to improve the efficiency, capacity, and resolution of forensic DNA typing. Furthermore, MPS enhances the capabilities of forensic DNA laboratories to solve crimes involving challenging samples, in particular of mixed specimens (Bruijns et al., 2018; Alonso et al., 2018). As noted previously, the ability to detect sequence variants of STR alleles that are of the same size and thus indistinguishable by capillary electrophoresis provides higher discrimination power, increased mixture resolution, and more accurate results. In addition, results can be obtained on degraded samples if the sequence analysis targets small regions (< 200 bp; Zhang et al., 2018). Finally, MPS provides the ability to deconvolute complex mixtures.

The analysis of forensics specimens that are mixtures remains one of the most problematic issues in forensics from both technical and statistical perspectives (see chapter 3). Until the introduction of NGS/MPS technology, forensic DNA sequencing was carried out by the Sanger method,[2] performed on automated capillary electrophoresis instruments using fluorescently labeled DNA fragments generated during a DNA synthesis reaction. Forensic sequencing was carried out primarily on the mtDNA since the length polymorphism of the STR loci could be analyzed faster and more cheaply by electrophoresis. The development of massively parallel and clonal sequencing systems represented a true paradigm shift for all DNA sequencing; for forensic sequencing in particular, it offered the potential for much more effective analysis of mixed samples (those with more than one DNA contributor). Each of the millions of parallel sequencing reactions generates a "sequence read." The clonal sequencing aspect of NGS provides the opportunity to recover different sequence reads for every genetic variant (allele) present in the mixture. Thus, the number of sequence reads recovered for a particular variant provides a digital readout and

---

1. These applications include the study of human disease genomics (Wu et al., 2018), cancer (Tan et al., 2018), immunology (Petersdorf & Ohuigin, 2018) microbiology and microbiomes (Cao et al., 2017; Ross et al., 2018), epidemiology (Gardy & Loman, 2018), evolutionary biology (Lea et al., 2017), plant genomics (Torkamaneh et al., 2018), crop genome evolution (Huang et al., 2018), metagenomics (Garrido-Cardenas & Manzano-Agugliaro, 2017), and forensic DNA (Borsting & Morling, 2016; Alonso et al., 2018; Bruijns et al., 2018; de Knijff, 2019).

2. This method, based on the incorporation of fluorescent nucleotides into a growing DNA strand complementary to the targeted DNA template, was developed by Fred Sanger, a British biochemist, in 1977. By synthesizing a series of fluorescently labeled DNA fragments, the sequence of the DNA template could be determined by analyzing the array of DNA fragments separated by capillary electrophoresis. Sanger received the Nobel Prize, along with Walter Gilbert and Paul Berg, in 1980 for this invention.

allows a quantitative analysis of the various contributors in a mixed sample within the limits of the technology.

INTRODUCTION TO MPS TECHNOLOGY

There are several different commercial MPS platforms that utilize different sequencing chemistries, giving them unique characteristics and limitations, but they are all capable of producing vast amounts of sequencing data in a more cost effective and timely manner than has previously been possible (McCombie et al., 2018). The common initial step for all MPS platforms is the creation of a *library*, that is, the conversion of DNA from the evidence or reference sample to a form that can be sequenced in each of the parallel sequencing reactions for a given MPS platform. For many platforms, this step involves the addition of short DNA sequences, known as *adapters*, to the ends of all the DNA molecules in the sample in an enzyme-mediated reaction known as *ligation*. These adapters contain sequences that allow each of the individual DNA fragments to be PCR amplified and to be bound to a location where the sequencing reaction will occur. These added sequences also contain short DNA sequences that function as "bar codes" (or "indices"), and the specific pair of bar codes indicates which DNA sample was the source of the library. Tagging each library with a unique pair of bar codes allows the sequencing of multiple libraries in the same MPS run. The mixture of many millions of sequence reads generated in the run can be deconvoluted and assigned by bioinformatic software to an individual sample (Lee, 2017; McCombie et al., 2019).

The MPS libraries can be either be "targeted" or "shotgun". Shotgun libraries are constructed from *all* the DNA present in the sample; this is the strategy for generating whole genome sequences (WGS).[3] Targeted libraries are focused on particular regions of interest; for forensic applications involving individual identification, the targeted regions are typically the highly polymorphic regions of the genome, such as the current standard STR panels, panels of SNPs, or the mtDNA genome. Since less than 1% of the human genome is polymorphic, virtually all forensic MPS applications involve targeted libraries.

Targeted libraries can be prepared in one of two ways. PCR amplification can be used to target specific regions of DNA and specific genetic markers. The PCR primers are designed to flank the region(s) of interest; amplification results in a targeted PCR product that contains the primers. In some methods, the adapter sequences are directly attached to the ends of PCR primers along with their bar codes. These adapter sequences then bind to complementary sequences tethered to the sequencing platform, typically beads or a slide (or flow cell) on which sequencing reactions occur. Alternatively, the adapters are connected in a ligation reaction to the amplified DNA.

---

3. WGS has the potential to distinguish even between monozygotic twins, using sequence variation that has been generated *after* the twins have separated from a single fertilized egg. See the discussion of the McNair case later in the chapter.

An alternative strategy for creating targeted libraries does not rely on selective PCR amplification but is based on the selective "capture" of the regions of interest from a shotgun library (Bose et al., 2018; Shih et al., 2018). The specific capture is accomplished by a set of probes, short synthetic DNA fragments, that hybridize (bind) to specific sequences in the shotgun library. These captured targeted regions can then be recovered and subsequently sequenced. This probe capture strategy, while slightly more cumbersome than the PCR targeted library system, offers some advantages for analyzing forensic samples. Many forensic samples contain highly degraded DNA; many of these short DNA fragments may not contain both of the intact PCR primer binding sites needed and therefore cannot be amplified by PCR. The probe capture–based library system, however, is often effective in recovering short DNA sequences from such samples (Bose et al., 2018; Shih et al., 2018). Another advantage of the probe capture strategy is that a given set of genetic markers (e.g., STRs and SNPs) can be selectively recovered from a shotgun library and sequenced and then, with a different set of probes, the mitochondrial genome can be captured and sequenced from the same library. Target enrichment strategies using probe capture have been applied recently to hair samples (Shih et al., 2018)These capabilities are particularly valuable for forensic samples with limited DNA.

Most current commercial NGS or MPS platforms rely on sequencing a clonal population generated by some form of DNA amplification from a single DNA molecule (Mardis, 2017; McCombie et al., 2019; de Knijff, 2019). The currently dominant commercial platforms are Illumina and the Ion Torrent technology (ThermoFisher), both of which use a "sequencing by synthesis" strategy. (Illumina has recently partnered with a venture capital firm to start an independent forensic genomics company, Verogen, to provide Illumina's forensic sequencing technology to forensic customers.) In both platforms, the target DNA to be sequenced serves as a template for a DNA synthesis reaction. Both systems sequence a clonal population of DNA molecules by detecting which of the four nucleotides is incorporated into a growing new DNA strand. The Ion Torrent system flows one of the four nucleotides in succession over the parallel sequencing reactions and determines which nucleotide is incorporated into a specific site in the new DNA strand by detecting a hydrogen ion (a proton) that is released during the incorporation event, using a highly sensitive pH meter.[4] The Illumina system uses four different fluorescent labels for each of the four nucleotides to determine which nucleotide is incorporated at each position of the growing DNA strand, using a camera to detect the fluorescence. Both the Illumina and Ion Torrent systems provide relatively short sequence reads, typically fewer than 300 (Illumina) or 400 (Ion Torrent) bases. Both strands of the double-stranded DNA template are sequenced; these sequence reads are termed *forward* or *reverse*. In the Illumina system, these "paired-end" reads from a single DNA molecule can be "merged" so that a DNA molecule of around 500 base pairs (bp) can be sequenced completely,

---

4. The first NGS system, the 454 platform, introduced in 2005, used a similar strategy by detecting a light flash that was generated during the nucleotide's incorporation into the new DNA strand. This system is no longer available.

with sequence reads in both directions. This length limitation may not have serious consequences for the analysis of STR regions of forensic samples, since the size of the targeted STR repeat regions falls within that size limit, and casework samples frequently contain degraded DNA templates. For reference samples or for database samples, longer sequence reads are useful for reconstructing the regions of the genome containing the polymorphisms within and outside the repeat region used for individual identification. Sequence read length limitations may create problems for STR genotyping targeting sites within and outside the repeats; determining the sequence of the entire tandem repeat region and the unique flanking DNA sequences in very large STR alleles may not be possible due to the sequence read length constraint. For MPS analysis of STRs, some STR loci, which have long alleles, have thus been excluded from the STR panels.

A few newer and less widely used MPS platforms, such as Pacific BioSciences and Oxford Nanopore technologies, do not require the clonal amplification step and sequence the individual DNA molecule. The Pacific BioSciences technology, which analyzes single DNA molecules without PCR amplification, provides far fewer sequence reads than Illumina or Ion Torrent technology, but the reads are substantially longer, often greater than 10 kilobases (Kb). The SMRT (Single Molecule Real Time) system converts the DNA from the sample into closed, single-stranded circular molecules and loads these single circular DNA molecules into an SMRT cell to serve as a template for hundreds of thousands of parallel DNA synthesis reactions in a well-like structure termed a zeromode wave guide (ZMV) (Gonzalez-Garay, 2015). As the bases are added to the growing DNA strand by DNA polymerase, a camera detects in real time light pulses corresponding to the different incorporated bases in the ZMV. Currently the Pacific BioSciences technology requires a large and expensive instrument; as of 2019, it has very rarely been used on forensic samples.[5]

Another more recent DNA sequencing technology, the MinION (Oxford Nanopore), also does not require amplification or fluorescent labels. Instead, it detects ion currents generated when a charged DNA molecule passes through a nanoscale pore in a membrane separating two chambers. Each nucleotide produces a unique current level, which allows the MinION to sequence the strand. The pocket-size MinION instrument is capable of producing very long sequence reads and has the unique advantage of being light and portable, offering the potential for use in the field. To be applied in the field, the small, compact MinIon will have to be used in conjunction with a system for extracting DNA from the crime scene evidence sample, as has been done in the Rapid DNA technology. However, the current MinIon system has the disadvantage of high error rates, that is, mistakes in the base calling especially for long stretches of the same base, known as homopolymers (Rang et al., 2018). A recent study demonstrated that STRs could be sequenced with the MinIon instrument (Cornelis et al., 2018).

For all these platforms, MPS has the potential to analyze both length and sequence polymorphism, so that a very large number of genetic markers are now potentially

---

5. In late 2018, Pacific BioSciences was acquired by Illumina.

available for forensic analysis. Genetic markers such as SNPs, depending on the distribution of alleles in different populations, can be used for individual identification, for ancestry, or for phenotypic characteristics, such as eye and hair color.[6] Data derived from evidence samples on ancestry and phenotype can be valuable in providing investigational leads, but it is the panel of genetic markers for individual identity that will be used to calculate the random match probabilities (RMPs) or likelihood ratios (LRs) that are presented to the jury.

COMMERCIAL FORENSIC MPS KITS

Currently, several commercial kits have been developed for various MPS platforms. The ForenSeq DNA Signature Prep Kit (Illumina, now known as Verogen) analyzes 58 STRs and 173 SNPs for identity, phenotypic, and biogeographic ancestry on the MiSeq MPS platform (Jäger et al., 2017). The HID Ion AmpliSeq Identity Panel, including 124 SNPs for identity and lineage, and the Precision ID GlobalFiler MPS STR Panel are both analyzed on the Ion Torrent MPS platform (ThermoFisher). However, these PCR-based MPS systems still require that intact primer binding sites be present on the DNA from the forensic sample. When DNA is degraded so that most DNA fragments are on the order of 100–150 bp, PCR-based methods are likely to result in "drop-out" (failure to amplify and be detected) of some or all loci or alleles. As noted previously, the strategy of target enrichment using probe capture to prepare MPS libraries may address this issue and has turned out to be more effective in recovering sequence reads from samples with degraded DNA (Bose et al., 2018; Shih et al., 2018).

The International Commission on Missing Persons, in partnership with Qiagen, designed a custom SNP panel targeting over 1,400 markers, using early versions of the Qiagen GeneReader MPS system. Qiagen is developing additional SNP panels for forensic use. Several other commercial vendors have MPS panels. Promega's PowerSeq 46GY System, designed to be compatible with the Illumina TruSeq and MiSeq, includes all CODIS and ESS STRs amelogenin and 23 Y-STR loci, and it has a recommended DNA input of 0.5 nanogram (ng). The PowerSeq Auto/Mito/Y System targets 22 autosomal STRs, including the CODIS STR extended panel; 23 Y-STRs; and 10 amplicons for mitochondrial control regions and amelogenin (Montano et al., 2018). A publication in 2018 reported on a National Institute of Justice–funded interlaboratory forensic MPS study comparing different instruments, panels, and

---

6. SNPs that have a relatively even distribution of alleles (close to 50-50) in many different populations are useful in individual identification. In addition, these SNPs should be statistically independent to allow the use of the product rule in calculating RMPs. SNPs that have very different allele distributions in different populations can be useful for ancestry investigation. For example, an A/T SNP might have a frequency of 98% A and 2% T in population X and 5% A and 95% T in population Y. Such SNPs are referred to as ancestry informative markers (AIMS). SNPs in genes that encode proteins involved in pigment metabolism may be associated with variation in eye, hair, and skin color.

software by Battelle.[7] Recent advances in forensic MPS have been covered in two recent reviews (Bruijns et al., 2018; Alonso et al., 2018[8]; McCord & Lee, 2018; Kayser 2018).

The special issue of *Forensic Science International: Genetics* entitled " Trends and Perspectives in Forensic Genetics 2018" contains several articles, including a perspective entitled "From Next Generation to Now Generation Sequencing" (de Knijff, 2019), an MPS database search algorithm for forensic mitogenome analysis (Huber et al., 2018), forensic Y chromosome STR analysis (Caliebe & Krawczak, 2018), epigenetics, tissue identification and beyond (Vidaki & Kayser, 2018), forensic DNA transfer and inference of activity levels (van Oorschot, 2019; Taylor et al., 2018), mixture interpretation (Coble & Bright, 2019), microhaplotype analysis (Oldoni et al. 2019), forensic microbiome analysis for PMI (Metcalf, 2019), and human ID (Woerner et al., 2019), and missing persons DNA identification (Parsons et al., 2019). Collectively, these articles provide up-to-date information and overviews of the state of MPS sequencing, validation, and applications in forensic genomics.

MPS PRIVACY, ETHICS, AND ADMISSIBILITY

Current STR analysis by capillary electrophoresis analyzes noncoding regions of the DNA, as noted previously, but some newer MPS applications, because they have the capability of analyzing sequence polymorphism, investigate coding regions that are related to phenotype and ancestry and may contain potentially medically relevant information, raising issues of privacy (see Chapter 15; Bradbury et al., 2019). Within the United States, individual states have the discretion to regulate these issues. Under current standards, Indiana, New Mexico, Rhode Island, and Wyoming specifically prohibit phenotyping of DNA held in CODIS; Texas specifically allows it.

Sequencing the whole mitochondrial genome, rather than just the hypervariable region, involves analyzing genes that may have disease-related information. This is a current issue of concern within the United States (Buford et al., 2018), although in Europe the more informative whole mitochondrial genome data can be used in the courtroom for individual identification.

As with all advances in technology, the use of MPS introduces a variety of logistical, legal, and ethical issues for forensic DNA analysis. According to a recent survey of 33 European laboratories, these issues include the current absence of a standardized nomenclature and reporting standards, as well as absence of

---

7. Battelle, "Next Generation Sequencing (NGS) Feasibility and Guidance Study for Forensic DNA" (special report for National Institute of Justice, February 9, 2019), https://www.ncjrs.gov/pdffiles1/nij/grants/252287.pdf.

8. The first contains a description of the first-, second-, and third-generation sequencing techniques along with an overview of the MPS STR and SNP technologies (Bruijns et al., 2018). The second includes reviews and summaries of forensic MPS STR validation and implementation studies, available panels, platforms, bioinformatics tools, population sequencing studies, and international projects and standardization group efforts to standardize nomenclature (Alonso et al., 2018).

compatibility with current databases and population data that can be referenced in statistical analysis. Additional concerns involve the lack of legislation and funding for MPS forensic applications as well as data management, analysis, and interpretation. Cloud computing is a potential solution for data storage, although safety and privacy remain critical considerations there as well.

MPS has the potential to have a huge impact on the criminal justice system. Given the history of forensic DNA technology in the courtroom (see chapter 13), a series of rigorous and contentious admissibility hearings on the reliability and on the general scientific acceptance of MPS can be anticipated. The first admissibility hearing for MPS was highly unusual in that it involved not only an assessment of the technology itself but a very novel application, namely the attempt to distinguish between monozygotic ("identical") twins: the *McNair* case in 2017.[9] In 2004, two men assaulted and raped two women in Suffolk County, Massachusetts. One of the women managed to hide a used condom and gave it to the police. DNA from the evidence, based on the standard STR profiles, matched Dwayne McNair, but it also matched his twin brother. The police found the second man, Anwar Thomas, who worked out a plea deal; he identified Dwayne as the perpetrator. In 2011, Dwayne McNair was charged with 12 counts of aggravated rape and 2 counts of armed robbery, but the conundrum of the DNA match with the twin brother complicated the prosecution.

In December 2013, a European genetics company, EurofinsScientific, claimed to be able to distinguish identical twins. They published the results of their twin study (Weber-Lehmann et al., 2014), in which they were able to identify by WGS mutations in one twin that were passed down to his child but were not present in his "identical" twin brother. According to their study, the twins must have been separated within five to nine days after fertilization in order for the WGS method to distinguish a difference between the two genomes. In order to establish definitively which of the McNair twins committed the crimes, the assistant district attorney contacted Eurofins to have the McNairs' genomes and the semen sample sequenced, at a cost of $120,000. Eurofins found that there were nine sequence differences between the two twins and that the whole genome sequence of the evidence was closer to Dwayne McNair; they claimed that he was several billion times more likely to have been the rapist.[10] However, during the admissibility hearing, the presiding judge granted McNair's motion to exclude the DNA evidence, finding that the Eurofins genome mapping did not meet the *Daubert* standard of admissibility. The initial test Eurofins conducted, in which they found that mutations between twins were passed down to their children, involved only one pair of twins. The judge noted that MPS had not been regularly employed in criminal cases and ruled that more tests, and more tests simulating crime scene conditions, would be needed before such evidence could be

---

9. *Commonwealth of Mass. v. Dwayne McNair*, docket no. 1484CR10768 (2017).

10. The prosecutor's language in this case was incorrect. The likelihood ratio refers to the probability of the observed DNA results given the prosecutor's hypothesis (the evidence was contributed by the suspect) divided by the defense hypothesis (someone else was the source of the evidence). The DNA evidence does not address the issue of guilt per se.

considered admissible. Based on other evidence, Dwayne McNair was convicted of the rape.

Thus, this admissibility hearing, which was focused on a specific novel and unvalidated application of WGS by MPS, did not really address the admissibility of MPS technology per se, and the scientific issues raised by MPS analysis of evidence samples still remain to be considered in future hearings. It should be noted, though, that the scientific foundations and methodologies of DNA sequencing and PCR (described previously in this chapter and in chapter 1) have been understood and accepted for many decades (Shendure et al., 2017; Heather & Chain, 2016; Dove, 2018) and have undergone several rounds of scientific evaluation, testing, and validation on casework samples for the past 30 years (Cormier et al., 2005). Sequencing of mtDNA using manual methods was accepted in 1996, over 20 years ago (Davis, 1998). Although additional MPS testing, validation, and adoption in the forensic DNA community in the United States is indicated, the scientific foundations of this technology are well established, and the forensics community has been engaged in MPS testing for over a decade (Holland, 2009).

Some critics of MPS in forensics have focused not on reliability or general acceptance (the *Daubert* and *Frye* admissibility standards) but on privacy concerns, arguing that allowing the state to search criminal DNA evidence against public, genomic, nonforensic databases is highly intrusive, requires affirmative informed consent, and can disclose secondary private genetic information of individuals and their relatives (see chapter 15).

RAPID DNA

Unlike MPS, the commercial Rapid DNA technology does not represent a technical innovation allowing the analysis of novel genetic markers or the more effective genotyping of complex mixtures; instead, it automates an existing system, the characterization of STRs by electrophoresis. Rapid DNA instruments can obtain a CODIS core STR profile from a reference cheek swab in less than two hours (Grover et al., 2017; Shackleton, 2019; Buscaino et al., 2018). The practical consequences of this automation and speed are very significant for law enforcement practices. These compact instruments can be deployed in police stations, border patrol stations, and potentially in the field (Morrison et al., 2018).

The ability to fully automate and speed up the standard STR genotyping method has been a dream and goal of the forensic community for many years. In recent years this dream has been realized (Buscaino et al., 2018; Moreno et al., 2017), and this capacity is now transforming how, when, and where forensic genotyping can be performed. Starting in 2012, several different companies have developed integrated microfluidic devices to automate DNA extraction from a cheek swab, PCR amplification of a panel of STR loci, electrophoresis of the PCR products, and data analysis. The companies offering these fully integrated devices (swab in, STR profile out) include ParaDNA, which amplifies 5 STRs and amelogenin directly from human and casework samples; ANDE, which uses PowerPlex 16 chemistry or a new 27 locus

STR system; FlexPlex, an automated allele calling expert system and RFID sample tracking; and RapidHit200, with PowerPlex 16 HS chemistry. An updated version, RapidHIT200, with the GlobalFiler Express, runs eight swabs simultaneously.

All of these commercial compact instruments are capable of providing an STR profile in less than 90 minutes. This short turnaround time is achieved, in part, by rapid PCR thermal cycling. Standard PCR amplification involves a series of cycles in which the temperature of the reaction is varied. A PCR of thirty 1.5-minute cycles will take 45 minutes. If each cycle can be reduced to 20 seconds, this PCR would take only 10 minutes. Gibson-Daw and coworkers were able to reduce the total time to amplify a 7 loci multiplex consisting of 6 STRs and amelogenin to 6.5 minutes (Gibson-Daw, 2017). The speed of these instruments and newly developed chemistries means that, in principle, these instruments could be deployed in police stations so that STR profiles of arrestees could be obtained and screened against criminal and evidence databases while the suspect was still at the booking station. Rapid DNA results that matched the profile of an evidence sample or a database profile would presumably lead to the arrestee being detained. rapidly generated profiles could also potentially prevent the detention of the wrongfully accused. This type of testing has been done in Bensalem, Pennsylvania, since 2010 and is now implemented in several police departments in Florida, at Palm Bay and Miami Beach. These instruments could also be used at the border to investigate claimed familial relationships in immigration cases (see chapter 10), but that raises issues of both consent and privacy.

These two applications have in common the analysis of high-quality DNA from blood or a cheek swab, unlike the degraded DNA typically found in forensic crime scene samples or in the human remains used in missing persons cases. In validation studies (Moreno et al., 2017; Shackleton et al., 2019), the performance of most of these Rapid DNA instruments, which have been optimized for reference samples, is quite good for reference samples but is less reliable for forensic evidence samples (Mapes et al., 2018; Alshehhi & Roy, 2015). In some of these validation studies, some degree of manual interpretation of the profiles generated from the instrument was necessary to achieve the correct result (i.e., concordant with a known sample profile). A recent validation study (Salceda et al., 2017) evaluated the RapidHIT ID system using the Global Filer Express chemistry on reference samples and reported a 98% success rate for a concordance data set. Another evaluation of the same instrument system (Date-Chong et al., 2016) found that some variability in the STR results was observed and that some samples required manual editing of the profiles; the study reported that 50% of the cheek swab samples would have been successfully genotyped without any need for manual review or editing. The authors concluded, on the basis of this 2016 study, that the system was not yet suitable for full hands-off usage.

One group has developed a system that concentrates the DNA from low template samples (Turingan et al., 2016) and has reported improved rapid DNA results from blood, buccal cells, chewing gum, bone, cell phones, cigarette butts, and other low template samples. Additional research and testing to improve results from challenging samples are needed.

The FBI guidelines for application of STR genotyping by Rapid DNA specify that it has been validated for reference samples, not casework evidence; accordingly, reference sample profiles can be uploaded to CODIS but currently not casework profiles. However, in spite of the FBI guideline and the reported performance issues on evidence samples (Alshehhi & Roy, 2015; Turingan et al., 2016), the Rapid DNA technology has been applied to such samples in some situations where speed is of the essence. There is a growing body of research to improve results on crime scene samples and to provide decision support guidance for Rapid DNA end users (Mapes et al., 2018).

During the Camp Fire in Butte County, California, in November 2018, the worst fire in California history, hundreds of people went missing, and many of the victims could not be identified by traditional methods due to extensive burning of the remains. ANDE offered Rapid DNA testing of the remains for free. Jim Davis, the chief federal officer with ANDE, said, "We've been able to get usable DNA data of about 80% of the remains we've tested. . . . [O]ur instrument works very well in collecting data from bones and teeth." This search for the missing required that family members get their cheek swabbed at the Butte County Sheriff's Office to provide reference samples. "We believe that we can bring closure to family members, quicker than has been done previously," Davis said. The use of Rapid DNA technology in the identification of missing persons, as in the analysis of the Camp Fire remains, does not raise many of the legal concerns that will inevitably attend its use in crime scene evidence and its introduction in the courtroom.

In some jurisdictions, Rapid DNA technology has in fact been used on casework samples. In the Orange County, California, District Attorney's Office (Hynds & Conitini, 2017), the desire to aid investigations led to the application of this technology to crime scene samples. The Orange County DA's office found searchable profiles on 65% of the samples tested by the RapidHit instrument. Obtaining genetic profiles from evidence samples quickly could force criminals into plea bargains instead of costly trials and can reduce jail time for exonerated inmates. Given the value of speed in analyzing evidence samples as well as the potential of these compact instruments to be used in the field, it is highly likely that the profiles generated by Rapid DNA analysis of casework samples will be used to search databases and be introduced into the courtroom. If and when this happens, admissibility challenges based on reliability and general scientific acceptance can be expected to follow very closely. The legal and ethical issues associated with Rapid DNA, however, are much broader than those addressed by the *Frye* and *Daubert* admissibility standards (see chapter 13). For example, the issue of arrestee DNA databases, a contentious and controversial application of forensic DNA technology, becomes a more pressing concern with the implementation of Rapid DNA technology.

The future envisioned by the FBI and many in law enforcement of Rapid DNA analysis at the booking station has been made possible by development and launch of these commercial instruments and by the passage of the Rapid DNA Act of 2017, which provides guidelines for the implementation of this new technology. This bill amends the DNA Identification Act of 1994 to require the FBI to issue standards and procedures for using these instruments. This vision involves obtaining STR profiles from cheek swabs using fully automated instruments for purposes of individual identification of arrestees and for uploading these profiles to CODIS and searching national criminal databases (National DNA Index System, NDIS). (CODIS refers to the software for searching, and NDIS is the name of the database.) This legislation specifies that only reference samples and no forensic samples should be analyzed by Rapid DNA technology. In particular, samples that require interpretation and samples with low quantity or degraded DNA are considered to fall outside the scope of Rapid DNA. The strategy for implementation is to introduce the technology first in accredited NDIS labs and then, once standards and requirements for police booking stations have been established by the Scientific Working Group on DNA Analysis Methods (SWGDAM), into the booking stations. SWGDAM has generated position statements for Rapid DNA.

The concerns about booking-station genetic profiling are logistical, technical, and legal. Although the Rapid DNA instruments to be deployed at police stations are intended to be fully automated, including interpretation, several issues will need to be addressed. These include the proper collection of samples; potential for contamination; technical proficiency that may be needed to interpret the data, should any complications arise; and legal issues. When the standard genetic profiles are introduced into the courtroom, the analyst presenting the data is subject to cross-examination. How will the data generated by an automated system be treated? Also, as previously noted, the issue of an arrestee database remains controversial. Some critics have argued that since the criteria for arrest are subjective (e.g., probable cause), such databases are subject to bias and lack the more objective foundations of a convicted felon database (i.e., based on conviction); others have noted that many of those arrested are never charged, raising issues of privacy and inappropriate governmental intrusion. As of 2019, 31 states are taking DNA from felony arrestees; the others do not. Among those states that do, some remove the profiles and destroy the DNA of arrestees who were never charged or were found innocent, while in other states these profiles remain part of the database.

The question of whether collection of DNA samples upon arrest, taken without a warrant and prior to a conviction, is constitutional under the Fourth Amendment was addressed by the US Supreme Court in *Maryland v. King* (King *v.* State of Maryland, 42a.3d 549, Md, 2012) in 2017. In an unusual 5–4 split (with Justice Breyer joining the majority, conservative wing and Justice Scalia, the liberal one), the Court upheld the Maryland DNA Collection Act (MDCA) and found that the warrantless collection of a cheek swab from Alonzo King in 2009 during booking did not violate his constitutional rights. The MDCA allows state and local law enforcement officers to collect DNA samples from individuals who have been arrested for a violent crime or burglary or for attempted crimes of violence or burglary. While he was under arrest for first- and second-degree assault charges, King's DNA was collected and genotyped, and the STR profile was logged into Maryland's DNA database. King's STR profile turned out to match DNA evidence from an unsolved 2003 rape, leading to his conviction for first-degree rape and a life sentence. The STR profile was the only evidence linking King to the rape. The trial judge denied King's motion to suppress the STR profile evidence, but the Court of Appeals of Maryland reversed this decision, ruling that the DNA sample taken under the MDCA was an unconstitutional violation of King's privacy rights.

In the Supreme Court's decision to uphold the MDCA, the majority argued that the cheek swab was less intrusive than a blood draw and distinguished between the noncoding STR profiles and other potential DNA sequence information. In particular, they argued that obtaining an STR profile at booking, like the taking of fingerprints, was a form of identification of the arrested individual. Although this argument prevailed, many legal scholars were skeptical. Four of the justices were also not convinced by the identification argument. As Scalia noted in his dissent, "DNA testing does not even begin until after arraignment and bail decisions are already made. The samples sit in storage for months and take weeks to test. When they are tested, they are checked against the Unsolved Crimes Collection—rather than the Convict and Arrestee Collection, which could be used to identify them" (Scalia, J., dissenting, *slip op.* at 12). Rapid DNA technology, however, is sufficiently fast that it could reasonably be considered a form of identification.

EMERGING FORENSIC GENETIC MARKERS

Advances in the field of forensic DNA include not only technological breakthroughs but also the addition of new, powerful genetic marker arrays and panels.

SNP Arrays: Another Genotyping Technology

The method of genotyping SNPs using hybridization to an array of immobilized probes is not a recently developed DNA technology, but in the context of familial searching of genealogy databases (see chapters 4 and 15), it is one that has been recently deployed to solve some very high-profile cases, such as the Golden State Killer. The technology, currently commercialized by the companies Illumina and Affymetrix, is based on arrays of short synthetic DNA fragments, known as *oligonucleotide probes*. These probes are designed to hybridize (bind) to the SNP region in the amplified DNA from the sample; depending on which probes bind to the DNA, the base present at the SNP position can be inferred. For example, for an Adenine/Thymine (A/T) SNP, some probes will bind to the DNA if the A is present, while other probes will bind if a T is present. Current SNP arrays contain probes for well over half a million SNPs, distributed throughout the genome. (Unlike the markers in forensic assays, these SNPs were not chosen to be statistically independent.) These SNP arrays have been widely used in research studies of many different diseases and in studies of human populations. Generating a profile with these arrays is less expensive and faster than MPS analysis, and the profiles with these SNP panels are highly informative (discriminating, in terms of individual identification) and are well suited to kinship analysis and for tracing familial relationships.

These SNP array profiles are the basis of the analyses conducted by many of the current ancestry companies, such as 23andMe and Ancestry.com, as well as the public database GEDMatch. Until very recently, this technology had not been applied to forensic samples (see chapter 1). In several high-profile, suspectless crimes, such as the case of the Golden State Killer, the profile of an evidence sample has been compared to the profiles in the GEDMatch database. The partial match obtained in this familial search eventually led to the identification and conviction of a distant relative of the individual in the genealogy database. (This case is discussed in more detail in chapters 4 and 15.) The genotyping technology used in all these genealogical searches is the SNP array. Once the suspect has been identified, however, and additional samples have been collected, the standard STR profile can be generated and compared to STR profiles of the evidence samples, and this result is what is presented to the jury.

Rapidly Mutating Y-STRs

Analyzing Y chromosome markers, such as the STRs included in the various commercial STR panels, is an efficient and informative strategy for genotyping male DNA in the presence of female DNA. As with other haploid lineage markers, such as mtDNA, a match with Y STRs is less discriminating than a match with autosomal STR markers. Unlike the case with autosomal STRs, the STR markers on the Y chromosome are linked in what is known as a *haplotype*. The probabilities of a match at each Y STR locus are therefore not statistically independent, so the product rule cannot be used in estimating RMPs. For these markers, the significance of the match and the RMP is estimated by the frequency of the Y STR haplotype (the series of linked STR alleles) in a population database. An additional limitation is that male relatives will share the same Y chromosome STR haplotype. A recent development that addresses this limitation and offers the potential for distinguishing among different male relatives is identification of "rapidly mutating" (RM) Y STR loci (Adnan et al., 2016; Hadi, 2016).

STR loci have a higher mutation rate than sequence polymorphisms like SNPs; additions or deletions of the repeat copy sequence occur more frequently during DNA replication than do base substitutions changing a SNP locus sequence. A subset of Y-STRs has been identified with a mutation rate that is 10–100 times higher than the standard Y-STR loci. The possibility that such a mutation may have occurred at one (or more) of these RM Y-STR loci may allow discriminating among different members (relatives) of the same Y chromosome lineage. Multiplex assays for analyzing 13 different RM Y-STRs have been developed (Alghafri et al., 2015), and RM Y STRs are now included in several commercial Y STR panels.

APPLICATIONS OF GENETIC MARKERS OF POTENTIAL INVESTIGATIONAL VALUE

In recent years genetic polymorphisms that influence visible physical attributes such as eye color, skin color, hair color, height, weight, tendency to baldness, and other appearance traits have been investigated (reviewed in Vidaki & Kayser, 2018). Some of these genetic variants, which are typically not 100% predictive but are still highly associated with traits like eye and hair color, can be used to analyze evidence samples to provide investigational leads. Companies now offer DNA tests of appearance for law enforcement (Parabon Nanolabs) or panels of genetic markers for ancestry, hair, and eye color (Verogen, ThermoFisher, and Promega).

The genetic markers that contain information about ancestry, such as the ancestry informative markers (AIMS) previously described, can also provide a prediction from an evidence sample of the appearance of the individual who left the sample. However, the analysis of the statistical significance of a match and the calculation of the RMPs or LRs that are presented to the jury are carried out with the standard genetic markers of individual identification.

Additional DNA assays that may provide potentially useful information for investigations are those that indicate the anatomical source of the evidence sample. Specific messenger RNAs may point to the tissue from which the DNA evidence was obtained (van den Berge & Sijen, 2017). Messenger RNA is not nearly as stable as DNA, so this approach has some significant limitations. Finding DNA sequences specific to certain bacteria may also indicate the nature or source of the evidence sample. For example, lactic bacteria (*lactobacilli*) are associated with vaginal tissue (Fleming & Harbison, 2010). Also, some bacteria are associated with specific environmental and geographic areas, information that might prove critical in an investigation. (The various forensic applications of analyzing microbiome DNA sequences are discussed in more detail in chapter 6.)

CONCLUSION

Innovation in the forensic applications of DNA technology continues at a high rate, but its implementation is constrained by the need to be "backward compatible" with existing DNA databases and the need for rigorous validation studies before the novel technology may be ruled fully admissible in the courtroom. These needs include the MPS STR universal allele nomenclature, standardized controls, validation protocols, and interpretation guidelines.

The standard STR genotyping by capillary electrophoresis will likely continue to be used on many, if not most, evidence samples into the near future, but interpreting the more challenging specimens, such as highly degraded samples, complex mixtures, and trace DNA samples will certainly benefit from the developments outlined in this chapter. By far the most consequential recent technical developments in DNA technology are the implementation of NGS technologies and the deployment of Rapid DNA instruments. Both of these technologies raise technical issues of validation and legal issues of privacy that must be addressed before they will be widely accepted in the courtroom. Nevertheless, these new technologies have the potential to transform what DNA analysis can contribute to the criminal justice system as well as to immigration and missing persons cases.

Another important recent technical innovation is the development of probabilistic genotyping, commercial software that can analyze STR profiles from mixed samples much more effectively than previous methods. Since these are software programs rather than new DNA technologies, they are not addressed in this chapter but in chapter 3, on mixtures. The use of familial searching in genealogical databases in suspectless crimes (discussed in chapters 4 and 15), while not new DNA technology, is certainly a new application of DNA technology, with many dramatic successes, but it remains somewhat controversial.

Although limited to investigational use, the genetic tests for ancestry and appearance may prove increasingly useful in providing objective data not subject to the vagaries of eyewitness accounts. DNA analysis of these genetic variants can truly be considered the "silent witness" invoked in the title of this book.

REFERENCES

Adnan, A., et al. (2016). Improving empirical evidence on differentiating closely related men with RM Y-STRs: A comprehensive pedigree study from Pakistan. *Forensic Science International: Genetics*, 25, 45–51. doi:10.1016/j.fsigen.2016.07.005.

Alghafri, R., et al. (2015). A novel multiplex assay for simultaneously analysing 13 rapidly mutating Y-STRs. *Forensic Science International: Genetics*, 17, 91–98.

Alshehhi, A., & Roy, R. (2015). Generating Rapid DNA profiles from crime scene samples commonly encountered in the United Arab Emirates. *Journal of Forensic Research*, 6(4), 296. doi:10.4172/2157-7145.1000296.

Alonso, A., et al. (2018). Current state-of-art of STR sequencing in forensic genetics. *Electrophoresis*, 39(21), 2655–2668. doi:10.1002/elps.201800030.

Ball, G., et al. (2015). Concordance study between the ParaDNA ® Intelligence Test, a Rapid DNA profiling assay, and a conventional STR typing kit (AmpFlSTR ® SGM Plus ®). *Forensic Science International: Genetics*, 16, 48–51. doi:10.1016/j.fsigen.2014.12.006.

Blackman, S., et al. (2015). Developmental validation of the ParaDNA ® Intelligence System—A novel approach to DNA profiling. *Forensic Science International· Genetics*, 17, 137–148. doi:10.1016/j.fsigen.2015.04.009.

Børsting, C., & Morling, N. (2017). Use of next-generation sequencing in forensic genetics. *ELS*, 1–9. doi:10.1002/9780470015902.a0027106.

Børsting, C., & Morling, N. (2015). Next generation sequencing and its applications in forensic genetics. *Forensic Science International: Genetics*, 18, 73–89. doi:10.1016/j.fsigen.2015.02.002

Borter, G. (2018, November 15). California taps war-zone DNA specialists after wildfire. *US News & World Report*. Retrieved from https://www.usnews.com/news/us/articles/2018-11-15/dna-firm-schooled-in-war-zones-helps-id-california-fire-victims.

Bose, N., et al. (2018). Target capture enrichment of nuclear SNP markers for massively parallel sequencing of degraded and mixed samples. *Forensic Science International: Genetics*, 34, 186–196. doi:10.1016/j.fsigen.2018.01.010.

Bradbury, C., Köttgen, A., & Staubach, F. (2019). Off-target phenotypes in forensic DNA phenotyping and biogeographic ancestry inference: A resource. *Forensic Science International: Genetics*, 38, 93–104. doi:10.1016/j.fsigen.2018.10.010.

Bruijns, B., Tiggelaar, R., & Gardeniers, H. (2018). Massively parallel sequencing techniques for forensics: A review. *Electrophoresis*, 39(21), 2642–2654. doi:10.1002/elps.201800082.

Budowle, B., Schmedes, S. E., & Wendt, F. R. (2017). Increasing the reach of forensic genetics with massively parallel sequencing. *Forensic Science, Medicine and Pathology*, 13(3), 342–349. doi:10.1007/s12024-017-9882-5.

Buford, T. W., et al. (2018). Mitochondrial DNA sequence variants associated with blood pressure among 2 cohorts of older adults. *Journal of the American Heart Association*, 7(18), 1–11. doi:10.1161/jaha.118.010009.

Bulbul, O., & Filoglu, G. (2018). Development of a SNP panel for predicting biogeographical ancestry and phenotype using massively parallel sequencing. *Electrophoresis*, 39(21), 2743–2751. doi:10.1002/elps.201800243.

Buscaino, J., et al. (2018). Evaluation of a rapid DNA process with the RapidHIT ® ID system using a specialized cartridge for extracted and quantified human DNA. *Forensic Science International: Genetics*, 34, 116–127. doi:10.1016/j.fsigen.2018.02.010.

Butler, J. M. (2015). The future of forensic DNA analysis. *Philosophical Transactions of the Royal Society B: Biological Sciences*, 370(1674), 20140252. doi:10.1098/rstb.2014.0252,

Caliebe, A., & Krawczak, M. (2018). Match probabilities for Y-chromosomal profiles: A paradigm shift. *Forensic Science International: Genetics*, 37, 200–203. doi:10.1016/j.fsigen.2018.08.009.

Cormier, K., et al. (2005). Evolution of DNA evidence for crime solving—a judicial and legislative history. *Forensic Magazine*, 2(3), 13–15. Retrieved from https://www.ncjrs.gov/App/Publications/abstract.aspx?ID=210455

Cao, Y., et al. (2017). A review on the applications of next generation sequencing technologies as applied to food-related microbiome studies. *Frontiers in Microbiology*, 8.

Coble, M. D., & Bright, J. (2019). Probabilistic genotyping software: An overview. *Forensic Science International: Genetics*, 38, 219–224. doi:10.1016/j.fsigen.2018.11.009.

Cornelis, S., et al. (2018). Forensic STR profiling using Oxford Nanopore Technologies MinION sequencer. https://www.biorxiv.org/content/10.1101/433151v1.full.pdf

Date-Chong, M., Hudlow, W. R., & Buoncristiani, M. R. (2016). Evaluation of the RapidHIT™ 200 and RapidHIT GlobalFiler ® Express kit for fully automated STR genotyping. *Forensic Science International: Genetics*, 23, 1–8. doi:10.1016/j.fsigen.2016.03.001.

Davis, C. L. (1998). *Mitochondrial DNA: State of Tennessee v. Paul Ware*. Retrieved from https://www.promega.com/-/media/files/resources/profiles-in-dna/103/mitochondrial-dna-state-of-tennessee-v-paul-ware.pdf?la=en.

de Knijff, P. (2019). From next generation sequencing to now generation sequencing in forensics. *Forensic Science International: Genetics*, 38, 175–180. doi:10.1016/j.fsigen.2018.10.017.

DNAscan 6C Rapid DNA Analysis System, (GE Healthcare DNAscan™ 6C Rapid DNA Analysis™ System) https://cdn.gelifesciences.com/dmm3bwsv3/AssetStream.aspx?mediaformatid=10061&destinationid=10016&assetid=17800

Dove, A. (2018). Technology feature | PCR: Thirty-five years and counting. *Science*, 360(6389), 673. doi:10.1126/science.360.6389.673-c. https://science.sciencemag.org/content/360/6389/673.3

Fleming, R. I., & Harbison, S. (2010). The use of bacteria for the identification of vaginal secretions. *Forensic Science International: Genetics*, 4(5), 311–315. doi:10.1016/j.fsigen.2009.11.008.

French, K. (2016, September 12). The path to legal standing. Proto???

Gangano, S., et al. (2013). DNA investigative lead development from blood and saliva samples in less than two hours using the RapidHIT™ Human DNA Identification System. *Forensic Science International: Genetics Supplement Series*, 4(1). doi:10.1016/j.fsigss.2013.10.022.

Gardy, J. L., & Loman, N. J. (2018). Towards a genomics-informed, real-time, global pathogen surveillance system. *Nature Reviews Genetics*, 19(1), 9–20.

Garrido-Cardenas, J. A., & Manzano-Agugliaro, F. (2017). The metagenomics worldwide research. *Current Genetics*, 63(5), 819–829. doi:10.1007/s00294-017-0693-8.

Gerth, O. (2017). Identical twins as a misnomer: How advancing technology protects the interest of justice in the courtroom. *Georgetown Journal of Legal Ethics, 30*, 783

Gettings, K. B., et al. (2018). Sequence-based U.S. population data for 27 autosomal STR loci. *Forensic Science International: Genetics, 37*, 106–115. doi:10.1016/j.fsigen.2018.07.013.

Gibson-Daw, G., Crenshaw, K., & Mccord, B. (2017). Optimization of ultrahigh-speed multiplex PCR for forensic analysis. *Analytical and Bioanalytical Chemistry, 410*(1), 235–245. doi:10.1007/s00216-017-0715-x.

Gonzalez-Garay, M. L. (2015). Introduction to isoform sequencing using Pacific Biosciences technology (iso-seq). In Wu J., ed., *Translational Bioinformatics Transcriptomics and Gene Regulation*, vol. 9, 141–160. Springer, Dord.

Goray, M., Pirie, E., & Oorschot, R. A. (2019). DNA transfer: DNA acquired by gloves during casework examinations. *Forensic Science International: Genetics, 38*, 167–174. doi:10.1016/j.fsigen.2018.10.018.

Grover, R., et al. (2017). FlexPlex27—highly multiplexed rapid DNA identification for law enforcement, kinship, and military applications. *International Journal of Legal Medicine, 131*(6), 1489–1501. doi:10.1007/s00414-017-1567-9.

Hadi, S. (2016). Analysis of rapidly mutating Y chromosome short tandem repeats (RM Y-STRs). *Methods in Molecular Biology, 1420*, 201–211. doi: 10.1007/978-1-4939-3597-0_16.

Heather, J. M., & Chain, B. (2016). The sequence of sequencers: The history of sequencing DNA. *Genomics, 107*(1), 1–8. doi:10.1016/j.ygeno.2015.11.003.

Hennessy, L. K., et al. (2013). Developmental validation studies on the RapidHIT™ Human DNA Identification System. *Forensic Science International: Genetics Supplement Series, 4*(1). doi:10.1016/j.fsigss.2013.10.003.

Holland, M., Makova, K., & Mcelhoe, J. (2018). Deep-coverage MPS analysis of heteroplasmic variants within the mtGenome allows for frequent differentiation of maternal relatives. *Genes, 9*(3), 124. doi:10.3390/genes9030124.

Holland, M., McQuillan, M., & Boese, B. (2009). *Next generation sequencing of forensic DNA loci using 454 Life Sciences technology*. Retrieved from https://www.promega.com/~/media/files/resources/conference%20proceedings/ishi%2020/oral%20presentations/holland.pdf?la=en.

Huang, Y., Liu, H., & Xing, Y. (2018). Next-generation sequencing promoted the release of reference genomes and discovered genome evolution in cereal crops. *Current Issues in Molecular Biology*, 37–50.

Huber, N., Parson, W., & Dür, A. (2018). Next generation database search algorithm for forensic mitogenome analyses. *Forensic Science International: Genetics, 37*, 204–214. doi:10.1016/j.fsigen.2018.09.001.

Hynds, & Contini. (2017). *The need for speed: swiftly solving orange county crime utilizing rapid technology and the OCDA DNA database* oral presentation at the 2017 international symposium of human identification. Retrieved from https://promega.media/-/media/files/products-and-services/genetic-identity/ishi-28-oral-abstracts/hynds-contini.pdf?la=en

Jäger, A. C., et al. (2017). Developmental validation of the MiSeq FGx forensic genomics system for targeted next generation sequencing in forensic DNA casework and database laboratories. *Forensic Science International: Genetics, 28*, 52–70. doi:10.1016/j.fsigen.2017.01.011.

Jobling, M., & Gill, P. (2004). Encoded evidence: DNA in forensic analysis. *Nature Reviews Genetics, 5*, 739–751. doi:10.1038/nrg1455.

Just, R. S., Irwin, J. A., & Parson, W. (2015). Mitochondrial DNA heteroplasmy in the emerging field of massively parallel sequencing. *Forensic Science International: Genetics, 18*, 131–139.

Kayser, M. (2017). Forensic use of Y-chromosome DNA: A general overview. *Human Genetics, 136*, 621–635. doi:10.1007/s00439-017-1776-9.

Kayser, M. (2019). Introduction to Trends and perspectives in forensic genetics 2018 [Special issue]. *Forensic Science International: Genetics, 38,* 254–255. doi:10.1016/j.fsigen.2018.11.007.

Kim, H., Erlich, H. A., & Calloway, C. D. (2015). Analysis of mixtures using next generation sequencing of mitochondrial DNA hypervariable regions. *Croatian Medical Journal, 56*(3), 208–217. doi:10.3325/cmj.2015.56.208.

Larue, B. L., et al. (2014). An evaluation of the RapidHIT® system for reliably genotyping reference samples. *Forensic Science International: Genetics, 13,* 104–111. doi:10.1016/j.fsigen.2014.06.012.

Lee, S. B., Varlaro, J., & Holt, C. (2016, June). Stepping into the future of forensic genomics: Developmental validation of a next-generation sequencing forensic DNA sample-to-answer system. *Forensic Magazine.* Retrieved from https://www.forensicmag.com/article/2016/07/future-forensic-genomics-developmental-validation-ngs.

Lea, A. J., Vilgalys, T. P., Durst, P. A. P., & Tung, J. (2017). Maximizing ecological and evolutionary insight in bisulfite sequencing data sets. *Nature Ecology & Evolution, 1*(8), 1074–1083.

Le Gallo, M., Lozy, F., & Bell, D. W. (2017). Next-generation sequencing. In L. Hedrick Ellenson, eds., *Molecular Genetics of Endometrial Carcinoma. Advances in Experimental Medicine and Biology,* vol. 943, (pp. 119–148). Springer, Cham. doi:10.1007/978-3-319-43139-0_5 https://link.springer.com/chapter/10.1007/978-3-319-43139-0_5

Liu, Y.-Y., & Harbison, S. (2017). A review of bioinformatic methods for forensic DNA analyses. *Forensic Science International: Genetics, 33,* 117–128. doi:10.1016/j.fsigen.2017.12.005.

Manna, A. D., et al. (2016). Developmental validation of the DNAscan™ Rapid DNA Analysis™ instrument and expert system for reference sample processing. *Forensic Science International: Genetics, 25,* 145–156. doi:10.1016/j.fsigen.2016.08.008.

Mapes, A. A., et al. (2018). Decision support for using mobile Rapid DNA analysis at the crime scene. *Science & Justice, 59*(1), 29–45. doi: 10.1016/j.scijus.2018.05.003.

Mardis, E. R. (2013). Next-generation sequencing platforms. *Annual Review of Analytical Chemistry* (Palo Alto Calif), 6, 287–303. doi:10.1146/annurev-anchem-062012-092628.

Mardis, E. R. (2017). DNA sequencing technologies: 2006–2016. *Nature Protocols 12*(2), 213–218. doi:10.1038/nprot.2016.182.

McCombie, W. R., McPherson, J. D., & Mardis E. R. (2019). Next-generation sequencing technologies. *Cold Spring Harbor Perspectives in Medicine, 9*(11), a036798. doi:10.1101/cshperspect.a036798.

McCord, B. R., & Lee, S. B., eds. (2018). Novel applications of massively parallel sequencing (MPS) in forensic analysis [Special issue]. *Electrophoresis, 39*(21), 2639–2641. doi:10.1002/elps.201870175.

Meakin, G., & Jamieson, A. (2013). DNA transfer: Review and implications for casework. *Forensic Science International: Genetics, 7*(4), 434–443. doi:10.1016/j.fsigen.2013.03.013.

Metcalf, J. L. (2019). Estimating the postmortem interval using microbes: Knowledge gaps and a path to technology adoption. *Forensic Science International: Genetics, 38,* 211–218. doi:10.1016/j.fsigen.2018.11.004.

Moreno, L. I., Brown, A. L., & Callaghan, T. F. (2017). Internal validation of the DNAscan/ANDE™ Rapid DNA Analysis™ platform and its associated PowerPlex® 16 high content DNA biochip cassette for use as an expert system with reference buccal swabs. *Forensic Science International: Genetics, 29,* 100–108. doi:10.1016/j.fsigen.2017.03.022.

Moreno, L. I., Galusha, M.B., & Just, R. (2018). A closer look at Verogen's Forenseq™ DNA Signature Prep kit autosomal and Y-STR data for streamlined analysis of routine reference samples. *Electrophoresis, 39*(21), 2685–2693. doi:10.1002/elps.201800087

Morrison, J., Watts, G., Hobbs, G., & Dawnay, N. (2018). Field-based detection of biological samples for forensic analysis: Established techniques, novel tools, and future innovations. *Forensic Science International, 285*, 147–160. doi:10.1016/j.forsciint.2018.02.002.

Murphy, E. (2015). *Inside the cell: The dark side of forensic DNA*. New York, NY: Nation Books.

Oldoni, F., Kidd, K. K., & Podini, D. (2019). Microhaplotypes in forensic genetics. *Forensic Science International: Genetics, 38*, 54–69. doi:10.1016/j.fsigen.2018.09.009.

Oorschot, R. A., et al. (2019). DNA transfer in forensic science: A review. *Forensic Science International: Genetics, 38*, 140–166. doi:10.1016/j.fsigen.2018.10.014.

Parsons, T. J., Huel, R. M., Bajunović, Z., & Rizvić, A. (2019). Large scale DNA identification: The ICMP experience. *Forensic Science International: Genetics, 38*, 236–244. doi:10.1016/j.fsigen.2018.11.008.

Petersdorf, E. W., & Ohuigin, C. (2018). The MHC in the era of next-generation sequencing: Implications for bridging structure with function. *Human Immunology, 80*(1), 67–78. doi:10.1016/j.humimm.2018.10.002.

Phillips, C., et al. (2018). Global patterns of STR sequence variation: Sequencing the CEPH human genome diversity panel for 58 forensic STRs using the Illumina ForenSeq DNA Signature Prep Kit. *Electrophoresis, 39*(21), 2708–2724. doi:10.1002/elps.201800117.

Phillips, C., García-Magariños, M., Salas, A., Carracedo, A., & Lareu, M. V. (2012). SNPs as supplements in simple kinship analysis or as core markers in distant pairwise relationship tests: when do SNPs add value or replace well-established and powerful STR tests? *Transfusion Medicine Hemotherapy, 39*(3), 202–210. doi:10.1159/000338857

Promega. (n.d.). *Target amplification and library prep*. Retrieved from https://www.promega.com/products/genetic-identity/mps-workflow/target-amplification-and-library-prep/.

Qiagen. (n.d.). *GeneReader platform*. Retrieved from https://www.qiagen.com/us/products/ngs/mdx-ngs-genereader/library-preparation/qiagen-genereader-platform/.

Qiagen. (2017, February 6) *Looking to the future of missing persons analysis with NGS*. Retrieved from https://corporate.qiagen.com/about-us/insights-magazine/2017/2017-06-icmp-ngs-development.

Rang, F. J., Kloosterman, W. P., & Ridder, J. D. (2018). From squiggle to basepair: Computational approaches for improving nanopore sequencing read accuracy. *Genome Biology, 19*(1), 90. doi:10.1186/s13059-018-1462-9.

Raza, S., & Ameen, A. (2017). Nano pore sequencing technology: A review. *International Journal of Advances in Scientific Research, 3*(8), 90. doi:10.7439/ijasr.v3i8.4333.

Romsos, E. L., & Vallone, P. M. (2015). Rapid PCR of STR markers: Applications to human identification. *Forensic Science International: Genetics, 18*, 90–99. doi:10.1016/j.fsigen.2015.04.008.

Ross, A. A., Muller, K., Weese, J. S., & Neufeld, J. (2018). Comprehensive skin microbiome analysis reveals the uniqueness of human skin and evidence for phylosymbiosis within the class Mammalia. *Proceedings of the National Academy of Science USA, 115*(25), e5786–5795. doi:10.1073/pnas.1801302115.

Salceda, S., et al. (2017). Validation of a rapid DNA process with the RapidHIT ® ID system using GlobalFiler ® Express chemistry, a platform optimized for decentralized testing environments. *Forensic Science International: Genetics, 28*, 21–34. doi:10.1016/j.fsigen 2017.01.005

Shackleton, D., et al. (2019). Development of RapidHIT® ID using NGMSElect™ Express chemistry for the processing of reference samples within the UK criminal justice system. *Forensic Science International, 295*, 179–188. doi:10.1016/j.forsciint.2018.12.015.

Shendure, J., et al. (2017). DNA sequencing at 40: Past, present and future. *Nature*, *550*(7676), 345–353. doi:10.1038/nature24286.

Shih, S. (2017). Characterization of germline heteroplasmy in mother-offspring pairs using next generation sequencing. Master's Thesis, Forensic Science. University of California, CA.

Shih, S., et al. (2018). Applications of probe capture enrichment next generation sequencing for whole mitochondrial genome and 426 nuclear SNPs for forensically challenging samples. *Genes*, *9*(1), 49. doi:10.3390/genes9010049.

Silva, D. S., et al. (2018). Genetic analysis of Southern Brazil subjects using the PowerSeq™ AUTO/Y system for short tandem repeat sequencing. *Forensic Science International: Genetics*, *33*, 129–135. doi:10.1016/j.fsigen.2017.12.008.

Spiewak, J. (2018). KUTV Utah's first guilty plea with rapid DNA technology. Retrieved from https://kutv.com/news/local/utahs-first-guilty-plea-with-rapid-dna-technology-took-just-weeks-to-close-case

Tan, O., Shrestha, R., Cunich, M., & Schofield, D. (2018). Application of next-generation sequencing to improve cancer management: A review of the clinical effectiveness and cost-effectiveness. *Clinical Genetics*, *93*(3), 533–544.

Taylor, D., Kokshoorn, B., & Biedermann, A. (2018). Evaluation of forensic genetics findings given activity level propositions: a review. *Forensic Science International: Genetics*, *36*, 34–49. doi:10.1016/j.fsigen.2018.06.001

Torkamaneh, D., Boyle, B., & Belzile, F. (2018). Efficient genome-wide genotyping strategies and data integration in crop plants. *Theoretical and Applied Genetics*, *131*(3), 499–511.

Tribble, N. D., Miller, J. A., Dawnay, N., & Duxbury, N. J. (2015). Applicability of the ParaDNA®Screening system to seminal samples. *Journal of Forensic Sciences*, *60*(3), 690–692. doi:10.1111/1556-4029.12742.

Turingan, R. S., et al. (2016). Rapid DNA analysis for automated processing and interpretation of low DNA content samples. *Investigative Genetics*, *7*(1). doi:10.1186/s13323-016-0033-7.

US Federal Bureau of Investigation (FBI). (2016). *Rapid DNA or Rapid DNA analysis.* Retrieved from http://www.fbi.gov/about-us/lab/biometric-analysis/codis/rapid-dnaanalysis.

van den Berge, M., & Sijen, T. (2017). Extended specificity studies of mRNA assays used to infer human organ tissues and body fluids. *Electrophoresis*, *38*(24), 3155–3160. doi:10.1002/elps.201700241

Vidaki, A., & Kayser, M. (2018). Recent progress, methods and perspectives in forensic epigenetics. *Forensic Science International: Genetics*, *37*, 180–195. doi:10.1016/j.fsigen.2018.08.008.

van Oorschot, R. A. H., Szkuta, B., Meakin, G. E., Kokshoorn, B., & Goray, M. (2019). DNA transfer in forensic science: A review. *Forensic Science International: Genetics*, *38*, 140–166. doi:10.1016/j.fsigen.2018.10.014

Weber-Lehmann, J., et al. (2014). Finding the needle in the haystack: Differentiating "identical" twins in paternity testing and forensics by ultra-deep next generation sequencing. *Forensic Science International: Genetics*, *9*, 42–46. doi:10.1016/j.fsigen.2013.10.015.

Weirather, J. L., et al. (2017). Comprehensive comparison of Pacific Biosciences and Oxford Nanopore Technologies and their applications to transcriptome analysis. *F1000Research*, *6*, 100. doi:10.12688/f1000research.10571.1.

Woerner, A. E., Novroski, N. M., Wendt, F. R., Ambers, A., Wiley, R., Schmedes, S. E., & Budowle, B. (2019). Forensic human identification with targeted microbiome markers using nearest neighbor classification. *Forensic Science International: Genetics*, *38*, 130–139. doi:10.1016/j.fsigen.2018.10.003.

Wu, I.-C., Liu, W.-C., & Chang, T.-T. (2018). Applications of next-generation sequencing analysis for the detection of hepatocellular carcinoma-associated hepatitis B virus mutations. *Journal of Biomedical Science*, *25*(1), 51–63.

Zascavage, R. R., Thorson, K., & Planz, J. V. (2018). Nanopore sequencing: An enrichment-free alternative to mitochondrial DNA sequencing. *Electrophoresis*, *40*(2), 272–280. doi:10.1002/elps.201800083.

Zhang, Q., et al. (2018). Evaluation of the performance of Illumina's ForenSeq™ system on serially degraded samples. *Electrophoresis*, *39*(21), 2674–2684. doi:10.1002/elps.201800101.

NEWS ARTICLES

https://scipol.duke.edu/content/bensalem-police-first-local-department-us-use-rapid-dna-testing

http://www.evidencemagazine.com/index.php?option=com_content&task=view&id=1618

https://www.nbcmiami.com/news/local/Miami-Beach-Police-Using-Rapid-DNA-Technology-to-Help-Solve-Crimes-Faster-440812163.html

https://www.nbcnews.com/news/us-news/dna-tests-separated-families-slammed-immigration-advocates-n889161

Asplen, C. (2014). Rapid advances in rapid DNA. *Evidence Technology Magazine*. May–June 2014. http://www.evidencemagazine.com/index.php?option=com_content&task=view&id=1618&Itemid=49

http://www.evidencemagazine.com/index.php?option=com_content&task=view&id=1618

Plasencia, Amanda. (2017). Miami beach police using rapid DNA technology to help solve crimes faster. South Florida 6 NBC News. Posted August 16, 2017. https://www.nbcmiami.com/news/local/Miami-Beach-Police-Using-Rapid-DNA-Technology-to-Help-Solve-Crimes-Faster-440812163.html

Quann, P. (2017). Bucks county courier times bensalem police first local department in U.S. to use rapid DNA testing. Posted March 16, 2017. https://www.buckscountycouriertimes.com/edb1505e-09a1-11e7-9b87-1fb8018118bb.html

https://scipol.duke.edu/content/bensalem-police-first-local-department-us-use-rapid-dna-testing

Silva, Daniella. (2018). DNA tests for separated families slammed by immigration advocates.

The government is reviewing the cases of some 3,000 children in its care who were separated from their parents, officials said. NBC News. Posted July 5, 2018 https://www.nbcnews.com/news/us-news/dna-tests-separated-families-slammed-immigration-advocates-n889161

https://www.nbcnews.com/news/us-news/dna-tests-separated-families-slammed-immigration-advocates-n889161

CHAPTER 6

Microbial Forensics

From Epidemiology to Crime Investigations

ANTTI SAJANTILA AND BRUCE BUDOWLE

The realization that humans are born with approximately 20,000 protein-encoding genes (Clamp et al., 2007; Ezkurdia et al., 2014; International Human Genome Sequencing, Consortium, 2004) but quickly gain over a million additional genes via their microbiome (Gensollen et al., 2016; Peterson et al., 2009) has expanded substantially the potential genetic markers associated with human beings and hence the range of applications in the field of microbial forensics (Schmedes et al., 2016). The foundation for these new possibilities for the forensic community lies in two extraordinary scientific advances. First is the sequencing of the entire human genome, which was simultaneously achieved by the publicly funded Human Genome Project (HGP; https://www.genome.gov/) (Lander et al., 2001) and by a privately funded group (Venter et al., 2001). Second, the ongoing Human Microbiome Project (https://www.hmpdacc.org/) (Peterson et al., 2009), which is based on the medical community's interest in the microbes associated with the human body and the important roles they play in health and disease, is collecting and characterizing the diversity and function of the human microbiome (Gevers et al., 2012; Human Microbiome Project, Consortium, 2012).

Microbial forensics was initially defined as the analysis of microbiological-based evidence, usually focusing on bioterrorism and biocrime, for attribution purposes— that is, determining either the source of the evidence or who committed the offense (Budowle et al., 2003). With advances in DNA technology over the last decade, particularly the advent of high throughput sequencing methods, the field has broadened from addressing the concerns of bioterrorism to other forensic and medico-legal issues. Microbial forensics can now be applied to human identification by identifying hosts—and in this context suspects or victims—by their "unique" microbiomes. The many strains/species of commensal microbial organisms that inhabit human bodies, not harming the host and possibly aiding health (e.g., *E.coli* and *Bifidobacterium*), constitute the human microbiome. In addition to human DNA markers, the human microbiome can be exploited for individual human identification. As well as the commensal microbes, analysis of those microbes transmitted among individuals causing a temporary or persistent infection can be utilized in a forensic context.

The use of microbes in forensic archaeology and mass-grave investigations was introduced in the 1990s (Hopkins et al., 2000; Janaway, 1996) but had no clear effect on the routine work. The use of microbes as weapons or for the purpose of terror is far older, dating back centuries (Hawley & Eitzen, 2001). However, it was the anthrax letter attacks of 2001 in the United States that generated considerable public alarm and prompted governments to develop better capabilities to respond to such threats (Budowle et al., 2003).

Today, almost 20 years later, microbial forensics has progressed significantly due to developments in a range of genetic technologies that now enable findings of significance to be made about the distribution, variation, and activity of microbes in the human body. These technologies, coupled with bioinformatic tools, are designed to analyze and classify human microbes as well as provide detailed phylogenetic analyses (i.e., evolutionary history and relationships among the different strains of microbes) of the molecular data used to identify the source of a particular microbe or community of microbes (Woerner et al., 2019)—all of which has made microbial forensics an attractive investigative tool for law enforcement. Microbiome analysis is also being applied in the field of public health to identify the etiologic agents and sources of disease outbreaks or bioterrorist attacks. This chapter describes the development of microbial forensics, its recent applications to human identification, and its potential in the field of forensic sciences and forensic medicine. It also illustrates the power of the new technologies in solving crimes by describing some notable cases that were investigated using microbial forensics.

DEFINITIONS AND QUESTIONS

The microbiota is an "ecological community of commensal, symbiotic and pathogenic microorganisms" found in and on all multicellular organisms from plants to animals. Single-cell microorganisms may harbor other microbes, such as bacteria and prophages, but the focus in this chapter is on multicellular organisms and their microbiomes. The microbiota includes the collection of bacteria, archaea, protists, fungi, and/or viruses. According to the Human Microbiome Project, *human microbiome* refers to the community of microorganisms that live in or on the human body. The *microbiota* and the *microbiome* can be considered the microbial community or assemblage and its complete set of genes (Spor et al., 2011)

Microbial forensics, in broad terms, is the genetic analysis of microbiological-based evidence for attribution purposes in criminal and civil cases. Attribution in this context is determining the source of a sample to the degree possible. Initially, microbial forensics was defined narrowly as "a scientific discipline dedicated to analyzing evidence from a bioterrorism act, biocrime, or inadvertent microorganism/toxin release for attribution purposes" (Budowle et al., 2003). However, the potential and scope of microbial forensics has recently been expanded to encompass a wide range of applications, including human identification, determining postmortem interval, human geolocation, and body fluid identification (Schmedes et al., 2016).

In microbial forensics, scientists typically seek to answer questions such as the following. (1) What is the agent (microbe) in question? In many cases, the identification of the agent is not challenging, but the time frame of exposure to the microbe is a key critical question. (2) Is it probative or relevant to the investigation in question? While certainty about the finding is important for law enforcement and public health consequences, a holistic understanding of the sample background may be needed to determine the source and whether its presence is an indication of an intentional crime. (3) Can the agent (microbe) be linked to a source? The power of methods used to discriminate the agent may be limited, and characterization should be described with a degree of confidence (or uncertainty). (4) What are the meaning and weight of the conclusion(s)? The degree of attribution (which could range from a generic conclusion such as species designation to individualization such as originating from a specific tube in a specific laboratory) may affect the course of action of law enforcement and/or public healthcare workers.

SPECIFIC INTEREST AREAS IN MICROBIOME RESEARCH AND MICROBIAL FORENSICS

The Human Microbiome Project (Peterson et al., 2009) laid an important foundation for microbial forensics by characterizing the microbiomes of 300 selected healthy human subjects at five major body sites: nasal passages, oral cavity, skin, gastrointestinal tract, and urogenital tract (Human Microbiome Jumpstart Reference Strains, Consortium, et al., 2010). Of particular interest has been the bacterial community in the human gut (Spor et al., 2011) and skin (Oh et al., 2016), specifically its association with human conditions such as obesity, ulcerative colitis, and diabetes (Blum, 2017; Paramsothy et al., 2017; Pedersen et al., 2016). The other body locations of course have also been of considerable interest (see figure 6.1). To some extent, human viruses have been studied in the same body regions (Reyes et al., 2010; Hannigan et al., 2015; Aggarwala et al., 2017; Oh et al., 2016).

While these initial targets are of interest to the academic and medical communities, the forensic community has had its own overlapping interests since even before the inception of the Human Microbiome Project. A relatively recent case that engaged the forensic and epidemiological communities—though well after the event—was the 1993 Kameido anthrax incident in a building owned by the Aum Shinrikyo cult in Japan (Keim et al., 2001). In this case, the group cultured and then sprayed *Bacillus anthracis* on the Tokyo resident population. However, no illnesses occurred, as the cult employed the Sterne strain, which is a vaccine strain. Members of this Japanese cult espoused tenets from Buddhism and Hinduism and were obsessed with the apocalypse, which they predicted would occur between 1999 and 2003, with the sole survivors being followers of the Aum. They also became infamous for dispersing the

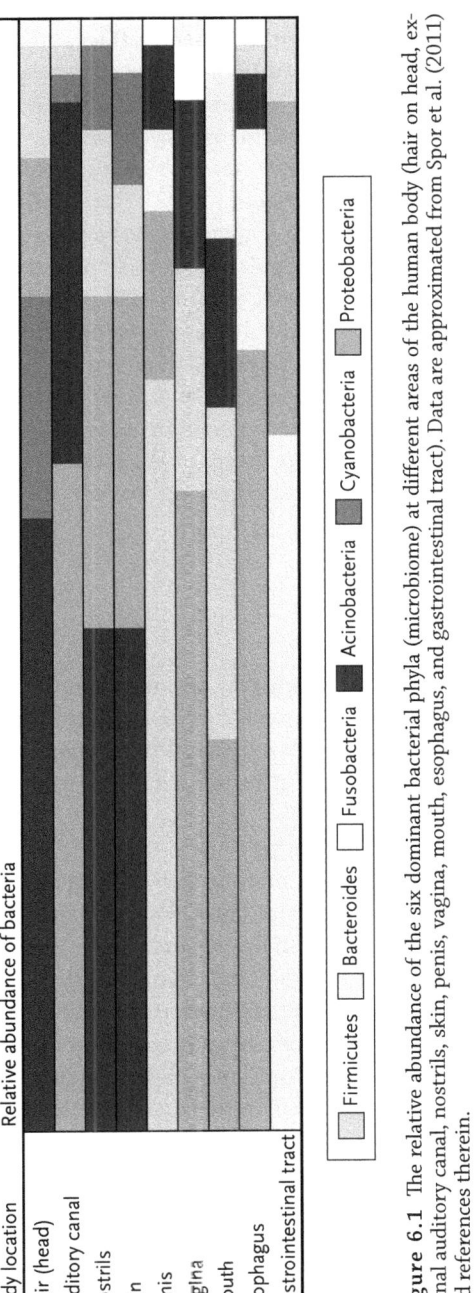

Figure 6.1 The relative abundance of the six dominant bacterial phyla (microbiome) at different areas of the human body (hair on head, external auditory canal, nostrils, skin, penis, vagina, mouth, esophagus, and gastrointestinal tract). Data are approximated from Spor et al. (2011) and references therein.

nerve agent sarin in the Tokyo subway system, killing 12 individuals and causing some 6,000 people to seek medical attention.

Eight years later, in 2001, the anthrax letter attacks in the United States demonstrated that bioterrorism could be perpetrated relatively easily and revealed our vulnerabilities and lack of preparedness to respond to such an attack. Since 2001 there have been many developments in strategies and technologies for detection of biological agents as a rapid response to terroristic mass-destruction attempts (Budowle et al., 2003). Dissemination of infectious agents, either deliberately or through negligence, has prompted the need for forensic investigations combined with epidemiology and microbiological analyses. Some cases have included patients infected with HIV (Ou et al., 1992) or HCV (Gonzalez-Candelas et al., 2013) by healthcare personnel and transmission of HIV via sexual assault (Albert et al., 1994), all of which are considered criminal acts in many countries. Beyond bioterrorism and other criminal investigations, the forensic community also is involved in civilian cases, which can vary from an outbreak of foot-and-mouth disease (Cottam et al., 2006) to tracing food-borne epidemics (Kotewicz et al., 2008) and emerging human infections (Howard & Fletcher, 2012).

METHODOLOGICAL EVOLUTION

While microbes are ubiquitous, and some may be quite abundant in a sample, many species and strains in a microbiome sample may be at relatively low abundance. As with human DNA, the target microbe's nucleic acids may need to be enriched to a level at which the analysis of the type, quantity, and other characteristics of the microbe is possible.

Microbial Cultures. A traditional method for amplifying a microorganism is culturing. In a culture, the microbial organisms reproduce in a specific medium under controlled (laboratory) conditions and can be isolated and identified morphologically or by a series of serological, immunological, or chemical-based techniques. The microorganisms can be cultured on a liquid medium contained in flasks (e.g., Lysogeny broth, LB), or, for example, agar in petri dishes or solid stab cultures in test tubes. These cultures often have different additives, which enable selective growth of bacterial species. Virus cultures, on the other hand, need specific host cells in which the virus can multiply. Several international microbial culture collections have been created to maintain and distribute strains among scientists (Boundy-Mills et al., 2015; Casaregola et al., 2016) for research and development endeavors. Culturing was, and still remains, a fundamental diagnostic tool for medically relevant microbial infections. However, culturing requires that the microbe be viable, and unfortunately, most microbial species cannot be cultured. For microbial forensics investigations, these limitations (the time required for cloning; the ability only to detect a priori known, culture-based microbes; only being able to culture select microbes) do not allow culturing to be a routinely practiced approach for microbial attribution.

Nucleic-Acid-Level Detection

For a more rapid diagnostic tool for microbe detection to meet the needs of epidemiology and/or forensic investigations, it is vital to have DNA (or RNA) level diagnostic capabilities. The value of analyzing nucleic acids is multifold. First, they can be enriched by direct amplification (by polymerase chain reaction, PCR) of specific target regions of the microbial genes or anonymous sequences, allowing presence or nonpresence type of diagnostics. Second, nucleic acid sequencing of target regions (barcoding) facilitates more detailed and hierarchical analysis of genomes. Third, advanced RNA/DNA sequencing technologies permit analysis of large portions or the whole genome, further detailing the relationships of the species find in a sample. Importantly, the enrichment strategy affords exquisite sensitivity of detection, allowing analysis of trace quantities of a sample.

Polymerase chain reaction (PCR). PCR is an in vitro enzymatic assay that generates millions to billions of copies of targeted DNA sequences. These abundant quantities of target copies afford great sensitivity, thus allowing minute quantities of an analyte (i.e., the target) to be detected. The specific amplified products of microbial DNA can be analyzed using various downstream technologies, such as gel electrophoresis (Mullis & Faloona, 1987), restriction-fragment length polymorphism analysis (Saiki et al., 1985), capillary electrophoresis (Schwartz et al., 1991), high-performance liquid chromatography (Oefner et al., 1994), dot blot/reverse dot blot arrays (Saiki et al., 1989), DNA chip array systems (Schena et al., 1995), oligonucleotide ligation assay (Landegren et al., 1988), primer extension/minisequencing assay (Syvanen et al., 1990), and mass spectrometry (Hillenkamp et al., 1991). Further, technologies that allow detection of microbial DNA directly in a closed PCR reduce sample manipulations as well as risk of contamination. These methods include PCR with intercalating dyes (Higuchi et al., 1992), 5′ exonuclease assay (TaqMan) (Holland et al., 1991), molecular beacons (Tyagi & Kramer, 1996), scorpion primers (Whitcombe et al., 1999), and dye-labeled oligonucleotide ligation (Chen et al., 1998).

Although PCR-based methods offer several advantages over traditional culturing, they also have some limitations. First, sample-related inhibition of PCR may occur, resulting in a need for extensive sample cleanup, high-volume samples, or enrichment of microbes prior to PCR (Vutukuru et al., 2016). Second, a microbial sample may include several species or heterogeneous strains that may not all be amplified by PCR relative to their abundance in the sample, making some metagenomic evaluations problematic. In addition, first-generation DNA sequencing technologies (Maxam & Gilbert, 1977; Sanger et al., 1977) do not have the throughput or resolution to characterize these complex samples. Third, and most important, all PCR-based methods have limited capabilities to analyze something that was not a priori characterized, and thus they, similarly to culturing, have limited capabilities for routine large-scale screening in cases of epidemiologic or forensic investigations.

Barcoding. One type of targeted PCR amplification is the DNA barcoding strategy (Hebert et al., 2003). DNA barcoding is a method that uses a designated portion of a specific gene or genes common to all members of a kingdom to identify an organism to the species level (Hebert et al., 2003; Savolainen et al., 2005). The barcoding idea has been both welcomed (Schindel & Miller, 2005) and criticized (Ebach & Holdrege, 2005). Nevertheless, it has grown into an international, internet-based informatics workbench system for taxonomists: the Barcode of Life Data System (BOLD; www.barcodinglife.org) (Ratnasingham & Hebert, 2007).

While the barcoding method was first developed to classify species in the animal kingdom, microbial barcoding developed in parallel based on microbial phylogeny and taxonomy using the 16S rRNA gene (Woese et al., 1975; Woese & Fox, 1977). Based on this early work, the Ribosomal Database Project (RDP; (http://rdp.cme.msu.edu) was created in the 1990s (Olsen et al., 1992) to provide ribosome-related data services to the scientific community. RDP offers online data analysis, 16S rRNA derived phylogenetic trees, and aligned and annotated 16S rRNA sequences. Today, this database has more than 3.5 million 16S rRNA sequences (Release 11, Update 5, September 30, 2016).

The 16S rRNA gene is highly conserved and serves as a DNA barcode for diverse bacterial species. Therefore, it has been the most widely studied gene for reconstructing bacterial phylogeny (Schloss & Handelsman, 2004). The 16S rRNA gene has also been described as a useful marker in medicine for analysis of clinical samples for infectious diseases (Clarridge, 2004), as well as in studies of somatic conditions, such as obesity (Ley et al., 2006). This gene also is a marker of interest for identifying human hosts based on their microbiomes for forensic applications (Fierer et al., 2010; Goga, 2012).

A similarly useful gene for barcoding of microbes, particularly bacteria, is a segment (658 base pair portion) of the cytochrome oxidase I (COI, COX, or COX1) gene within the mitochondrial genome. Using this COI marker, a DNA barcode has been developed for 22 pathogenic species that are food borne (Jones et al., 2013).

New DNA sequencing technologies. The most reliable way to determine microbial species in any sample is to analyze its nucleic acid sequence in full or the parts of its genome that separate it from other close species. The first successful DNA sequencing experiments were reported in 1968 (Wu & Kaiser, 1968), but it was not until 1977 that two different methods made it possible to realistically sequence DNA. Up until the late 1970s, DNA was the most difficult biopolymer to analyze. Since then, and with the advances over the last decade, DNA has become the easiest biopolymer to sequence. These methods transformed the field of molecular genetics: Sanger and his co-workers (1977) used a chain terminator procedure (1977), and Maxam and Gilbert (1977) used chemical cleavage, to determine the sequences of the bases along the DNA strand. These first-generation methods both relied on electrophoretic separation of different length DNA strands produced in the assay and radioactive labeling for detection of the fragments. Although a great boon to molecular genetics, these procedures were slow, tedious, and hazardous. The Sanger method soon became the method of choice, primarily due to the advent of fluorescently labeled sequencing primers (Ansorge et al., 1986; Smith et al., 1985) or terminators (Prober et al., 1987) and capillary electrophoresis (Swerdlow et al., 1990; Swerdlow & Gesteland, 1990). However, by today's standards, Sanger sequencing is still a slow and low throughput process that makes it difficult to sequence large genomes such as those from microbes and eukaryotes. To put this in perspective, at the beginning of the twenty-first century, it took more than a decade at a cost of over $3 billion and 40 institutions to sequence one human genome, approximately 3 billion nucleotides (Lander et al., 2001; Venter et al., 2001). It is no wonder whole genome analyses were not practical just a few years ago.

A little over a decade ago, the landscape of molecular genomics changed dramatically. Several novel high-throughput DNA sequencing technologies were developed. These technologies were commonly called the "next-generation" or "second-generation" sequencing methods. This generation of sequencing technologies is also referred to as massively parallel sequencing (or MPS). In contrast to the Sanger sequencing method, the next generation sequencing (NGS) technology is highly scalable, allowing the entire genome of a variety of species to be sequenced in a single analysis. Generally, the genome was fragmented into small pieces, and each fragment was cloned individually (but simultaneously) in a massively parallel fashion in vitro. Subsequently, the clones, again in a massively parallel fashion, were sequenced using one of a number of sophisticated downstream chemistries (Shendure et al., 2017). In contrast, Sanger sequencing could only sequence one fragment at a time. If there were multiple fragments in a sample, all were sequenced together in a single reaction, resulting in a garbled sequence. In contrast, with NGS, a multitude of fragments are individually sequenced but simultaneously, then the sequences of the randomized fragments are "patched" together through the use of bioinformatics.

It should be noted that a third trend of DNA sequencing technologies has been developed or is underway that differs from the second-generation NGS or MPS approaches. These third-generation chemistries and platforms, such as PacBio (Eid et al., 2009; Levene et al., 2003) and Nanopore sequencing (Branton et al., 2008; Deamer et al., 2016), offer real-time detection of single DNA molecules. These latest technologies can eliminate the need for DNA template amplification and thus, in theory, reduce copy-errors, sequence-dependent biases, and information loss related to the MPS platforms (cf. Heather & Chain, 2016; Mardis, 2017; Shendure et al., 2017). However, single-molecule chemistries, such as those used by Nanopore, are nascent and still have relatively high error rates. It is expected that with continued development the error rate is likely to drop toward a tolerable rate (Heather & Chain, 2016; Shendure et al., 2017).

RNA molecules also can be sequenced using NGS modified chemistries (Marioni et al., 2008), which lately have been applied to allow single-cell RNA sequencing (Tang et al., 2009). RNA sequencing has not yet made it to the mainstream of epidemiology and forensic microbiology, but its usefulness can be anticipated. There are a number of devastating human infectious agents for which RNA sequencing may prove important in major epidemiological investigations, including Ebola hemorrhagic fever, SARS, rabies, the common cold, influenza, hepatitis C, West Nile fever, polio, and measles, which are all RNA viruses.

The DNA/RNA MPS technologies are coupled with phylogenetic and machine-learning analyses to determine the origin of the donor of the samples in a forensic setting and also the screening of "the unknown" using metagenomic approaches (Eisen, 2007; Gill et al., 2006). For example, Schmedes and colleagues (2017) developed a human identification microbial to characterize skin microbiomes and used supervised learning to attribute skin microbial signatures to their respective individual hosts. The initial data for their panel were obtained from metagenomic data sets generated from skin microbiome samples collected from 14 body sites on 12 individuals, sampled over three time points over the course of ~three-year period (Oh et al., 2016). They supervised identified features from the abundant skin bacterium *Propionibacterium acnes*, enabling selection of a subset of markers for human attribution purposes. Indeed, skin microbiomes from the foot, manubrium, and hand and their respective human hosts were associated with up to 92%, 96%, and 100% accuracy, respectively (Schmedes et al., 2018). The largest microbial genome data set was reported in 2019; it contained more than 9,000 metagenomes from various human populations, geographies, lifestyles, body sites, and ages (Pasolli et al., 2019). From these samples, over 150,000 genomes were reconstructed with close to 5,000 species-level genome bins of bacteria and archaea, 77% of which were not present in public databases.

APPLICATIONS AND CASES

Microbial investigation traditionally has been used in investigations of bioterrorism or biocrime, as previously noted, but microbial analysis in conjunction with epidemiology has also been commonly utilized in tracing transmission of infectious agents in suspected sexual assaults or medical malpractice cases. Microbial forensics also has applications in postmortem settings. The technology can be used to estimate postmortem interval (the time since death) in investigation of the sequence of postmortem events in cadavers found in particular environments (e.g., graves or water), for molecular diagnostics in cause-of-death investigations, and to estimate the geographical origin of unknown cadavers.

Tracing Infectious Agents in Living People

Bioterrorism . According to the Centers for Disease Control and Prevention (CDC), bioterrorism is a biological attack with the intentional release of viruses, bacteria, or other germs that can sicken or kill people, livestock, or crops (www.cdc.gov/anthrax/bioterrorism/). The need for microbial forensics became evident in 2001 as a result of the anthrax letter attacks in the United States. Soon after the terrorist attacks on the World Trade Center and Pentagon on September 11, 2001, a previously healthy 63-year-old man died and was found to be infected with Gram-positive bacilli, *B. Anthracis*, in his blood and cerebral spinal fluid (Bush et al., 2001). This man was the first person diagnosed related to the anthrax attack; altogether there were 22 infections and 5 deaths from an outbreak that was definitively attributed to bioterrorism. The ease of disseminating a bioweapon by readily accessible means—via the postal service—exposed the vulnerability and demonstrated the unpreparedness of various governmental agencies in the United States for bioweapon attacks.

Rape and medical malpractice. The basis of molecular epidemiological investigations of transmission events of human infectious agents is the use of molecular phylogenies. The sequence data from a source are compared to that of the target(s) (suspect(s) and/or victim(s) and other reference samples). The questions that can be answered in such cases depend naturally on the infectious agent (DNA/RNA sequences), the number of suspects, and the number of victims. Typical questions asked in these investigations are presented in table 6.1.

The first cases reported using molecular analysis were on transmission of human immunodeficiency virus (HIV) in suspected malpractice (Metzker et al., 2002; Ou et al., 1992), rape or sexual assault cases (Albert et al., 1994; Lemey et al., 2005; Machuca et al., 2001), and other criminal activities (Birch et al., 2000). These early investigations involved a relatively few transmission events in a short period of time. Although strong support for relatedness between HIV samples obtained from the suspect and victims was attained, the direction of transmission between the evidence sample pairs often was unclear. Later, the use of phylogenetic methods to establish the direction of transmission was better established (Scaduto et al., 2010).

An epidemiological investigation of a suspected infection of 270 patients with hepatitis C virus (HCV) by an anesthetist in Spain was reported in 2013 (Gonzalez-Candelas et al., 2013). The origin of transmissions was relatively old, 25 years at the time of the case; more than 300 candidate patients potentially associated with the original source of infection had undergone a minor surgery in one of two hospitals and subsequently were infected with HCV. For the first time a molecular clock, a

Table 6.1 EXAMPLES OF QUESTIONS POSED TO THE MOLECULAR EPIDEMIOLOGIST USING MOLECULAR PHYLOGENETIC METHODS

1. Is the suspect the source responsible for the infection or outbreak?
2. Which of the victim(s) has been infected from a common source (i.e., suspect's strain)?
3. Alternatively, which of the victims could have been infected from other sources?
4. Can alternative sources be excluded (existence of different but simultaneous outbreak)?
5. Can the duration of the outbreak be determined?
6. Can infection for each patient in the outbreak be dated?
7. Can the date of infection of the source (i.e., suspect's strain) be determined?

technique using the mutation rate of a biomolecule (DNA, RNA, protein) to deduce the time when two or more forms of a microbe diverged—in other words, the initial infection time frame—was used to estimate the date of various transmission events. HCV, like HIV, is a fast-evolving virus, making its genetic analysis challenging. Thus, even individuals harboring the virus with the same origin will have differences in their quasi-species population of viral RNA sequences, a population of related viruses that derive from a parental virus(es). However, those sequences from the same source are likely to be more similar than those from epidemiologically unassociated patients. Therefore, phylogenetic approaches could prove informative. In court, these data were compared with those obtained from the nonmolecular investigation (documents and testimonies), and in two out of three cases, the molecular data estimates were congruent with other epidemiologic (and forensic) data.

It is important to note that often the molecular epidemiological data, as in many other DNA investigations in a legal setting, are used to associate evidence with an alleged source, not to establish guilt or innocence per se. Rather, microbial evidence in concert with other evidence must be weighed to determine guilt or innocence.

Postmortem Forensic Microbiology

In addition to the cases involving living people, microbiological investigations have been useful in postmortem settings. These applications include estimating the postmortem interval, cause-of-death investigations, and assessment of ancestry of cadavers.

Estimation of Postmortem Interval (PMI)

Estimation of the PMI is at the core of cause-of-death investigations in cases of unwitnessed deaths. In forensic medicine, PMI is emphasized especially in homicide investigations, where tracking down when the death has occurred may help direct the investigation along fruitful paths and include or exclude potential perpetrators. Some early textbook chapters suggested the use of soil microbes in the analysis of PMI in the forensic archaeology context (Evans, 1963; Janaway, 1996). Relatively recent studies have established that postmortem microbial and biochemical activity results in a series of decomposition stages, with a reproducible microbial succession across human corpses (Hyde et al., 2013; Pechal et al., 2014). When viewed as an ecosystem, the expired human body is a suite for a range of bacteria, insects, and fungi. Some of these microbes are obligate: they require host cells to reproduce. As such, decomposition is a mosaic process with an intimate association between biotic and abiotic factors, including the cadaver itself, intrinsic and extrinsic microbes, and insects, along with weather, climate, and soil. In order to be useful for forensic purposes, this decomposition process should be analyzable and measurable as a function of a specific biological succession with an ecological impact. In this process, bacteria have been credited as a driving force (Meyers & Foran, 2008). Metcalf and colleagues (2016) have shown that microbial diversity in soil is positively associated with the decomposition of corpses, and this phenomenon is repeatable and generalizable across selective ecosystems, seasons, and host taxa. However, even with the advances of molecular methods for understanding the mosaic ecosystem of decomposition, a holistic approach utilizing microbiology, entomology, and chemistry is still needed to estimate the PMI.

Cause-of-Death Investigation

Use of microbiology in the investigation of the cause of death is not new. Microbiological samples have been taken from the deceased since simple methods for detection of infective agents were available for detection of *Mycobacterium tuberculosis*, influenza measles, polio, and so forth. However, since postmortem microbial succession affects the evaluation of the diversity and causality to death of true antemortem microbes, it is of utmost importance to characterize this succession immediately after death and before the autopsy in cause-of-death investigations. Hurtado and colleagues (2018) showed that diagnostic use of microbial methods in autopsy material is reliable at least up to 24 hours, and even beyond, keeping in mind that the proportion of some bacteria, such as *Enterobacteriaceae* and *Pseudomonas* spp. as etiological agents of infections leading to death, may be overestimated. In their study, they observed that increasing postmortem intervals were associated with an increase in the number of different bacteria identified.

Aside from infection as an underlying cause of death, investigation of other types of death of medico-legal interest may also benefit from microbial forensics. One example is bodies found in aqueous media, which represent a diagnostically difficult task even for an experienced medico-legal specialist (Piette & De Letter, 2006). Similar to the findings in cadavers from graves, where the soil microbiome affects the postmortem appearance and abundance of microbes, cadavers found in various aqueous environments, typically in fresh water versus saltwater, in lakes or seas, respectively (Lunetta et al., 2002; Lunetta et al., 2004), present possibilities for microbial analysis. The earliest microbiological testing for putative drowning was the "diatom test" (Lunetta et al., 1998; Lunetta et al., 2013), which originally was based on the chemical or enzymatic disruption of the cellular material of the diatoms, a major group of algae found in oceans, waterways, and soil, and the identification of diatom species based on their characteristic shells. It is likely that enzymatic or chemical analysis of diatoms, combined with barcoding and NGS-based sequencing, will become the standard for cause-of-death diagnostics in drowning, allowing investigation of the potential conditions and geographical place of drowning.

Ancestry Estimation

Persistent human DNA viruses have been proposed for use in assessing ancestry, alone or as an adjunct to human DNA genetic markers. In addition to our (human) genes, humans carry biological signatures acquired from birth through childhood, and these signatures—as geographical hallmarks—can be informative about the geography where people grew up. The microbial signatures provide a grossly untapped resource of data for identification and for the assessment of the geographical origins of human remains. Systematic studies have yet to show their utility in tracing the origins of human remains at large, though promising examples already exist (Kolb et al., 2013; Toppinen et al., 2015; Zheng et al., 2007; Zhong et al., 2009).

CONCLUSION

As understanding of the human and environmental microbiomes as a whole progresses with the advances of molecular genetic technology, microbial forensics is likely to become part of the standard suite (Schmedes et al., 2016) of specialized forensic laboratories. For these laboratories to act in a concerted way to meet the demands of law enforcement and public health, several issues related to quality and standards of technology will need to be addressed (Budowle et al., 2008; Budowle et al., 2014; Morse & Budowle, 2006). Readiness to analyze specimens from potential biocrime or bioterrorism acts is one of the main tasks for such laboratories, as will be investigation of transmission of infectious agents in sexual assault and medical malpractice. Medico-legal institutes may have an interest in developing such techniques in their laboratories for delineation of various phases and scenarios in cause-of-death investigations. In all these areas, collaboration between practitioners and academia will be critical in implementing these new strategies and making their contribution to law enforcement.

NOTE TO THE READER

In addition to the specialized articles in scientific journals listed in the References, several publications may be helpful in understanding the importance of microbial forensics in a larger context; see Budowle et al. (2003); Budowle & Harmon (2005); Budowle et al. (2005); Morse & Budowle (2006); and National Research Council (2014).

REFERENCES

Albert, J., et al. (1994). Analysis of a rape case by direct sequencing of the human immunodeficiency virus type 1 pol and gag genes. *Journal of Virology, 68*(9), 5918–5924.
Wu, R., & Kaiser, A. D. (1968). Structure and base sequence in the cohesive ends of bacteriophage Lambda DNA. *Journal of Molecular Biology, 35*, 523–537.
Ansorge, W., et al. (1986). A non-radioactive automated method for DNA sequence determination. *Journal of Biochemical and Biophysical Methods, 13*(6), 315–323.

Birch, C. J., et al. (2000). Molecular analysis of human immunodeficiency virus strains associated with a case of criminal transmission of the virus. *Journal of Infectious Disease, 182*(3), 941–944.

Blum, H. E. (2017). The human microbiome. *Advances in Medical Science, 62*(2), 414–420.

Boundy-Mills, K., et al. (2015). The United States Culture Collection Network (USCCN): Enhancing microbial genomics research through living microbe culture collections. *Applied and Environmental Microbiology, 81*(17), 5671–5674.

Branton, D., et al. (2008). The potential and challenges of nanopore sequencing. *Natures Biotechnology, 26*(10), 1146–1153.

Budowle, B., et al. (2003). Public health: Building microbial forensics as a response to bioterrorism. *Science, 301*(5641), 1852–1853.

Budowle B., & Harmon, R. (2005). HIV legal precedent useful for microbial forensics. *Croatian Medical Journal, 46*, 514–521.

Budowle, B., et al. (2005). Towards a system of microbial forensics: from sample collection to interpretation of evidence. *Applied and Environmental Microbiology, 71*, 2209–2213.

Budowle, B., et al. (2008). Criteria for validation of methods in microbial forensics. *Applied and Environmental Microbiology, 74*(18), 5599–5607.

Budowle, B., et al. (2014). Validation of high throughput sequencing and microbial forensics applications. *Investigative Genetics, 5*, 9.

Bush, L. M., et al. (2001). Index case of fatal inhalational anthrax due to bioterrorism in the United States. *The New England Journal of Medicine, 345*, 1607–1610.

Casaregola, S., et al. (2016). An information system for European culture collections: The way forward. *Springerplus, 5*(1), 772.

Chen, X., Livak, K. J., & Kwok, P. Y. (1998). A homogeneous, ligase-mediated DNA diagnostic test. *Genome Research, 8*(5), 549–556.

Clamp, M., et al. (2007). Distinguishing protein-coding and noncoding genes in the human genome. *Proceedings of the National Academy of Sciences of the USA, 104*(49), 19428–19433.

Clarridge, J. E. 3rd. (2004). Impact of 16S rRNA gene sequence analysis for identification of bacteria on clinical microbiology and infectious diseases. *Clinical Microbiology Reviews, 17*, 840–862.

Cottam, E. M., et al. (2006). 'Molecular epidemiology of the foot-and-mouth disease virus outbreak in the United Kingdom in 2001. *Journal of Virology, 80*(22), 11274–11282.

Deamer, D., Akeson, M., & Branton, D. (2016). Three decades of nanopore sequencing. *Nature Biotechnology, 34*(5), 518–524.

Ebach, M. C., & Holdrege, C. (2005). DNA barcoding is no substitute for taxonomy. *Nature, 434*(7034), 697.

Eid, J., et al. (2009). Real-time DNA sequencing from single polymerase molecules. *Science, 323*(5910), 133–138.

Eisen, J. A. (2007). Environmental shotgun sequencing: its potential and challenges for studying the hidden world of microbes. *PLoS Biology, 5*(3), e82.

Evans, W. E. D. (1963). *The chemistry of death*. Springfield, IL: Charles C. Thomas

Ezkurdia, I., et al. (2014). Multiple evidence strands suggest that there may be as few as 19,000 human protein-coding genes. *Human Molecular Genetics, 23*(22), 5866–5878.

Fierer, N., et al. (2010). Forensic identification using skin bacterial communities. *Proceedings of the National Academy of Sciences of the USA, 107*(14), 6477–6481.

Gensollen, T., et al. (2016). How colonization by microbiota in early life shapes the immune system. *Science, 352*(6285), 539–544.

Gevers, D., et al. (2012). The Human Microbiome Project: A community resource for the healthy human microbiome. *PLoS Biology, 10*(8), e1001377.

Gill, S. R., et al. (2006). Metagenomic analysis of the human distal gut microbiome. *Science, 312*(5778), 1355–1359.

Goga, H. (2012). Comparison of bacterial DNA profiles of footwear insoles and soles of feet for the forensic discrimination of footwear owners. *International Journal of Legal Medicine, 126*(5), 815–823.

Gonzalez-Candelas, F., et al. (2013). Molecular evolution in court: analysis of a large hepatitis C virus outbreak from an evolving source. *BMC Biology, 11*, 76.

Hawley, R. J., & Eitzen, E. M., Jr. (2001). Biological weapons—a primer for microbiologists. *Annual Review of Microbiology, 55*, 235–253.

Heather, J. M., & Chain, B. (2016). The sequence of sequencers: The history of sequencing DNA. *Genomics, 107*(1), 1–8.

Hebert, P. D., et al. (2003). Biological identifications through DNA barcodes. *Proceedings: Biological Sciences, 270*(1512), 313–321.

Higuchi, R., et al. (1992). Simultaneous amplification and detection of specific DNA sequences. *Biotechnology* (New York), *10*(4), 413–417.

Hillenkamp, F., et al. (1991). Matrix-assisted laser desorption/ionization mass spectrometry of biopolymers. *Analytical Chemistry, 63*(24), 1193A–1203A.

Holland, P. M., et al. (1991). Detection of specific polymerase chain reaction product by utilizing the 5'—3' exonuclease activity of Thermus aquaticus DNA polymerase. *Proceedings of the National Academy of Sciences of the USA, 88*(16), 7276–7280.

Hopkins, D. W., Wiltshire, P. E. J., & Turner, B. D. (2000). Microbial characteristics of soils from graves: An investigation at the interface of soil microbiology and forensic science. *Applied Soil Ecology, 14*, 283–288.

Howard, C. R., & Fletcher, N. F. (2012). Emerging virus diseases: Can we ever expect the unexpected? *Emerging Microbes & Infections, 1*(12), e46.

Human Microbiome Jumpstart Reference Strains, Consortium, et al. (2010). A catalog of reference genomes from the human microbiome. *Science, 328*(5981), 994–999.

Human Microbiome Project, Consortium. (2012). Structure, function and diversity of the healthy human microbiome. *Nature, 486*(7402), 207–214.

Hurtado J. C., et al. (2018). Postmortem interval and diagnostic performance of the autopsy methods. *Scientific Reports, 8*, 16112.

Hyde, E. R., et al. (2013). The living dead: Bacterial community structure of a cadaver at the onset and end of the bloat stage of decomposition. *PLoS One, 8*(10), e77733.

International Human Genome Sequencing, Consortium. (2004). Finishing the euchromatic sequence of the human genome. *Nature, 431*, 931.

Janaway, R. C. (1996). The decay of buried human remains and their associated materials. In J. Hunter, C. Roberts, & A. Martin, eds., *Studies in crime: An introduction to forensic archaeology* (pp. 58–85). London, UK: B. T. Batsford Ltd.

Jones, Y. L., et al. (2013). Potential use of DNA barcodes in regulatory science: Identification of the U.S. Food and Drug Administration's "Dirty 22," contributors to the spread of foodborne pathogens. *Journal of Food Protection, 76*(1), 144–149.

Keim, P., et al. (2001). Molecular investigation of the Aum Shinrikyo anthrax release in Kameido, Japan. *Journal of Clinical Microbiology, 39*(12), 4566–4567.

Kolb, A. W., Ane, C., & Brandt, C. R. (2013). Using HSV-1 genome phylogenetics to track past human migrations. *PLoS One, 8*(10), e76267.

Kotewicz, M. L., et al. (2008). Optical mapping and 454 sequencing of Escherichia coli O157: H7 isolates linked to the US 2006 spinach-associated outbreak. *Microbiology, 154*(Pt. 11), 3518–3528.

Landegren, U., et al. (1988). A ligase-mediated gene detection technique. *Science, 241*(4869), 1077–1080.

Lander, E. S., et al. (2001). Initial sequencing and analysis of the human genome. *Nature, 409*(6822), 860–921.

Lemey, P., et al. (2005). Molecular testing of multiple HIV-1 transmissions in a criminal case. *AIDS, 19*(15), 1649–1658.

Levene, M. J., et al. (2003). Zero-mode waveguides for single-molecule analysis at high concentrations. *Science, 299*(5607), 682–686.

Ley, R. E., et al. (2006). Microbial ecology: human gut microbes associated with obesity. *Nature, 444*(7122), 1022–1023.

Lunetta, P., Penttila, A., & Hallfors, G. (1998). Scanning and transmission electron microscopical evidence of the capacity of diatoms to penetrate the alveolo-capillary barrier in drowning. *International Journal of Legal Medicine, 111*(5), 229–237.

Lunetta, P., Penttila, A., & Sajantila, A. (2002). Circumstances and macropathologic findings in 1590 consecutive cases of bodies found in water. *American Journal of Forensic Medicine and Pathology, 23*(4), 371–376.

Lunetta, P., et al. (2004). Unintentional drowning in Finland 1970–2000: A population-based study. *International Journal of Epidemiology, 33*(5), 1053–1063.

Lunetta, P., et al. (2013). False-positive diatom test: a real challenge? A post-mortem study using standardized protocols. *Legal Medicine* (Tokyo), *15*(5), 229–234.

Machuca, R., et al. (2001). Molecular investigation of transmission of human immunodeficiency virus type 1 in a criminal case. *Clinical and Diagnostic Laboratory Immunology, 8*(5), 884–890.

Mardis, E. R. (2017). DNA sequencing technologies: 2006–2016. *Nature Protocols, 12*(2), 213–218.

Marioni, J. C., et al. (2008). RNA-seq: An assessment of technical reproducibility and comparison with gene expression arrays. *Genome Research, 18*(9), 1509–1517.

Maxam, A. M., & Gilbert, W. (1977). A new method for sequencing DNA. *Proceedings of the National Academy of Sciences of the USA, 74*(2), 560–564.

Metcalf J. L., et al. (2016). Microbial community assembly and metabolic function during mammalian corpse decomposition. *Science, 35,* 158–162.

Metzker, M. L., et al. (2002). Molecular evidence of HIV-1 transmission in a criminal case. *Proceedings of the National Academy of Sciences of the USA, 99*(22), 14292–14297.

Meyers, M. S., & Foran, D. R. (2008). Spatial and temporal influences on bacterial profiling of forensic soil samples. *Journal of Forensic Sciences, 53*(3), 652–660.

Morse, S. A., & Budowle, B. (2006). Microbial forensics: Application to bioterrorism preparedness and response. *Infectious Disease Clinics of North America, 20*(2), 455–473, xi.

Mullis, K. B., & Faloona, F. A. (1987). Specific synthesis of DNA in vitro via a polymerase-catalyzed chain reaction. *Methods in Enzymology, 155,* 335–350.

National Research Council. (2014). *Science needs for microbial forensics: Developing initial international research priorities.* Washington, DC: The National Academies Press. Retrieved from https://doi.org/10.17226/18737.

Oefner, P. J., et al. (1994). High-performance liquid chromatography for routine analysis of hepatitis C virus cDNA/PCR products. *Biotechniques, 16*(5), 898–899, 902–908.

Oh, J., et al. (2016). Temporal stability of the human skin microbiome. *Cell, 165*(4), 854–866.

Ou, C. Y., et al. (1992). Molecular epidemiology of HIV transmission in a dental practice. *Science, 256*(5060), 1165–1171.

Paramsothy, S., et al. (2017). Multidonor intensive faecal microbiota transplantation for active ulcerative colitis: A randomised placebo-controlled trial. *Lancet, 389*(10075), 1218–1228.

Pasolli, E., et al. (2019). Extensive unexplored human microbiome diversity revealed by over 150,000 genomes from metagenomes spanning age, geography, and lifestyle. *Cell, 176*(3), 649–662.

Pechal, J. L., et al. (2014). The potential use of bacterial community succession in forensics as described by high throughput metagenomic sequencing. *International Journal of Legal Medicine, 128*(1), 193–205.

Pedersen, H. K., et al. (2016). Human gut microbes impact host serum metabolome and insulin sensitivity. *Nature, 535*(7612), 376–381.

Peterson, J., et al. (2009). The NIH Human Microbiome Project. *Genome Research, 19*(12), 2317–2323.

Piette, M. H., & De Letter, E. A. (2006). Drowning: still a difficult autopsy diagnosis. *Forensic Science International, 163*(1–2), 1–9.

Prober, J. M., et al. (1987). A system for rapid DNA sequencing with fluorescent chain-terminating dideoxynucleotides. *Science, 238*(4825), 336–341.

Ratnasingham, S., & Hebert, P. D. (2007). Bold: The Barcode of Life Data System (http://www.barcodinglife.org). *Molecular Ecology Notes, 7*(3), 355–364.

Saiki, R. K., et al. (1985). Enzymatic amplification of beta-globin genomic sequences and restriction site analysis for diagnosis of sickle cell anemia. *Science, 230*(4732), 1350–1354.

Saiki, R. K., et al. (1989). Genetic analysis of amplified DNA with immobilized sequence-specific oligonucleotide probes. *Proceedings of the National Academy of Sciences of the USA, 86*(16), 6230–6234.

Sanger, F., Nicklen, S., & Coulson, A. R. (1977). DNA sequencing with chain-terminating inhibitors. *Proceedings of the National Academy of Sciences of the USA, 74*(12), 5463–5467.

Savolainen, V., et al. (2005). Towards writing the encyclopedia of life: An introduction to DNA barcoding. *Philosophical Transactions of the Royal Society B: Biological Sciences, 360*(1462), 1805–1811.

Scaduto, D. I., et al. (2010). Source identification in two criminal cases using phylogenetic analysis of HIV-1 DNA sequences. *Proceedings of the National Academy of Sciences of the USA, 107*(50), 21242–21247.

Schena, M., et al. (1995). Quantitative monitoring of gene expression patterns with a complementary DNA microarray. *Science, 270*(5235), 467–470.

Schindel, D. E., & Miller, S. E. (2005). DNA barcoding a useful tool for taxonomists. *Nature, 435*(7038), 17.

Schloss, P. D., & Handelsman, J. (2004). Status of the microbial census. *Microbiology and Molecular Biology Reviews, 68*(4), 686–691.

Schmedes, S. E., Sajantila, A., & Budowle, B. (2016). Expansion of microbial forensics. *Journal of Clinical Microbiology, 54*(8), 1964–1974.

Schmedes, S. E., Woerner, A. E., & Budowle, B. (2017). Forensic human identification using skin microbiomes. *Applied and Environmental Microbiology, 83*(22). e01672–17.

Schmedes, S. E., et al. (2018). Targeted sequencing of clade-specific markers from skin microbiomes for forensic human identification. *Forensic Science International: Genetics, 32*, 50–61.

Schwartz, H. E., et al. (1991). Analysis of DNA restriction fragments and polymerase chain reaction products towards detection of the AIDS (HIV-1) virus in blood. *Journal of Chromatography, 559*(1–2), 267–283.

Shendure, J., et al. (2017). DNA sequencing at 40: Past, present and future. *Nature, 550*(7676), 345–353.

Smith, L. M., et al. (1985). The synthesis of oligonucleotides containing an aliphatic amino group at the 5' terminus: synthesis of fluorescent DNA primers for use in DNA sequence analysis. *Nucleic Acids Research, 13*(7), 2399–2412.

Spor, A., Koren, O., & Ley, R. (2011). Unravelling the effects of the environment and host genotype on the gut microbiome. *Nature Reviews Microbiology, 9*(4), 279–290.

Swerdlow, H., & Gesteland, R. (1990). Capillary gel electrophoresis for rapid, high resolution DNA sequencing. *Nucleic Acids Research, 18*(6), 1415–1419.

Swerdlow, H., et al. (1990). Capillary gel electrophoresis for DNA sequencing. Laser-induced fluorescence detection with the sheath flow cuvette. *Journal of Chromatography, 516*(1), 61–67.

Syvanen, A. C., et al. (1990). A primer-guided nucleotide incorporation assay in the genotyping of apolipoprotein E. *Genomics, 8*(4), 684–692.

Tang, F., et al. (2009). mRNA-Seq whole-transcriptome analysis of a single cell. *Nature Methods, 6*(5), 377–382.

Toppinen, M., et al. (2015). Bones hold the key to DNA virus history and epidemiology. *Scientific Reports*, *5*, 17226.

Tyagi, S., & Kramer, F. R. (1996). Molecular beacons: Probes that fluoresce upon hybridization. *Nature Biotechnology*, *14*(3), 303–308.

Venter, J. C., et al. (2001). The sequence of the human genome. *Science*, *291*(5507), 1304–1351.

Vutukuru, M. R., et al. (2016). A rapid, highly sensitive and culture-free detection of pathogens from blood by positive enrichment. *Journal of Microbiological Methods*, *131*, 105–109.

Whitcombe, D., et al. (1999). Detection of PCR products using self-probing amplicons and fluorescence. *Nature Biotechnology*, *17*(8), 804–807.

Woerner, A. E., et al. (2019). Forensic human identification with targeted microbiome markers using nearest neighbor classification. *Forensic Science International: Genetics*, *38*, 130–139.

Woese, C. R., & Fox, G. E. (1977). Phylogenetic structure of the prokaryotic domain: the primary kingdoms. *Proceedings of the National Academy of Sciences of the USA*, *74*(11), 5088–5090.

Woese, C. R., et al. (1975). Conservation of primary structure in 16S ribosomal RNA. *Nature*, *254*(5495), 83–86.

Zheng, H. Y., et al. (2007). Relationships between BK virus lineages and human populations. *Microbes and Infection*, *9*(2), 204–213.

Zhong, S., et al. (2009). Distribution patterns of BK polyomavirus (BKV) subtypes and subgroups in American, European and Asian populations suggest co-migration of BKV and the human race. *Journal of General Virology*, *90*(Pt. 1), 144–152.

PART II
Human Rights and Humanitarian Disasters

In 2013, more than 30 years after Salvadoran soldiers abducted Serapio Cristian Contreras Recinos, at the age of two, during an attack on his village and gave him away for illegal adoption, he meets and embraces his biological mother, María Maura Conteras. The *reencuentro*, or reunification, was made possible through DNA analysis. (Photo: Pro-Búsqueda)

CHAPTER 7

The Living Disappeared

Forensic DNA Typing and the Search for Argentina's Stolen Children

MARIANA HERRERA PIÑERO, ERIC STOVER, MELINA TUPA, AND VÍCTOR B. PENCHASZADEH

Argentines awoke on the morning of March 24, 1976, to learn that the nation's military leaders had deposed President María Estela "Isabel" Martínez de Perón in a coup hours earlier, placed her under arrest, and flown her to a lodge in the Andes (Dias, 1976; Joyce & Stover, 1991). The coup was one of approximately 30 military takeovers that Argentina had endured since the popular president Hipólito Yrigoyen was deposed in 1930. In the intervening years, only a few elected presidents were able to finish their terms. Among them was Juan Domingo Perón, husband of Isabel, a member of the military and founder of the populist Peronist movement, who had been elected in 1946 and served several terms before being ousted and forced to leave the country.

During Perón's exile, the Argentine military solidified its power and set out to repress demands for political freedom and socioeconomic rights. Leftist political labor organizations responded by organizing armed groups, the most active of which were the Montoneros (of Peronist ideology) and the Ejército Revolucionario del Pueblo (ERP), or People's Revolutionary Army (of Marxist/Guevarist ideology), which carried out armed attacks on military and police installations and kidnappings of wealthy businessmen for ransom.

In March 1973, the military finally allowed free elections to be held again. The winner, a Peronist named Héctor Cámpora, allowed Juan Perón to return from exile, then called for new elections that fall. With his wife as running mate, Perón won by a landslide, only to die in office 10 months later, effectively leaving his wife, who at the time was vice president, the presidency.

During Isabel Perón's presidency, Argentina was a deeply divided country with its economy in shambles. Tens of thousands of industrial workers were either on strike or had cut production more than half through work stoppages to protest economic policies. The Montoneros, who by then had become a massive popular organization with an armed wing, had gone underground. Meanwhile, the right-wing death squad Triple A (Argentine Anticommunist Alliance), set up by Juan Peron in 1973 and led by the minister of social welfare, José Lopez Rega, had increased its campaign of killing "left-wing subversives" (Gambini, 2007).

The day after the March 1976 coup that deposed Isabel Perón, the armed forces installed a three-man military junta headed by General Jorge Rafael Videla, who announced on public television that he and his fellow junta members would henceforth exercise all judicial, legislative, and executive power. Within weeks, the junta had dissolved Congress, banned political parties and labor unions, replaced members of the Supreme Court, and appointed retired and active military officers as university rectors.

As for the guerrilla groups and their sympathizers, the military junta broadened its definition of *terrorist* to include "not only someone who plants bombs but a person whose ideas are contrary to Western Christian civilization." It soon became apparent that the junta's "anti-subversive campaign" was aimed not just at radical leftists but at anyone who opposed the military government. "First we will kill all the subversives, then their collaborators, then their supporters, then those who remain indifferent and finally the timid ones," announced General Ibérico Manuel Saint-Jean, the governor of the province of Buenos Aires, at an official dinner in 1977.

Relying on an extensive intelligence network, military and police death squads, known as *grupos de tareas*, or task forces, began to operate out of a labyrinth of hundreds of secret detention centers. Death squad members, driving cars with their license plates covered or missing, abducted "suspected subversives" as they stepped off buses, walked home from work or school, or gathered in parks or university campuses to demonstrate against the regime. Regular military units, often numbering in the hundreds, raided private homes and safe houses where groups of guerrillas or student activists and trade unionists lived in hiding. Most of the abductees were tortured and killed, then buried in unmarked graves or dropped from planes into the Atlantic Ocean or the estuary of the Rio de la Plata (National Commission on the Disappeared, 1986).

Thousands of people disappeared during the seven years of military rule that ended in 1983 (Snow & Bihurriet, 1992; National Commission of the Disappeared, 1986).[1]

---

1. Argentine human rights organizations, including the Abuelas de Plaza de Mayo and the Madres de Plaza de Mayo, place the figure as high as 30,000. In a report released in late 1984, the National Commission of the Disappeared provided the names of 8,960 people who had disappeared. However, at the time Ernesto Sábato, the chair of the commission, cautioned that the true number of disappeared was likely to be much higher, as many families were still afraid to report a disappearance for fear of reprisals. The commission submitted to the courts more than 1,800 cases for investigation of possible criminal charges for human rights violations (National Commission on the Disappeared, 1986).

Of these, nearly a third were women. Some were abducted with their small children. Others were pregnant at the time of their abductions or became so while in detention, usually through rape by guards and torturers (Goldman, 2012). Many of the pregnant detainees were taken to makeshift maternity wards at the Escuela de Mecánica de la Armada (ESMA), the Navy Mechanics School, or the Army's Campo de Mayo Hospital, which was disguised as an epidemiology unit. Women held at these wards were often given medication to accelerate birth or Caesarean sections (National Commission on the Disappeared, 1986). Once a woman gave birth, she was removed from the facility and executed at a secret location, and her child was given to "politically acceptable" parents—usually families with some connection to the regime (Joyce & Stover, 1991; Penchaszadeh, 1992).

All told, as many as 500 kidnapped newborns and young children—referred to by their abductors as *botín de guerra*, or war booty—were given to childless military and police couples and others favored by the regime (Nosiglia, 1986; Penchaszadeh, 1992, 1997; Arditti, 1999). Using falsified documents, the "adoptive families" would then register the children as their biological sons and daughters (Vegh Weis, 2017). In a groundbreaking study of the military junta, *Lexicon of Terror*, Marguerite Feitlowitz wrote that the military considered the children "seeds of the tree of evil," who, by being placed with "decent" and "patriotic" families, would be saved from becoming the next generation of subversives (Feitlowitz, 2011).

Argentine scholars would later trace the military junta's justification for appropriating the children of the disappeared to a theory developed in 1938 by Antonio Vallejo-Nágera, a Spanish psychiatrist and Franco loyalist (Federación Estatal de Foros por la Memoria, 2013). Vallejo-Nágera claimed that members of the Republican Army and their supporters in the International Brigades—all of whom opposed the Spanish dictator Francisco Franco—possessed a Marxist "red gene" that turned them into psychopaths and "social imbeciles." The only way to prevent the spread of the red gene among the Spanish people, he argued, was to take the children away from their Marxist mothers (Vallejo-Nágera, 1938). By 1943, tens of thousands of children had been seized and handed over to orphanages or families loyal to the Franco regime. In some cases, birth records were destroyed and the children's names changed to prevent any further contact with their parents (Fotheringham, 2011). Decades later, Spanish prosecutors revealed that what may have begun as a form of political retaliation mutated into a trafficking business that continued after Franco's death in 1975, in which Spanish doctors, nurses, and also nuns colluded with criminal networks (Minder, 2018).

THE SEARCH FOR LIFE

One afternoon in April 1977, 14 middle-aged and elderly women wearing white kerchiefs embroidered with the names of their disappeared children converged on the Plaza de Mayo, directly across from the presidential palace. Since it was forbidden to gather in public spaces, the women formed a single line on the plaza's cobbled walkway and began walking silently in a circle.

The vigil marked the creation of the Madres de Plaza de Mayo (Mothers of the Plaza de Mayo). From then on, the women met weekly at the plaza to demand that their children be found alive. Among the demonstrators were dozens of women who protested the disappearance not only of their sons and daughters but of their grandchildren as well. In October 1977, a group of these women, led by an art history teacher named María Isabel "Chicha" Chorobik de Mariani, formed a second organization called the Abuelas de Plaza de Mayo (Grandmothers of the Plaza de Mayo).

From the outset, the Abuelas de Plaza de Mayo (hereafter the Abuelas) had three goals. The first was to locate and identify their kidnapped grandchildren and restore their "right to identity." The second was to work with the courts to return their grandchildren to their biological families—a process they referred to as "restitution" (Arditti, 1999). The third goal was to bring to justice those responsible for the child theft (Gandsman, 2012; Grandsman, 2019; Avery 2004).

In 1989, "Chicha" Mariani was succeeded as president of the Abuelas by Estela de Carlotto. A 48-year-old schoolteacher, Carlotto had learned through a military contact in early 1978 that her daughter, Laura, and her companion, a drummer in a rock band and a member of the Montoneros guerrilla organization, had been abducted and taken to a secret detention center called La Cacha. Several months later, Carlotto and her husband were summoned to a police station, where they received Laura's body and were given an autopsy report stating that she had died of "multiple gunshot wounds."[2] As Carlotto would find out years later, while in detention Laura had given birth to a boy whom she named Guido, after his grandfather. Two months later, Laura had been dragged out of the camp to an unknown location, where soldiers staged an armed confrontation. When Laura's body was turned over to her parents, she appeared to have been shot multiple times, including in the face. Her baby, meanwhile, had been taken away and given up for illegal adoption (Goldman, 2012; Goñi, 2015).

Unlike most of the disappeared, who were buried in mass graves, Laura's body was given to her family, who later buried her in a single grave in a local cemetery. In April 1985, after the military had left power, the American forensic anthropologist Clyde Snow, with the assistance of the Argentine Forensic Anthropology Team, exhumed Laura's skeletonized remains. In his forensic report to the judge, he determined that Laura had not been killed in a shootout but while she was on the ground, lying face down. Closely examining the pelvic bones, he found a groove in the preauricular sulcus, a small, shallow trench immediately in front of the sacroiliac joint, which indicated that the she had given birth to a term or near-term infant.[3]

---

2. A copy of the autopsy report, dated August 29, 1978, is on file at the Human Rights Center, School of Law, University of California, Berkeley.

3. Clyde Snow's report is on file at the Human Rights Center, School of Law, University of California, Berkeley.

CITIZEN DETECTIVES

Estela de Carlotto recalled how she felt her first day in 1978 marching with the Abuelas in the Plaza de Mayo: "I was very frightened. . . . I saw so many weapons, police mounted on horses, assault dogs, and water cannons . . . all of [which] made me want to leave. But the more experienced Grandmothers told me: 'Walk Estela, nothing will happen to us, I promise'" (Tupa, 2015).

Solidarity, as Carlotto soon learned, was at the heart of the Abuelas' struggle. It not only gave the women courage and hope; it also inspired them to become clever detectives. Grandmothers spied on couples they suspected had adopted missing children and even went undercover in their search for clues. "Whenever we got information that a family had adopted a child illegitimately, we would follow the family closely," explained Mariani. "There were even cases in which some of us offered ourselves as domestic housekeepers to gain entrance into a household we suspected had one of our grandchildren" (Chelala, 1986). One grandmother, having learned that a pedicurist might have adopted a kidnapped child, paid a visit to the woman's shop. As the woman pedicured her feet, the grandmother peppered her with questions about her family (Arditi, 1999). Still another grandmother parked her car in front of schools with the hood up, as if it were broken down, and photographed children who were believed to have been illegally adopted (Vegh-Weis, 2017).

Working with their legal team, the Abuelas regularly submitted habeas corpus petitions to judges inquiring about the whereabouts of their missing children and grandchildren. Judges forwarded these petitions to municipal police departments, the Ministry of the Interior, and the military command. Invariably, these agencies denied that the person in question was under their control, forcing the judge to close the case (Arditti, 1999; Avery, 2004; Vegh-Weis, 2017).

The Abuelas had their first breakthrough in March 1980. The case involved two sisters, seven-year-old Tatiana and three-year-old Laura Britos, who were living with Carlos and Inés Sfiligoy, a couple who had adopted the two girls from an orphanage in 1978. In 1977 a death squad had abducted and killed the girls' parents and then abandoned the sisters in a park on the outskirts of Buenos Aires. Acting on a tip, the Abuelas contacted the judge who had presided over the girls' adoption. He, in turn, met with one of the Abuelas, an elderly woman who claimed she was their paternal grandmother, and asked her to identify her granddaughters. Photographs of the girls were produced, but the judge said they were insufficient proof of the putative grandmother's relation to Tatiana and Laura. In the end, the judge ruled that the sisters should remain with the Sfiligoys (Abuelas de Plaza de Mayo, 2018).

By June 1984, the Abuelas had located 25 of their kidnapped grandchildren. Still, as was the case with the Britos sisters, they faced the challenge of finding a an identification method that would persuade a judge to return the children to their biological families. "The quandary we faced had been born in the uncertainty of how we would identify the [Britos sisters]," Mariani recalled. "We had photos and other evidence, but that wasn't enough. . . . It had been three years since the girls had been kidnapped, and they were taller and, of course, they had aged. And so we asked ourselves, 'What are we going to do about the children who were born in captivity?'

In some cases, we didn't even know the sex of the child or to whom they belonged" (Joyce & Stover, 1991).

Then one day Mariani arrived at the headquarters of the Abuelas with an article from *El Día*, her hometown newspaper. The article was short, but it would change the course of history for the grandmothers. As Estella de Carlotto explained: "[The article] was about a man who had denied his own paternity. But then blood tests were ordered, and the boy in question turned out to be the man's [son]. This case made us wonder if maybe our own blood could help us [identify our grandchildren]" (Tupa, 2015).

With this question as their polestar, the Abuelas traveled out into the world in search of an answer.

SEARCHING FOR AN ANSWER

A group of Abuelas first posed this question to Arnault Tzanck, a pioneer in blood transfusions, in his office at the Hôpitaux Universitaires Pitié-Salpêtrière in Paris. Intrigued, Tzanck left the women in his office and wandered throughout the hospital, conferring with his colleagues. Returning several hours later, he entered the room, shaking his head. "I'm sorry," he said. "But I can't answer your question" (Nosiglia, 1986).

Undeterred, the group went on to meet with geneticists in 12 European countries, only to leave each meeting empty handed. Then, in November 1982, they received the first indication that their question could be answered. The occasion was a meeting of the General Assembly of the Inter-American Commission on Human Rights of the Organization of the American States in Washington, D.C. While in the capital, the Abuelas met with Isabel Mignone, the daughter of Emilio Mignone, founder of the Center of Legal and Social Studies in Buenos Aires. Isabel, who was aware of their quest, put the group in contact with Víctor Penchaszadeh, an Argentine geneticist living in New York and a coauthor of this chapter. Years later, Penchaszadeh recalled the meeting:

> I met with Chicha Mariani and Estela de Carlotto in a hotel in Manhattan. And, of course, the first thing they asked me was: "How can we identify our kidnapped grandchildren?" They knew I was a geneticist. They knew I was Argentine and had been exiled. And so they asked me directly: "What could be more important for an Argentine geneticist than this topic?" It was as if they had thrown me a challenge. I told them: ". . . I'll need to study it. And I'll need to pull together a group of experts. But it can be done." (Tupa, 2016)

Penchaszadeh contacted Fred Allen, a renowned American hematologist who, until his retirement, had directed the New York Blood Center. Allen was intrigued by the question but warned his Argentine colleague it would require, as a first step, adapting the statistical formula for determining paternity and applying it to grandpaternity.

While Allen worked on the formula, Mariani and her group traveled to Washington, D.C., in October 1983, to meet with Eric Stover, another author of this chapter. At the time, Stover was the director of the Science and Human Rights Program of the American Association for the Advancement of Science (AAAS). During the military dictatorship, Stover had made several trips to Argentina to gather information about disappeared scientists and health professionals. During one visit the Abuelas, fearing that the military might raid their offices, arranged for Stover to smuggle duplicates of case files on kidnapped children out of the country for safekeeping.

In Stover's office, Mariani and Carlotto posed the same question they had asked Penchaszadeh. Stover then consulted his colleague, Cristián Orrego, a Chilean biochemist working at the National Institutes of Health, who relayed the query to Mary-Claire King, a geneticist then at the University of California, Berkeley. King, in turn, consulted with Penchaszadeh and Allen and assembled a group of population and molecular geneticists, mathematicians, and statisticians; as the Abuelas had hoped, they came up with an answer in a few months (Banco Nacional de Datos Genéticos, 2017; Snow et al., 1984).

The team of geneticists found grandpaternity, like paternity, could be determined to a high degree of certainty thanks to numerous and highly specific genetic markers, including human leukocyte antigens (HLA), blood groups, red-cell enzymes, and plasma proteins. Working with these genetic markers, geneticists could establish that a child shares specific genetic variants with a specified set of grandparents because the child is, in fact, their grandchild, compared to the probability that the child and the grandparents share similar genetic variants only by chance (Di Lonardo et al., 1984).

The question now was how this cutting-edge science could be brought to Argentina.

THE RETURN TO DEMOCRACY

By June 1982, the writing was on the wall for Argentina's military junta. That month, British forces succeeded in ending Argentina's two-and-a-half month occupation of the Falkland Islands, or Islas Malvinas, a cluster of windswept islands that had been the object of a 150-year-old dispute between the two countries. The junta, now faced with violent antimilitary demonstrations, lifted the six-year ban on political activity and promised elections in late 1983.

On October 30, 1983, Argentines went to the polls to elect Raúl Ricardo Alfonsín, a small-town lawyer and leader of the Radical Party, as president. Within weeks of taking office, Alfonsín had retired dozens of generals; ordered the prosecution of the junta leaders; and established the National Commission on the Disappearance of Persons (CONADEP), chaired by the novelist Ernesto Sábato, to probe the fate of the disappeared. The commission was comprised of 13 members and 5 secretaries, who interviewed victims and relatives, visited secret detention centers, and reviewed government documents in an effort to reconstruct a comprehensive picture of what had happened to the disappeared and who was responsible for these crimes (Joyce & Stover, 1991).

In February 1984, the Abuelas and other human rights organizations met with Sábato and urged him to contact Stover at the AAAS, which he did, asking for scientific assistance to the commission. In June Stover flew to Argentina with geneticist Mary-Claire King and a forensic team, including forensic anthropologist Clyde Snow. Over the next 10 days, the forensic team traveled throughout the country, meeting with judges, morgue workers, human rights activists, and relatives of the disappeared.

King remained in Buenos Aires, where the Abuelas introduced her to Ana Maria Di Lonardo, then head of the Department of Immunology at the Hospital Durand in Buenos Aires (Beckwith, 1987; Arditti, 1999). As the grandmothers waited anxiously, the two women refined the mathematical formula for the *índice de abuelidad* (grandparentage index) and developed a test for grandpaternity based on results from HLA antigens, blood groups, red-cell enzymes, and plasma proteins.[4] Soon after King left Argentina, a judge agreed that the test could be applied in the case of a suspected kidnapped child (Di Lonardo et al., 1984; Penchaszadeh, 1997).

PAULA EVA LOGARES

The case involved an eight-year-old girl named Paula Eva Logares, who was living with a former police commissioner, Rubén Lavallén, and his wife, Raquel Teresa Leiro. In court, the couple claimed that Paula was their biological daughter. To prove it, they produced a birth certificate signed by a police surgeon on October 29, 1978, bearing the name Paola Lavallén. A fellow policeman testified that he was the owner of the house where the birth had taken place.

But the Abuelas knew differently. Paula, they argued, wasn't born in 1978, as the Lavalléns claimed, but in a private clinic two years earlier, on June 19, 1976, to Mónica Grinspon and her husband, Claudio Logares. Later that year, the couple fled with Paula to Montevideo, Uruguay. On the afternoon of May 18, 1978, heavily armed men kidnapped Paula, then 23 months old, and her parents as they stepped off a bus. The family was taken to a detention center in Buenos Aires, where Paula's parents were disappeared and Paula given to Rubén Lavallén and his wife (Arditti, 1999; Abuelas de Plaza de Mayo, 2009).

When Paula's maternal grandmother, Elsa Pavón de Aguilar, learned what had happened that night in Montevideo, she quit her job and joined the Abuelas, dedicating herself to finding her daughter, son-in-law, and granddaughter. The first glimmer of hope came in April 1980, when the Brazilian organization CLAMOR (Human Rights Defense Committee for the Southern Hemisphere) turned over an unmarked package to the Abuelas. In it were photos of a small girl bearing a remarkable resemblance to Paula. An address included with the photos indicated that she

---

4. These HLA tests, like all HLA tests at the time, were based on antibodies detected in the blood. Now HLA testing is based on PCR and DNA testing, like the testing done in the first DNA forensic case (*Colin Pitchfork*) discussed in chapter 1.

was living in the home of Rubén Lavallén near the botanical gardens in Buenos Aires (Joyce & Stover, 1991).

Elsa Pavón would stand for hours across the street from the Lavallén's apartment building, hoping to catch a glimpse of Paula. One day, in a park near the apartment, she spotted a girl walking with a middle-aged woman; she was sure it was Paula. When Pavón returned several weeks later, a "For Rent" sign was posted on the Lavalléns' apartment door. Fearing the Lavalléns might flee the country, the Abuelas launched a campaign to find Paula. Posters of the little girl were tacked onto telephone poles and community bulletin boards all over the city. Then one day Pavón received an anonymous call from someone who apparently lived in the same building as the Lavalléns: the Lavalléns, Pavon was told, were planning to spend the Christmas holiday in Uruguay.

The Abuelas immediately petitioned a court to prohibit the couple from taking Paula out of the country. The court accepted the petition. When the Lavalléns appeared before the court, they argued that Paula was their biological daughter and provided a birth certificate signed by a police surgeon. Another police officer testified that he was the owner of the house where Paula was born. The court ordered Paula to undergo paternity testing. A blood sample was obtained from Paula, but the Lavalléns refused to be tested, which at the time was a right guaranteed to them under Argentine law. As Rubén Lavallén later told a journalist: "I was unfamiliar with this kind of analysis, [and] when it became clear that it was a humiliation for us and our own daughter, I decided I wouldn't go through with it. I won't allow anybody, anybody, to question my fatherhood, especially since I've seen this child leave my wife's womb. No way will I submit to such tests" (Lavallén, 1985).

Stymied, the court turned to Di Lonardo and King. Di Lonardo obtained blood samples from three of Paula's putative grandparents, including Elsa Pavón, and three of her maternal uncles. Communicating by telephone and fax, the two geneticists, one in Argentina, the other in the United States, established a 99.9% certainty on the basis of HLA testing that Paula was related to the grandparents who claimed her (Di Lonardo et al., 1984).

The court accepted the scientific finding and ordered that Paula not leave the country, and that custody of her be granted to her grandmother, Elsa Pavón. But the Lavalléns appealed the case, arguing that removing Paula from their home would be too traumatic for her. Finally, in December 1984, after the military junta had left power, the Supreme Court ruled in favor of Paula's grandmother and ordered that Paula remain in her custody. The court's ruling also prompted Teodoro Puga, then health secretary for the Municipality of Buenos Aires, to place Ana María Di Lonardo in charge of conducting genetic tests of future cases of kidnapped children that appeared before the justice system.

The Abuelas, with the help of science, had won a historic victory—one that would galvanize their efforts to find more kidnapped children. Yet success, as it often does, brought a new set of challenges. While blood testing had established Paula Logares's biological identity with a high degree of certainty, it had left her torn, at least initially, between two identities. As an eight-year-old, she now had to come terms with the fact that her "adopted" father was complicit in her abduction and the death of her parents. She also had to wrestle with the fact that her "adoptive parents" had falsified her birth date, changed her name, and kept her origins a secret from her. In short, her whole life had been built on a false identity.

Years later, Paula recalled how she felt when she met her grandmother for the first time in the judge's chambers:

> I was on the other side of the table from her, and [the judge] wanted me to approach her. But I didn't want to be near her. She showed me photos: one of me as a baby, and another of my mother and father, and I started to cry.... I mean, think what it is like to be an eight year old and to be told the only parents you have ever known are not your [real] parents, not your true family. And then to be told, "It's a lie, this is your real family." (Banco Nacional de Datos Genéticos, 2017)

Paula's integration into her family has taken years. In an interview in 1998, Elsa said that only recently had she and Paula been "able to laugh, fight, and cry together" (Arditti, 1999). Today, Paula and Elsa regularly appear together at rallies and conferences in support of the Abuelas.

JUSTICE DELIVERED AND DENIED

On September 20, 1984, the National Commission on the Disappearance of Persons delivered "Nunca Mas," its report on the crimes of the military dictatorship, to President Alfonsín. In his introduction to the report, Ernesto Sábato wrote:

> The vast majority of the [disappeared] were innocent not only of any acts of terrorism, but even of belonging to the fighting units of guerrilla organizations: these latter chose to fight it out, and either died in shoot-outs or committed suicide before they could be captured. Few of them were alive by the time they were in the hands of repressive forces. (National Commission of the Disappeared, 1986)

The 50,000-page report laid the groundwork for the trials and imprisonment of the nine generals and admirals who had made up the three successive juntas that ruled Argentina after the 1976 coup.

But it also angered the top military brass.

In December 1986, Alfonsín, under pressure from the military, passed Law 23.492, *Punto Final*, or "Full Stop," which set a 60-day deadline for opening prosecutions of members of the armed forces, police, and prison officials accused of past human rights abuses. The only cases excluded from this law were those concerning rape, theft, and the abduction and concealment of minors. Though outraged by the new law, the Abuelas continued pressing for charges against the individuals—or "appropriators"—who had illegally adopted their stolen grandchildren or registered them falsely as biological offspring (Arditti, 1999).

The following year, in the wake of a military takeover of an army base near Buenos Aires, Alfonsín sent Congress another amnesty law, this one granting immunity to a large number of subordinate officers (Avery, 2004). But even Alfonsín's two amnesty laws were not enough for the military. Two more uprisings took place, resulting in more negotiations and compromises. Finally, in July 1989, Alfonsín relinquished the presidency five months ahead of schedule, and Carlos Saúl Menem, who had been elected in May of that year, became president. Five months later, Menem pardoned and released the imprisoned junta leaders, claiming that it was time to unify the country and move toward reconciliation, even though 80% of the population opposed the pardons (Arditti, 1999; Joyce & Stover, 1991).

OVERCOMING OBSTACLES

During this period of political turmoil, the Abuelas continued to face a host of legal and social challenges. Some judges, lacking experience dealing with baby-theft cases, had declared themselves incompetent and passed on their cases to other judges, causing lengthy delays. Others failed to question the legitimacy of the adoptions or even the false birth certifications that were produced. A few judges had permanently closed cases when testing indicated the genetic match of a child with a particular family was negative, effectively ending the possibility of comparing the child's blood sample to that of other putative families. Meanwhile, some appropriators were refusing to let the children identified as falsely registered as biological offspring or supposedly "adopted" be tested. And to make matters worse, a growing number of children, now entering their late teens and early twenties, were themselves refusing to be tested for fear that their "adopted parents" would be subjected to criminal charges (Gandsman, 2008). Finally, some of the older grandparents were dying, with no formal means of preserving their blood samples for analysis in future cases.

Still another challenge was Argentina's "closed adoption system," which prohibited nonparental relatives (grandparents, uncles or aunts, and siblings) from becoming interested parties in custody cases. Unlike open adoption systems, which allow for contact and information to be shared between the biological and adoptive families, closed adoption systems involve sealed records, so that adopted children do not know their genetic forebears (Appel, 1995). In other words, unless the Abuelas obtained information, usually in the form of a tip, about a stolen child from the period of military rule, they were prevented from gaining access to adoption records that might reveal an illegal adoption and potentially the child's biological identity.

Until the law was changed, the Abuelas argued, couples who had knowingly—and *illegally*—appropriated children during the military dictatorship would continue to get away with their crime, while the children would never learn their true identities (Avery, 2004). This secrecy was potentially damaging to children who had been *legally* adopted during military rule. By now, publicity about the Abuelas and their mission was widespread, and some of these children would surely wonder about their own origins and the fate of their biological parents (Arditti, 1999).

The Abuelas took these challenges in stride and began lobbying Congress. Their first success was passage of a law, in 1987, creating the National Genetic Data Bank, or Banco Nacional de Datos Genéticos (BNDG), which would offer state-of-the-art services without charge to the relatives of kidnapped children and to anyone whose identity was in question. Once operational, the BNDG established a permanent repository, or biobank, for genetic profiles of grandparents and other family members, which it promised to keep until the year 2050. It also hired psychologists to provide counseling to recovered children. Finally, the law that created the data bank provided that whenever biological identity was in dispute, a court could require all concerned parties to submit to genetic testing; refusal to comply would be construed as evidence of involvement in kidnapping. By 1998, the BNDG had collected blood samples from more than 2,000 individuals, representing about 175 family groups, and identified more than 20 grandchildren (Arditti, 1999; Banco Nacional de Datos Genéticos, 2017).

In 1992 the Abuelas and BNDG received a further boost to their work through the establishment of the government-appointed National Commission for the Right to Identity (Comisión Nacional por el Derecho de Identidad de las Personas, 1992), known as CONADI Its members were given the authority to conduct interviews with survivors of illegal detention centers and comb through government records from the years of military rule to identify pregnant women who had passed through the secret detention system and were later executed. With this knowledge, CONADI would approach parents who were unaware that their daughters were pregnant at the time of their detention or if they had been raped and become pregnant in detention and ask them to provide blood samples to the BNDG. Starting in 1998, children who thought they had been illegally adopted could also approach CONADI, which in turn would refer them directly to the BNDG. By 1996, the commission's investigations had uncovered 40 additional cases of disappearances of pregnant women and children that had not been reported previously (Banco Nacional de Datos Genéticos, 2017, p. 91; Arditti, 1999, p. 151).

The Abuelas next took on Argentina's closed adoption system and successfully lobbied Congress to pass a new law, in 1997, that gave adopted children "the right to know their true biological identity and [to] have access to their adoption file once they have reached the age of eighteen" (Ley No. 24.779, 1997). In effect, all adopted children age 18 or older now had the right to know their places of origin, the conditions surrounding their births, and the identities of their birth parents. The law also specified that illegality in the adoption process was sufficient grounds for annulment, and individuals who had participated in illegal adoptions could be arrested and prosecuted (Avery, 2004).

JUSTICE RESTORED: THE CASE OF CLAUDIA POBLETE HLACZIK

In 1999, the Abuelas filed a complaint in Argentine federal court that would set the groundwork for reopening prosecutions of past human rights offenders. The case involved a 22-year-old computer analyst named Claudia Poblete Hlaczik. In 1978 Claudia, then eight months old, and her parents, José Liborio Poblete and Gertrudis María Hlaczik, were detained and taken to a secret detention center in Buenos Aires known as El Olimpo. José and Gertrudis were later killed, and Claudia was given to Ceferino Landa, an army intelligence officer, and his wife, Mercedes Beatriz Moreira, along with a false birth certificate signed by a military doctor named Julio César Cáceres Monié.

For 22 years, Claudia's grandmother, Buscarita Roa, searched for her granddaughter with the assistance of the Abuelas. In 1999, Buscarita received information that Claudia was living with Landa and his wife in Buenos Aires. Soon thereafter, Claudia was summoned to the BNDG for a genetic test. She resisted at first, fearing what might happen to the couple who had raised her, but eventually changed her mind: "At the time, I realized if I refused I would end up being a big media story, so I went to the Hospital Durand and they took my DNA.... The truth is I never thought I was a daughter of the disappeared. For me, the story ended with the Mothers of the Plaza de Mayo looking for their children, and they were all old crazy women."

DNA tests confirmed that Claudia was the child of José Liborio Poblete and Gertrudis María Hlaczik "They called me from the courthouse, so I went there to look at the report," Claudia recalled. As she turned the pages, she came across a blurry photo of José and Gertrudis and one of a baby frowning at the camera. "Clearly I was that baby. And, at that very moment, the world fell out from under me"[5] (Ginzberg, 2015).

Claudia's appropriators were sentenced to seven years in prison. Under the amnesty laws, however, the two police agents responsible for the abduction, torture, and murder of her biological parents couldn't be prosecuted. One of the officers, Julio Héctor Simón, also known as "El Turco Julián," was an avowed fascist who carried a key chain fashioned out of a swastika and had served as an interrogator at El Olimpo and other detention centers. In the late 1980s, Simón had been prosecuted for 58 cases of torture but was pardoned under the Full Stop and Due Obedience laws. However, the presiding judge, Gabriel Cavallo, argued in a 188-page ruling that both amnesty laws conflicted with Argentina's obligation, under international law, to bring to justice those responsible for crimes against humanity.

Cavallo's ruling helped persuade the Argentine Congress to repeal the two amnesty laws in 2003 and the Supreme Court to declare them unconstitutional in 2005. A year later, a federal court sentenced Simón to 25 years for the Poblete abductions and murders. (The other police officer, also charged with the crime, died in a prison hospital before the verdict was issued.) Simón's conviction opened the floodgates, and cases against police and military personnel suspected of abuses during the military dictatorship began stacking up like firewood (Goldman, 2012; Ginzberg, 2015).

---

5. Claudia moved to Venezuela and fell in love with a colleague at work. They soon wed and had a baby. Marriage changed Claudia's life: "It helped me put into perspective both my relationship with my appropriators and with my biological family.... Forming my own 'family' helped me construct my own identity and allowed me to look towards the future, not the past" (Qassim, 2007; Drapkin, 2010).

LAW CATCHING UP WITH SCIENCE: THE CASE OF EVELYN VÁZQUEZ

Since the early 1990s, advances in human genetic identification had enabled scientists to extract DNA from a person's clothing and personal items, such as hair and toothbrushes. This development caught the attention of Alcira Ríos, a lawyer with the Abuelas for more than 20 years and a former political prisoner. By the late 1990s, the Abuelas were concerned that a growing number of grandchildren, now in their twenties or early thirties, were refusing to be tested. Some, like Claudia Poblete, were terrified of being responsible for the imprisonment of the only parents they'd ever known. Others feared that their appropriators might harm them. At least one appropriator had held a gun to a child's head to stop her from going to the BNDG to give blood (Goldman, 2012).

Ríos took it upon herself to share these new genetic methods with several Argentine judges, and before long a number of them were forgoing orders for mandatory blood testing in favor of less invasive methods. One of the earliest cases involved Evelyn Vázquez, who, according to the Abuelas, was born in the maternity ward of the Navy Mechanical School in September 1977.[6] Evelyn's putative mother, Susana Pegoraro, was executed after giving birth, and her baby was given to Policarpo Vázquez, a former naval intelligence officer, and his wife, Ana María Ferra.

In 1999, the Abuelas filed a case against Policarpo Vázquez and his wife before federal judge María Servini de Cubría. During the proceedings, Policarpo Vázquez admitted that Evelyn, now 22, was not his biological daughter, and that he had known she was a child of one of the disappeared but chose to do nothing about it. As he told the press at the time: "God put the baby in my hands, and so it was my responsibility to raise her" (Meyer & Ginzberg, 1999).

During the trial, Judge Servini de Cubría ordered Evelyn to submit to a blood test at the BNDG, but she refused. When an appellate court upheld the judge's decision, she appealed to the Supreme Court. Evelyn argued that ordering her to undergo a DNA test amounted to a violation of not only her privacy, but her dignity, as she did not wish "to betray intense emotional ties" with the people she saw as her parents. The Supreme Court, in a 7–1 decision, ruled in Evelyn's favor, arguing that an adult, unlike a minor, could not be forced to undergo a blood test, as it would amount to a violation of privacy. However, the justices placed no prohibitions on obtaining genetic material by other means (Valente, 2002).

The Supreme Court ruling notwithstanding, Evelyn Vázquez's case was taken up by another court. After her appropriators confessed to their crime, the judge ordered the police to raid Evelyn's upper-floor apartment. While officers guarded

---

6. During the proceedings, it was revealed that Jorge Luis Magnacco, the military doctor in charge of the maternity ward at the naval academy, had delivered Evelyn. Magnacco was later convicted of three cases of kidnapping and illegal adoption, as well as for his participation in a plan (Plan Sistemático) implemented at the highest levels to kidnap, hide, and rename the children of people detained and disappeared during the military dictatorship (Abuelas de Plaza de Mayo, 2017).

the windows to prevent her from jumping out, a BNDG technician confiscated her gym bag, stuffed with sweaty clothes, and a toothbrush. DNA testing of these items proved that Evelyn was the daughter of Susana Pegoraro and Santiago Bauer. Policarpo Vázquez and Ana María Ferra were later sentenced to 14 and 10 years in prison, respectively, while Justina Cáceres, the midwife who had falsified Evelyn's birth certificate, was sentenced to 7 years ("Condenas de catorce y diez años para un apropiador y una apropiadora," 2011; Goldman, 2012). Evelyn eventually changed her last name to Bauer, and she soon established a relationship with her biological grandmother but still remained close to her appropriators.

With the Vázquez case resolved in their favor, the Abuelas decided to lobby Congress to pass a law codifying the extraction of DNA from personal items. Passed in November 2009, Law 26.549 provided for the extraction of "DNA by means other than body inspection, such as the requisitioning of personal objects containing cells already detached from the body, by means of a house search or personal searches."[7] In 2008, the Supreme Court reversed its earlier ruling in the Vázquez case by allowing that the state, when required to solve crimes against humanity such as forced disappearance and suppression of identity, had the obligation to obtain DNA for testing, and if the possible victim declined, DNA could be obtained by noninvasive means (Penchaszadeh, 2015).

Taken together, Law 26.549 and the 2008 Supreme Court ruling reaffirmed what the Abuelas and other Argentine human rights activists had been arguing for years: that the right to privacy is superseded by rights and obligations of a higher status, such as the obligation of the state under international law to investigate crimes against humanity, find and punish its perpetrators, and provide reparations to victims. In addition, the two legal measures confirmed the right of society and of families of the disappeared to learn the truth about the fate of their missing relatives and the identity of their appropriated offspring (Méndez, 1997).

In purely practical terms, the two legal measures, similar to the 1997 adoption law, greatly enhanced the ability of the Argentine courts and CONADI—and by extension, the BNDG—to gain access to genetic material for identification purposes. To meet this new demand, the BNDG assembled a technical team, developed a protocol for the extraction and analysis of genetic material from personal effects, and held training workshops for law enforcement on the proper methods for gathering personal items (Banco Nacional de Datos Genéticos, 2017).

THE NATIONAL GENETIC DATA BANK

The BNDG, which is under the direction of Mariana Herrera Piñero, a PhD biologist specializing in forensic genetics and a coauthor of this chapter, currently includes

---

7. See Congreso Argentino, Codigo Proceso Penal, Ley 26.549, November 26, 2009. Retrieved from http://servicios.infoleg.gob.ar/infolegInternet/anexos/160000-164999/160779/norma.htm.

in its database profiles of 24 nuclear autosomal short tandem repeats known as microsatellites or STR markers, from more than 300 families, complemented with mitochondrial DNA (mtDNA) sequences and chromosome Y haplotypes (Penchaszadeh, 2015.) Each month the lab analyzes the profiles of some 100 people who believe they may be offspring of one of the disappeared.

To date, the BNDG has used DNA analysis to identify 79 of the 129 recovered children (Banco Nacional de Datos Genéticos, 2017; Ginzberg, 2017). Of the 50 children not identified by the BNDG, 35 had been kidnapped with their parents and were eventually located and recognized by their relatives. Some of the remaining 15 missing children were unborn fetuses whose skeletons were found in the grave comingled with the remains of their mothers. In other cases, the cemetery records showed that the woman was buried as NN on a date compatible with a recent pregnancy.

BNDG: The Story of Appropriated Children in Four Figures

Evolution of Techniques

Figure 7.1 shows the evolution of genetic techniques that have been applied in the search for Argentina's stolen children since 1984. New genetic technologies that have been developed around the world have had an enormous impact on the work of the BNDG. One of these developments enabled geneticists to determine which specific DNA sites (or DNA markers) along the genome could be analyzed to generate genetic profiles unique to each individual for identification purposes (Schneider, 2007). Prior to the analysis of autosomal and chromosome Y DNA markers, the discovery that DNA located in the mitochondria (cytoplasmic organelles transmitted to

National Genetic Data Bank: Evolution of Techniques

STAGE	PERIOD	DESCRIPTION
EXPLORATION	1984–1992	HLA, Restriction fragment length polymorphism, first use of Polymerase chain reaction.
STABILIZATION	1992–2005	Mithochondrial DNA sequencing, microsatellites (STRs) standardization, Chromosome Y haplotypes, quality standars.
GROWTH	2005–2019	Chromosome X microsatellites. Automation, DVI and MPI (*) massive searching softwares. ISO/IEC 17025 standars for forensic genetic laboratories.
SOPHISTICATION	Present and future	New tools / Next Generation Sequencing (NGS)

(*)DVI: *Identification of victims in disasters.*
MPI: *Missing persons identification*

Figure 7.1 National Genetic Data Bank: Evolution of Techniques

the offspring exclusively through maternal lineage) was extraordinarily variable and thus useful for human identification was the first complement to HLA and blood-group studies (King, 1991; Skolnick, 1993; see chapter 1). A further development came from the emerging field of *bioinformatics*, a subdiscipline of biology and computer science concerned with the acquisition, storage, display, and analysis of complex biological and genetic information, which is used to find genes, examine their functions, and analyze genetic variation. Bioinformatics increased exponentially the capability of storing and analyzing vast amounts of complex probabilities in human genetic identification and familial relationships (National Human Genome Research Institute, 2015).

Processing DNA samples

Figure 7.2 shows the workflow of the BNDG. Young (or not-so-young) people presumed to be children of the disappeared provide reference samples—comprised of blood, buccal swabs, or personal items—to the BNDG either by judicial order (20%) or on their own, having approached the CONADI (80%) to have a test performed. Once a reference sample is received from a presumed offspring of a disappeared person, the youth's mtDNA is compared to the sequences of all the maternal family groups in the database. Next, the autosomal profile from different variable STR markers or microsatellites is compared with all the family groups that share the same mtDNA to calculate a likelihood ratio (LR) of a match to each of these families. If a positive familial match is made with a biological family group, the results are reported to CONADI or to the courts, which will review the case file documents to verify the preliminary information and inform the young person and the biological family members of the finding. The Abuelas will at that time also announce the identification at a press conference (Banco Nacional de Datos Genéticos, 2017).

The BNDG is constantly grappling with the need to broaden the number of genetic profiles—and thus create stronger statistical values—for each of the family groups in its database. Of the more than 300 family profiles in the database, approximately

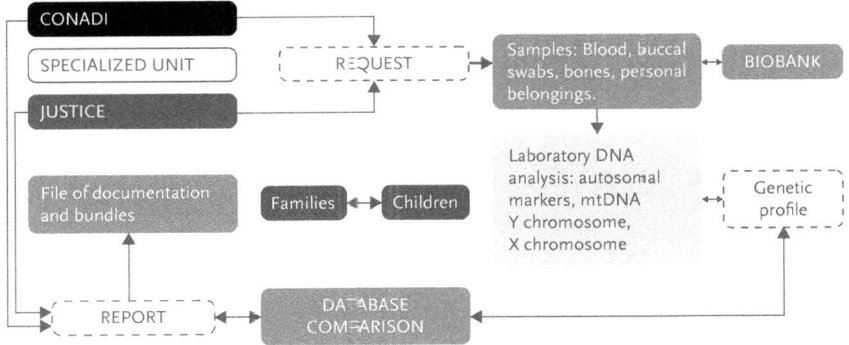

Figure 7.2 National Genetic Data Bank: Processing of samples

30% are incomplete and without enough statistical power because of the high number of family members who disappeared during the military dictatorship, or because the grandparents had already died at the time their children were disappeared. To overcome this challenge, the BNDG established the Forensic Anthropology Unit in 2015. The unit conducts exhumations of relatives of the disappeared to obtain and type DNA from the skeletal remains so the information can be added to the profile of a particular family group. The unit's work has greatly improved the statistical power of each family group by increasing the probability of exclusion in negative cases and by overcoming defined threshold LRs and thus increasing the probability of inclusion (i.e., LR > 10,000) and the certainty of the DNA identification method in positive cases (Banco Nacional de Datos Genéticos, 2017).

The rate of identifications began to increase in 1998 due to (1) the close working relationship developed between BNDG and CONADI; (2) the growing number of grandchildren who had reached the age of 18, which gave them the legal right to request a genetic test without the permission of their appropriators; and (3) the acquisition by the BNDG of automated technologies, which enabled it to process reference samples more quickly and efficiently (Banco Nacional de Datos Genéticos, 2017).

Missing Children Identified

Only 9 of the children who are known to have disappeared with their parents continue to be missing and are being searched for. Of the 129 identified children, 89 were born in captivity and appropriated as babies, either by the military themselves or sold to clinics associated with the clandestine detention centers. Of those, 70% were falsely registered as biological offspring with fraudulent birth certificates, resulting in suppression of identity. Twenty-eight percent were adopted legally, and 2% were placed in foster care. Of the remaining 40, 25 identifications are of children who were abandoned in the street or taken up by neighbors when their parents were abducted or left in orphanages, and 15 were unborn fetuses. (See figure 7.3.)

Figure 7.3 Fate of appropriated children identified to date (n=90)

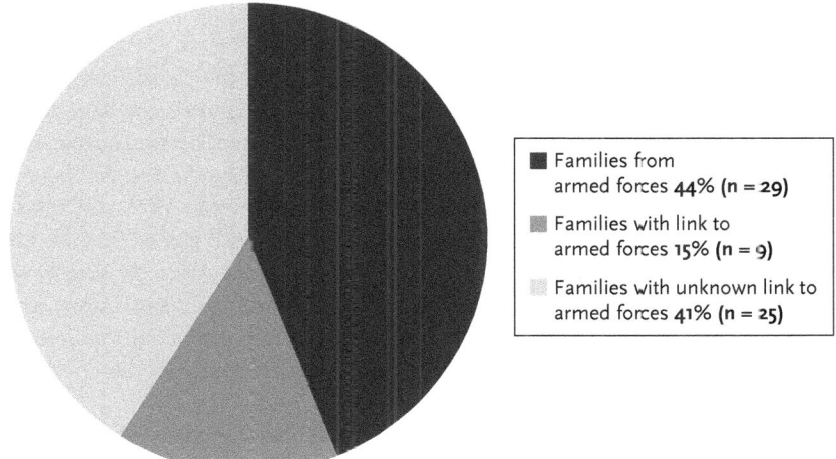

Figure 7.4 "Stolen children registered as biological offspring" (n=63)
In 2013, more than 30 years after Salvadoran soldiers abducted Serapio Cristian Contreras Recinos, at the age of two, during an attack on his village and gave him away for illegal adoption, he meets and embraces his biological mother, María Maura Conteras. The *reencuentro*, or reunification, was made possible through DNA analysis. (Photo: Pro-Búsqueda)

Appropriating Families

Of the 63 stolen children who were falsely registered as a couple's biological offspring, about 46% were appropriated by couples in which at least one member was an active or retired member of the armed forces, 14% by nonmilitary couples with connections through relatives and friends of military personnel, and 40% by nonmilitary couples with no known connections to the armed forces. (See figure 7.4.)

These data support early findings of the National Commission on the Disappearance of Persons that the theft and illegal adoption of children of the disappeared was not a sporadic activity carried out by rogue military units but a centrally coordinated plan reaching up into the top echelons of the armed forces (Feitlowitz, 2011; Avery, 2004; National Commission of the Disappeared, 1986). Most significantly, in July 2012 a federal court convicted the first junta leader, Rafael Videla, and eight other military officers of crimes against humanity for organizing and implementing a "Systematic Plan for the Appropriation of Minors," which included the "widespread practice of abduction, retention, and concealment of minors . . . [by] altering or suppressing their identity." Videla was sentenced to 50 years in prison, while the other defendants received between 5 and 40 years for these crimes.[8] The high visibility of the trial, with the continuing appearances of former high-ranking officers accused and convicted of stealing babies and killing their mothers, cast a new light

---

8. The verdict can be accessed, through the Centro de Información Judicial, at http://www.cij.gov.ar/nota-9445-Lesa-humanidad--condenaron-a-50-a-os-de-prisi-n-a-Jorge-Rafael-Videla-por-robo-de-beb-s.html

on the brutality of the military dictatorship, stirring up public opinion and giving new impetus to the search for the missing grandchildren.

Figures 7.3 and 7.4 are based on data compiled by the Specialized Unit in Appropriations of the Attorney General's Office.[9] Taken together, the data demonstrate that appropriators often obtained falsified birth certificates, usually signed by doctors or midwives linked to the military or trafficking syndicates. For example, the Attorney General's Office uncovered evidence that between 1976 and 1978 an obstetrician by the name of Juana Elena Arias de Franicevich signed 93 false birth certificates involving disappeared and trafficked children. Arias de Franicevich, who died in 1995, had a private hospital on the outskirts of Buenos Aires, where she worked with Jorge Bergés, another obstetrician and convicted torturer who served in clandestine detention centers in Buenos Aires province ("Love Overcomes Hate,", 2017).

Today, the BNDG serves as a regional model for the identification of the disappeared in Peru, Colombia, and El Salvador. In Peru armed conflict from 1980 to 2000 resulted in about 20,000 people being missing. The BNDG is currently training Peruvian professionals in forensic genetic techniques and the use of DNA databases. In Colombia, the BNDG is training professionals in mtDNA analysis and has taken part with CONADI and the Office of the Special Prosecutor for Crimes of Appropriation in workshops to help their Colombian counterparts and family organizations achieve common objectives for the identification of their missing relatives. So far, the Colombian authorities have registered more than 30,000 unmarked graves believed to contain the remains of persons missing because of civil war and political repression since the mid-1960s (Gutman, 2017). The BNDG is also training technicians and university students affiliated with the Asociación Pro-Búsqueda de Niñas y Niños Desaparecidos (Pro-Búsqueda) in El Salvador. Pro-Búsqueda is an organization of families who are looking for thousands of children who were separated from their parents during El Salvador's civil war from 1980 to 1992 and given up for adoption in Central America, Europe, and the United States (see chapter 8).

CONCLUSION

Today the Abuelas are an inspiring example of citizen activism throughout the world. As the Argentine writer Rita Arditti put it: "[The Abuelas are] a group of women with no scientific background who used common sense and prodded scientists to develop the tools that would put their work on a firmer ground. They offer an outstanding example of lay citizens enlisting scientists to work for human rights" (1999, p. 76). Simply put, the Abuelas have always been ahead of the curve. Not only did they recognize that genetic testing would be the linchpin for the restitution of their missing grandchildren; they also appreciated that, given their advancing age, a permanent

---

9. More information about the Specialized Unit for Appropriation Offences is available at https://www.fiscales.gob.ar/unidad-de-apropiacion/.

genetic data bank was needed to preserve their DNA for future identifications. Building partnerships with like-minded activists, judges, scientists, and governmental officials, they led the effort to establish the BNDG and CONADI and successfully lobbied for sweeping legal reforms, including an end to Argentina's closed adoption system.

As for the recovered grandchildren, many have gone through the complex and painful process of finding out and coping with who their biological parents really were and what happened to them. For many of the grandchildren, knowledge of the truth, though at times painful, has helped liberate them from the violence that surrounded their births and the lies that defined their lives as children (Hollander, 1992). The outcome in each case, however, has been unique and distinct. It has depended on who the biological parents were and why and how they disappeared and were killed; where the children were born and under what circumstances; who appropriated them, how they were raised, and what they were told about their identity; how old they were when they learned their true origins; and the nature of the process by which they came to discover their true biological identity (ranging from a legally enforced testing of a child to a voluntary search by a young adult who was suspicious of his or her true ancestry) (Penchaszadeh, 2011, 2015).

Today the recovered grandchildren, now in their late thirties and early forties, have an emblematic presence in Argentine society. Many of them hang out together, call each other brother and sister, and help those newly recovered cope with the mixed emotions of having a commingling of past and present identities. Some work with the Abuelas or in governmental agencies that deal with human rights issues, claiming it is their way of honoring the social activism of their disappeared parents. A few have entered politics and serve as congressmen and congresswomen.

One day in August 2014, Estela de Carlotto, still the current president of the Abuelas de Plaza de Mayo, ended her 36-year search for her missing grandson. That morning, she learned that DNA tests had identified him as Ignacio Hurban, a musician who taught piano and played in several bands in the Buenos Aires area. By nightfall, his identification—the 114th since 1984—and his reunification with his grandmother were the leading news story throughout the country.

Once again, Argentina and the world were reminded of a past that should never have happened—nor should ever be forgotten. As director of the BNDG Mariana Herrera Piñero has written: "Each identification of a stolen grandchild not only impacts the person who recovers his or her identity, but it also puts our society's own identity in front of the mirror, allowing us to recover our own memory and those parts of our history we were denied" (Banco Nacional de Datos Genéticos, 2017).

REFERENCES

Abuelas de Plaza de Mayo. (2009). *Las abuelas y la genética: El aporte de la ciencia en la búsqueda de los chicos desaparecidos*. Buenos Aires, Argentina: Talleres Gráficos Gutten Press.

Abuelas de Plaza de Mayo. (2017, December 6). *El medico represor Jorge Luis Magnacco fue excarcelado a causa de us bajas penas*. Retrieved from https://www.abuelas.org.ar/

noticia/el-medico-represor-jorge-luis-magnacco-fue-excarcelado-a-causa-de-sus-bajas-penas-922.

Abuelas de Plaza de Mayo. (2018). *Casos resultos: Laura Malena Jotar Britos*. Retrieved from https://www.abuelas.org.ar/caso/jotar-britos-laura-malena-229.

Appel, A.R. (1995). Blending families through adoption: Implications for collaborative adoptive law and practice. *Boston University Law Review*, 75, 997.

Arditti, R. (1999). *Searching for life: The grandmothers of the plaza de mayo and the disappeared children of Argentina*. Berkeley, CA: University of California Press.

Avery, L. (2004). A return to life: The right of identity and the right to identify Argentina's "living disappeared". *Harvard Women's Law Journal*, 27, 235–272.

Banco Nacional de Datos Genéticos. (2017). *Una pregunta, 30 años: Memoria escrita del Banco Nactional de Datos Genéticos*. Buenos Aires, Argentina: Ministerio de Ciencia, Tecnología e Innovación Productiva.

Beckwith, B. (1987). Science for the people: Using genetic screening and forensic science to find Argentina's disappeared. *Science for the People*, 19, 6.

Butler, J. M. (2010). *Fundamentals of forensic DNA typing*. Burlington, MA: Elsevier.

Chelala, C. (1986, October 6). Grandmothers of the disappeared. *Christian Science Monitor*, 47.

Comisión Nacional por el Derecho de Identidad de las Personas. (1992). El trabajo del Estado en la recuperación de la identidad de jóvenes apropiados en la última dictadura militar. *Informaciones*, (35), 20–22.

Condenas de catorce y diez años para un apropiador y una apropriadora: Con énfasis en el daño psicológico. (2011, September 24). *Pagina 12*. Retrieved from https://www.pagina12.com.ar/imprimir/diario/elpais/1-177467-2011-09-24.html.

Di Lonardo, A. M., Orrego, C., Darlu, P., King, M-C., & Baur, M. (1984). Human genetics and human rights: Identifying the families of kidnapped children. *The Journal of Forensic Medicine and Pathology*, 5, 339–347.

Dias, J. (1976, March 24). Mrs. Peron overthrown by military in Argentina and reported arrested. *The New York Times*, 1.

Drapkin, J. K. (2010, November 11). Torn between identities in Argentina. *Public Radio International*. Retrieved from https://www.pri.org/stories/2010-11-11/torn-between-identities-argentina.

Federación Estatal de Foros por la Memoria. (2013, November 7). *El mapa de la ideología Franquista que funcionó como cimiento de la dictadura*. Retrieved from http://www.foroporlamemoria.info/2013/11/el-mapa-de-la-ideologia-franquista-que-funciono-como-cimiento-de-la-dictadura/.

Feitlowitz, M. (2011). *A lexicon of terror: Argentina and the legacies of torture*. Oxford, UK: Oxford University Press, 2011.

Fotheringham, A. (2011, January 2). The 30,000 lost children of the Franco years are set to saved from oblivion. *The Independent*. Retrieved from https://www.independent.co.uk/news/world/europe/the-30000-lost-children-of-the-franco-years-are-set-to-be-saved-from-oblivion-2173996.html.

Gambini, H. (2007, February 19). Perón, creador de la Triple A. *La Nacion*. Retrieved from https://www.lanacion.com.ar/opinion/peron-creador-de-la-triple-a-nid884744.

Gansdman, A.E. (2008). *Reclaiming the past: The search for the kidnapped children of Argentina's disappeared* Doctoral thesis, McGill University, Montreal.

Gandsman, A. E. (2012). Retributive justice, public intimacies and the micropolitics of the restitution of kidnapped children of the disappeared in Argentina. *The International Journal of Transitional Justice*, 6, 423–443.

Gandsman, A. E. (2019). "A prick of a needle can do no harm": Compulsory extraction of blood in the search for the children of Argentina's disappeared. *Journal of Latin American and Caribbean Anthropology*, 14, 162–184.

Ginzberg, V. (2015, June 14). La década juzgada. *Página 12*. Retrieved from https://www.pagina12.com.ar/diario/elpais/1-274879-2015-06-14.html.

Ginzberg, V. (2017, May 30). La identidad es mucho mas que los genes, restituimos historia. *Pagina 12*. Retreived from https://www.pagina12.com.ar/40920-la-identidad-es-mucho-mas-que-los-genes-restituimos-historia.

Goldman, F. (2012, March 19). Children of the dirty war: Argentina's stolen orphans. *The New Yorker*. Retrieved from https://www.newyorker.com/magazine/2012/03/19/children-of-the-dirty-war.

Goñi, U. (2015, June 7). A grandmother's 36-year hunt for a child stolen by the Argentinian junta. *The Guardian*. Retrieved from https://www.theguardian.com/world/2015/jun/07/grandmothers-of-plaza-de-mayo-36-year-hunt-for-stolen-child.

Gutman, D. (2017, August 28). Colombia mira a Argentina para identificar a sus desaparacidos. Inter Press Service. Retrieved from http://www.ipsnoticias.net/2017/08/colombia-mira-a-argentina-para-identificar-a-sus-desaparecidos/.

Hollander, N.C. (1992). Psychoanalysis and state terror in Argentina. *American Journal of Psychoanalysis*, 52, 173–289.

Joyce, C., & Stover, E. (1991). *Witnesses from the grave: The stories bones tell*. Boston, MA: Little, Brown.

King, M. C. (1991). An application of DNA sequencing to a human rights problem. *Journal of Molecular Genetic Medicine*, 1, 117–131.

Lavallén, R. (1985). Interview on "All Things Considered." National Public Radio

Ley No. 24.779. (1997, February 28). *Adopción: Código civil, título IV de la sección segunda: El Senado y Cámara de Diputados de la Nación Argentina*. Retrieved from http://servicios.infoleg.gob.ar/infolegInternet/anexos/40000-44999/42438/texact.htm.

Love overcomes hate. (2017, December 9). *Buenos Aires Times*. Retrieved from http://www.batimes.com.ar/news/argentina/love-overcomes-hate-always.phtml.

Méndez, J. E. (1997). Derecho a la verdad frente a las graves violaciones a los derechos humanos. In *La Aplicación de los tratados sobre derechos humanos por los tribunales locales*. Buenos Aires Argentina: Editorial Del Puerto, 517–540.

Meyer, A., & Ginzberg, V. (1999, March 16). La red del robo de bebés al desnudo. *Pagina 12*. Retrieved from https://www.pagina12.com.ar/1999/99-03/99-03-16/pag03.htm.

Minder, R. (2018, October 3). In Spain's "stolen babies" scandal, doctor Escapes punishment. *New York Times*. Retrieved from: https://www.nytimes.com/2018/10/08/world/europe/spain-stolen-babies-ines-madrigal.html.

National Commission on the Disappeared. (1986). *Nunca más*. London, UK: Faber and Faber.

National Human Genome Research Institute, National Institutes of Health. (2015, March 5). *Bioinformatics*. Retrieved from https://www.genome.gov/25019999/understanding-bioinformatics-and-sequencing/.

Nosiglia, J. E. (1986). *Botín de guerra*. Buenos Aires, Argentina: Cooperativa Tierra Fértil.

Penchaszadeh, V. B. (1992) Abduction of children of political dissidents in Argentina and the role of human genetics in their identification. *Journal of Public Health Policy*, 13, 291–305,

Penchaszadeh, V. B. (1997) Genetic identification of children of the disappeared in Argentina. *Journal of the American Medical Women's Association*, 52, 16–27.

Penchaszadeh, V. B. (2011) Forced disappearance and suppression of identity in children of Argentina: Experiences after genetic identification. In: S. Gibbon, R. Ventura Santos, & M. Sans, eds., *Racial identities, genetic ancestry and health in Latin America* (pp. 213–243). New York, NY: Palgrave Macmillan.

Penchaszadeh, V. B. (2015) Ethical, legal and social issues in restoring genetic identity after forced disappearance and suppression of identity in Argentina. *Journal of Community Genetics*, 6, 207–213.

Qassim, A. (2007, August 3). Broken bonds. *The Guardian*. Retreived from https://www.theguardian.com/lifeandstyle/2007/aug/04/familyandrelationships.family2.

Schneider, P. M. (2007). Scientific standards for studies in forensic genetics. *Forensic Science International, 165,* 238–243.

Skolnick, A. A. (1993). Mitochondrial DNA studies help identify lost victims of human rights abuses. *Journal of the American Medical Association, 269,* 1911–1913.

Snow, C. C., & Bihurriet, M. J. (1992). An epidemiology of homicide: Ningun nombre burials in the province of Buenos Aires 1970 to 1984. In T. B. Jabine & R. P. Claude, eds., *Getting the record straight* (pp. 328–363). Philadelphia, PA: University of Pennsylvania Press.

Snow, C. C., et al. (1984). The investigation of the human remains of the 'disappeared' in Argentina. *The Journal of Forensic Medicine and Pathology, 5,* 297–299.

Tupa, M. (2015, November). Interview with Estela de Carlotto, Buenos Aires, Argentina.

Tupa, M. (2016, January). Interview with Víctor Penchaszadeh, Buenos Aires, Argentina.

Valente, M. (2003, October 1). Argentina: Identity dispute for child of disappeared. Inter Press Service. Retreived from http://www.ipsnews.net/2003/10/argentina-identity-dispute-for-child-of-the-disappeared/.

Vallejo-Nágera, A. (1938). Psiquismo del fanatismo Marxista. *Semana Médica Española, 6,* 172–180.

Vegh-Weis, V. (2017). The relevance of victims' organizations in the transitional justice process: The case of the grandmothers of the plaza de mayo in Argentina. *Intercultural Human Rights Law Review, 12,* 11.

CHAPTER 8

Disappeared, Not Lost

Finding El Salvador's Missing Children

ANDREA LAMPROS, MONTSERRAT MARTÍNEZ GÓMEZ,
CRISTIÁN ORREGO BENAVENTE, AND
PATRICIA VÁSQUEZ MARÍAS

When Salvadoran soldiers formed a ring around their village on August 25, 1982, and began to open fire, Mariá Maura Contreras and Fermín Recinos Ayala grabbed their three small children and ran from their home through the hilly, uneven land (Contreras, 2018). "We started running so that we would be killed running instead of captured," recounted Maura.

Unable to carry all three children, Maura held their four-month-old baby Julia Inés as four-year-old Gregoria Herminia and nearly two-year-old Serapio Cristián lagged slightly behind. Maura looked back to see a soldier grab her daughter Gregoria by the hair. When she stumbled with Julia Inés, a soldier snatched the baby from her arms (Contreras, 2018; Inter-American Court of Human Rights, 2011).

Maura and Fermín lost three children that day, and three children lost their parents and each other. It was a day like other days in the course of the 12-year Salvadoran armed conflict, when the horrors and indignities of war were amplified by the forced disappearance of children from their families. Attacks like this one on Maura's community were part of a "scorched earth" policy intended to target the civilian population that supported the leftist Farabundo Martí National Liberation Front (FMLN).

Maura didn't know then that this act of war wasn't a singular, isolated incident and that scores of other mothers and fathers in El Salvador were experiencing a similar horror. She started searching for her children in the midst of the war, even when it was terribly dangerous to do so. She went to every organization she could think of to seek assistance, including the Red Cross. "But the people closed the door," she said. "Nobody helped me." Nobody listened, she said, until she found "Padre Jon"

(Contreras, 2018). Known more formally as the Jesuit father Jon de Cortina, Padre Jon served in several communities that happened to support the FMLN during the 1980–1992 war. He listened to the stories of grieving loved ones, but soon he would do more.

In 1994, two years after the war's end, with the help of Dutch human rights worker Ralph Sprenkels and a cadre of families who had lost children, Padre Jon established the Asociación Pro-Búsqueda de Niñas y Niños Desaparecidos (the Association in Search of Disappeared Children), commonly referred to as Pro-Búsqueda. The organization soon pulled geneticists and human rights investigators into the slipstream, merging traditional investigative work with modern forensic science in a search for Salvadoran children that would last for decades and continues still.

"IN THE NAME OF GOD, STOP THE REPRESSION"

The attack on Maura's village—two years into the 12-year armed conflict in El Salvador—exemplified the Salvadoran military's tactics to retain control of the country and the socioeconomic status quo by any means they decided were necessary. El Salvador historically had the most unequal distribution of income in all of Latin America, with 78% of farmland in the hands of 10% of the landowners (Bonner, 1984, p. 19). In the 1970s, Salvadoran trade unionists had joined with farmers in a *movimiento popular*, or popular movement, that embraced the Vatican II virtues of "liberation theology," based on the idea that ordinary people did not have to wait for heaven to share in the bounty of the world (Blachman et al., 1986). The long-standing institutional violence and disparity in wealth in El Salvador sparked a grassroots, community-based opposition movement, including the formation of the coalition of leftist guerillas, the FMLN (Blachman et al., 1986). The coalition took its name from the rebel Agustín Farabundo Martí, who had led an indigenous peasant uprising in 1932 and was executed by a firing squad during what is known as La Matanza (the slaughter) (Bonner, 1984). At least 30,000 peasants were killed in the massacre (Bonner, 1984, p. 23).

Nearly 50 years after La Matanza, El Salvador's bloodier and more protracted conflict was triggered on March 24, 1980, when a car stopped outside of the chapel of the Hospital of the Divine Providence in the center of San Salvador. A lone gunman stepped out, rested his rifle on the car door, and shot Salvadoran archbishop Oscar Romero through the heart with a single bullet (Gibb, 2000). The day before, Archbishop Romero, an outspoken critic of the country's oligarchy and its use of death squads, had called on the government to stop its violence: "In the name of God, in the name of this suffering people whose cries rise to heaven more loudly each day, I implore you, I beg you, I order you in the name of God: stop the repression" (Bonner, 2016).

Romero's murder galvanized communities across the country that supported the guerrillas—especially in the rural departments of Chalatenango, Morazán, San Vicente, and San Miguel—and as a result, they would bear the brunt of the military's brutal response. That response included massacres like the one at El Mozote, where approximately 1,000 men, women, and children were killed with guns and machetes in December 1981 (Binford, 1996). It also included the widespread capture and "disappearance" of activists, union members, and students who were suspected supporters of the FMLN.

In the context of the Cold War, the US Ronald Reagan administration moved vigorously to keep the Salvadoran government and military, no matter how unsavory, in power. Receiving approximately $2 million a day in military and economic aid, as well as military training at the School of the Americas (opened in the Panama Canal Zone and later based in Fort Benning, Georgia), the Salvadoran government had strong backing from the United States. In the first year of Reagan's administration, the United States sent $82 million to El Salvador; by 1982, the tiny country was the fourth largest recipient of US aid (Bonner, 1984, p. 12). Reagan's 1980 presidential campaign had hammered incumbent Jimmy Carter for being soft on communism. Except for a handful of diplomatic admonishments, the Reagan and then George H. W. Bush administrations turned a blind eye to human rights and the rule of law (Blachman et al., 1986) in their support of the Salvadoran government's campaign to quash the Left at any cost—seeking to avoid another Cuba or Nicaragua in Latin America.

"I AM ALIVE"

In the 1980s, Jon de Cortina taught engineering during the week at the University of Central America (UCA) in San Salvador and returned to his parishioners on the weekends, splitting his time between towns that supported the FMLN (including San José Las Flores) and towns under military control (including San Francisco Lempa). One weekend in mid-November 1989, Padre Jon was in San José Las Flores when the FMLN launched a major military offensive in the capital. The following Monday, his parishioners begged him not to make the 68-mile drive to the UCA; they felt it would be too risky for him to travel. They also feared the military might in his absence seize the opportunity to attack FMLN strongholds in San José Las Flores, Guarjila, and Los Ranchos, towns that had been made vulnerable when FMLN soldiers left to fight in the San Salvador offensive (Sprenkels, 2001). Padre Jon heeded their warning and stayed in Guarjila.

Four days later, when Padre Jon would ordinarily have been in the capital, the FMLN-controlled station Radio Venceremos brought the chilling news that soldiers from the US-trained Atlacatl Battalion had entered the grounds of the UCA, dragged six Jesuit priests, their housekeeper, and her daughter from their dormitory, and shot them all in the rectory's rose garden—a swift and brutal response to the FMLN's military offensive (UN Truth Commission, 1993). In its broadcast, the radio station erroneously reported that Padre Jon was among the dead. Indeed, he might have been. After hearing the news on the radio many miles from the rectory's rose garden, Padre Jon shouted: "Yo estoy vivo" (I am alive) (Sprenkels, 2001).

Some years later, Padre Jon befriended Ralph Sprenkels, who had come to El Salvador as the war was winding down. Sprenkels was studying Spanish and history at the University of Guadalajara in Mexico when he began working with a solidarity group that supported the FMLN and refugees from the Salvadoran conflict (Sprenkels, 2018). He moved to El Salvador in 1992 and, with Padre Jon's assistance, joined a human rights team that was documenting the experiences of thousands of residents of Chalatenango who had witnessed massacres and other atrocities during the armed conflict.

Padre Jon and Sprenkels were in Guarjila, Chalatenango, in 1992 when United Nations representatives—who were in the country to report on wartime atrocities and help negotiate the peace—first learned that children had been forcibly taken from their families by the military as a tactic of war. For hours in a dusty community center, survivors of attacks in May and June 1982—just a few months before the attack on Maura's village—told their stories about the Guinda de Mayo.[1] Planes had circled overhead, dropping bombs as soldiers flooded their communities, they said. One woman said she lost 51 relatives during the attack (University of Washington, 2013). Others recounted moving through dark, rain-soaked ravines for days on end, clasping their hands over their children's mouths, praying the soldiers wouldn't hear them (SWPictures, 2017). Still others told of crossing the treacherous Río Sumpul and watching some of their friends and loved ones drown along the way as they tried to flee military repression (Sprenkels, 2001).

During the Guinda de Mayo, soldiers moved from house to house, rounding up more than 200 people. Mothers described children being ripped from their arms and watching, distraught and impotent, as soldiers tossed them into military helicopters and sped them away to unknown fates (University of Washington, 2013).

The UN Truth Commission's final report, *From Madness to Hope: The 12-Year War in El Salvador*, released on March 15, 1993,[2] detailed many stories of violence perpetrated during the war, concluding that 70,000 people were killed (a figure later revised to 75,000). The report blamed the government for systematically violating the rights of its citizens and culpability for most of the deaths and other atrocities. Yet curiously, the report did not make any reference to the systematic disappearance of children. Their names were included only in the lengthy list of victims offered in the report's annex (Sprenkels, 2018).

Both Padre Jon and Sprenkels were deeply troubled by the commission's failure to address the question of the El Salvador disappeared children in its final report. This omission may have strengthened their resolve to honor the testimony of survivors by ramping up efforts to find the children lost in the Guinda de Mayo—and perhaps nationwide. As Sprenkels later explained: "We settled on the need to investigate the

---

1. In Spanish, *guinda* means "flight" and *Mayo* is May. This term signifies a retreat or flight from violence that happened in May.
2. Three days after the commission published its report, President Alfredo Cristiani signed into law the General Amnesty Law for the Consolidation of Peace, which gave amnesty to all those who had engaged in political crimes between 1980 and 1992 (Orrego Benavente, 2014).

kidnapped children because we felt . . . these children might be alive and that the family members were really committed to [finding them]. It was not a hard choice for us."

FINDING THE FIRST CHILDREN

Two years after this testimony, in 1994, Padre Jon followed his first clues to the whereabouts of children from the Guinda de Mayo.

A man who worked near the Aldeas Infantiles SOS orphanage (SOS Children's Village) outside of San Salvador told him that several children he suspected could be from Chalatenango had been brought there by women volunteering with the Salvadoran Red Cross. Like other orphanages in El Salvador, the SOS Children's Village was organized into group homes in which about 10 children lived together with a woman who was hired to be their "mother" (Sprenkels, 2018). While passing by the orphanage one day, the man had noticed a disabled girl he believed to be his cousin, Andrea, who had lost an arm during a bombing of Chalatenango in 1981.

Intrigued by this news, Padre Jon sent Sprenkels in the town's community van—packed with family members from Guarjila—to visit the orphanage. The community had lost many children in the attack (the exact number of killed or forcibly disappeared has not been determined), and they had no idea how many they might find at the orphanage. "Everybody wanted to go," said Sprenkels. Although they desperately wanted to see the children, the families agreed to meet only with the orphanage director and other staff and to review the facility's files. After the meeting, the families were convinced that these were their children—stolen years earlier during the Guinda de Mayo.

After visiting the SOS Children's Village himself, Padre Jon arranged for the children to visit Guarjila 10 days later, coincidentally on the second anniversary of the signing of the Salvadoran peace accords (Sprenkels, 2001). Sprenkels wrote about this first reunion:

> The young people got out [of the microbus] with their heads down. They started to look out at the multitude of people watching them. The silence lasted on a fraction of a second. It was interrupted by the screams, the running, the hugs, and the tears: My son! My daughter! All of the families rushed at once. . . . Padre Jon and everyone present cried upon hearing the first cries of reunion. (Sprenkels, 2002, p. 88)

For some of the children, the *reencuentro*—the reunion of families separated during the war—was emotionally difficult, especially as the managers of the orphanage had told them their parents were dead or communists who had abandoned them (Sprenkels, 2002). For others, it was a mixture of pure joy and disbelief.

Andrea, the teenager first spotted at the orphanage, spent time looking through her family's belongings after the reunion: the kitchen, the horse, the chairs, and finally the family photo album. She saw herself in these images, and childhood memories trickled back (Sprenkels, 2001).

While powerful, this first reunion begged for more than anecdotal evidence. Some of the families that survived this hardship, especially those who had been separated when their children were very young, needed scientific inquiry to determine *and believe* the kinship.

DNA TYPING AND THE SEARCH FOR EL SALVADOR'S MISSING CHILDREN

In summer 1994, following this first *reencuentro*, Padre Jon, Sprenkels, and several relatives of missing children visited the offices of Reed Brody, a human rights lawyer and UN official monitoring the implementation of El Salvador's peace accords (Riordon, 2014). At the meeting, Padre Jon asked if Brody could find a way to conduct DNA testing to match children with their parents.

That same day, Brody called Eric Stover, at the time executive director of Physicians for Human Rights in Boston, Massachusetts. Since the mid-1980s, Stover and his colleague Cristián Orrego Benavente, a coauthor of this chapter who died in 2018, had overseen the engagement of geneticists in the search for hundreds of children who had disappeared during Argentina's military dictatorship in the 1970s. Stover then tapped Cook County medical examiner Robert Kirschner to help with testing.

The first child to be tested with DNA was a boy named Juan Carlos, who was believed to have been abducted from the Guinda de Mayo as a baby and was found in the SOS orphanage. At an emotionally charged press conference in San Salvador in 1995, Padre Jon announced that geneticists in the United States had found a connection between Juan Carlos's DNA and that of Maria Magdalena Ramos, a woman who had lost her baby in the Guinda de Mayo and thought this quiet boy could be hers. News of the match filled headlines around the world as the first use of DNA confirming the forced disappearance of Salvadoran children during the armed conflict (Crossette, 1995).

Along with Padre Jon, founder of Pro-Búsqueda, Stover and Kirschner then received permission from the head of the SOS orphanage to extract blood samples from more children. They traveled to Guarjila, where they slept in hammocks in a house the priest shared with a community member (Stover, 2018).

In a radio interview broadcast throughout the community, Stover announced that to help establish the children's identities, he and Kirschner would be at the local church the next day collecting blood samples from relatives who suspected that their children had been taken during the Guinda de Mayo. Early the following morning, they laid out their instruments for extracting blood on the church altar and waited. Within an hour, family members streamed into the church. Some walked, while others arrived on horseback, winding their way down the dusty roads that crisscrossed the surrounding hills. Once they were seated in the church, Padre Jon explained how DNA testing would be performed and that anyone wishing to have blood drawn would need to sign a consent form. He also warned that finding a match to a missing child could take months or years—or never occur at all. In many of the Guinda de Mayo cases, the consistency of the children's names from birth as well as the strong family resemblance was enough; the DNA test would just add certainty.

"WE CAN DO THIS!"

The tiny staff of Pro-Búsqueda (which originally consisted of only Padre Jon and Sprenkels) expanded in 1995 to include family members of missing children who, much like the Grandmothers of the Plaza de Mayo in Argentina, turned themselves into skilled investigators. Pro-Búsqueda's early years focused on gathering testimony from relatives about the missing person, the disappearance, family members, and the person opening the case. Five people on average were interviewed about each case, and occasionally family reference samples were taken for DNA analysis (Orrego Benavente, 2014).

The Pro-Búsqueda team soon learned that not all of the children brought to their attention had been kidnapped. Rather, some mothers, who had been desperately impoverished or made homeless by the war, had given up their children to adoption for, they hoped, their children's own safety and well-being. Some thought they had given up their children temporarily, only to realize later that the adoptions were irreversible and that their children were nowhere to be found (Martínez Gómez, 2018). In some cases, families had given their children to FMLN members who operated "security houses" in the capital. The children would essentially be human camouflage for the FMLN operatives, creating the façade of an ordinary family while spying on the activities of the military (Sprenkels, 2018).

As news of successful reunions began to spread across the country, more testimonies and blood samples arrived at the offices of Pro-Búsqueda in San Salvador.[3] It became apparent that the organization needed to establish a data bank

---

3. Pro-Búsqueda supported one of the first legal actions from the families of the disappeared: the case of the Serrano-Cruz sisters, victims of the Guinda de Mayo (Parmar et al., 2010). This case bolstered the use of forensic science in the search for the disappeared. María Victoria Cruz Franco, mother of Ernestina and Erlinda Serrano Cruz, first filed a criminal complaint against the Atlacatl Battalion in the Chalatenango Trial Court shortly after the UN Truth Commission's report was released in 1993 and then with the Salvadoran Supreme Court in 1995 (Inter-American Court of Human Rights, 2005). Unsuccessful in national courts, Cruz and Pro-Búsqueda eventually took the case to the Inter-American Court and won a victory. The court required the Salvadoran government to conduct criminal investigations against alleged perpetrators in the disappearance of children, create a website to trace disappeared children, and build a genetic data system to aid in the search and finally to establish a national commission to find disappeared children (UNICEF, 2010). Although the ruling was issued in 2005, the Salvadoran government only began to comply under President Mauricio Funes, a member of the FMLN who was elected in 2009. Funes took responsibility for the crimes the military had committed decades earlier and issued a formal apology (Reuters, 2010). Following a later Inter-American Court ruling in 2011, the government provided reparations to the families of the victims

for the storage of the family reference samples and to develop a standardized process for collecting those samples from young people who believed they might be among El Salvador's missing children. Many of the children had been adopted in El Salvador by US and European couples; all told, researchers found that children were sent through forced or irregular adoptions to 11 countries (Orrego Benavente, 2014).

By 2003, Orrego had become a trainer in forensic genetics at the California Department of Justice's Richmond DNA lab (Cal DOJ), now called the Jan Bashinksi DNA Lab, in Richmond, California. As part of his job, he helped to educate his colleagues about global issues related to the forensic sciences. He asked Stover, now with UC Berkeley's Human Rights Center, to speak to the California DOJ staff about the work of the International Commission of Missing Persons (ICMP) and its use of DNA typing to identify the remains of victims exhumed from mass graves in the aftermath of the recent wars in the former Yugoslavia (see chapter 9). During his presentation, Stover mentioned the work of Pro-Búsqueda and its quest to reunite families torn apart by the war, and the possibility of developing a DNA database to aid in the effort (Gima, 2018).

"We can do this!" said Lance Gima, then director of Cal DOJ, recounting his response to the presentation. During the 1990s, Gima had already led the development of a groundbreaking DNA database that collected reference samples from convicted felons in California and compared them to samples from crime scenes. Gima wanted to help with the El Salvador effort but knew he couldn't simply use the Richmond lab and those among the staff willing to volunteer after hours for the project without getting approval from California's attorney general at the time, Bill Lockyer.

Gima and Orrego met with Lockyer's staff and received the green light for lab staff to volunteer on nights and weekends to test Pro-Búsqueda's DNA reference samples. One of the first to volunteer was Brian Harmon, then a 32-year-old criminalist and DNA analyst who had just begun his career at the laboratory (Harmon, 2018). Using vacation time, in 2004 Harmon traveled to El Salvador with Orrego and other scientists from the lab to meet with the Guinda de Mayo families and explain how the Richmond lab would collect and analyze their reference samples. The scientists explained how the process worked and that family members would simply have to use a tongue depressor and swab their cheeks "ocho veces (eight times) to gather the DNA.

Working into the night and on weekends back in Richmond, the California DOJ scientists punched the paper at the end of plastic tongue depressors (which contained the saliva sample) onto a plastic plate that could hold up to 84 samples. They soaked the samples in chemicals, some of which would burst open the cells. They then used an enzyme to digest the proteins in the sample and spun the samples quickly in a centrifuge. This served to separate the components, with the heavy parts sinking to the bottom and lighter parts (such as the DNA) floating.

and publicly acknowledged its involvement in those crimes. As part of the reparations program, plaques were unveiled in the presence of the six victims—including Maura Contreras's children—in schools and community centers where the children had been abducted (Contreras, 2018; Inter-American Court of Human Rights, 2011). Funes, however, did not open the military archives that could have shed more light on the abductions (Associated Press, 2013).

Before analyzing the purified DNA, the DOJ scientists would copy it multiple times using polymerase chain reaction (PCR) (see chapter 1) because the instruments weren't sensitive enough to analyze low quantities of DNA (Calloway, 2018; Harmon, 2018). They would determine the genetic profile by using short tandem repeats, or STRs, a framework to analyze an individual's DNA and create a profile. The results would then be analyzed with a unique tool developed by forensic mathematician Charles Brenner called DNA-View, a system that performs statistical analyses and DNA comparisons for kinship using these genetic profiles (Dreifus, 2000).

The California DOJ team worked with pioneers like Brenner and also with Berkeley scientist George Sensabaugh, Oakland-based scientist Henry Erlich, and other leaders in the field to make the best use of forensic DNA for investigations of missing children. "If we can determine who these children are and where they are from, it becomes a fact," Harmon has said. "It [doesn't] in itself create justice. [But] it's that piece that, as a scientist, I thought, I could help. . . . It spoke to my core beliefs."

After the creation of the DNA database, Pro-Búsqueda set out to populate it with data—as many family reference samples as possible. These samples were collected from the closest biological relatives that could be found of the disappeared person and were stored together. The number of family members required to give samples for any particular case would depend on the relationships within the family, with every case being different. Pro-Búsqueda would use a point system with each type of relative (parent, sibling, aunt, uncle, etc.) being worth a different value depending on the degree of kinship. The investigators would strive for a 10. A parent would be worth 10 points, while a grandparent was only worth 2 because of the limited amount of DNA shared and the corresponding statistical confidence (Orrego Benavente, 2014).

The newly created database meant that Pro-Búsqueda would be changing the way it used DNA information, Harmon pointed out: "What was proposed was something more radical: instead of using DNA at the end of the process [to confirm kinship that had already been suspected], it would be used at the beginning."

To aid in the DNA collection process, UC Berkeley/UC San Francisco Joint Medical student Elizabeth Barnert spent the summer of 2005 in El Salvador as a UC Berkeley Human Rights Center fellow supporting Pro-Búsqueda in collecting reference samples. Just 25 years old at the time, Barnert traveled around the country with Pro-Búsqueda investigators, who took buccal swabs while she documented the experiences of family members whose children were missing. In all, the team took 600 samples at that time. Thanks to Cristián Orrego, Barnert also carefully documented the *chain of custody*, the chronological documentation or paper trail that records the sequence of custody, control, transfer, analysis, and disposition of physical and electronic evidence (Barnert, 2018).

Once back in Berkeley, Barnert was anxious to hand off to Orrego the samples she had carried in her black duffel bag. "I wanted him to come right away," said Barnert. "I had the feeling that these swabs were so incredibly precious. What if my house burned down?"

Following the creation of the DNA database, the Californians and Salvadorans realized it was time to shift the forensic expertise from North to South (Gima, 2018).

This effort, however, would be completely dependent on securing significant funding and close cooperation between Pro-Búsqueda and UC Berkeley's Human Rights Center. Unlike Argentina, where the civilian government of Raul Alfonsín had established a national genetic bank to aid the work of the Grandmothers of the Plaza de Mayo, the Salvadoran government had done nothing to assist Pro-Búsqueda.

PRO-BÚSQUEDA'S FIRST FORENSIC GENETICIST

Transfer of scientific know-how from North to South gained momentum during a visit by Padre Jon to the Richmond lab in 2004. Together, he and Orrego mapped out a plan to relocate the DNA effort to El Salvador and hire a forensic geneticist to oversee its operations. "Cristián was the one who made [it] happen," Barnert recalled. "He was laying the groundwork for it. [And] from a distance, I watched that process succeed."

UC Berkeley's Human Rights Center, led by Eric Stover, launched a Forensic Program and hired Orrego to lead it. The program was dedicated to supporting the work of Pro-Búsqueda in finding and documenting the cases of forced disappearances. A Chilean who had moved with his family to Bloomington, Indiana, in 1961, Orrego brought both scientific expertise—he had studied biochemistry at Brandeis University—and a deep human commitment to the effort. In talking about his passion for finding El Salvador's lost children and reuniting families, Orrego told an interviewer: "I only have to think of the strength and determination of the families, who carry on this struggle for decades in the face of so much official indifference, greed, and laziness" (Riordon, 2014).

In 2007, the Human Rights Center's Forensic Program, under Orrego's leadership, won the first of two grants from the US State Department to establish a DNA database at Pro-Búsqueda and to support the work of a physician and forensic geneticist. "Finally, it had come down to a true partnership," said Stover.

The funding secured by the Human Rights Center enabled Pro-Búsqueda to hire Patricia Vásquez, one of only three trained forensic geneticists in El Salvador. Vásquez, a coauthor of this chapter, graduated with a medical degree in 1999, then worked briefly at a free community clinic in San Salvador. Although she liked caring for people, she wanted to pursue her dream of studying "legal medicine," a goal that eventually took her to Spain, where she earned an advanced degree in forensic genetics in 2003. Her mother had attended Archbishop Romero's funeral, along with about 250,000 fellow Salvadorans, in the Plaza Libertad where dozens had been killed killed (Miglierini, 2010; Vásquez, 2018). As university students, Patricia and her brother, Mauricio, had been on campus when the military occupied the National University. And during the FMLN's offensive in the late 1980s, the war came into her San Marcos neighborhood as tanks filled the streets. As military planes dropped bombs nearby, Vásquez and her brother hid under the bed.

After returning to El Salvador, Vásquez taught science at a private university in the capital; there, a colleague heard her lecture about forensic genetics and mentioned an opening at Pro-Búsqueda. "This was new for me," Vásquez later recalled. "I thought

I would only use [DNA analysis] for paternity or criminal investigations. I had no idea of the human rights applications."

Orrego, a member of Pro-Búsqueda's interview committee, and the other committee members were immediately taken with Vásquez and offered her the position. (Indeed, Orrego was so taken with Vásquez that he asked her to marry him a few years later.)

Vásquez took up her post at Pro-Búsqueda with some trepidation. After all, she was now in charge of a DNA database containing more than 350 family reference samples and would deal with families with high expectations. To help with the transition, Vásquez was given training at the Richmond DNA laboratory and maintained ongoing consultations with lab scientists there. But still the work was challenging. In Spain, with one biological sample, she had only been looking at one genetic marker (called a genetic locus) at a time. With Pro-Búsqueda, she had to analyze 14 different genetic markers in one biological sample. In spite of the challenge, the work of reuniting families, Vásquez found, could be thrilling.

PRO-BÚSQUEDA TODAY

Today Pro-Búsqueda is a bustling community organization, based in a modest office in San Salvador and comprised mainly of people who lost family members during the armed conflict. The armed conflict may have ended decades ago, but there is a palpable urgency to the work and a dogged commitment among the staff. At midday, one of the mothers, Inés, who gave up three children under duress during the war, provides a simple lunch for staff members. The organization serves as a hub for all types of grassroots meetings: one day representatives of the Mothers of the Disappeared are visiting, and on another, members of the newly formed government commission for the El Salvador disappeared. The walls are adorned with larger-than-life photos or posters of Padre Jon, who died in 2005 after suffering a stroke.

Six family members—five mothers and a father—joined Padre Jon and Ralph Sprenkels to formally launch Pro-Búsqueda in the summer of 1994. One of them was Margarita Zamora, who at the age of 17 lost four siblings and her mother, Elva Josefina, in the Guinda de Mayo. "We don't say 'lost,' we say 'disappeared,'" explains Zamora, who believes at least one of her siblings could be alive. "In reality, they were not *lost*. They were *disappeared*" (Zamora, 2018).

According to one of her colleagues, Zamora, who is Pro-Busqueda's lead investigator, is "like Wikipedia" because she seems to remember every detail of every case she's handled. Although Pro-Búsqueda's methodology has evolved over time, Zamora details the three constant components: testimonial, documentary, and scientific evidence. "One is not more important than another," she says (Zamora, 2018).

In the early years, the investigators relied almost entirely on testimonial evidence, interviewing family members to gather details about their missing children. As they gained greater access to orphanages and government offices, they began combing through files to supplement their interviews. At times they have been denied access to certain government documents.

Eduardo García, the current director of Pro-Búsqueda, is a Spaniard from Cádiz, Andalusia. When he told his mother he wanted to go to El Salvador to act in solidarity with the people struggling for human rights instead of finishing college, she said absolutely not. So he finished college and then came to El Salvador in 1993, soon after the war had ended. He never went back.

García details the steps Pro-Búsqueda investigators take: from searching the topography of the area where a child was kidnapped (even though it's been decades since the crime) with the aim of finding every clue possible to explain what transpired, to working with family members, hospital administrators, and local government officials to obtain birth certificates and relevant adoption records. There were reportedly some 30,000 adoptions during the armed conflict, and tracing records of an individual adoptee can be quite daunting. Prior to the use of forensic genetics, Pro-Búsqueda had located some of the missing children and reunited them with their families, but doubts always remained (García, 2018). "Now we are able to reconstruct what happened with scientific evidence," said García. "Before this there wasn't always certainty—especially because a lot of the documents are fake."

Efforts to bolster the DNA database with family samples have contributed to a more scientific and proactive approach by the organization, García said. With help from Berkeley's Human Rights Center, which launched an international outreach campaign to find the Salvadoran children who had been adopted in the United States, Europe, Canada, and other Central American countries, the database was infused with family samples (of grandparents, parents, aunts, uncles, and children). Under the leadership of Forensic Project director Cristián Orrego, the Human Rights Center designed and posted posters with a child's photo and the words "Donde están?" (Where are they?) in immigrant community centers and US consulates. They created radio ads that aired in cities around the country and established a texting "hotline" for anyone who wanted immediate information about how they could provide a DNA sample and seek their loved ones.

García tells the story of a young man (we'll call him Salvador) who requested DNA testing at Pro-Búsqueda's office after hearing about the organization on the radio. Salvador explained that he suspected he had been adopted during the war and raised by a military soldier, but he wasn't sure. He had birth records that seemed to indicate that he was the biological son of the soldier. However, once tested, Salvador's DNA sample showed a strong connection to other family members in Pro-Búsqueda's data bank—people unknown to him—who had contributed DNA samples in the hopes of one day finding their lost loved one. When Pro-Búsqueda's investigators looked deeper into the case, they discovered that Salvador's birth records had been falsified and were actually the papers of the soldier's biological son, who had died early on in the armed conflict. Salvador had been given the deceased boy's identity.

In this case of falsified documents, testimonial and scientific evidence was critical to discovering the truth (García, 2018). "There are records, but the question is: are they right?" said García. Rural families will often have no birth certificates or paper trails to go on. Young people who know they are adopted will give DNA samples to the database in hopes of a match with a family already in the records. Testimony and often documentary evidence in these cases are much less important than the hard science (García, 2018).

Pro-Búsqueda estimates that at least 994 Salvadoran children were abducted and given up for illegal adoption during the armed conflict. Hundreds—and possibly thousands—more were given up under duress because of the war. To date, of those abducted and given up for illegal adoption, 443 have been found in Central America, the United States, and Europe, while more than 551 are yet to be located. During El Salvador's armed conflict, the United States issued more than 2,300 adoption visas to American citizens, allowing them to bring Salvadoran children to the United States (Physicians for Human Rights, n.d.). However, nobody knows how many of these adoptions happened illegally or under irregular or coerced circumstances. Because of privacy laws, the US State Department (which oversees its nation's international adoptions) will only give adoption details to those who are directly affected and who make inquiries themselves and does not proactively inform parents about the possibility of irregular adoptions (Orrego Benavente, 2013).

Long after the end of the armed conflict, the question persists: Why are parents still searching for their missing children, and why are missing children still searching for their biological families? Montserrat Martínez Gómez, an investigator who has worked with Pro-Búsqueda for nearly a decade and is an author of this chapter, says on the children's side, it's because the children, now young adults, want to know the story of why they were given up for adoption. Adoptees are also driven by a curiosity to meet their biological families and to know that they are safe. It is not as common, however, for adoptees to seek out their parents because they want to know their genetic history (Martínez, 2018).

A Pro-Búsqueda investigation usually begins with a family member who believes a young relative was abducted or given up for adoption at the time of the conflict. The family member is interviewed and then asked to provide a DNA sample (Orrego, 2014). A DNA kit is dispatched, and the person is asked to swab the inside of his or her cheek (for as long as it takes to sing the verse of a song) and send it back to Pro-Búsqueda. If the query comes from a US resident, the kit is returned to UC Berkeley's Human Rights Center. Whenever possible, Human Rights Center staff will hand-deliver samples to Pro-Búsqueda.

Once a case is opened, Pro-Búsqueda's investigators search for whatever documentary evidence they can find. They scour their own digital database for birth certificates and adoption records, for example. Often an investigator will go to the court and ask for the child's file. When it is supplied, it can be an invaluable resource with a wealth of information about where the child was found and even the name given by the birth parents and a summary of circumstances (Martínez Gómez, 2018).

Sometimes the files themselves reveal vast irregularities. In one case, some 150 children were registered as being the children of a single man (García, 2018). In another case, an adopted boy's file showed two different birth names. After tracing evidence to a woman who was believed to be the boy's mother DNA testing showed that the child, now a young adult, was not, in fact, her biological son.

Even when DNA analysis is done, it can be hard to prove kinship. Sometimes investigators will have a 90% probability of a match but cannot say with certainty that the parties are connected. Usually it's because the DNA profiles being compared are from a grandparent to a grandchild or sibling to sibling or without the DNA from the father. Pro-Búsqueda seeks a 99.9% probability to conclude that two samples are genetically related. It can be hard for families to understand that more DNA samples and a wider analysis are needed to hit that high percentage. "Sometimes it can be difficult to balance the investigation in the field with the DNA process," said Martínez.

When Pro-Búsqueda makes a match, it's exciting for everyone involved. But it's not the end of the story and certainly not closure either for the children or their families (García, 2018). "You are dealing with intense stories, related or not with the war," said Martínez. "Every case is intense. You can see and [sense] a lot of feelings on [both] sides. You can be close to pain and love and hate. It's very intense."

THE COMPLEXITIES OF FAMILY REUNIFICATION

While the use of DNA to reunite families can bring answers and certainty, it can be fraught with complexity. The experiences of families that lost children and adoptees can be vastly different, as life experiences can be divided by culture, economics, and language barriers. Moreover, some of the children were raised by the soldiers responsible for killing members of their birth families.

The study "Long Journey Home: Family Reunification Experiences of the Disappeared Children of El Salvador," conducted between 2005 and 2009, interviewed 26 young adults who had been reunited with their biological families after irregular or forced adoptions in El Salvador. The study unveiled the complexity and difficulty of family reunions for the children (Barnert, 2015).

Pro-Búsqueda has always paid attention to these psychosocial factors, attempting to manage the expectations of everyone involved. When a DNA match is made, the organization contacts the adoptees to let them know that their biological family has been found. They do not put pressure on them to communicate with their biological families but rather provide a path via the organization for them to do so. And for the family members who lost children decades ago they help to manage expectations. Pro-Búsqueda staff broker any relationship between the children and parents and do not put them directly in touch unless the children consent. Some missing children, now young adults, may not want to meet them, either virtually or in person, or may want to connect only once with their biological families and never be in touch again. Some, especially those living in Europe and the United States, may not even speak Spanish.

Rebeca Durán was an eight-year-old girl with four siblings while on the move with her family during the *guindas* of the early 1980s. "The military was bombing us and chasing us," she recalls. She says her mom became very ill and had to leave the countryside for the city with her younger siblings. After years apart and desperate to see her mother, Durán got news of her mother's existence from a businessman who would sell things in her village. She went with him back to San Salvador and found her mother in a hovel and with only her little sister, not her two brothers. Her mother, María Inés, had given the boys to a lawyer, who said they would be returned after six months. The boys were sick at the time, and she wanted them to receive care. She didn't realize that she might never see them again (Durán, 2018).

The boys—now men named Álvaro and Andrés—were adopted by two couples in Italy and later located by Pro-Búsqueda. Only one of them, Andrés, after many years of refusing to do so, consented to meet. Rebeca says that although she and her siblings write long messages to Andrés, they only receive short, sporadic replies. Still, it's better than nothing, she says. "There's a lot of pain, much sadness, much frustration, but also much happiness," says Durán. "With my brother, we know where he is, where he lives, that he exists." Unfortunately, she says, her other brother doesn't want to be in touch.

Angela Fillingim, now a professor at Western Washington University and the mother of two, offers the equally complex perspective of an adoptee. She was a UC Davis student when she visited El Salvador for the first time in 2005 and had a chance meeting with Ester Alvarenga, who worked at Pro-Búsqueda at the time and was later its director. Raised by a family in Berkeley, California, Fillingim knew she was adopted but was never on a quest to find her biological family. Alvarenga's persistence, along with seeing copies of her adoption papers and meeting Human Rights Center fellow Elizabeth Barnert, led Fillingim to investigate her birth and adoption. Pro-Búsqueda investigators followed geographic leads, took DNA samples, and found members of her biological family. It turned out that Fillingim's mother, Blanca, who worked in a sweatshop during the war, had given her up as a means of protecting her from the violence.

Not all of the Salvadorans shepherding her through this process fully understood the additional complexities of international adoptees, she says. Because she and others like her didn't live through the war in the same way as those raised in El Salvador, the often casual conversations about massacres, torture, and kidnapping were especially shocking; the war had been normalized. Fillingim recalls an occasion before the DNA testing when she overheard a Pro-Búsqueda staff person saying that Fillingim's family members had probably been executed because they lived in the conflict zone of Suchitoto, Cuscatlán.

Fillingim describes the lead-up to and day of the *reencuentro* in the town of Ilobasco as "weird," "surreal," "beautiful," and "the most awkward blind date." She kept thinking as she met her birth mother and relatives, "What would my life have been like [for them]?" Because a PBS *Newshour* crew filmed the reunion, it was somewhat distinct from other *reencuentros*. But she was supported by one of Pro-Búsqueda's psychologists, who had managed her expectations and provided "emotional scaffolding," as well as by others from her Berkeley community. The Human Rights Center's Eric Stover, Cristián Orrego, and Barnert, who had become a friend, traveled to El Salvador with her.

Fillingim returned to El Salvador a few years after the *reencuentro* to work with Pro-Búsqueda and build relationships with her biological family members— relationships that were sometimes complicated by money. Her birth mother, who died seven years after the reunion, often asked for money, apparently assuming that because Fillingim was raised in the United States, she must be wealthy. This was far from the case. Despite such complications, Barnert has commented, "Through the process, Angela came to better understand her identity, which I think brought her a lot of peace and resolution."

MAURA'S STORY

And what of Maria Maura Contreras, Fermín Recinos Ayala, and their three children, stolen from them during the attack known as La Conacastada, the description of which begins this chapter? Maura's oldest daughter, Gregoria Herminia, recounted the night of her disappearance in Inter-American Court testimony:

> They captured us and told me to look after my little sister, and they asked me: What about your parents? And I told them that they were there, and then they followed them and told me that they had killed them.... That was really hard because it was something that I didn't want to hear because I loved my parents.... The day that we camped was the last time [I saw my siblings].... I told them not to separate us, but they didn't want to leave them with me; they didn't want us to be together. (Inter-American Court of Human Rights, 2011)[4]

Gregoria later told the media that she was taken by a Salvadoran soldier who physically and emotionally abused her (Associated Press, 2013). "That soldier stole everything from me," Contreras told the Associated Press. "He took away my parents, he took away my siblings, he took away my identity. I couldn't live like a girl because he never gave me the love of a father and he was always abusing me, even raping me." Years later, Gregoria ran away from that soldier's home and lived with a member of his family in Guatemala (Contreras, 2018).

Maura was among the first family members to work directly with Pro-Búsqueda in the search for disappeared children. When she heard about the families of the Guinda de Mayo, she tracked down Padre Jon to ask if there could be other children and parents who had suffered a similar horror. In the course of her detective work, Maura interviewed a soldier with knowledge of the attack on her community, who confessed that her daughter Gregoria Herminia was taken by the military. He confirmed what Maura felt in her heart: her oldest daughter was alive.

Maura learned Gregoria had been taken to Guatemala. She petitioned the Guatemalan officials for information and eventually found her girl. Maura was reunited with Gregoria in 2006, twenty-four years after her disappearance. DNA extracted from mother and daughter confirmed their relationship. Sadly, the reunion happened a year after Padre Jon died.

---

4. The Inter-American Court of Human Rights ruled in 2011 in the case of *Contreras et al v. El Salvador*: "Invoking statements it had made during a hearing held before the Inter-American Commission on Human Rights in another case, the State acknowledged that, 'in the context of the armed conflict that took place in the country between 1980 and 1991, there was a systematic pattern of forced disappearances of children and adolescents in different areas, especially in those most affected by armed combat and military operations.'"

The stunning finding of Serapio Cristián involved one of the most exciting moments of the geneticist Patricia Vásquez's career. While poring over data on her computer at the Pro-Búsqueda office in 2012, Vasquez discovered a rare familial *match*,[5]—a link between one person's DNA and another indicating a genetic relationship with a high degree of certainty.

Vazquez had unexpectedly matched the DNA of Maura Contreras with the second of her three children, Serapio Cristian. Both mother and son had given DNA samples to the Pro-Búsqueda database. This rare "cold hit" was the first of its kind in the group's work.

Serapio Cristián, who grew up as Mario Ulises Carballo, had been taken by a military family and raised by an officer who had participated in the 1982 attack on his village. Unlike his sister Gregoria, he was not mistreated and maintained a relationship with the family that raised him. Ironically, he grew up a short distance from his biological parents.

The DNA match discovered by Vásquez led in 2012 to a joyful *reencuentro* with Serapio Cristián, who embraced his mother Maura and his sister Gregoria Herminia in front of their community after 30 years of separation. His biological father, Fermín Recinos Ayala, had died before he could be reunited with his son.

While Maura expresses gratitude for finding two of her three children so many years after their violent disappearance, her voice falls when she mentions the baby, Julia Ines. After so much time and searching, this child, who would be in her thirties as of this writing, still has not been found.

PRO-BÚSQUEDA'S LIVING LEGACY

Now decades since the end of the armed conflict in El Salvador, Pro-Búsqueda's DNA database can help reveal the truth about the disappeared children, even if parents and other family members have died. Long delayed by politicians of all parties who feared reopening of wartime wounds and potential efforts to hold the state accountable, and sidetracked by El Salvador's brutal gang war, the creation of the Comisión Nacional de Búsqueda de Personas Desaparecidas en El Salvador (National Commission on the Search for Disappeared Persons, or CONABÚSQUEDA) in 2017 may lead to the investigation of thousands of adults who disappeared during the war. The 1993 UN Truth Commission report noted at least 5,000 disappeared, but human rights organizations have documented the number to be closer to 10,000. Investigations by the national commission may lead to the exhumation of mass graves of the disappeared throughout the country and in turn to the application of advanced DNA methods to identify the human remains recovered from these sites.

In 2018, Pro-Búsqueda investigators, a student from UC Berkeley, and a student from UC Davis traveled throughout El Salvador to lay the groundwork for a new

---

5. In the context of identifying relatives, *match* refers to evidence of relatedness and not the idea that the two samples have identical profiles.

DNA project commissioned by Cristián Orrego and scientists at Children's Hospital Oakland Research Institute (CHORI)—what Orrego called the Population Registry of Forensic DNA of El Salvador. This registry is intended to describe the genetic composition, for forensic purposes, of Salvadorans, but not to be a population database of relatives of the disappeared.

Scientists say this will be an important resource for Pro-Búsqueda and many others interested in using forensic genetics to investigate human rights violations from the distant and not-so-distant past. Given the crisis on the US/Mexico border that unfolded in 2018 with the separation of families—many of them from El Salvador—this type of database could become a critical resource for reuniting families separated for three months or three decades.

Pro-Búsqueda continues to search for—and find—children separated from their families during the war. Investigator Zamora says that Pro-Búsqueda receives on average 10 to 20 new cases each year. Each year, as the war recedes further into the past and the parents and grandparents who lost children and grandchildren die, the search becomes ever more challenging, and the need for advanced uses of DNA becomes pressing to solve cases. But as Zamora and other Pro-Búsqueda staff explain, the use of scientific evidence to reunite families separated during El Salvador's armed conflict is just one function of human rights organizations. Another function is to commemorate and thereby permanently implant the facts and horrific pain of those facts in the collective memory.

To this end, every year Pro-Búsqueda commemorates massacres like the Guinda de Mayo, in which mothers literally had their children ripped from their arms. Family members come from surrounding towns and hamlets and form a procession, walking the land where they lost loved ones and picking flowers to float in the Río Sumpul to represent the children and adults who drowned trying to escape.

ACKNOWLEDGMENTS

The authors wish to acknowledge the tremendous assistance Ralph Sprenkels provided us as we researched this chapter. Ralph died on September 14, 2019. We will always remember him for his enduring commitment to human rights in El Salvador. We also thank Lili Spira, a UC Berkeley student, for her research and translation assistance.

REFERENCES

Associated Press. (2013, February 22). Soldiers stole children during El Salvador's war. Retrieved from https://www.usatoday.com/story/news/world/2013/02/22/soldiers-children-el-salvador/1940533/.

Barnert, E. (2018, April). Interview with Elizabeth Barnert, assistant professor of pediatrics, David Geffen School of Medicine, University of California, Los Angeles, by Andrea Lampros.

Barnert, E., Stover, E., Ryan, G., & Chung, P. (2015). Long journey home: Family reunification experiences of the disappeared children of El Salvador. *Human Rights Quarterly*, 37(2), 2015.

Binford, L. (1996). *The El Mozote massacre: Anthropology and human rights*. Hegemony and experience. Tucson, AZ: The University of Arizona Press.

Blachman, M., Leogrande W., & Sharpe, K. (1986). *Confronting revolution: Security through diplomacy in Central America*. New York, NY: Pantheon.

Bonner, R. (1984). *Weakness and deceit: U.S. policy and El Salvador*. Times Books.

Bonner, R. (2016, April 15). Time for a US apology to El Salvador. *The Nation*. Retrieved from https://www.thenation.com/article/archive/time-for-a-us-apology-to-el-salvador/.

Calloway, S. (2018, May). Interview with Oakland Children's Hospital Research Institute scientist Sandy Calloway by Andrea Lampros.

Contreras, M. (2018, June). Interview with Maura Contreras, mother of three children taken during the attack at La Conacastada, by Andrea Lampros.

Crossette, B. (1995, January 21). DNA test reunites Salvadoran mother and child. *The New York Times*. Retrieved from https://www.nytimes.com/1995/01/21/world/dna-test-reunites-salvadoran-mother-and-child.html.

Dreifus, C. (2000, August 8). A conversation with Charles Brenner: A math sleuth whose secret weapon is statistics. *The New York Times*. Retrieved from https://www.nytimes.com/2000/08/08/science/conversatipon-with-charles-brenner-math-sleuth-whose-secret-weapon-statistics.html.

Durán, R. (2018, June). Interview with Rebecca Durán by Andrea Lampros.

El Salvador's Funes apologizes for civil war abuses. (2010, January 16). *Reuters*. Retrieved from https://www.reuters.com/article/us-elsalvador/el-salvadors-funes-apologizes-for-civil-war-abuses-idUSTRE60F26M20100116.

Escalante, A. (2018). Interview with Ana Julia Escalante, Director of pyschosocial support for Pro-Búsqueda, by Andrea Lampros.

Fainaru, S. (1996, July 14). A country awakes to the reality of its "disappeared" children. *Boston Globe*. Retrieved from http://poundpuplegacy.org/node/29658.

Fillingim, A. (2018, June). Interview by Andrea Lampros.

García, E. (2018, June). Interview with Eduardo García, director, Asociación Pro-Búsqueda, by Andrea Lampros

Gibb, T. (2000). The killing of Archbishop Oscar Romero was one of the most notorious crimes of the cold war: Was the CIA to blame? *The Guardian*. Retrieved from https://www.theguardian.com/theguardian/2000/mar/23/features11.g21.

Gima, L. (2018, June). Interview with Lance Gima, Former director of the California Department of Justice and director of Forensic Training Program at Conference of Western Attorneys General and Jan Bashinski DNA Laboratory, by Andrea Lampros.

Harmon, B. (2018, June). Interview with Brian Harmon, criminalist manager, California Department of Justice, Jan Bashinski DNA Laboratory, by Andrea Lampros.

Inter-American Court of Human Rights. (2005). Serrano-Cruz Sisters v. El Salvador.

Inter-American Court of Human Rights. (2011). Contreras et al. v. El Salvador. University of Minnesota, Human Rights Library.

Martínez Gómez, M. (2018, April and May). Interviews with Montserrat Martínez Gómez, staff member at Pro-Búsqueda, by Andrea Lampros.

McKinley, J. (2006, December 22). Separated by war, reunited through DNA. *The New York Times*. Retrieved from https://www.nytimes.com/2006/12/22/us/22salvador.html.

Michaels, S. (2007). DNA testing reunites families separated by war. *PBS Newshour*. Retrieved from https://www.pbs.org/newshour/show/dna-testing-reunites-families-separated-by-war.Miglierini, J. (2010, March 24). El Salvador marks Archbishop Oscar Romero's murder. BBC News, San Salvador. Retrieved from http://news.bbc.co.uk/2/hi/8580840.stm.

Órgano Judicial Corte Suprema de Justicia and Pro-Búsqueda (2016). *Jurisprudencia sobre desaparición forzada de niñas y niños*. San Salvador, El Salvador.
Orrego Benavente, C. (2013). Secretary of State Kerry and Salvadoran president Funes: Help find El Salvador's missing children. *Huffington Post*. Retrieved from https://www.huffpost.com/entry/el-salvador-missing-children_b_2885174.
Orrego Benavente, C. (2014). *Pro-Búsqueda: El Salvador Family Reunification Manual*; distributed by Asociación Pro-Búsqueda de Niñas y Niños Desaparecidos, San Salvador, El Salvador.
Parmar, S., Roseman, M., Siegrist, S., & Sowa, T., eds. (2010). *Truthtelling, accountability, and reconciliation*. UNICEF Innocenti Research Centre, Human Rights Program, Harvard Law School. Cambridge, MA: Harvard University Press.
Physicians for Human Rights. (n.d.). *Where we work: El Salvador work*. Retrieved from https://phr.org/countries/el-salvador/.
Reuters. (2010, January 16). El Salvador's Funes apologizes for civil war abuses. Retrieved from https://www.reuters.com/article/us-elsalvador/el-salvadors-funes-apologizes-for-civil-war-abuses-idUSTRE60F26M20100116.
Riordon, M. (2014). Bold scientists: Dispatches from the battle for honest science. Toronto, Ontario: *Between the Lines*.
Rohter, L. (1996). El Salvador's stolen children face a war's darkest secret. *The New York Times*. Retrieved from https://www.nytimes.com/1996/08/05/world/el-salvador-s-stolen-children-face-a-war-s-darkest-secret.html.
Sprenkels, R. (2001). *El día más esperado: Buscando a los niños desaparecidos de El Salvador*. San Salvador, El Salvador: Asociación Pro-Búsqueda de Niñas y Niños Desaparecidos and UCA Editores.
Sprenkels, R. (2002). *Historias para tener presente*. San Salvador, El Salvador: Asociación Pro-Búsqueda de Niñas y Niños Desaparecidos and UCA Editores.
Sprenkels, R., Hernández, L. C., and Villacorta, C. (2002). *Lives apart: Family separation and alternative care arrangements during El Salvador's civil war*. Save the Children. Retrieved from https://resourcecentre.savethechildren.net/library/lives-apart-family-separation-and-alternative-care-during-el-salvadors-civil-war.
Sprenkels, R. (2009). Caminar con el Pueblo: Entrevista con Jon Cortina. Retrieved from https://www.researchgate.net/publication/330193789_Caminar_con_el_pueblo_Entrevista_con_Jon_Cortina.
Sprenkels, R. (2018, May and June). Interviews with Ralph Sprenkels, cofounder of Pro-Búsqueda and lecturer, Department of History and Art History, Ultrecht University, by Andrea Lampros.
Stover, E. (2018, April). Interview with Eric Stover, faculty director, Human Rights Center, UC Berkeley, by Andrea Lampros.
SWPictures. (2017). *El Salvador's lost children* [Video]. Retrieved from www.swpictures.co.uk.
UN Truth Commission on El Salvador (1993, March 15). From madness to hope: The 12-year war in El Salvador. Retrieved from http://www.derechos.org/nizkor/salvador/informes/truth.html.
University of Washington. (2013). *Unfinished sentences: The "May Guinda"*. The Unfinished Sentences Testimony Archive, a project of the University of Washington Center for Human Rights in Seattle, Washington, and the Human Rights Institute of the "José Simeon Cañas," Central American University in San Salvador, El Salvador.
Utrecht University. (2019). *In memorium Dr. Ralph Sprenkels (9 March 1969–14 September 2019)*. Retrieved from https://www.uu.nl/en/news/in-memoriam-dr-ralph-sprenkels-9-march-1969-14-september-2019.
Vásquez, P. (2018, April and June). Interviews with Patricia Vásquez, Pro-Búsqueda's forensic geneticist, by Andrea Lampros.
Zamora, M. (2018, June). Interview with Margarita Zamora, Pro-Búsqueda investigator, by Andrea Lampros.

CHAPTER 9
Large Scale Identification of the Missing

Experiences and Perspectives of the International Commission on Missing Persons

ANDREAS KLEISER AND THOMAS J. PARSONS

Because of its rigorous scientific underpinnings, forensic genetics can now in many cases provide a conclusive means of identifying the remains of missing persons in a range of contexts, from the aftermath of armed conflict to humanitarian disasters (National Research Council, 2009). This does not mean that DNA is the only way that missing persons can be accurately identified, or that human identification should necessarily be based on a single means of identification or divorced from investigatory context. Nonetheless, it was the power of DNA kinship matching that enabled the International Commission on Missing Persons (ICMP) to identify nearly 18,500 persons missing from the conflicts in the former Yugoslavia from 1991 to 1995 (Sarkin et al., 2014).[1]

The ICMP was created at the instigation of US president Bill Clinton in 1996 at the G-7 Summit in Lyon, France, to help account for the tens of thousands of persons who were missing as a result of the fighting in the western Balkans. From the beginning ICMP has worked with all sides in the conflict to locate the missing by encouraging them to cooperate and by asserting the right of families of the missing and others to an effective investigation, as well as by supporting the processes and

---

1. Overall, ICMP's contribution in the former Yugoslavia has helped account for ~27,000 (70%) of the 40,000 persons reported missing as a result of the conflict, including ~8,500 individuals identified by authorities through traditional means prior to the implementation of DNA testing.

agencies that played a significant role in the broader effort to rebuild a war-torn society. In 2003, ICMP's mandate and sphere of activity were extended to address the issue of missing persons globally and persons going missing from a broad range of circumstances, including disasters. In this chapter we examine ICMP's "DNA-led" approach to identifying the missing on a large scale in the former Yugoslavia and other regions of the world. This approach combines expert archaeological excavation and anthropological examination and DNA sampling, as well as public outreach and engagement for reference sample collection, together with cooperation with and strengthening of national mechanisms of medical legal death investigation.

Even before the cessation of hostilities in the former Yugoslavia, the United Nations had established the International Criminal Tribunal for the former Yugoslavia (ICTY) in May 1993 to investigate and prosecute violations of international humanitarian law, including genocide, crimes against humanity, and war crimes occurring during an armed conflict characterized by "ethnic cleansing" and the intentional killing of large numbers of civilians. One of the worst massacres was the systematic execution of ~8,000 men and boys associated with the 1995 fall of the UN Safe Areas of Srebrenica and Zepa to Bosnian Serb forces. Soon thereafter, the international tribunal deployed forensic teams to exhume mass graves and collect evidence of potential war crimes. This initiative, and other forensic efforts, resulted in thousands of mostly skeletonized bodies and body parts being raised above ground, documented, and stored in an ad hoc manner. No provision was initially made for the identification of the remains of the majority of the missing and their return to families, a circumstance that intensified the trauma felt by families desperate to know the fate of their loved ones.

Between 1996 and 2001, the ICMP and others made dedicated attempts to identify the missing, following a traditional model of antemortem and postmortem comparison of biological profiles from human remains and missing persons reports. With limited success in the face of such large numbers, relatively uniform demographics, and a dearth of medical or dental records, DNA was sometimes used to test *presumptive* hypotheses of identity, a term used when evidence short of scientific identifications indicates a potential identity. However, this effort, applied on a small scale with narrowly directed DNA comparisons, often resulted in exclusions rather than identifications.

In 2001 the ICMP decided to employ forensic DNA analysis on a large scale, in a manner that relied on high-throughput nuclear short tandem repeat (STR) testing of skeletal remains. These results were then subjected to blind, computerized comparisons to STR profiles of family members of the missing collected on a regional level. This process has resulted in a large number of kinship DNA reports with high evidentiary strength that have resulted in a large number of identifications. In addition to enabling the return of human remains to families for proper burial, the scientific identifications played a large role in prosecutions by the ICTY, including verdicts of genocide and other crimes against the Bosnian Serb general Ratko Mladic and Radovan Karadzic, former president of the Republika Srpska.

This chapter considers some of the policy developments that have led to the systematic application of forensic sciences, and in particular of DNA-led human identification processes, in a range of settings. These developments include a marked shift in some areas of the world to the rule of law and human rights reference framework for missing persons investigations that involve judicial bodies and related institutions, as well as the active participation of families of the missing, as part of advancing state responsibility for investigating the fate of the missing.

THE MISSING: A RESPONSIBILITY OF STATES

Since the Second World War, the laws of war, also known as international humanitarian law, and especially human rights, have been recognized in a wide range of international instruments and treaties, beginning with the *Universal Declaration of Human Rights* of 1948 and the 1949 Geneva Conventions. The Geneva Conventions in particular made several important advances. Among them are the identification of certain violations of the laws of war as "grave breaches" (war crimes that warrant prosecution and punishment) and a requirement that all parties to the Conventions be "under the obligation to search for persons alleged to have committed, or to have ordered to be committed, such grave breaches" and to bring them to trial in their own courts or those of another country (Reisman & Antoniou, 1994).

Further formal human rights protections came about in 1951 with the adoption of the Geneva Convention on the Status of Refugees and its Protocol, and the establishment of the UN High Commissioner for Refugees in 1950. Since that period the International Covenant on Civil and Political Rights (ICCPR) and its regional counterparts, including the European Convention on Human Rights and Fundamental Freedoms (ECHR), the American Convention on Human Rights (ACHR), and the African Charter on Human and Peoples' Rights (ACHPR), have greatly advanced these protections.

However, new and complex issues have come to challenge the fundamental guarantees of human rights. These include implications of digitalization and the rights to privacy and freedom of expression, as well as threats posed by global warming and political instability to the right to security, food, water, and housing. At the same time, mass migration and refugee flows in many parts of the world have rendered millions homeless, stranded in camps, and often missing (Neier, 2012).

To secure the future protection of human rights, it is critical to enhance the feasibility and credibility of enforcement through more effective human rights investigations. A major push in this direction took place in the 1990s with the establishment of the ICTY, the International Criminal Tribunal for Rwanda (ICTR), and the International Criminal Court (ICC). These three tribunals increased the international community's ability to enforce human rights protections and investigate violations of serious international crimes in a more effective manner. In 2014, the *Agreement on the Status and Functions of the International Commission on Missing Persons*, which conferred on the ICMP full and permanent international legal status 18 years after its establishment as a quasi-intergovernmental body, further strengthened the human rights dimension of addressing the issue of persons going missing, along with creating more effective investigative capacities.

The ICMP, as well as many domestic institutions dedicated to the investigation of missing persons cases, has benefited greatly from scientific and technological advances over the past 20 years. These advances include forensic science methods and procedures forming part of investigations of war crimes and violations of human rights. Invoking state responsibility for accounting for missing persons is a more recent phenomenon generally, although there is some historical precedent pointing to the relevance of accounting for the dead and missing in democratic society.

In the *History of the Peloponnesian War*, Thucydides recorded that in a funeral procession of those who had fallen in battle for Athens, "one empty bier was carried decked for the missing, that is, for those whose bodies could not be recovered" (Thucydides, 2009). The procession was a public manifestation, woven into the fabric of democratic society, according to Pericles, who is recorded as speaking at one of these occasions. A link to democracy and the state still resonates in Abraham Lincoln's Gettysburg Address of November 1863: "[W]e can not hallow this ground. The brave men, living and dead, who struggled here, have consecrated it" (Nicolay & Hay, 1890). Lincoln's celebrated speech, which may have been inspired by Pericles's funeral oration (McPherson, 1992), was delivered following one of the most devastating battles of the American Civil War, involving thousands of missing combatants. In respect of them, Lincoln said, "[F]rom these honored dead we take increased devotion to that cause . . . that government of the people, by the people, for the people, shall not perish from the earth." In respect of state responsibility specifically concerning the missing, Lincoln officially appointed Clara Barton to undertake the task of searching for the missing, eventually through the establishment of the Bureau of Records of Missing Men of the Armies of the United States.

However, the link between accounting for the missing (and the unidentified dead) and democratic society is not a given. Earlier that same year when Lincoln had spoken at Gettysburg, a Swiss businessman, Henry Dunant, appalled by the treatment of the wounded on the battlefield of Solferino, had noted that the dead—and even some of the badly wounded—were buried in trenches and declared "missing" to avoid the formalities of returning their personal belongings to family members. The missing were not his primary concern, though, and in his book about the battle, they warranted only a footnote. Dunant did not make the link to state responsibility and democracy, but proposed that private societies be formed to alleviate the suffering of the maimed and dying. His proposal fell on fertile soil in his native Geneva, where it became the seed of the Red Cross Movement (Dunant, 1862), as part of which the aforementioned Clara Barton launched the American Red Cross Society in 1880.

To Dunant's disappointment, however, his proposal was not embraced by Napoleon III, the victor of Solferino on the side of Piedmont-Sardinia. When the German kaiser accepted Dunant's idea of private societies to alleviate suffering in war as a way of boosting support for an unpopular war against France in 1870, France characterized the rallying of civil society as "war preparations." It appeared therefore that Dunant's sternest critic, Florence Nightingale, was proved right. After returning from the Crimean War of 1853–1856, she had objected bitterly to Dunant's proposal of civil society intervening on behalf of the suffering on the battlefield. In her view, these responsibilities "really belong to the governments of each country," and relieving them of these duties "would render war more easy" (Hutchinson, 2018; Kahn, 2013).

Since the proliferation of international legal instruments and human rights mechanisms in the 1970s, the context and meaning of addressing the issue of missing persons have changed dramatically (Ishay, 2004). Today, the legal and institutional framework no longer prioritizes commemoration and tribute to the missing. There is also greater recognition that cases of persons going missing often involve multiple human rights abuses of the persons themselves and their families, and that as human rights abuses they require an official response (Council of Europe, Commissioner on Human Rights, 2016).

Since the mid-1980s, when forensic investigations of the disappeared were first launched in Argentina, the focus has shifted from ritual commemoration and affirmation for societal purposes to accountability and the rights of family members to know the fate of their loved ones (see chapter 7). This rights-based approach mandates the conduct of official, transparent, and effective investigations (Chevalier-Watts, 2010) capable of establishing the facts of the alleged crimes and the circumstances that led to the disappearance of a missing person or persons.

A rights-based investigation should seek to educate family members about the scientific methods used to identify victims and to determine the manner and cause of death, as well as the official procedures that must be followed (OHCR Working Group, 2011). In the absence of effective and official investigations, disappearances represent violations of the rights of surviving relatives and others, including violations of due process, the prohibition of torture, and violation of the right to a family life and the right to recognition as a person before the law. The Rome Statute of the International Criminal Tribunal states that enforced disappearance is a crime against humanity, with a corresponding obligation to officially investigate disappearances.[2]

FORENSIC INVESTIGATIONS

Forensic science consists of a variable range of activities, theoretical and applied, that are accepted as capable of establishing facts, traditionally to serve as evidence in court (Fraser, 2009). Investigations serve many other purposes, including establishing the severity and scale of human rights violations and abuses, and contribute to the reinstatement of the rule of law more generally (Humphrey, 2003).

The impact of criminal investigations, including forensic investigations, on the broader objectives of dealing with the past violations of serious international crimes has been questioned, however. In particular, contributing to justice in a broader societal sense can sometimes appear to take second place to investigating specific crimes (Fonderbrider, 2002). The sequencing of priorities, such as determining cause and

---

2. The Rome Statute of the International Criminal Court can be accessed at https://www.icc-cpi.int/resourcelibrary/official-journal/rome-statute.aspx.

manner of death before identifying victims, or the division of tasks among various agencies, such as between international investigators and domestic law enforcement, may play a role in shaping that perception (La Vaccara, 2019). In some instances, investigation of the crimes takes second place to the broader objective. For example, efforts to identify the victims of the Turkish invasion of Cyprus in 1974 remain to this day a strictly humanitarian effort, not also a criminal investigation or one of circumstances of persons having gone missing. As a result, the European Court of Human Rights has not accepted the UN-mandated process in Cyprus as commensurate with human rights guarantees under the European Convention on Human Rights, to which both Turkey and Cyprus are parties (ECtHR, 2001).

When UN Security Council Resolution 827 established the ICTY in May 1993 as an ad hoc response to violations of international law in the region, the newly formed body was given the authority to conduct investigations, including exhumations. ICTY rules and standard operating procedures instructed crime scene investigators to secure evidence of crimes and document leads that could assist in identification of the dead. Nevertheless, the ICTY's field investigations in Bosnia and Herzegovina left behind a legacy of more than 6,000 sets of unidentified human remains, because no provision was made for large-scale identification (Sarkin et al., 2014). The relationship between the families of the missing and the ICTY was hence rightly described as "far from symbiotic" when the tribunal completed field investigations in 2001 (Stover & Shigekane, 2002).

However, when Radovan Karadzic was finally arrested and put on trial in 2009, more than a decade after the fall of the UN Safe Areas of Srebrenica and Zepa to Bosnian Serb forces, 90% of the 7,751 persons reported missing to the ICMP had been located and conclusively identified. As part of this process, over 20,000 relatives of the missing provided personal data on the missing and themselves, including DNA references, with the largest reporting volume taking place from 2001 to 2003. The transition from the ICTY's criminal and forensic investigations to the forensic work conducted by ICMP may hence be regarded as seamless in respect of the international, state-led effort that ICMP represented and the large number of relatives of the missing who participated. And it seems fair to say that it created, if somewhat belatedly, a symbiosis with the ICTY. Radovan Karadžić, Radko Mladić, and others were convicted of genocide, crimes against humanity, and war crimes based, inter alia, on evidence gathered through the DNA-led process.

Yet ICMP's success was viewed with apprehension in some quarters. In particular, some feared that international criminal investigations and the deployment of forensic methods on a large scale in the future would deter parties to the armed conflict from providing information on the whereabouts of the missing (Sassòli & Tougas, 2002). However, the primary engine pushing governments and the international community to account for the missing is civil society, especially associations of the families of the missing. These efforts have in large measure ensured that state institutions take responsibility for identifying the missing. The contribution of relatives of the missing to the case against Radovan Karadžić illustrates this shift in perspective (ICTY, 2013).

Asked by the ICTY chief prosecutor and the accused alike to produce evidence of the fate of the missing after the Srebrenica massacre, the ICMP sought the consent of more than 1,200 relatives of the missing to submit more than 9,000 pages of genetic and associated personal data to the trial chamber. The ICMP's efforts to obtain consent from the families were critical, as the commission had originally collected and processed these personal data, including genetic data, only for the purposes of locating and identifying the missing not for criminal investigations.

Of the relatives who were asked to participate, 95% provided written consent for use of their personal data in criminal trials, about 4% could no longer be located, and fewer than 1% declined. A motion by the accused to exclude all DNA evidence was rejected by the Trial Chamber on the grounds that "in giving the accused the opportunity to retest a large number of ICMP DNA identifications . . . the Chamber has done its absolute utmost to ensure that the accused is able to exercise his right under Article 21(4)(e) of the Statute" (ICTY, 2013).

The overwhelming participation of families of missing persons in providing permission to use DNA evidence in the case of Radovan Karadžić was at odds with the results of a countrywide opinion survey in Bosnia and Herzegovina that the ICMP had commissioned in 2011. The survey had found that only about 58% of respondents felt comfortable providing their genetic and other personal data to courts as evidence in criminal trials for war crimes, crimes against humanity, and genocide. The results of the survey, 58%, and the participation rate in the case of Radovan Karadžić, 95%, can be reconciled.

The countrywide survey addressed all families having reported missing relatives. The outreach to families in the case of Radovan Karadžić addressed only the group of families whose missing relatives had been located and identified based on completed DNA matching and proper judicial procedures, and who had been officially informed by the domestic courts that their missing relatives had been found. The fact that practically all of the families of the missing permitted their personal data to be used as evidence in the case of Radovan Karadžić may be seen as indicative of heightened confidence in judicial institutions. It also testifies to families of the missing seeking not only "closure," but also justice.

STRENGTHENING INVESTIGATIVE CAPACITIES

Effective investigations, especially when underpinned by forensic genetics, can serve as benchmarks for the credibility of justice not only in the aftermath of armed conflict but also following humanitarian disasters. In the Mediterranean region, accounting for missing migrants through effective investigations is becoming recognized as important for upholding human rights (ICMP, 2018). In addition, the Dutch investigation into the downing of Malaysian Airlines flight MH 17, including the identification of all victims by DNA analysis, is widely seen as an affirmation of the Netherlands' commitment to both the criminal justice process and the rule of law generally (Ter Haar, 2017).

The application of forensic DNA analysis to the identification of missing persons has been driven by cultural, social, economic, and legal factors, as well as by scientific advances (Fraser, 2009). Placing forensic science—because of its "technicity"— in opposition to humanism (Wagner, 2008) risks removing both from the societal contexts from which they derive meaning. However, as forensic science cannot fall below the standard of science, it has raised the bar for the credibility of efforts to uphold fundamental rights through investigations. In developed countries, investigating the fate of missing persons invariably requires the use of forensic science, but if forensics, and practically DNA analysis, is not becoming accessible to all, it may also cause new inequities between more- and less-developed countries.

Inequities between those who can afford forensic science and those who cannot affect the universal application of identical human rights standards in respect of the missing and may consequently call into doubt the feasibility of universal human rights (Rosenblatt, 2010). The search for minimum standards in the face of the increasing frequency of disasters and emergencies worldwide can then be read as an effort to dissociate missing persons scenarios somewhat from human rights standards, so as not to raise the bar too much. Doing so would risk weakening protections for human rights through a proliferation of divergent standards (Fitzpatrick, 1994). Rather than one standard building on or reinforcing the other, humanitarianism may simply become the poor man's human rights, competing with other urgent needs, to the detriment of advancing human rights as a universal guarantee.

The issue is not a trivial one, as attitudes toward migrants who go missing in the Mediterranean region and on the US-Mexico border have shown. And the issue is not simply one of lower investigative standards being applied to a migrant vessel capsizing off Lampedusa, for example, than to an aviation disaster in the French Alps. The issue is also one of the capacity and willingness of developed countries to deploy resources effectively and beyond the realm of domestic scenarios.

ADVANCING ACCESS TO SCIENTIFIC, DNA-LED MISSING PERSONS PROCESSES

There are two key technical limitations in the DNA identification of missing persons: the frequent need to perform human identifications through kinship analysis and the highly degraded DNA contained in many postmortem samples. The need for kinship analysis means there is a corresponding need to collect reference samples, generally from multiple closely related individuals. A consequence of degraded DNA is that specialized and lengthy techniques are required to process samples such as bone; this in turn necessitates taking measures that will avoid contamination. These considerations drive up cost, reduce throughput, and consequently limit access to effective investigations.

These limitations are in many cases interlinked: degraded DNA may give rise to incomplete STR profiles, which will then limit the power of kinship analysis. A 12-locus partial profile, when compared, for example, to a single sibling as a reference residing in a large DNA database of family reference profiles, may very well not be detected as a kinship association.

The need for DNA reference samples from close relatives is a severely limiting factor in many important missing persons contexts. To take the example of migrants lost at sea in the Mediterranean, their relatives may live in distant countries and may be difficult to trace. Similarly, victims who disappeared during periods of political repression in Albania in the 1960s are unlikely to have surviving close relatives. Even if contact can be made with family members, it is not assured that first- or second-degree relatives are still alive. Therefore, devising a method that provides enough genetic data to make matches with a single, distant relative with high evidentiary strength is critical in DNA identification of the missing. In addition, such a DNA test would also have to be suitable for application to highly degraded DNA samples.

Next generation sequencing (NGS), or massive parallel sequencing (MPS), has revolutionized many areas of the life sciences, increasing the throughput and decreasing the cost of obtaining DNA sequence information by many orders of magnitude (see chapters 1 and 5; Goodwin et al., 2016). In recent years, commercial adaptations of NGS methods to forensic applications have been developed primarily targeting sequence-based analysis of existing STR markers, but also including single nucleotide polymorphisms (SNPs) selected to increase the discrimination power of identity testing and to provide information on biogeographic ancestry and/or appearance (Montano et al., 2018; Jäger et al., 2017; Wang et al., 2017). However, in currently available assays, recovery of sizable fragments of DNA is still needed to span STR markers, while the number of SNPs targeted does not provide for a power of kinship analysis in excess of the STR markers. For full profiles combining both STRs and SNPs, however, the power for resolution of kinship cases is significantly enhanced, although these systems still have to be tested in large-scale missing persons cases where their practical utility can be evaluated.

In our opinion, the maximum utility for NGS methods in missing persons cases will be realized through targeting large numbers of carefully selected SNP loci. SNP loci are ideal for targeting degraded DNA, as they represent the smallest possible target. For many forensic applications, there is a requirement for backward compatibility to existing STR databases, but new events or projects involving missing persons do not have this limitation. Indeed, the power of SNP typing has recently been highlighted in high-profile cases where SNP profiles involving the hundreds of thousands of SNPs produced by commercial assays from companies such as 23andMe and Ancestry.com have been used to trace criminals through the profiles of their distant relatives (Murphy, 2018). While the methods used by these companies are proprietary, they are based on chip-hybridization assays that require large amounts of DNA that are not available in many forensic case samples relevant to missing persons.

Geneticists working with ICMP, in collaboration with others[3]—for example, the University of Santiago de Compostela, Linkoping University, the Swedish National

---

3. Chris Phillips, University of Santiago de Compostela; Andreas Tillmar, Linkoping University & Swedish National Board of Forensic Medicine; and Kenneth Kidd, Yale University and QIAGEN Corportation.

Board of Forensic Medicine, and Yale University—have developed a novel SNP-based assay, conceived specifically for missing persons applications. The current iteration of this assay, undergoing final optimization at the time of writing, targets 1,241 autosomal and 29 X-chromosomal tri-allelic SNPs, carefully selected from the 1,000 Genomes Project as having high heterozygosities in global populations. These are complemented by 45 selected micro-haplotype loci (Kidd et al., 2014a) and 55 ancestry informative SNP markers (Kidd et al., 2014b). These loci are targeted in a single assay using Qiagen's QiaSeq chemistry, originally developed for clinical diagnostics (Peng et al., 2015). Continuing trials on degraded skeletal remains extracts indicate nearly full or high-partial profiles from samples with less than a nanogram of total DNA.

These loci offer exceptionally high power for kinship analysis, with kinship simulations on the autosomal SNP loci alone showing typical sibling index values on the order of 10 to the 200th power, and likelihood ratios for first cousins on the order of 10 to the 40th. At the time of writing, ICMP has just issued MPS DNA match reports on three siblings identified by kinship matching to first-cousin references.

NGS assays, such as the one being developed by ICMP, clearly have the potential to redefine the strategic approach to reference sample collection in missing persons applications. They could therefore alter the range of application of forensic genetic investigations in the overall human rights context.

While NGS methods that provide phenomenal power of kinship matching on highly degraded samples now exist, issues of cost and throughput still limit timely and economical application. In theory, readdressing NGS assay systems from the ground up to permit very high sample simultaneous testing, and implementation on much greater capacity instruments than the benchtop models currently in use, could, with concomitant bioinformatics tools, increase throughput and lower cost. This would, however, require testing in one or several large, highly specialized testing centers. Such an exercise would represent a significant contribution to an internationally supported mechanism to secure the rights of families to effective forensic DNA investigations. A model that provides a very large standing capacity for international response may in fact be effective, as the specialization and sophistication of methods involved may be difficult to implement in a large number of laboratories, especially when considered from the standpoint of data security.

Alternatively, attention could be focused on developing new NGS techniques, since miniaturization and portability, together with standing mechanisms of data connectivity, could permit rapid deployment of instrumentation with minimal upstream DNA processing. The type of capability offered in principle by Oxford Nanopore Technologies, with single-strand direct sequencing on instruments small enough to be held in the hand, is very attractive in this regard. At present, however, this technology has a sequencing error rate and template length requirements that greatly limit its applicability to forensic genetic systems (Cornelis et al., 2018).

Though still at the conceptual stage, novel approaches that take full advantage of rapidly emerging technologies are likely to produce the next revolution in the application of DNA testing to missing persons. ICMP's experience has been that innovation, research, and dedication can contribute to a more effective approach to accounting for missing persons, and this in turn will contribute more broadly to the pursuit of justice and the advancement of human rights globally.

ICMP'S EXPERIENCE

In 2001, when ICMP embarked on large-scale DNA identification efforts in the western Balkans (Parsons et al., 2019), the prospects for success were uncertain. Until then, most DNA testing on degraded skeletal remains had involved mitochondrial DNA (mtDNA), primarily due to its high copy number per cell, which increased the chances of successful PCR amplification (see, e.g., Holland et al. 1993; Edson et al. 2004; chapter 1). However, there were some 30,000–40,000 persons missing in the conflict. In addition, the Srebrenica mass killings were characterized by large mass graves, containing many hundreds of individuals, that were subsequently exhumed by perpetrators and distributed in more than 90 secondary mass graves in a process that badly fragmented and commingled the mortal remains (Sarkin et al., 2014). It was clear that the resolving power of mtDNA would not be sufficient, especially considering the systematic lack of distinctive medical or dental records and a dearth of helpful contextual information with which mtDNA matching could be combined.

Early forays into the use of DNA analysis in the western Balkans, with samples sent to external laboratories, involved testing presumptive hypotheses of identity for confirmation. However, in many instances the presumptive identifications based on "traditional" identifiers (mostly anthropological characteristics of skeletal remains and/or family recognition of clothing) proved incorrect, and there was little to establish identity hypotheses. What was needed was a DNA system that utilized the power of autosomal STRs and centralized DNA database capabilities to make matches independent of non-DNA information.

Optimization of DNA extraction methods based on silica purification yielded a high DNA profiling success rate (Davoren et al., 2007), permitting adoption of a modular, high-throughput laboratory workflow. Steady-state capacities of the laboratory varied over time, but for many years averaged 65 bone or tooth samples per working day, with a short-term maximum of 105 samples per day. The multiplex STR kit chosen for the primary database was Promega PowerPlex 16, with 15 autosomal STRs and the amelogenin sex-indicating marker. As time progressed, DNA extraction methods and validated low-copy-number amplification methods more than kept pace with the increasing degradation of samples that were sequentially exhumed over several years (Milos et al., 2007; Huel et al., 2012; Amory et al., 2012).

In order to obtain family reference DNA profiles, massive public outreach campaigns were instituted, with various collection centers located throughout the region and mobile collection teams that acted within the western Balkans and in targeted campaigns in Europe and North America. In this way, families lodged missing persons reports with ICMP and at the same time provided genetic reference samples under informed consent, as well as information for contacting additional family members. Several first- or second-degree relatives were sought for each missing person, in order to be confident that a DNA match could be made

The DNA databases grew rapidly, calling for development of informatic systems capable of archiving the data in a reliable way and performing kinship matching. Initially, pairwise database screens were performed between each postmortem DNA profile and each family reference DNA profile, with an output of parentage indices, sibling indices, and direct matching (among postmortem profiles for purposes of reassociation). Indications of family associations found in these screens were then subject to full pedigree kinship calculations using DNAView software (Brenner, 1997). DNA match reports were issued for posterior probabilities of 99.95% or greater, with prior probabilities taken as the number of missing persons associated with a particular event or region. Because multiple family references were collected, in most cases the reported certainty far exceeded the 99.95% threshold.

Currently, DNA matching is conducted as a module within ICMP's Identification Data Management System (iDMS). The iDMS has multiple modules spanning all aspects of forensic work, from field recoveries, to mortuary examination and case management, to DNA sample tracking and matching. The iDMS houses the Missing Persons and Relatives modules and links to the Online Inquiry Center, where families may provide and receive information on the missing and official partners can track the progress of submitted cases in real time.

As part of its global mandate, ICMP has conducted DNA testing and in many instances DNA matching on numerous occasions. In post-conflict work, this includes Bosnia and Herzegovina, Serbia, Kosovo, Cyprus, Iraq, and Libya. In relation to political disappearances, work has included cases from Chile, Colombia, South Africa, Albania, and Brazil. ICMP has engaged in large-scale disaster victim identification (DVI) in relation to, among others, the Southeast Asian tsunami in 2004, hurricane Katrina (United States) in 2005, an aircraft incident in Cameroon in 2007, typhoon Frank (Philippines) in 2008, and the downing of Malaysian Air Flight 17 over Ukraine in 2014. At the time of writing, the ICMP database holds more than 101,189 family reference DNA profiles and 49,638 postmortem DNA profiles and has issued DNA match reports on 20,034 individuals.

CONCLUSION

The world has seen significant capacity building to implement human rights more effectively through international technical provisions. In regard to missing persons, this has been done by creating international DNA capacities and personal data processing solutions. DNA analysis in the context of effective investigations constitutes just one of three so-called primary identifiers, the others being fingerprints and dental records (see chapter 10). Any one of these is enough to perform human identifications with a level of certainty that is generally accepted (Black & Bikker, 2017).

The concept of primary identifiers originates from the DVI context, specifically the Interpol *Disaster Victim Identification Guide* (Interpol, 2014). However, as discussed previously, there is no reason to suppose that a lesser investigative standard is justified, whether it be for a mass grave investigation or the investigation of a plane crash. The increasingly widespread use of these primary identifiers can itself be regarded as a significant advancement of the right to an effective investigation. Processes relying on visual examinations of remains and antemortem data comparisons have proven unreliable in many instances. For example, one-third of the victims of the 2002 Bali bombing were reportedly wrongly identified by visual means (Bikker, 2017). Fingerprints and dental records can provide a reliable means of human identification in cases where surviving next of kin are not available to provide antemortem data.

To secure the protection of human rights in the future, it is critical to enhance the feasibility and credibility of human rights enforcement through better and broader access. Advances in technical capacity can help to sustain international policymaking and thus universal human rights. There is a particular and very clear value to DNA analysis as a primary means of human identification. Identifications based on fingerprints require, among other things, that postmortem fingerprints or footprints can still be taken and that antemortem reference prints exist or can be located. The same applies to the use of dental records, which also tend to be available only for older people in more developed countries.

DNA's obvious advantage is that everybody has it. That fact, coupled with recent advancements in NGS analysis and database management systems, strongly suggests that forensic DNA typing has emerged as the predominant means of identifying the remains of missing persons. Such progress will be further enhanced if scientific advances continue to lower cost and increase availability, and if available capacities for cost-effective DNA identification systems continue to grow.

NOTE TO THE READER

The views expressed in this chapter are of those of the authors and are not to be attributed to the International Commission of Missing Persons

REFERENCES

Amory, S., et al. (2012). Automatable full demineralization DNA extraction procedure from degraded skeletal remains. *Forensic Science International: Genetics*, 6, 398–406.
Bikker, J. (2017). Disaster victim identification. In A. Greene, ed., *Missing persons* (pp. 200, 203). New York, NY: Springer.
Black, S., & Bikker, J. (2017). Forensic identifications. In A. Greene, ed., *Missing persons* (pp. 188, 190). New York, NY: Springer.
Brenner, C. H. (1997). Symbolic kinship program. *Genetics*, 145(2), 535–542.
Chevalier-Watts, J. (2010, August). Effective investigations under Article 2 of the European Convention on Human Rights: Securing the right to life or an onerous burden on a state?, *European Journal of International Law*, 3, 701–721.
Cornelis, S., et al. (2018). Forensic STR profiling using Oxford Nanopore Technologies' MinION sequencer. *bioRxiv*. https//doi.org/10.1101/433151.

Council of Europe, Commissioner on Human Rights. (2016). *Missing persons and victims of enforced disappearance in Europe*. Retrieved from https://rm.coe.int/missing-persons-and-victims-of-enforced-disappearance-in-europe-issue-/16806daa1c.
Davoren, J., et al. (2007). Highly effective DNA extraction method for nuclear short tandem repeat testing of skeletal remains from mass graves. *Croatian Medical Journal, 48*, 478–485.
Dunant, H. (1862). *Un souvenir de Solférino*. Geneva, Switzerland: Jules-Guillaume Fick.
ECtHR. (2001). Cyprus v. Turkey. Application no. 25781/94, p. 9, 115 f.
Edson, S. M., et al. (2004). Naming the dead—Confronting the realities of rapid identification of degraded skeletal remains. *Forensic Science Review, 16*, 63–90.
Fitzpatrick, J. (1994). *Human rights in crisis: The international system for protecting rights*. Philadelphia, PA: University of Pennsylvania.
Fonderbrider, L. (2002). Reflections on the scientific documentation of human rights violations. *IRRC, 84*(848), 885.
Fraser J., & Williams R. (2009). *Handbook of forensic science*. Devon, UK: Willan Publishing.
Goodwin, S., McPherson, J. D., & Crombie, M.R. (2016). Coming of age: 10 years of next generation sequencing technologies. *Nature Reviews Genetics, 17*, 333–351.
Holland, M. M., et al. (1993). Mitochondrial DNA sequence analysis of human skeletal remains: Identification of remains from the Vietnam War. *Journal of Forensic Science, 38*, 542–553.
Huel, R., et al. (2012). DNA extraction from aged skeletal samples for STR typing by capillary electrophoresis. In *DNA electrophoresis protocols for forensic genetics*. Methods in molecular biology, Vol. 830. Totawa, NJ: Humana Press.
Humphrey, M. (2003). International intervention, justice and national reconciliation: The role of the ICTY and the ICTR in Bosnia and Rwanda. *Journal of Human Rights*, 2(4), 496.
Hutchinson, J. (2018). *Champions of charity: War and the rise of the Red Cross*. New York, NY: Routledge.
ICMP. (2018). *Joint process to account for persons missing as a result of migration in the Mediterranean region—joint statement*. Retrieved from https://www.icmp.int/press-releases/developing-a-joint-process-on-the-issue-of-missing-migrants-in-the-mediterranean-region/.
ICTY. (2013, April 14). *Decision on the accused's motion to exclude DNA evidence*. Retrieved from https://icty.org/x/cases/karadzic/tdec/en/130416.pdf.
Interpol. (2014). *Disaster victim identification guide*. Lyon, France: Interpol.
Ishay, M. (2004). Promoting human rights in the era of globalization and interventions: The changing spaces of struggle. *Globalizations, 22*, 181–193.
Jäger, A. C. et al (2017). Developmental validation of the MiSeq FGx Forensic genomics system for targeted next generation sequencing in forensic DNA casework and database laboratories. *Forensic Science International: Genetics, 28*, 52–70.
Kahn, D. E. (2013). *Das rote Kreuz: Geschicte einer humanitaren bewegun*. Munich, Germany: C. H. Beck.
Kidd, K. K., et al. (2014a). Current sequencing technology makes microhaplotypes a powerful new type of genetic marker for forensics. *Forensic Science International: Genetics, 12*, 215–224.
Kidd, K. K., et al. (2014b). Progress toward an efficient panel of SNPs for ancestry inference. *Forensic Science International: Genetics, 10*, 23–32.
La Vaccara, A. (2019). *When the conflict ends, while uncertainty continues: Accounting for missing persons between war and peace in international law*. Paris, France: Editions A. Pedone; Oxford, UK: Hart Publishing.
McPherson, J, M. (1992, July 16). The art of Abraham Lincoln. *The New York Review of Books*. Retrieved from https://www.nybooks.com/articles/1992/07/16/the-art-of-abraham-lincoln/.
Milos, A., et al. (2007). Success rates of nuclear short tandem repeat typing from different skeletal remains. *Croatian Medical Journal, 48*, 486–493.

Montano, E.A., et al. (2018). Optimization of the Promega PowerSeq™ Auto/Y system for efficient integration within a forensic DNA laboratory. *Forensic Science International: Genetics*, 32, 26–32.

Murphy, H. (2018, June 27). Genealogists turn to cousins' DNA and family trees to crack five new cold cases. *The New York Times*. Retrieved from https //www.nytimes.com/2018/06/27/science/dna-family-trees-cold-cases.html.

National Research Council. (2009). *Strengthening forensic science in the United States—A path forward*. Washington, DC: National Academies Press.

Neier, A. (2012). *International human rights movement: A history*. Princeton, NJ: Princeton University Press.

Nicolay, J. G., & Hay, J. (1890). *Abraham Lincoln: A history*. New York, NY: A Century Company.

OHCR Working Group on Enforced or Involuntary Disappearances. (2011). *General comment on the right to the truth in relation to enforced disappearances*. (A/HRC/16/48). Retrieved from https://www.ohchr.org/Documents/Issues/Disappearances/GC-right_to_the_truth.pdf.

Parsons, T., Huel, R., Bajunovic, Z., and Rizvic, A. (2019). Large scale DNA identification: The ICMP experience. *Forensic Science Int: Genetics*, 38, 236–244.

Peng, Q., et al. (2015). Reducing amplification artifacts in high multiplex amplicon sequencing by using molecular barcodes. *BMC Genomics*, 16, 589.

Reisman, W. M., & Antonio, C. T., eds. (1994). *The laws of war: A comprehensive collection of primary documents on international law governing armed conflict*. New York, NY: Vintage Books.

Rosenblatt, A. (2010). International forensic investigations and the human rights of the dead. *Human Rights Quarterly*, 32, 921–950.

Sarkin, J., Nettelfield, L., Matthews, M., & Kosalka, R. (2014). *Bosnia and Herzegovina: Missing persons from the armed conflicts of the 1990s: A stocktaking*. International Commission on Missing Persons (ICMP). Retrieved from https://www.icmp.int/wp-content/uploads/2014/12/StocktakingReport_ENG_web.pdf.

Sassòli, M., & Tougas, M. (2002). The ICRC and the missing. *IRRC*, 84(848), 278.

Stover, E., & Shigekane, R. (2002). The missing in the aftermath of war: When do the needs of victims' families and international war crimes tribunals clash? *IRRC*, 84(848), 847.

Ter Haar, B. (2017). *Lessons of the MH17 disaster opinion (revisited)*. Barend ter Haar, Netherlands: Netherlands Institute of International Relations Clingendael.

Thucydides. (2009). *The history of the Peloponnesian War* (R. Crawley, Trans.). North Chelmsford, MA: Digireads.com Publishing.

Wagner, S. (2008). *To know where he lies*. Berkeley, CA: University of California Press.

Wang, Z., et al. (2017). Massively parallel sequencing of 32 forensic markers using the Precision ID GlobalFiler™ NGS STR Panel and the Ion PGM™ System. *Forensic Science International: Genetics*, 31, 126–134.

CHAPTER 10
Tracing Windblown Seeds

Genetic Information as a Biometric for Tracking Migrants

SARA H. KATSANIS

> Human borders mean nothing to air, water, windblown soil or seeds or migrating fish, birds or mammals.
> —David Suzuki (2009)

Nongovernmental organizations and governments have used genetic information to identify the missing in the aftermath of political violence, war, and natural disasters since the late 1980s. But only in recent years have governments begun to collect such information as a means of monitoring and controlling immigration. And nowhere has this trend been more prevalent than in the Untied States.

Immigration authorities may use genetic information to protect national security, prevent illegal immigration, detect human trafficking, and expose cases of fraud within the immigration system itself. As we have learned from other chapters in this volume, the use of DNA in any of these scenarios is in itself neither "good" nor "bad," but rather laced with ethical and legal questions regarding privacy and definitions of identity and family. In the immigration context, use of DNA information also opens up questions of race and nationality, potentially to the exclusion or stigmatization of people with particular genetic backgrounds or relationships. With increased use of DNA in immigration contexts in the United States, particularly with its use to identify children displaced from their families and to detect human trafficking, a nuanced understanding of how DNA information is applied is essential. We need to understand why DNA-based "biometric" information has become the go-to technology of choice at the border and what considerations must be examined to ensure that such information is used in ways that do not harm individuals or social systems.

Sara H. Katsanis, *Tracing Windblown Seeds* In: *Silent Witness*. Edited by: Henry Erlich, Eric Stover and Thomas J. White, Oxford University Press (2020). © Oxford University Press. DOI: 10.1093/oso/9780190909444.003.0011

To watch the American news media, one would think that the United States is being overrun by migrants, refugees, and foreign-born criminals. In reality, the United States has experienced a drop in immigration since 2005, with far fewer immigrants in proportion to the population than in the preceding years (see figures 10.1a and 10.1b). Prior to the 1960s, however, half of the immigrants entering the United States were from European countries, whereas today Europeans comprise less than 10% of immigrants. What has changed, then, are the proportions of immigrants who have come to the United States from different parts of the world (see figure 10.1c).

In our post-9/11 world, many Americans have become weary and even fearful of immigrants. Xenophobia and racism fuel some of these fears, including apprehension about a changing way of life and fear of possible terrorist attacks by extremists. In recent years, US immigration officials have intensified their scrutiny of visa applicants and other persons coming into the United States.

A hundred years ago, in the days of Ellis Island (where some of my own ancestors arrived in the nineteenth century), the screening of certain immigrants, especially from southern European countries, could be just as invasive as it is today, although they were treated more as diseased individuals than as criminals. The anti-immigration sentiment was as strong then as it is in current times, targeting similarly impoverished and disadvantaged populations. Back then, immigrants who

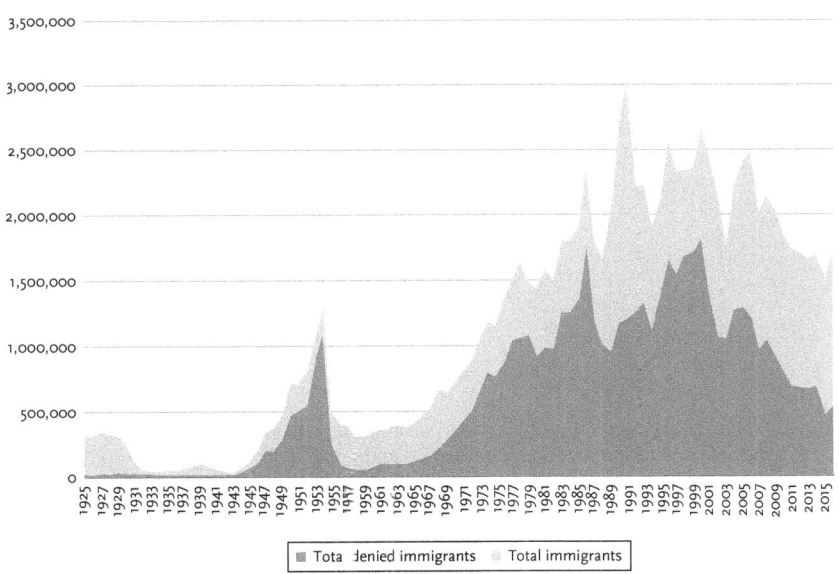

Figure 10.1a Total immigrants to the United States, 1925–2015. The trend for number of immigrants and those deported tends to track global and economic events. Denied immigrants are those who illegally entered the United States and were deported and those that were legally in the United States and then deported. *Sources:* US Department of Homeland Security (2016) and US Census Bureau (2016). Graph by Sara H. Katsanis.

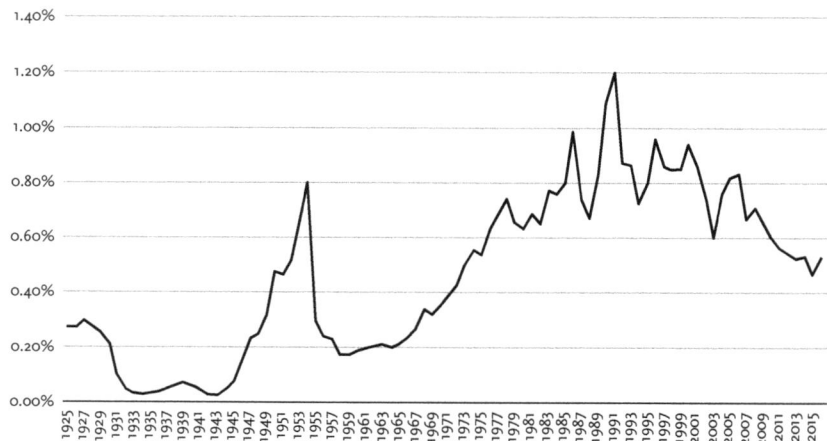

Figure 10.1b Ratio of immigrants to US population, 1925–2015. Number of immigrants includes both legal immigrants and those rejected and deported. *Sources:* US Department of Homeland Security (2016) and US Census Bureau (2016). Graph by Sara H. Katsanis.

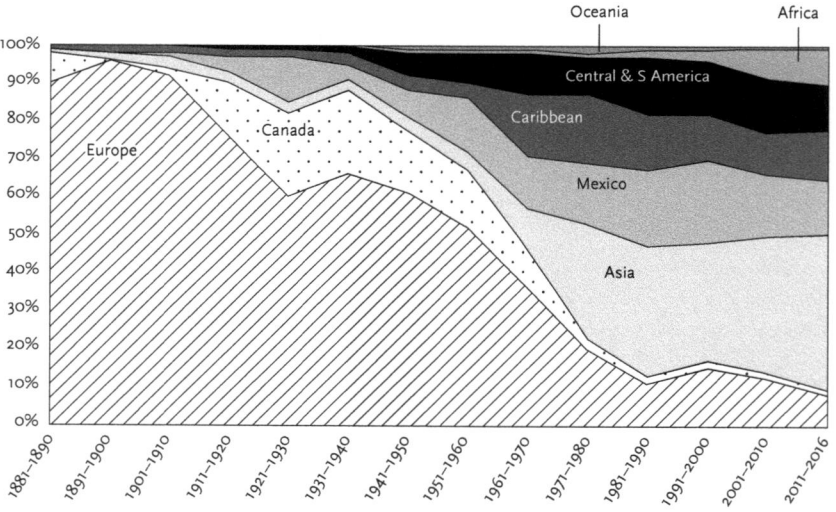

Figure 10.1c Immigration by region, 1881–2016. Historic figures showed more proportion of immigrants from Europe. The figure is adapted and updated from Fairchild (2004); data from US Department of Homeland Security (2016). Graph by Sara H. Katsanis.

failed to qualify for entry were sent back to their countries of origin—expenses paid. Today they are locked up, and if they are rejected, they are expelled with orders not to return. Further, migrants are often labeled as either "legal" or "illegal," implying a criminality to the act of crossing a border.

In November 2002, just over a year after the 9/11 attacks, President George W. Bush established the Department of Homeland Security (DHS), shifting the screening of migrants from a quality-control approach to a defensive tactic (Pub. L. No. 107-296). Rather than screening newcomers for illnesses and moral standing (such as "a lunatic, idiot, or any person unable to take care of him or herself without becoming a public charge" as described in the Immigration Act of 1882), preference was to be given to persons who might benefit the country, while those who were believed to be a security threat were denied entry. Taking this a step further, President Donald Trump signed an executive order soon after his inauguration in January 2017 restricting the entry of travelers from seven Middle Eastern countries (Executive Office of the President, 2017).

In April 2018, President Trump doubled down on his harsh immigration policies, giving US immigration officials the authority to separate children from their accompanying adult(s) who were detained while trying to cross into the United States. Adults were incarcerated, while the children were placed in holding facilities operated by the US Immigration and Customs Enforcement (ICE) and then transferred to foster care[1] by the Department of Health and Human Service's Office of Refugee Resettlement (ORR). Within weeks of implementing the new policy, it became clear that the poorly funded ORR foster care system was incapable of tracing the whereabouts of thousands of displaced children. In June 2018, in response to widespread public outrage, President Trump reversed this policy and allowed children to remain in the same detention center as their parents (Gambino & Laughland, 2018).

These increasingly harsh policies are pushing the US immigration system toward adopting stiffer identification requirements for migrants who wish to enter the United States. And what better means is there for individual identification than DNA evidence, particularly when it comes to families? While fingerprints are an apt marker for identity, genetic information is far more valuable for its ability to connect people who are biologically related. And so it stands to reason that biometric technology is now being incorporated into travel and other identity documents as a natural expansion of the procedures used over the last century to trace migrants entering the United States (see table 10.1 for relevant immigration laws; Thomas, 2006; Greenwald, 2013).

PROOF OF IDENTITY

Proof of identify (POI) at US borders includes three main forms of identification: biometric (e.g., fingerprints), attributed (e.g., full name), and biographical (e.g., education or employment history). Traditionally, POI has relied upon attributed and biographical data, such as that recorded at Ellis Island, but this approach is waning given the ability to compromise POI documents easily (Rodger et al., 2008).

---

1. The Homeland Security secretary at the time, Kirstjen Nielsen, defended the administration's policy: "That's no different than what we do every day in every part of the United States—when an adult of a family commits a crime," she told NPR. "If you as a parent break into a house you will be incarcerated by police and thereby separated from your family" (Burnett & Gonzalez, 2018).

Table 10.1 LANDMARK LAWS AND POLICIES FOR IMMIGRATION AND BIOMETRICS

Policy	Content
Naturalization Act of 1790 (1 Stat. 103)	Set requirement of 2 years' residence and limited to free white persons.
Page Act of 1875 (Sect. 141, 18 Stat. 477)	Prohibited certain groups of Asian immigrants from entry by inspection at port.
Chinese Exclusion Act of 1882 (Pub. L. No. 47-126, 22 Stat. 58)	Restricted entry of Chinese migrants and excluded Chinese already in the United States from citizenship. Required documentation from Chinese government that migrants were qualified to emigrate.
Immigration Act of 1891 (Ch. 551, 26 Stat. 1084)	Modified 1882 act to restrict certain medical conditions. Required medical exam upon entry into the United States to screen for contagious people.
Asiatic Barred Zone Act of 1917 (Pub. L. No. 301, 39 Stat. 874)	Restricted additional medical conditions and "moral" conditions such as illiteracy, alcoholism, and vagrancy.
Emergency Immigration Act of 1921 (Pub. L. No. 67-5, 42 Stat. 5)	Introduced numerical quotas based on country of origin and on the basis of the 1790 act.
Immigration Act of 1924 (Pub. L. No. 68-139, 43 Stat. 153)	Amended the 1921 act to prioritize immigrants from northern and western European countries, affecting primarily southern and eastern European, Jewish, and Chinese immigrants. Required medical exam to occur abroad rather than in the United States (Bateman-House & Fairchild 2008).
Immigration and Nationality Act of 1952 (Pub. L. No. 82-414)	Abolished race restrictions but retained the quota system.
Immigration and Nationality Act of 1965 (Pub. L. No. 89-236, 79 Stat. 911)	Ended 1921-instated quotas and instead developed a formula permitting a percentage of applicants from any particular country, not including immediate relatives of US citizens and certain refugees.
Immigration Reform Act of 1986 (Pub. L. No. 99-603, 100 Stat. 3359)	Provided amnesty to immigrants entering the United States prior to 1982.
DNA Identification Act of 1994 (42 U.S.C. § 14132; Pub. L. No. 103-322)	Authorizes creation of CODIS.
Illegal Immigration Reform and Immigrant Responsibility Act of 1996 (Pub. L. No. 104-208, 110 Stat. 3009-546)	Prevents re-entry of deported immigrants and authorizes a border fence.

Table 10.1 CONTINUED

Policy	Content
INS Memorandum July 14, 2000 (US Department of State, Immigration and Naturalization Service, 2000)	Provides guidance on parentage testing for family-based immigrant visa petitions.
USCIS Memorandum March 20, 2006 (US Department of Homeland Security, Office of the Citizenship and Immigration Services, 2006a)	Recommends acceptance of DNA test results as secondary evidence of family relationship and grants authority to require DNA testing and to initiate a DNA testing pilot project to study the effect of requiring DNA testing as evidence of family relationships.
USCIS Memorandum July 5, 2006 (US Department of Homeland Security, Office of the Citizenship and Immigration Services, 2006b)	Response to recommendation for DNA testing, concluding that the benefits of testing would justify the costs.
DOJ Regulation, December 10, 2008 (73 F.R. 74932)	Directs federal agencies to collect DNA from individuals who are arrested, facing charges, or convicted, and from non-US persons who are detained under the authority of the United States, subject to certain limitations and exceptions.
September 8, 2010 Public Comment Request for P-3 DNA collection (75 F.R. 54690)	Notice of request for public comment on plan to require an affidavit of relationship (AOR) to establish qualification for access to the Priority 3 (P-3) admissions program. The AOR also informs the anchor relative that DNA evidence of all claimed parent-child relationship will be required as a condition of access to P-3 processing and who is responsible for the costs. Program was renewed in 2015 (80 F.R. 18923).
November 14, 2014 CAM DNA collection (US Department of State, Bureau of Population, Refugees and Migration, 2014)	Announced a program to provide visas to Central American migrant minors joining families, requiring DNA verification of claimed relationships. Program was halted in 2017.
2017 Executive Order 13769 (Executive Office of the President, 2017)	Restricts entry of travelers from 7 Middle Eastern countries.
Rapid DNA Act of 2017 (Pub. L. No. 115-50)	Establishes ability to process DNA for CODIS without a laboratory involvement.
USCIS Memorandum April 17, 2018 (US Department of State, 2018)	Provides guidance on sibling testing for family-based immigrant visa petitions.

Scrutiny of immigrants to the United States prior to World War I involved a perfunctory examination of passengers, then newcomers were provided with immigration ID cards by the US government (McLaughlin, 1905). Immigration procedures included a medical exam to detect any suspected contagious diseases but also to quickly scrutinize a person's apparent character: whether someone was an alcoholic, a vagrant, or otherwise unfit to be granted entry. It wasn't until after World War I that government-issued passports were instated, and they remained optional for US citizens until 1952 (Immigration and Nationality Act, 1952, § 215(b), 8 U.S.C. § 1185). Passports were not required of immigrants seeking entry to the United States, but by the 1950s some form of government attestation, such as a letter from an embassy, was required as proof of a person's country of origin.

From a government's perspective, POI is useful for immigration authorities because it helps them document who is arriving and departing. It also helps to screen out known criminals and terrorists. From a migrant's perspective, POI is useful for attaining citizenship and residency privileges and necessary in many countries for access to jobs or to marry. By law, the United States cannot deport individuals without knowing their country of origin (8 U.S.C. § 1158(a)(2)(A)), and many other countries have similar laws. If they lack POI, asylum seekers may be sent back to their states of origin, but those states might use uncertainty about the background of failed asylum seekers to justify refusal to accept their return (Ramirez et al., 2015).

Since many asylum seekers may have lost or destroyed their travel documents, biometric information might be a useful means of documenting identity. Table 10.2 outlines some of the databases with biometric information used by the United States. While there are numerous ways to track migrants through paperwork, biometric attributes are the most efficient means for border officials to verify a person's claimed identity and to identify persons of interest, such as known drug dealers or human traffickers. Documenting a unique and permanent profile of each person crossing a border ensures efficient border controls and is far more reliable than the subjective judgment of border guards (Ramirez et al., 2015). Recent technological advances have enabled an arsenal of new tools that are portable, discrete, and ruggedized. Commonly used biometrics include face recognition, iris imaging, palm prints, hand geometry, and voice recognition (Lynch, 2012). Most biometrics are minimally invasive in nature, like a fingerprint, iris scan, or DNA buccal swab, and others are collected at a distance, with or without the subject's knowledge, such as photographs and videos (Lynch, 2012).

In recent years, the US Visitor and Immigration Status Indicator Technology program (US-VISIT), administered by DHS, has expanded its traditional process of reviewing written or verbal information provided by tourists and immigrants to include biometric data (US Department of Homeland Security, 2010). These data are used by a range of law enforcement and intelligence agencies, including ICE, the Customs and Border Protection (CBP), the Transportation Security Administration (TSA), and the Citizenship and Immigration Services (USCIS) (US Department of Homeland Security, 2006).

Table 10.2 IMMIGRATION DATABASES

Administrated by target population	Database	Biometrics	Status
Department of Homeland Security (DHS) Noncitizen Americans Foreign visitors	IDENT/US-VISIT Automated Biometric IDENTification System US Visitor and Immigration Status Indicator Technology	Fingerprints Digital facial images Iris scans Palm prints	IDENT was established in 1994 US-VISIT was established in 2004 220 million individuals as of 2017 Interoperable with IAFIS/NGI
	HART Homeland Advanced Recognition Technology	Digital facial images DNA profiles	Under development Intended to cross federal departments
Department of Justice (DOJ), Federal Bureau of Investigation (FBI) Criminal offenders Immigrant detainees Arrestees	IAFIS/NGI Integrated Automated Fingerprint Identification System Next generation identification	Fingerprints Digital facial images Iris scans Palm prints	IAFIS was established in 1999 Expanded to NGI to encompass other biometrics in 2011 70 million individuals as of 2018 Interoperable with IDENT
	CODIS Combined DNA Index System	DNA profiles	Established in 1998 17 million individuals as of 2018 Only accessible to CODIS users

Data collected and analyzed by US-VISIT are used to identify repeat border crossers or individuals who might pose a security risk to the United States (National Science and Technology Council, 2008). The "e-passports" that have been issued by the United States since August 2007 include digital photographic face recognition (Katsanis & Kim, 2014) and unique chip identification numbers and digital signatures (Goth, 2008). In 1994, immigration authorities started scanning fingerprints in IDENT, the Automated Biometric IDENTification System, as part of the US-VISIT effort. At first, IDENT and the federal criminal fingerprint database, Integrated Automated Fingerprint Identification System (IAFIS), did not share data, but in 2006 a pilot project considered the technical enhancements needed to allow two-way sharing of data. IAFIS is now termed IAFIS/NGI as the Federal Bureau of Investigation (FBI) develops its next generation identification (NGI) to expand the fingerprint database to other biometrics.

By early 2018, IDENT held biometric data from over 200 million people coming into the United States. Some of the early data, in particular fingerprints, were reportedly fraudulent, exhibiting keyboard errors and discrepancies in names of contributors (Sternstein, 2012); hence the need for a broader array of biometrics that could be more robust than fingerprints, like DNA.

GENETIC INFORMATION AS A BIOMETRIC

Immigration officials can use genetic information in three ways: *biometric identity*, *relationship testing*, and *biogeographic ancestry testing*. Each of these approaches provides information that can help immigration officials detect document falsification, identify human traffickers, and even facilitate deportation procedures (Wanem, 2008).

Biometric identity is essentially the collection and typing of DNA from a person in order to store the DNA profile in a database for future matching. For instance, use of DNA-based biometrics might deter migrants from using false identities and detect repeat border-crossers with false documents (Ramirez et al., 2015).

Relationship testing involves the comparison of DNA profiles from two or more individuals, which could be used in immigration to verify or refute claimed familial relationships. Genetic information stands apart from other biometric identifiers in that it can be used to identify biological relationships. Comparative DNA profiles can be used in some cases to verify claimed kinship for placement of children with sponsors in the United States or for reunification with relatives in their originating countries (Katsanis, 2015). Verifying relationships of claimed relatives of displaced children is essential to prevent children being placed with traffickers or returned across the border to smugglers.

Biogeographical ancestry tests, commonly called ancestry DNA tests, are based on the analysis of a person's biological ancestral origins as indicated by particular markers in the genome. Biogeographical data within our genomes can provide hints about people's world origins, which in turn can be useful for making claims of citizenship. Examples of these are discussed further later in the chapter.

DATA COLLECTION FROM IMMIGRANT DETAINEES

In 2008 the FBI announced its plan to begin collecting DNA from migrants arrested or detained under the authority of the United States for inclusion in the federal criminal database known as the Combined DNA Index System (CODIS) (73 F.R. 74932, 42 U.S.C. § 14135a). This expansion of CODIS meant that the hundreds of thousands of apprehended aliens each year would be added to the detainee index of the National DNA Index System (NDIS). As of August 2017, NDIS had gathered about 25,000 DNA profiles of detainees, at a rate of approximately 1,000 per year. Clearly, with only 25,000 profiles in the database after 10 years, the US immigration authorities are not collecting DNA from every detainee, but it is unclear why this is the case. It is possible that DNA is collected only from detainees who are charged with the crime of crossing the border illegally. Or perhaps border agents find other biometrics to be more efficient to collect and use, or are not aware of the law, not trained in DNA collection, or lack the manpower and resources to collect the specimens.

In due time, the resources will undoubtedly catch up with the legislation. DHS, for instance, has developed Rapid DNA instrumentation useful for quickly processing detainees' samples (US Department of Homeland Security, Science & Technology, 2015). The Rapid DNA Act of 2017 makes it possible to upload DNA profiles from a Rapid DNA instrument into CODIS, making it conceivable to detain immigrants and check their DNA against CODIS for crimes in which they might have been involved within a couple of hours (Pub. L. No. 115-50). The technology (described in more technical detail in chapter 5) is gradually gaining acceptance and is ready for use for processing samples from arrestees and detainees at booking stations (ANDE Corporation, 2018).

CODIS is restricted—by design and purpose—for use by law enforcement only. The 13 markers originally chosen for CODIS were not well-suited for detecting biologically related profiles, which has been a challenge both for missing persons casework (see chapter 11), and for familial searching (Kim et al., 2011). In 2016, the number of CODIS markers was expanded to 20, with the additional markers chosen in part to improve relationship detection (Katsanis & Wagner, 2013).

As mentioned previously, CODIS already contains profiles of immigrants detained by US immigration agents. Soon the DHS-based IDENT database will also include DNA profiles for immigrants and visitors to the United States (Boyd, 2017). Once the interchange of data is feasible between IDENT and CODIS, cross-comparison of biometric markers between DHS and FBI might enable immigration officials to identify criminals, terrorists, or known traffickers attempting to travel under false identities.

In 2011, DHS initiated efforts to replace US-VISIT and IDENT with a new system called Homeland Advanced Recognition Technology (HART) (US Government Accountability Office, 2017). HART was expected to contain by 2020 DNA data, as well as face recognition and other biometrics (Lynch, 2012). In addition, recent developments in blockchain technology and its integration into identification systems might make sharing of genetic data more secure than in the past (Roberts, 2017). The next phase for DNA biometric databases could allow data to be exchanged between CODIS and HART, just as the criminal and immigration fingerprint databases, IAFIS/NGI and IDENT, once separate, are now compatible across purposes.

VERIFYING CLAIMED RELATIONSHIPS

When US law abolished the immigration quota system that prioritized certain countries and races, legislators replaced it with a system that instead prioritizes visas and citizenship for family members of citizens. The family reunification programs that were started in the 1960s relied on birth and marriage certificates to navigate relationships. By the 1980s, ABO blood typing was used as biological evidence of relationship claims, but without much specificity. But by the end of the twentieth century, documents had become easy to fake, prompting the government to require more specific genetic information to verify claimed relationships.

International Adoptions

International adoptions became the first test case arena for using DNA markers for verification of biological parents. In the wake of the conflicts in prior decades, poverty was particularly widespread in Guatemala in the early 1990s. This meant many children were abandoned or orphaned, prompting a number of American families to seek Guatemalan adoptions, which were reportedly quicker than adoptions from other countries (Siegel, 2011). An expanding demand for babies and young children from Guatemala sparked underground "baby farming" enterprises and kidnappings (Sherwell, 2008). In some underground circles in Guatemala, women were paid to conceive and give up their babies for adoption or were held hostage to do the same. In other trafficking rings, pregnant women were drugged during birth and told after birth that their babies had died. Some children were even kidnapped from their homes and placed in orphanages for adoption (International Commission against Impunity in Guatemala, 2010).

By 1996, reports of such shocking practices prompted the use of DNA testing through commercial laboratories in the United States to verify the maternal relationship of any Guatemalan child being relinquished for adoption. By 2008, however, child traffickers in Guatemala had caught on and were paying substitute mothers and children to undergo the DNA test and paying the medical examiners in Guatemala to look the other way (Siegel, 2011). Once the fraud was detected, the US embassy began handling all DNA testing directly rather than through Guatemalan medical examiners (Associated Press, 2008). But by then it was too late. From 2000 to 2008, US citizens had adopted nearly 30,000 Guatemalan children (Joint Council on International Children's Services and National Council for Adoption, 2007). Most of these adoptions were legal and probably above board. But because of ongoing investigations of fraud, including the use of DNA, US authorities ended the adoptions in June 2008, in hopes of breaking the trafficking rings and the underground market for children.

Family-Based Immigration Programs

By 2000, US immigration authorities had welcomed the voluntary use of DNA parentage tests to support relationship claims that are central to family reunification programs (US Department of State, Immigration and Naturalization Service, 2000). In 2018, USCIS expanded the policy to include sibling relationship testing as well (US Department of State, Immigration and Naturalization Service, 2018). Using DNA, immigrants and asylum seekers could provide credible, immutable evidence of their biological claims of relationship (Ramirez et al., 2015). Such testing is conducted by commercial laboratories accredited by the American Association of Blood Banks (AABB) (US Citizenship and Immigration Services, 2008) and paid for by the applicant.

The process is fairly straightforward. For example, if my husband and I were to decide to bring his Greek mother to live with us in the United States, he would order a DNA test, submitting both his own sample and that of his mother seeking to immigrate. The visa application office would then review the laboratory report to determine the authenticity of the relationship claim (Taitz et al., 2002). If the relationship being tested is that of a parent and a child (like my husband and his mother), then the two people should share half of the genetic markers reported by the laboratory. A DNA test might be used in other circumstances as well: If a claimed relationship is not a biological relationship, like a husband and a wife, for example, then the couple should not share any markers, except those by chance. The use of DNA tests in this way is meant to support any relationship claims in the application, not to exclude family members who are not biologically related.

The difficulty with using DNA relationship tests in this manner is in the varying cultural translations of what it means to be a family. A sister in one culture might not be a biologically related sister by American standards. Or what is considered to be a parent in another culture might be what Americans term a guardian, godparent, or honorary parent. Moreover, the restrictions of family laws in the United States might exclude true family members in other countries. For instance, US law allows a man to bring only one wife into the United States (8 U.S.C. § 1101(a)(35)), thus prohibiting the immigration of secondary wives. These cultural differences prompt families to make false (or seemingly false) statements of relationships in order to bypass US immigration restrictions.

Against this backdrop, relationship testing laboratories began noticing what they termed *genotype recycling*: the submission of the same person(s) DNA profile multiple times for verification (Wenk, 2010). Laboratories with a careful eye can notice discrepancies in paperwork or see obvious errors in genetic results, such as differences between the claimed sex and the genetic findings. In this way, laboratories noted that applicants would sometimes provide DNA samples of related persons in place of unrelated persons (Wenk, 2011), similar to what happened in Guatemalan adoptions.

According to one laboratory director I interviewed in 2010, men with multiple wives were sending in a daughter's or sister's specimens multiple times and claiming the additional wives as sisters or daughters. He described one example: "The male-petitioner said he was bringing in his daughter to the United States but the [sex chromosome test] shows a male. . . . Now here's a poor immigrant, and instead of saying 'you already have my file there, can you just test the new children,' he doesn't do it that way. He brings in each alleged child as a new case, and he's paying for himself over and over again. He brings in 3 or 4 children at a time. So I begin to look at the whole alleged pedigree and I see that in the case where he is trying to bring in a daughter and she has a male type, that male type matches [the DNA profile of a child he had already tested]. The genotypes are identical. . . . I go through the 16 kids [he had already tested] and he had done this 4-5 times. What's clearly needed is, like the FBI has its own database of profiles, the State Department and USCIS need their own computer [so] that [they] can look for identical profiles."[2] In most cases, such fraudulent behavior would not be so easily detectable if the laboratory wasn't specifically looking for discrepancies. Moreover, perpetrators could also use multiple commercial laboratories, rather than a common government facility.

Once the US Department of State (DOS) learned of these fraudulent practices, it launched a pilot program to test the use of biometric measures, primarily DNA typing, to detect fraud in family relationship claims. By 2006, the USCIS Ombudsman had recommended that local offices have the authority to require DNA testing in specific circumstances, such as when fraud is suspected or where the applicant lacks evidence of a claimed family relationship, such as birth certificates or school records (US Department of Homeland Security, Office of the Citizenship and Immigration Services, 2006). With this new authority, the USCIS launched a pilot program in 2008 requiring DNA testing for the Priority 3 (P-3)[3] family reunification refugees (8 C.F.R. 204.2(d)(2)(vi),10). The 2008 pilot program was aimed at East African refugees applying for entry into the United States (Esbenshade, 2010). The pilot results using DNA testing seemed to confirm the prior reports of high levels of fraud among East African family refugee applications (US Department of State, Bureau of Population, Refugees, and Migration, 2008).

Soon thereafter, scholars pointed out that the methodology was flawed in reporting fraud because the pilot report included cases of people refusing to take a DNA test (Dove, 2012). The DOS grouped these refusals with those cases in which the DNA test result did not match the relationship claim. However, there was no accompanying social science data to indicate why people refused the tests, nor whether the families had legitimate cultural reasons that could explain why the DNA tests did not show the expected results. As a result of the pilot study, all applications under the P-3 program were halted until the DOS developed a DNA testing plan (Bruno, 2017). In 2010, the DOS passed new rules requiring DNA testing for international refugees seeking to reunify with their families in the United States (Holland, 2011). As a result, it took another two years to reinstate the P-3 program (Dove, 2012). The DNA testing aspect of

---

2. Statements made during the author's unpublished interviews with relationship-testing laboratory directors in 2010.

3. The DOS looks at refugees to be considered for admission to the United States who are referred by the United Nations High Commissioner for Refugees (UNHCR) under three priorities: Priority 1, individual cases referred by designated entities to the program by virtue of their circumstances and apparent need for resettlement; Priority 2, groups of special concern designated by the DOS as having access to the program by virtue of their circumstances and apparent need for resettlement; and Priority 3, individual cases from designated nationalities granted access for purposes of reunification with family members already in the United States (US Department of State, 2017a).

the program was reaffirmed by public comment in 2015 (80 F.R. 129). Although the actual numbers are relatively small, the P-3 program now requires DNA testing to verify claimed relationships of all visa applicants. The P-3 and other refugee programs are capped each year, dropping from a 1,150 ceiling in FY2013 to 760 in FY2018 (Worth, 2015). These family reunification cases are only a fraction of those refugee cases that are considered for settlement in the United States; however, both the ceilings for refugees and the actual overall number of refugees obtaining permission to settle in the United States have dropped precipitously in recent years (Bruno, 2018). Whereas in FY2017 the proposed ceiling was 110,000, the FY2019 ceiling was 30,000 (Bruno, 2018), and whereas in 2016 the United States admitted almost 85,000 refugees, in 2018 it admitted just shy of 22,500 (Bruno, 2018).

Mandatory testing for DNA in immigration marks a new dimension in biometrics, carrying sociopolitical and legal implications. Edward S. Dove, a law professor at University of Edinburgh, has noted: "The refugee's body has become a site of evidence where truth lies and fraud lurks. 'Family' has regressed from relational understanding to biological, corporeal understanding" (Dove, 2012). Moreover, some of the cases of fraud detected in the P-3 program and in the genotype recycling cases might be cases of miscommunication, and some might be the result of families circumnavigating the too stringent rules regarding relationships. But some also might be attempts to traffic persons.

Unaccompanied Migrant Minors and Family Separations at the US-Mexico Border

In response to extreme violence and economic instability in their home countries in Central America and Mexico, children and families increasingly are migrating north, largely to reunite with relatives and friends already in the United States (Hicks, 2015). In many cases, a parent will migrate first in order to establish a home for the family, then the minors may travel unaccompanied to join their parent. In other cases, a family cannot afford to travel together, but with a minor at risk of violence in their home community, such as gang threats, the minor will travel alone to escape the violence. Unaccompanied children from Central America and Mexico arrived at the Texas border in 2014 in considerably greater numbers than in previous years (Pierce, 2015). This sudden increase sparked new policies (US Government Accountability Office, 2015), political debates, and media inquiries into the safety and welfare of children who arrive at the border without relatives. Policies to divert child migrants staved off another potential surge in 2015, although the numbers still remained high (Pierce, 2015). Since the initial crisis, the number of unaccompanied children has remained steady, at about the same levels as in 2014 (see figure 10.2). Not only have the Trump administration's policies aimed at curbing immigration been unsuccessful at preventing migration, but managing children separated from their parents has become increasingly more expensive, such that authorities have diminishing resources for vetting the identities of the custodians with whom children are placed as they await their hearing or deportation (Stillman, 2015).

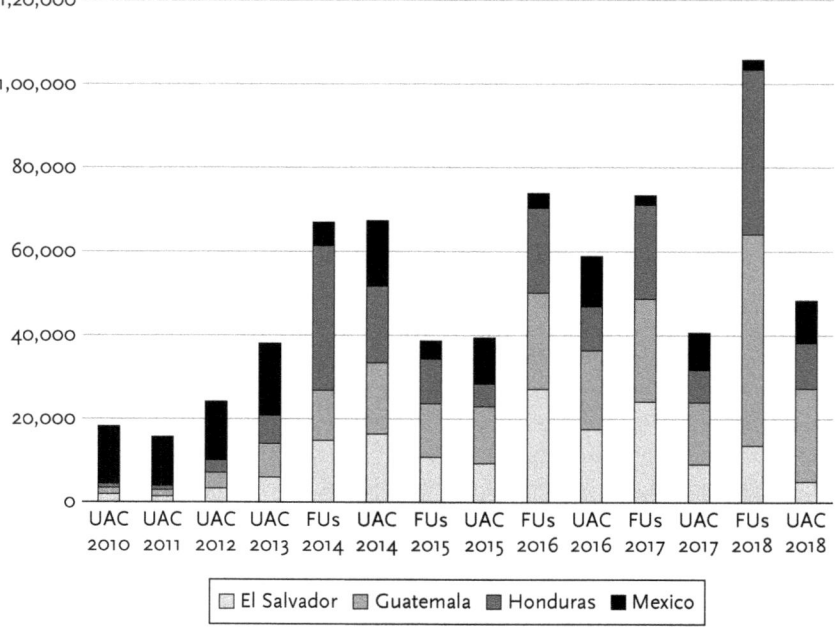

Figure 10.2 Southwest border apprehensions of unaccompanied minors (officially termed unaccompanied alien children, UAC) and family units (FUs). In 2014, an increasing number of children crossed into the United States, turned themselves over to authorities, and were then placed with sponsors in the United States pending a hearing. In parallel, family units increasingly arrived, with at least one child. The fiscal year (FY) runs September though October. *Source:* US Customs and Border Protection (2019). Graph by Sara H. Katsanis.

Human trafficking of children crossing the border alone is also a persistent problem. In the midst of the 2014 surge, at least one group of eight teenagers from Guatemala was placed with traffickers to work on an egg farm in Ohio (Hennessy-Fiske, 2015). Possibly more children have been trafficked, but data are lacking (Portman, 2016). This event brought to light the inadequacy of screening of adults hosting unaccompanied migrant children, whether they are relatives, family friends, or foster parents. Most children (~60%) are placed with their parents, whether or not the parents have legal status themselves, and another 32% are placed with a relative (US Government Accountability Office, 2016). The vetting of the sponsors varies, but none of them are required to provide documentation of the relationship. This oversight is one opportunity for traffickers to claim children as their own.

For the most part, the current DNA testing for child migrants is ad hoc. When an unaccompanied child is too young to verbally verify the relationships with a sponsor-to-be, the child's attorney or social worker might purchase a $400 to $1,000[4] DNA test through one of the commercial AABB-accredited laboratories to verify if a

---

4. Costs vary from one laboratory to the next. These figures are estimated from Dove (2012) and Barata et al. (2015).

particular person is indeed a relative. DNA tests might be used in cases where a social worker suspects child trafficking. Still, DNA tests are not used routinely to vet biological relationships.

The 2014 surge of unaccompanied minors, however, introduced an occasion to systemize the vetting of biological relationships, similar to the P-3 program. The Obama administration instated the Central American Minors Refugee/Parole program (CAM) to provide a legal mechanism for children fleeing violence to apply for sponsorship while still in their originating countries (US Department of State, Bureau of Population, Refugees and Migration, 2014; Hicks, 2015). To be eligible for sponsorship, a child's parents had to be lawfully in the United States, while other relatives, such as siblings and aunts, were ineligible to be sponsors for children. Visa applicants were required to submit to DNA testing to demonstrate the claimed parentage. Two years later, the Trump administration rescinded the program but court battles have sought to ensure that CAM petitions that were under consideration at the time the program ended will be processed through completion (Gonzales, 2019).

In 2018 several media outlets reported that the ORR's flawed child placement program had resulted in US immigration having "lost" thousands of migrant children after placement (Wang, 2018; Harmon, 2018). It was also reported that the processes for finding appropriate sponsors and housing for children and for following up with them for their immigration hearings were deeply flawed and inconsistent (US Government Accountability Office, 2016; 2018).

Beginning in April 2018, the Trump administration increased detention of migrants crossing illegally into the United States, creating a human rights crisis as thousands of children were separated from their parents, relatives, or other adults traveling with them and placed into the ORR unaccompanied minors program (Blake, 2018). This massive influx into an already flawed system resulted in the ORR being unable to trace the placement of children (Katsanis & Wagner, 2018). In June 2018, President Trump issued an executive order reversing his family separation policy. The new policy allowed children to stay with their parents (albeit in detention or prior to deportation). DNA testing was already called for to verify relationships of children crossing with potential human traffickers. After the policy reversal, with the haste to reunify families, came an outpouring of offers from consumer genomics companies to use DNA tests to assist efforts to reunify families (Molteni, 2018; Song, 2018; Ray, 2018).

However, concerns about privacy of genetic information and the vulnerability of the children prevailed, and many groups working with migrant families declined use of the tools offered by consumer genomics companies (The American College of Medical Genetics and Genomics, 2018; Peters, 2018). The ORR, meanwhile, proceeded with DNA testing to screen for trafficking, using an undisclosed laboratory and unknown methods (Moghe & Chavez, 2018). At the time of writing, the ORR has not publicly disclosed what procedures have been followed to ensure consent for voluntary samples, what tools have been applied to ensure accuracy and security of data and specimens, and what measures have been undertaken to protect the families from misuse of their genetic information (Ray, 2018). In fact, the ORR's lack of transparency regarding its approach and methods have in part fueled a backlash against the US government, as well as privacy concerns about how genetic information for immigration purposes in general is obtained, shared, and applied (Richards, 2018).

Rapid DNA instruments used by DHS at border sites could potentially have been used in these family reunification cases in 2018 to verify biological relationship claims (see chapter 5). They could have been used also to identify human traffickers in their attempts to use false identities (Ramirez et al., 2015). In fact, in 2019 a pilot run by the ICE examined whether systematic use of Rapid DNA instruments would be effective at detecting fraudulent relationships (Molteni, 2019). But Rapid DNA instruments need not be relegated to use only by government officials and law enforcement. Rapid DNA instruments could be used by nongovernmental organizations (NGOs) working with immigrant families to ensure that children are placed with relatives and to similarly verify claims of biological relationships.

The 2018 border crisis, in which families were separated needlessly, opened up broad discussions among geneticists and policymakers on whether DNA tests to screen for trafficking might be one measure to ensure relationship claims (Moghe & Chavez, 2018). The dialogue appeared to be split, with anti-immigration activists embracing the use of DNA in this manner, and pro-immigration activists split in advocating for or against the use of DNA (Wagner et al., 2019). Certainly privacy is a concern, and it is essential that any government use of genetic information for reunification be restricted from secondary uses, including law enforcement. It is a real and ongoing bioethical challenge to judge consent among people who are under duress, or in fact whether consent is even possible. But these debates about rights to privacy pale in comparison to the immediate and devastating trauma that can be inflicted upon a child who has been separated from his or her family and housed indefinitely with strangers (Katsanis & Wagner, 2018). Context is essential when weighing the ethical challenges of any scientific approach, and engagement with both experts and the public is an important part of developing policies that minimize public harm and maximize public benefit of technological tools like DNA (Farahany, 2018).

Incremental versus Universal DNA Databases

Governments that wish to monitor their citizens using biometric data can either expand criminal and immigration databases or create a universal DNA database comprised of the DNA profiles of an entire population. In 2015, following an ISIS bombing, Kuwait became the first country in the world to consider establishing a universal DNA database, passing legislation requiring all citizens, residents, and visitors to submit DNA samples (AFP, 2015; see also chapter 15). The justification was the need to protect national security and assist identification of victims following a mass disaster (Sperry, 2017). This provision was heavily criticized, with claims that genetic fingerprinting is a "massive infringement on human rights" and unlikely to prevent terrorist attacks (Whitson, 2015; Coghlan, 2016). Eventually the legislation was withdrawn (Al-Hamoud Al-Seyassah, 2017).

According to these critics, the acquisition by governments of any massive amounts of sensitive personal data from citizens goes against democratic principles. Even so, law enforcement and government agencies have increasingly gained access to genetic information from a broad swath of the population. In the United States, for example, CODIS has been expanded to include DNA samples obtained from people convicted of nonviolent crimes and misdemeanors and from people merely arrested for crimes, on the premise that it will improve law enforcement's ability to solve more crimes. As previously mentioned, US immigration authorities have justified an increase in DNA collection as a way to prevent undocumented immigrants and criminals from entering the country. Such a move to capture DNA from individuals who are merely detained could potentially increase DNA collection from lower income families and people of color.

The trend to focus on collection of DNA from nonwhite immigrants was reflected in 2016 when US members of Congress proposed legislation that would require DNA verification of any claimed relationships in immigrant petitions, in addition to abolishing the possibility of visas for immigrants from Iran, Iraq, Libya, Somalia, Syria, Sudan, and Yemen (Visa Integirty and Security Act, 2016). The bill died in committee, but Trump's executive order restricted the issuance of visas to individuals from these same seven countries (Executive Office of the President, 2017). Since then, newly proposed immigration bills have included provisions requiring genetic testing at the discretion of the DOS (Katsanis & Wagner, 2018).

As with the universal DNA database proposed in Kuwait, critics claim such systems are a gross intrusion of privacy at best and a human rights violation at worst (Sperry, 2017). A DNA database of a population subset can be used to discriminate against populations through genetic surveillance, as appears to be the case for the DNA data collected from Muslim Uighur people in Xinjiang, China (Cyranoski, 2017). But universal databases can also be considered an equalizer by eliminating the disparities that currently exist in our criminal justice and immigration systems. Kirsten Dedrickson has argued: "A universal DNA database's benefits in efficiently and effectively solving crimes, exonerating the innocent, and decreasing racial disparities in law enforcement ... make such a database immensely appealing from a public safety and criminal justice perspective" (Dedrickson, 2018). The DNA databases of federal agencies, such as CODIS and soon HART, are comprised of people who enter these systems and thus are a reflection of disparities.

Statelessness

Many migrants travel without documented POI. In the case of migrant children from Latin America, they often choose to travel without documentation to protect the names and contacts of their direct family members on both sides of the border from both immigration authorities and traffickers. Once migrants are in the hands of an immigration authority, however, without any POI, their asylum claims can be disputed. Documentation of some type is essential for consideration, and if asylum seekers are denied entry into the United States, they might, without the necessary documentation to prove otherwise, become stateless, which happens around the world to thousands of migrants.

In 2014, the United Nations High Commissioner for Refugees (UNHCR) implemented a global action plan to address poverty and crime associated with statelessness, particularly the risks to individuals who are not necessarily refugees but are otherwise rendered stateless through changing borders or who have lost their citizenship after a prolonged period outside of their homeland (United Nations High Commissioner for Refugees, 2014b). The UN noted that identification of individuals is important for documenting their rights to statehood, and with statehood, to provide them with access to privileges and fulfillment of human rights. But the outlined mechanisms to trace individuals are population-based approaches through birth registration and nationality documentation. With poor documentation in developing countries in particular, advocates are turning to iris scans, digital fingerprints, and other biometrics for accurate identification (Ensur, 2016).

An estimated 10 million people cannot claim citizenship in any country because they lack a nationality. In addition to changing borders or personal absence rendering a person stateless, many people also are lacking nationality documents due to gender discrimination or discrimination against particular ethnic or religious groups (United Nations High Commissioner for Refugees, 2014b). As a result, they face difficulties in accessing basic privileges, like getting married, opening a bank account, or getting a job (United Nations High Commissioner for Refugees, 2014b). One such group is the ethnic highlanders in the northern parts of Thailand. Known as the *chao khao* in Thai, which means "hill people," they comprise one of the largest stateless populations in the world.[5] The *chao khao* have resided in Thailand for generations, yet nearly half a million cannot prove their claims to Thai citizenship because they have been sequestered in mountainous rural regions without access to modern means of documentation, like birth certificates at hospitals (Flaim, 2017). Various local, national, and international governmental organizations and NGOs have sought to address the precarious status of ethnic minorities and refugees in Thailand by demonstrating their relationship with Thai citizens through DNA testing. Relationship DNA tests were used in citizenship adjudication cases by the Thai state, as well as by the DOS in verifying claims to asylum by refugees seeking resettlement in the United States (Flaim, 2017).

However, science is not always sufficient. The hill people of Thailand have long been a persecuted ethnic minority. In some cases, when courts are presented with DNA test reports documenting a person's relationship and clear Thai lineage, the judge would state, "He doesn't look Thai to me" and deny the application. One district official interviewed by Amanda Flaim of Michigan State University stated, "Even DNA cannot always be trusted. I can only really trust DNA tests with a person's mother. Why? Because you are born from your mother. We are not born of fathers. When someone submits a DNA test with their father without other proof, how can I know this is not the bastard child of a Burmese prostitute?" (Flaim, 2017). In these

---

5. The *chao khao* consist of the Karen, Hmong, Lahu, Lisu, Akha, Khamu, and Lua.

cases genetic information that clearly demonstrates a biological relationship with a parent was entirely ignored in favor of traditional xenophobia and racism.

Genetic tools might indeed empower stateless migrants who lack paper documentation to support their cases for asylum. But we lack data on the societal implications of using DNA as evidence of race, ethnicity, and identity, especially in vulnerable populations. Although using ancestry DNA as a matter of policy to determine nationality has been condemned (as discussed later in this chapter in the context of the United Kingdom's 2010 attempted pilot), stateless migrants might find personal utility in it on a case-by-case basis. For example, in 2016 I interviewed a young man I will call Nyan, who was residing in a Scandinavian country. Nyan's mother told him he was Karen, an ethnic Burmese minority, but she died when he was age four, leaving him in the care of his mother's Thai friend in Thailand. After his foster mother died, Nyan, who was a teenager at the time, smuggled himself onto a ship bound for Scandinavia. Upon his arrival, he was granted temporary asylum, pending documentation of his country of origin. The Scandinavian country likely will grant him asylum if he is a refugee from Myanmar, but not if he is Thai. But neither the Myanmar nor Thai governments have birth records for him. If he can demonstrate he is of Burmese heritage through biogeographical DNA testing, he might be granted citizenship.[6]

ETHICAL CHALLENGES WITH GENETIC INFORMATION

The historic lessons of eugenics and concerns (both historic and modern) about abuse of power have thus far limited the broad implementation of DNA as a biometric in the United States. While DNA technology is valuable for detecting fraud in claimed relationships, especially for detecting human trafficking at borders, developing policies that favor biological relationships over other community links might not be in the best interest of migrants, nor in line with American ideals. Further, biometric and genetic data collected for immigration might also be used for surveillance of a population or to track migrants suspected of being a threat to the state (Currion, 2015).

In 2008, the UNHCR noted that legitimate privacy concerns about the use of DNA information meant that "DNA testing must be carefully regulated" (United Nations High Commissioner for Refugees, 2008). Data collected for criminal justice purposes in the United States are governed under the Privacy Act of 1974 (5 U.S.C. § 552), but that act only provides protections for citizens and legal residents. Thus, the Privacy Act that governs the federally permitted data collected in the United States does not apply to the majority of data collected for immigration purposes. The implication is that DNA or other biometric data collected by a federal agency from non-US persons can be shared with other agencies across the government. There also

---

6. This is an ongoing case study that the author is pursuing to document the utility of ancestry DNA testing. The name has been changed to protect his identity.

are no protections for what kind of information can be elicited for immigration. The legal regulations restrict CODIS to markers that only confer identity, but no such restrictions exist outside of law enforcement (Katsanis & Wagner, 2013).

Even if an agency does not have a mandate to collect genetic information from detained persons or immigrants, there are no provisions to protect anyone from surreptitious DNA collection, the collection of abandoned DNA specimens, such as from chewing gum or water bottles. Some states restrict testing of abandoned DNA, but the laws are sporadic and almost all exclude law enforcement purposes and nonresidents from the restrictions.

Race and Nationality Are Social Constructs, Not Biological Ones

As immigration policies emerge to exclude visas for persons with particular nationalities, a new trend is emerging to use biological ancestry to verify nationalities (Khandaker, 2018). Genomic markers that indicate a person's ancestral roots are increasingly robust as more and more consumers and researchers provide data on discrete population differences. As in the case of Nyan's ancestry, which may be Burmese or Thai, genetic markers may provide some clues to his origins.

But nationality, like race, is not equivalent to ancestry, since nationality is a social construct, a product of cartographers and politicians. Race as a biological concept dates back to theories of polygenism versus monogenism (human races arising from multiple origins verses a single origin) in the eighteenth century, and phrenology in the nineteenth century. Eugenic policies sought to define human character based on genetic traits, and the application of genetic difference to racial differences resulted in horrific abuses (Black, 2003). Even today, genome scientists stratify data by race, calculate population frequencies of genetic traits by race, and develop genome-guided treatment by race. As a social construct, race cannot classify humans or determine a person's nationality, yet geneticists find race to be a useful, if imprecise, proxy for ancestry (Howard, 2016). Movements within genome studies to substitute biological ancestral markers—those single-nucleotide polymorphic markers that can trace ancestral roots in a person—have advantages over race for determining genetic heritage that might inform health risks (Yudell et al., 2016). However, this use of ancestral genetic markers to infer race and social belonging is uncharted territory. For example, we do not know if the use of ancestry DNA to determine Native American tribal enrollment and rights has increased notions of identity, family, and heritage among native peoples (Tallbear, 2013).

In 2009 the United Kingdom piloted the use of ancestry DNA tests in combination with isotope analysis of hair or fingernails from persons claiming to be Somali refugees. Following the civil war in Somalia, the United Kingdom was seeing a rapid increase in Somali asylum applications (Casciani ,2006). But given that most Somalian refugees were arriving without POI, the Home Office was finding it challenging to distinguish Somalian from Kenyan, Nigerian, or other Africans. The UK Border Force developed a systematic pilot to use biogeographical DNA markers and isotope analysis to determine nationality of refugees. For a few months, the program tested applicants' mtDNA and Y chromosome, hypothesizing that they could distinguish Somalian from Kenyan or other applicants based on the genetic haplotypes that differ among populations (Dove, 2012). However, scientists and ethicists alike criticized the program, claiming it was premature for border use (Travis, 2009). They contended that such ancestral markers were not precise enough for predicting subgroups and that biological roots were not a substitute for nationality. Further, by adopting scientific technologies to screen Somalian applicants, the screeners bypassed review of testimonies that might have revealed persecution and thus legitimate grounds to be considered for entry (Tutton et al., 2014). The UK pilot program was halted abruptly but left several questions unanswered. For example, could DNA ancestry tests be useful in place of POI? And if so, what policies might prevent DNA tests being used to exclude population groups from state rights?

Scientifically, genetic testing can provide only information about probable place of origin and cannot serve as a reliable guide to nationality (Tutton et al., 2014). As a *Nature* editorial put it following the UK scandal, "the idea that genetic variability follows man-made national boundaries is absurd" (Editorial, 2009). While that might be true, use of genetic information when other traditional POI is not available should be considered for the most vulnerable stateless people, as in Nyan's case.

Indeed, systematic application of biological interpretations of race, ethnicity, and nationality as a metric for immigration can be used to discriminate against or stigmatize entire populations (Katsanis & Kim, 2016). The tightening of immigration policies in the United States and in Europe means that asylum seekers will use more clandestine methods of entry (Schuster, 2011). In turn, refugees and migrants are likely to seek out smugglers and forge documentation in hopes of gaining passage (Tutton et al., 2014). False claims of nationality are not a new challenge, but using DNA to detect these false claims could open new opportunities to discriminate against and stigmatize groups of people.

Family Is a Social Construct, Not a Biological One

While genetic information is useful for verifying identity and informing biological relations, the long-standing prejudice that family must be biological is institutionalized by requiring DNA testing as a matter of policy. The authorities that require DNA as verification of a family relationship negate the nontraditional family constructs that are common and at last acceptable in this century, including families with same-sex or transgender parents, families with donor-conceived children, and those with adopted children. The Affidavit of Relationship form used for the P-3 refugee program lists almost 50 categories of relationship in their attempt to be as inclusive as possible. But the definitions of these relationships vary greatly from one culture to the next. For refugees, the family unit might fluctuate over time, rendering categories nonsensical (Dove, 2012). For example, in some cultures with village-oriented parenting, the term *mother* might apply to multiple women responsible for child care. And in many cultures, family extends to a network of relatives and clan structure not detectable by typical DNA typing (Dove, 2012). For example, a genetic test result that appears to a DNA analyst to be a misattributed relationship might actually be a miscommunication of the terms to define relationships in a family (Parker et al., 2013).

A recent case of twins born to a same-sex couple in California highlights these struggles. One of the fathers is an American, the other Egyptian. The married couple had their sons through a Canadian surrogate mother, who carried twins, one related to each of the fathers. Once they were born in Canada, the couple brought their sons home to California and applied for citizenship papers for the twins. One boy was granted US citizenship through his biological American father, Andrew, but the other twin was not because his biological father, Elad, is not a US citizen (Cruz, 2017). As reported by Brian Melley of *The Chicago Tribune*, "The consular official told them she had discretion to require a DNA test to show who the biological father was of each boy and without those tests neither son would get citizenship. The men knew that Andrew was Aiden's biological father and Elad was Ethan's but they had kept it a secret and hadn't planned on telling anyone" (Melley, 2018).

This case has since been resolved through court battles (Flynn, 2019), but it highlights how nontraditional families can be adversely affected by basing policies solely on biological assumptions. This is a tragic byproduct of the misuse of genetic data that should be used for informing immigration decisions, not as the deciding information itself. In this case, the family knew that their relationships were not traditionally biological, and they had hoped to keep the biologically attributed fatherhoods private. But in other cases, family members might not know their biological relationships.

Misattributed parentage is far more common than most of us think. Usually paternity is in question, but the use of donor eggs and embryos has also complicated genetic relationships (Hercher & Jamal, 2016). Infidelity and sexual assault can cause pregnancies that are unwanted, and the biological origins of children are sometimes masked. Sometimes this is because women are raped and want to mask the child's birth father to protect the child and themselves from stigmatization and further violence. Sometimes the woman is not aware of the true parent, having had multiple partners, by choice or through violence. Rates of misattributed paternity range from 1 to 20%, depending on the population (Bellis, 2005).

Immigration genetic tests to verify biological relationships are bound to reveal misattributed parentage (Sheets et al., 2016). In medical testing, genetic counselors have long debated whether to disclose unexpected biological relationships, generally advising that findings be disclosed only to the mother, in private. In human rights contexts, such as the identification of war dead or victims of a mass disaster, ethicists generally advise not disclosing unexpected relationship findings, but rather testing as many relatives as possible to limit the likelihood of one person being excluded as biologically related (Parker et al., 2013). The revelation of a lack of a biological relationship that is believed to exist can have serious ramifications for family relationships and even put the woman in danger. In the worst of circumstances, DNA tests revealing nonpaternity could be used as evidence against a rape victim.

CONCLUSION

If this chapter leaves you confused about whether or not genetic data are useful for immigration, the author is in the same position. In some cases, using DNA might be the most efficient way to detect fraud and to prevent human trafficking. DNA as a technological tool is far better than human minds in affirming claimed truths. Yet some critics are concerned that immigration data containing DNA profiles could also be used for investigating crimes unrelated to a given person's immigration status. If DNA data for immigration can be protected from misuse, such as stigmatizing a given population, then DNA could be a useful biometric. If DNA can be used to expedite laborious border processes, then the United States could develop protections for privacy of sensitive data and consent and assent processes for migrants undergoing testing, whether voluntary or required.

But social constructs of nationality, race, and family cannot be measured using biological data alone. DNA should never be used to deny privileges, including entry to a country. DNA data collected from migrants should not be used to screen populations or surveille future threats, nor should DNA be used to stigmatize or exclude populations from earning citizenship or residency.

REFERENCES

AFP, Kuwait City. (2015, July 1). Kuwait makes DNA tests mandatory after ISIS bombing. *Al Arabiya*. Retrieved from https://english.alarabiya.net/en/News/middle-east/2015/07/01/Kuwait-makes-DNA-tests-mandatory-after-ISIS-bombing-.html

Al-Hamoud Al-Seyassah, J. (2017, June 10). High court rules against controversial law on DNA—"articles violate Constitution". *Arab Times*. Retrieved from http://www.arabtimesonline.com/wp-content/uploads/pdf/2017/oct/06/ATKWT20171006.pdf.

The American College of Medical Genetics and Genomics. (2018, June 29). *Use of genetic testing to reunify families separated as a result of immigration enforcement policies: A statement from the American College of Medical Genetics and Genomics (ACMG)*. Retrieved from https://www.prnewswire.com/news-releases/use-of-genetic-testing-to-reunify-families-separated-as-a-result-of-immigration-enforcement-policies-a-statement-from-the-american-college-of-medical-genetics-and-genomics-acmg-300674668.html.

ANDE Corporation. (2018, June 4). Press release: ANDE Corporation's Rapid DNA Identification System first to receive FBI approval under new standards. *PR Newswire*. Retrieved from https://www.prnewswire.com/news-releases/ande-corporations-rapid-dna-identification-system-first-to-receive-fbi-approval-under-new-standards-300659210.html.

Associated Press. (2008, November 23). To save adopted girl, U.S. couple gives her up. *USA Today*. Retrieved from http://archive.boston.com/news/nation/articles/2008/11/23/to_save_adopted_daughter_calif_couple_gives_her_up/.

Barata, L. P., et al. (2015). What DNA can and cannot say: Perspectives of immigrant families about the use of genetic testing in immigration." *Stanford Law & Policy Review*, 26, 597–638.

Bateman-House, A., & Fairchild, A. (2008). Medical examination of immigrants at Ellis Island. *AMA Journal of Ethics*, 10, 235–241.

Bellis, M. A., Hughes, K., Hughes, S., & Ashton, J. R. (2005). Measuring paternal discrepancy and its public health consequences. *Journal of Epidemiology: Community Health*, *59*, 749–754.
Black, E. (2003). *War against the weak: Eugenics and America's campaign to create a master race*. Westport, CT: Dialog Press.
Blake, M. (2018, June 20). How a child moves through a broken immigration system. *The Wired*. Retrieved from https://www.wired.com/story/broken-immigration-system-family-separation/.
Boyd, J. (2017). *Leveraging DNA information in support of customer needs*. (Rapid DNA Technology Forum). Alexandria, VA: Forensic Technology Center of Excellence.
Bruno, A. (2017). *Refugee admissions and resettlement policy*. (CRS Report.) Washington, DC: Congressional Research Service.
Bruno, A. (2018, December 18). *Refugee admissions and resettlement policy*. (CRS Report). Washington, DC: Congressional Research Service. Retrieve from https://crsreports.congress.gov/product/pdf/RL/RL31269/23.
Burnett, J., & Gonzalez, R. (2018, May 10). Homeland Security secretary defends separating families who cross border illegally. National Public Radio. Retrieved from https://www.npr.org/2018/05/10/609480137/homeland-security-secretary-defends-separating-families-of-illegal-border-crosse.
Casciani, D. (2006, May 30). Somalis' struggle in the UK. BBC News. Retrieved from http://news.bbc.co.uk/2/hi/uk_news/magazine/5029390.stm.
Coghlan, A. (2016, October 21). Kuwait to change law forcing all citizens to provide DNA samples. *New Scientist*. Retrieved from https://www.newscientist.com/article/2109959-kuwait-to-change-law-forcing-all-citizens-to-provide-dna-samples/.
Cruz, C. (2017, September 6). DNA dilemma: Baby can't get US citizenship. CBS Los Angeles. Retrieved from https://losangeles.cbslocal.com/2017/09/06/baby-cant-get-us-citizenship/.
Currion, P. (2015, August 26). Eyes wide shut: The challenge of humanitarian biometrics. *IRIN*. Retrieved from https://www.thenewhumanitarian.org/opinion/2015/08/26/eyes-wide-shut-challenge-humanitarian-biometrics.
Cyranoski, D. (2017). China expands DNA data grab in troubled western region. *Nature*, *545*, 395–396.
Dedrickson, K. (2018). Universal DNA databases: A way to improve privacy? *Journal of Law and Biosciences*, *4*, 637–647.
Dove, E. S. (2012). Back to blood: The sociopolitics and law of compulsory DNA testing of refugees. *University of Massachusetts Law Review*, *8*, 1–42.
Editorial. (2009). Genetics without borders. *Nature*, *461*, 697.
Ensur, C. (2016, February 22). Biometrics in aid and development: Game-changer or trouble-maker. *The Guardian*. Retrieved from https://www.theguardian.com/global-development-professionals-network/2016/feb/22/biometrics-aid-development-panacea-technology.
Esbenshade, J. (2010). *Special report: An assessment of DNA testing for African refugees*. San Diego, CA: Immigration Policy Center.
Executive Office of the President. (2017, February 1). *Protecting the nation from foreign terrorist entry into the United States*. Executive Order 13769 of January 27, 2017. 82 F.R. 8977-8982. Washington, DC: The White House.
Fairchild, A. L. (2004). Policies of inclusion: Immigrants, disease, dependency, and American immigration policy at the dawn and dusk of the 20th century. *American Journal of Public Health*, *94*, 528–539.
Farahany, N., Chodavadia, S., & Katsanis, S. H. (2019). Ethical guidelines for DNA testing in migrant family reunification. *American Journal of Bioethics*, *19*(2), 4–7.
Flaim, A. (2017). Problems of evidence, evidence of problems: Expanding citizenship and reproducing statelessness among highlanders in Northern Thailand. In B.

N. Lawrence & J. Stevens, eds., *Citizenship in question: Evidentiary birthright and statelessness* (pp. 226–254). Durham, NC: Duke University Press.

Flynn, M. (2019, February 22). One twin was a citizen, the other undocumented: A victory in court for their same-sex parents rebukes the State Department. *The Washington Post*. Retrieved from https://www.mercurynews.com/2019/02/22/one-twin-was-a-citizen-the-other-undocumented-a-victory-in-court-for-their-same-sex-parents-rebukes-the-state-department/.

Gambino, L., & Laughland, O. (2018, June 20). Donald Trump signs executive order to end family separations. *The Guardian*. Retrieved from https://www.theguardian.com/us-news/2018/jun/20/donald-trump-pledges-to-end-family-separations-by-executive-order.

Gonzales, R. (2019, April 12). Trump administration to allow 2,700 Central American children into the U.S. National Public Radio. Retrieved from https://www.npr.org/2019/04/12/712948933/trump-administration-to-allow-2-700-central-american-children-into-the-u-s.

Goth, G. (2008). Biometrics could streamline border crossings. *Computing Now*, 4, 2.

Greenwald, G. (2013, June 6). NSA collecting phone records of millions of Verizon customers daily. *The Guardian*. Retrieved from https://www.theguardian.com/world/2013/jun/06/nsa-phone-records-verizon-court-order.

Harmon, A. (2018, May 28). Did the Trump administration separate immigrant children from parents and lose them? *The New York Times*. Retrieved from https://www.nytimes.com/2018/05/28/us/trump-immigrant-children-lost.html.

Hennessy-Fiske, M. (2015, November 15). Is Ohio case of migrant youth trafficking evidence of a "systemic problem"? *Los Angeles Times*. Retrieved from https://www.latimes.com/nation/la-na-ohio-immigrant-sponsor-20151115-story.html.

Hercher, L., & Jamal, L. (2016). An old problem in a new age: Revisiting the clinical dilemma of misattributed paternity. *Applied & Translational Genomics*, 8, 36–39.

Hicks, J. (2015, April 3). This new but little-known immigration program seeks to reunite Central American families in US. *The Washington Post*. Retrieved from https://www.washingtonpost.com/news/federal-eye/wp/2015/04/03/little-known-immigration-program-seeks-to-reunite-central-american-families-in-u-s/.

Holland, E. (2011). Moving the virtual border to the cellular level: Mandatory DNA testing and the U.S. refugee family reunification program. *California Law Review*, 99, 1635–1682.

Homeland Security Act of 2002. Pub. L. No. 107-296, 116 Stat. 2135.

Howard, J. (2016, February 9). What scientists mean when they say "race" is not genetic. *The Huffington Post*. Retrieved from https://www.huffpost.com/entry/race-is-not-biological_n_56b8db83e4b04f5b57da89ed.

Immigration Act of 1882. 22 Stat. 214.

Immigration and Nationality Act of 1952. 8 U.S.C. § 1185, Pub. L. No. 82-414, 66 Stat. 163.

International Commission against Impunity in Guatemala (CICIG). (2010). *Decree 77-2007: Report on players involved in the illegal adoption process in Guatemala since the entry into force of the adoption law*. Guatemala City, Guatemala.

Joint Council on International Children's Services and National Council for Adoption. (2007, September 29). *U.S. adoptions from Guatemala to halt as of January 1, 2008*. Adoption Social Work New York.

Katsanis, S. H. (2015, March 11). Humanitarian crisis at border calls for an unusual tool: DNA testing. *News and Observer*. Retrieved from https://www.newsobserver.com/opinion/op-ed/article13558967.html.

Katsanis, S. H., & Kim, J. (2014). DNA in immigration and human trafficking. In D. Primorac & M. Schanfield, eds., *Forensic DNA applications: An interdisciplinary perspective* (Chapter 22). New York, NY: CRC Taylor and Francis.

Katsanis, S. H., & Kim, J. (2016). Privacy challenges with genetic information. In S. J. Morewitz & C. Sturdy Colls, eds., *Handbook of missing persons* (Chapter 25). New York, NY: Springer.

Katsanis, S. H., & Wagner, J. K. (2013). Characterization of the standard and recommended CODIS markers. *Journal of Forensic Sciences, 58*(Suppl. 1), S169–172.

Katsanis, S. H., & Wagner, J. K. (2018, June 25). Why aren't we taking DNA instead of taking children. *The Herald Sun.* Retrieved from https://www.heraldsun.com/opinion/article213759989.html.

Khandaker, T. (2018, July 26). Canada is using ancestry DNA websites to help it deport people. *Vice News.* Retrieved from https://news.vice.com/en_ca/article/wjkxmy/canada-is-using-ancestry-dna-websites-to-help-it-deport-people?utm_campaign=sharebutton.

Kim, J., Mammo, D., Siegel, M. B., & Katsanis, S. H. (2011). Policy implications for familial searching. *Investigative Genetics, 1*, 2–22.

Lynch, J. (2012). *From fingerprints to DNA: Biometric data collection in U.S. immigrant communities and beyond.* Washington, DC: Immigration Policy Center, American Immigration Council.

McLaughlin, A. (1905, February). How immigrants are inspected. *The Popular Science Monthly,* 6.

Melley, B. (2018, January 23). Lawsuit filed after one twin boy born to gay couple granted U.S. citizenship, the other denied. *The Chicago Tribune.* Retrieved from https://www.chicagotribune.com/nation-world/ct-gay-couple-twins-citizenship-20180123-story.html.

Moghe, S., & Chavez, N. (2018, July 5). DNA testing being done on separated migrant children and parents, official says. CNN. Retrieved from https://www.cnn.com/2018/07/05/politics/dna-testing-migrant-family-separation/index.html.

Molteni, M. (2018, June 22). Family DNA testing at the border would be an ethical quagmire. *The Wired.* Retrieved from https://www.wired.com/story/family-dna-testing-at-the-border-would-be-an-ethical-quagmire/.

Molteni, M. (2019, May 2). How DNA testing at the US-Mexico border will actually work. *The Wired.* Retrieved from https://www.wired.com/story/how-dna-testing-at-the-us-mexico-border-will-actually-work/.

National Science and Technology Council. (2008). *Biometrics in government post-9/11: Advancing science, enhancing operations.* Washington, DC: National Science and Technology Council.

Parker, L. S., London, A. J., & Aronson, J. D. (2013). Incidental findings in the use of DNA to identify human remains: An ethical assessment. *Forensic Science International: Genetics, 7*, 221–229.

Peters, A. (2018, June 26). Why immigration groups said no to using DNA to reunite separated kids. *Fast Company.* Retrieved from https://www.fastcompany.com/40589668/why-immigration-groups-turned-down-23andmes-offer-of-dna-tests.

Pierce, S. (2015). *Unaccompanied child migrants in U.S. communities, immigration court, and schools.* (U.S. Immigration Policy Program, Migration Policy Institute Report). Migration Policy Institute.

Portman, R. (2016, January 28). *Statement of Chairman Rob Portman to US Senate Permanent Subcommittee on Investigations hearing.* 115th Congress. .

Ramirez, W., McKenna, M., & Somers, A. (2015). Repatriation and reintegration of migrant children. In K. Musalo, P. Ceriani Cernadas, & L. Frydman, eds., *Childhood and migration in Central and North America: Causes, policies, practices and challenges* (Chapter 12). San Francisco, CA: UC Hastings.

Rapid DNA Act of 2017. Pub. L. No., 115-50.

Ray, T. (2018, July 5). Reported DNA testing on migrants raises questions, concerns. GenomeWeb. Retrieved from https://www.genomeweb.com/policy-legislation/reported-dna-testing-migrants-raises-questions-concerns#.XuZwJ2pKiis.

Richard, S. E. (2018, July 17). Why there's a deep cultural aversion to DNA testing, even when it can reunite separated immigrant families. *Time*. Retrieved from https://time.com/5340278/dna-testing-immigration-family-separation/.

Roberts, J. J. (2017, June 19). Microsoft and Accenture unveil global ID system for refugees. *Fortune*. Retrieved from https://fortune.com/2017/06/19/id2020-blockchain-microsoft/.

Rodger, J., Winchester, S., Stephens, G., & Smith, S. (2008). *Developing a conceptual framework for identity fraud profiling* [Paper presentation]. 16th European Conference on Information Systems, Galway, Ireland.

Schuster, L. (2011). Turning refugees into "illegal migrants": Afghan asylum seekers in Europe. *Ethnic and Racial Studies 34*, 1392–1407.

Sheets, K., Baird, M., & Berger, D. (2015). Navigating DNA testing in immigration cases. *Bender's Immigration Bulletin, 21*, 719–726.

Sherwell, P. (2008, July 26). Guatemalan mother reunited with baby stolen and sold for adoption by US couple. *The Sunday Telegraph*. Retrieved from https://www.telegraph.co.uk/news/worldnews/centralamericaandthecaribbean/guatemala/2461557/Guatemalan-mother-reunited-with-baby-stolen-and-sold-for-adoption-by-US-couple.html.

Siegel, E. (2011). *Finding Fernanda: Two mothers, one child, and a cross-border search for truth*. Oakland, CA: Cathexis Press.

Song, K. (2018, June 29). A lesser-known DNA test that can help reunite immigrant parents with detained children. *CNBC*. Retrieved from https://www.cnbc.com/2018/06/28/the-least-bad-way-to-use-dna-test-to-match-immigrant-parents-children.html.

Sperry, B. P., Allyse, M., & Sharp, R. R. (2017). Genetic fingerprints and national security. *The American Journal of Bioethics 17*(5), 1–3.

Sternstein, A. (2012, August 23). Fingerprint records reveal 825,000 immigrants with multiple names, inconsistent birth dates. Nextgov.com. Retrieved from https://www.nextgov.com/cio-briefing/2012/08/fingerprint-records-reveal-825000-immigrants-multiple-names-inconsistent-birth-dates/57620/

Stillman, S. (2015, April 27). Where are the children? *New Yorker Magazine*. Retrieved from https://www.newyorker.com/magazine/2015/04/27/where-are-the-children.

Suzuki, D. (2009, December 31). *The Right Livelihood Award, acceptance speech*. Retrieved from https://www.rightlivelihoodaward.org/speech/acceptance-speech-david-suzuki/.

Taitz, J., Weekers, J. E., & Mosca, D. T. (2002). DNA and immigration: The ethical ramifications. *Lancet, 359*, 794.

Tallbear, K. (2013). *Native American DNA: Tribal belonging and the false promise of genetic science*. Minneapolis, MN: University of Minnesota Press.

Thomas, R. A. L. (2006). Biometrics, international migrants and human rights. *European Journal of Migration and Law, 7*, 377–411.

Travis, J. (2009). Forensic science: Scientists decry isotope, DNA testing of "nationality". *Science, 326*, 30–31.

Tutton, R., Hauskeller, C., & Sturdy, S. (2014). Suspect technologies: Forensic testing of asylum seekers at the UK border. *Ethnic and Racial Studies, 37*, 738–752.

UK Home Office. (2017, May 24). *Citizenship grants by previous country of nationality*. Table cz 06. Retrieved from https://www.gov.uk/government/publications/immigration-statistics-october-to-december-2016/citizenship#long-term-trends-in-applications.

United Nations High Commissioner for Refugees. (2008). *UNHCR note on DNA testing to establish family relationships in the refugee context*. Retrieved from https://www.refworld.org/docid/48620c2d2.html.

United Nations High Commissioner for Refugees. (2014a). *Ending statelessness*. Retrieved from http://www.unhcr.org/pages/49c3646c155.html.

United Nations High Commissioner for Refugees. (2014b, November 4). *Press release: UNHCR launches 10-year global campaign to end statelessness*. . Retrieved

from http://www.unhcr.org/afr/news/latest/2014/11/545797f06/unhcr-launches-10-year-global-campaign-end-statelessness.html.
US Census Bureau. (2016). *History*. Retrieved from https://www.census.gov/history/.
US Citizenship and Immigration Services. (2008, March 19). *Memorandum: Genetic relationship testing; suggesting DNA tests revisions to the adjudicators field manual (AFM)*. Chapter 21 (AFMUpdate AD07-25). Washington, DC: U.S. Citizenship and Immigration Services.
US Customs and Border Protection. (2018). *U.S. Border Patrol Southwest border apprehensions by sector FY2018*. Retrieved from https://www.cbp.gov/newsroom/stats/usbp-sw-border-apprehensions.
US Customs and Immigration Services Senior Policy Council. (n.d.). *Expanding DNA testing in the immigration process*. Washington, DC: US Customs and Immigration Services Senior Policy Council.
US Department of Homeland Security. (2006). *Privacy impact assessment for the automated biometric identification system (IDENT)*. Washington, DC: US Department of Homeland Security.
US Department of Homeland Security. (2010). *Biometric standards requirements for US-VISIT, version 1.0*. Washington, DC: US Department of Homeland Security.
US Department of Homeland Security. (2016). *Yearbook of immigration statistics*. Retrieved from https://www.dhs.gov/immigration-statistics/yearbook.
US Department of Homeland Security, Office of the Citizenship and Immigration Services. (2006a, April 12). *Memorandum from Prakash Khatri, CIS ombudsman to Dr. Emilio T. Gonzalez, director, USCIS*. Washington, DC: US Department of Homeland Security, Office of the Citizenship and Immigration Services.
US Department of Homeland Security, Office of the Citizenship and Immigration Services. (2006b). *Memorandum from Dr. Emilio T. Gonzalez, director, USCIS to Prakash Khatri, CIS ombudsman*. Washington, DC: U.S. Department of Homeland Security, Office of the Citizenship and Immigration Services.
US Department of Homeland Security, Office of the Citizenship and Immigration Services. (2017, November 9). *In-country refugee/parole program for minors in Honduras, El Salvador, and Guatemala (Central American minors—CAM)*. Retrieved from https://www.uscis.gov/CAM.
US Department of Homeland Security, Science & Technology. (2015). *DHS Science and Technology directorate—Rapid DNA*. Washington, DC: U.S. Department of Homeland Security. Retrieved from https://www.dhs.gov/sites/default/files/publications/Rapid_DNA_508.pdf.
US Department of State. (2017, October 4). *Proposed refugee admissions for fiscal year 2018*. Retrieved from https: //www.state.gov/wp-content/uploads/2018/12/Proposed-Refugee-Admissions-for-Fiscal-Year-2018.pdf.
US Department of State. (2018, April 17). *Policy memorandum PM-602-0106.1*. Washington, DC: US Department of State, Immigration and Naturalization Service.
US Department of State, Bureau of Population, Refugees, and Migration. (2008). *Fact sheet: Fraud in the refugee family reunification (priority three) program*. Washington, DC: US Department of State.
US Department of State, Bureau of Population, Refugees and Migration. (2014, November 14). *In-country refugee/parole program for minors in El Salvador, Guatemala, and Honduras with parents lawfully present in the United States* [Fact sheet]. Retrieved from https://reliefweb.int/report/united-states-america/country-refugeeparole-program-minors-el-salvador-guatemala-and-honduras.
US Department of State, Immigration and Naturalization Service. (2000, July 14). *Memorandum from Michael D. Cronin, acting executive associate commissioner, Office of Programs to all regional directors*. Washington, DC: US Department of State, Immigration and Naturalization Service.
US Government Accountability Office. (2015). *Improved evaluation efforts could enhance agency programs to reduce unaccompanied child migration: Report to Congressional*

Requesters (GAO 15-707) (GAO Report). Washington, DC: US Government Accountability Office.

US Government Accountability Office. (2016). *Unaccompanied children: HHS can take further actions to monitor their care: Report to Congressional Requesters* (GAO 16-180) (GAO Report). Washington, DC: US Government Accountability Office.

US Government Accountability Office. (2017). *Homeland Security acquisitions: Earlier requirements definition and clear documentation of key decisions could facilitate ongoing progress* (GAO Report). Washington, DC: US Government Accountability Office.

US Government Accountability Office. (2018). *DHS and HHS have taken steps to improve transfers and monitoring of care, buy actions still needed.* (GAO 18-506T) (GAO Report). Washington, DC: US Government Accountability Office.

Visa Integrity and Security Act of 2016. H.R. 5203, 114th Cong.

Wagner, J. K., Madden, D., Oray, V., & Katsanis, S. H. (2019, December 13). Conversations surrounding the use of DNA tests in the family reunification of migrants separated at the US-Mexico border in 2013. *Front Genetics*. Retrieved from https://pubmed.ncbi.nlm.nih.gov/31921289/.

Wanem, R. E. (2008). *Immigration fraud: Policies, investigations, and issues* (CRS Report for Congress). Washington, DC: Congressional Research Service.

Wang, A. B. (2018, May 27). The U.S. lost track of 1,475 immigrant children last year: Here's why people are outraged now. *The Washington Post*. Retrieved from https://www.washingtonpost.com/news/post-nation/wp/2018/05/27/the-u-s-lost-track-of-1500-immigrant-children-last-year-heres-why-people-are-outraged-now/.

Wenk, R. E. (2010). Sporadic genotype recycling fraud in relationship testing of immigrants. *Transfusion*, 50, 1852–1853.

Wenk, R. E. (2011). Detection of genotype recycling fraud in U.S. immigrants. *Journal of Forensic Science*, 56(Suppl. 1), S243–246.

Whitson, S. (2015). *Kuwait: New counterterror law sets mandatory DNA testing*. Human Rights Watch: Human Rights Watch. Retrieved from www.hrw.org/news/2015/07/20/kuwait-new-counterterror-law-sets-mandatory-dna-testing.

Worth, K. (2015, October 19). For some refugees, safe haven now depends on a DNA test. *Frontline*. Retrieved from https://www.pbs.org/wgbh/frontline/article/for-some-refugees-safe-haven-now-depends-on-a-dna-test/.

Yudell, M., Roberts, D., DeSalle, R., & Tishkoff, S. (2016). Taking race out of human genetics. *Science*, 351, 564–565.

CHAPTER 11

Preventing a Third Death

Identification of Missing Migrants at the US-Mexico Border

SARA H. KATSANIS AND KATHERINE M. SPRADLEY

There are three deaths. The first is when the body ceases to function. The second is when the body is consigned to the grave. The third is that moment, sometime in the future, when your name is spoken for the last time.
David Eagleman, *Sum: Forty Tales from the Afterlives (2009)*

Over the past 10 years, hundreds of human remains have been found along the US-Mexico border (Spradley, 2014). Though they are nameless, they are believed to be the remains of migrants who died while attempting to enter the United States. Most migrants enter the United States with the hope of finding work or, increasingly, to escape violence in their home countries. They may be children, parents, spouses, or siblings—or even smugglers, drug or human traffickers, rapists, or murderers. No matter the cause of death, every person deserves a name. Though far from their homes, migrants are not forgotten; their families and friends continue to search for them—at least until that third death.

Migrants die all along the nearly 2,000-mile southwestern border of the United States, but within the past 10 years, the Rio Grande valley has become the major migrant corridor into Texas. The majority of migrants who cross the border in Texas come from Central America, followed by Mexico. Others come from South America and as far away as Haiti, India, Bangladesh, and Africa (Zavis, 2016).

One of the most dangerous routes for migrants is through Brooks County, Texas, which has a US Customs and Border Protection checkpoint that migrants attempt to evade. Located 70 miles from the Mexican border, Brooks County has no mountains for orientation, only live oak trees, mesquite trees, and scrub brush that look the same from every direction. In the summer the mercury regularly tops 100 degrees

F. Migrants can easily get lost, roaming the same square mile for days, exhausted and dehydrated, until they succumb to the elements. After the migrants make it past the checkpoint, they are picked up in a vehicle and head north. A treacherous route, Highway 281, located in Brooks County, is one of two major highways in the Rio Grande valley that lead north.

Brooks County has the most recorded deaths within the Rio Grande valley. Texas ranchlands can be large; the famous King Ranch is just shy of one million acres. In order to search for human remains of people reported as missing in Brooks County, permission must be granted by the landowners, and not all of them grant it in which case no search takes place.[1] Anyone on a ranch who finds human remains is legally required to call the authorities, and a justice of the peace will decide if the death will be investigated as a crime. Unfortunately landowners sometimes decide not to call police,[2] and if they do call too often the authorities assume the remains are of a migrant and forgo any formal or criminal investigation into the cause of death or the identity of the deceased.

International and national efforts are emerging to trace missing migrants like those found in Brooks County (International Organization for Migration, 2017), yet as of this writing, use of DNA to identify them remains inconsistent and disconnected, using multiple databases that are not interchangeable (Katsanis & Faith, 2016). As a result, the vast majority of human remains recovered at the Mexican border are either buried unidentified or left to languish indefinitely in forensic laboratories. The processes in place to identify them and assist postmortem repatriation are riddled with legal and logistical challenges (Laczko et al., 2017). In recent years several nongovernmental organizations (NGOs) have emerged to assist with preventing migrant deaths, locating the missing, and identifying the dead. Small and large NGOs, international organizations, academics, and government officials are working to find short-term solutions for DNA data sharing to identify remains, but so far no organized process for coordinating these efforts across county, state, and international borders has been developed.

In many cases family members of migrants do not file a missing persons report, unsure whether to report to authorities in their country of origin or the country of destination or fearing they may further endanger their migrant relative if that person is still alive. Undocumented family members in the United States expecting a relative to join them fear immigration authorities themselves, so they neglect to report missing persons to law enforcement (Reineke & Halstead, 2017). Families have even reported being turned away by law enforcement when trying to report a missing person (Reineke & Halstead, 2017). Many families work with a smuggler (or "coyote"), who is paid to smuggle a family member across the Mexican border, but

---

1. There is an exception: if the search is for living individuals who are presumed endangered, landowner permission is not necessary, but a courtesy notification is provided to the landowner.
2. The second author has had conversations with relatives of a landowner and reported that the landowner did not call the authorities when human remains were discovered on that person's property.

many such smugglers are notoriously corrupt and abusive. Some turn their clients into drug or weapons mules or traffic them for sex or labor. So when a family member goes missing, the family might opt to contact the smuggler, who might in turn extort or threaten the family.

Even if a family reports a missing migrant and eventually provides a DNA sample, making a kinship match with a set of remains might take years or never happen at all. DNA analysis of human remains can take months to years to run through a standard laboratory, depending on the quality of the specimens, available technology, and the resource allocation of the laboratory. With few resources, laboratories must prioritize casework. And most laboratories prioritize DNA typing of sexual assault or homicide cases over typing of unidentified migrant remains. Ideally, DNA analysis of human remains includes analyzing both mitochondrial DNA (mtDNA) and short tandem repeats (STRs). But given the high costs of completing both analyses, only one form of typing may be conducted.

Difficulties with DNA analysis are further compounded by the emergence of private DNA laboratories and lack of coordination with law enforcement laboratories. For example, remains recovered in Texas or Arizona might be submitted first to a private laboratory for comparison to the laboratory's database of family reference samples (FRSs) before being turned over to a law enforcement laboratory (normally, a laboratory associated with the Combined DNA Index System; see chapter 10 for a discussion of CODIS). A private laboratory might maintain FRSs from both US and non-US residents, whereas CODIS is limited to cases submitted by US-based law enforcement. Efforts to develop a comprehensive database to assist migrant identifications have been slow to bear fruit.[3] While CODIS, the US-based criminal justice tool for connecting DNA from crime evidence to other crimes and to suspects, includes a missing persons index, the protocol does not permit comparison of DNA data in CODIS to non–law enforcement databases.[4] The result of this broken communication among stakeholders is a patchwork of efforts to identify the migrant remains that is fragmented, inefficient, and only occasionally successful. Equipo Argentino de Antropología Forense (EAAF) reported in 2017 having collected DNA on 1,082 cases and having made identifications on 69 of these, for a rate of ~6% (Doretti et al., 2017); Operation Identification reported that 6/143 (4.2%) cases

---

3. The first effort was the Reuniting Families Project out of Baylor University, started in 2003 and reported by Baker & Baker (2008).

4. According to NDIS protocol (US Federal Bureau of Investigation, 2017, p. 14), CODIS laboratories can submit DNA data only to the NDIS indices and are not permitted to maintain a separate database: "The generation of DNA data and/or a DNA database for dissemination beyond the purposes authorized by the Federal DNA Act [42 U.S.C. § 14132(b)(3)] shall be considered an unauthorized use of the CODIS software. The generation of DNA data and/or a DNA database consisting of such DNA data for dissemination to individuals, entities, agencies or laboratories other than NDIS participating laboratories shall be considered an unauthorized use of CODIS software." This policy is designed to prevent fragmentation of databases. At the same time, CODIS laboratories can only process samples that have been collected by law enforcement or associated with a law enforcement case. This policy precludes the laboratories from uploading DNA data obtained by NGOs or private citizens.

were identified through CODIS, and 17/143 (11.9%) through EAAF, for a total of 16% identified (Nelson, 2017). As a result, many cases remain unresolved for years (Doretti et al., 2017).

BURIED UNKNOWN

For many years, the unidentified remains of migrants recovered in Texas were buried (Spradley, 2014) in graves without autopsy (Kimmerle et al., 2010), and from about 2007 to 2013 without DNA collection (Frey, 2015). Although these burials have been referred to as mass graves, the majority of bodies that have been exhumed are from single graves, with a small number of burials having two bodies layered on top of each other. It is unknown for how long counties in south Texas have not been properly investigating unidentified deaths prior to 2007. However, based on the Forensic Border Coalition Cemetery Survey project (www.forensicbordercoaltion.org), this improper burial practice has occurred in other south Texas counties, including remains found in Cameron, Jim Hogg, and Starr Counties, and still continues in some areas (Spradley et al., 2017). It is unknown to what extent this practice continues elsewhere in the United States. Indeed, it is unclear to what extent migrants are buried without autopsy in Mexico, or for that matter anywhere in the world.

Prior to 2013, remains found in Brooks County, Texas, which does not have its own medical examiner, were sent to a funeral home rather than to a medical examiner in another county. If the funeral home failed to make an identification, the remains were then buried in an anonymous grave in the Sacred Heart Cemetery (see figure 11.1), under the auspices of a local Catholic church. During this process, no DNA samples were collected and, contrary to state law, no procedure was taken to record where the remains were buried in the cemetery. Temporary aluminum markers, some bearing an electronic death record number, were placed next to the graves, but they tended to get knocked out of place by lawn mowers, as they were not flush with the ground. While the electronic death record might seem useful, it is assigned by the funeral home, and recordkeeping has been so poor that it is not possible to correlate the electronic death record number with incident reports at the sheriff's office; it is thus of little use in accessing case information that could be helpful in locating a missing person (i.e., date discovered, location discovered, circumstances surrounding death). Recent forensic archaeological excavations in Sacred Heart Cemetery found 16 individuals who did not have markers. A groundskeeper at the cemetery pointed out locations based on memory recall. In March 2018, Texas State University faculty and students conducted a geophysical survey of the cemetery and found other areas that likely contain migrant remains, with no markers.

More populated counties along the Mexican border, like Hidalgo and Webb Counties, which have a forensic pathologist or medical examiner, are able to autopsy remains within their jurisdictions and often from other counties (Stern, 2015). However, the Forensic Border Coalition found in a 2015–2016 survey of six nearby counties without medical examiners (Spradley et al., 2017) that while a few counties send remains for autopsy (Cameron and Kenedy), a DNA sample may or may not be

Figure 11.1 Unidentified graves in the Sacred Heart Cemetery, Brooks County, Texas, awaiting exhumation. Image courtesy of Sara H. Katsanis.

taken, and the burial location may or may not be marked or be marked only with a temporary marker. Therefore if remains are identified, it may be impossible to locate the burial. Remains in Starr County were buried with no investigation and no DNA sampling. Brooks and Jim Hogg Counties would send remains to a funeral home for identification, and if the remains were not identified, they were buried in the Sacred Heart Cemetery. Hidalgo County has a contracted forensic pathologist and was the only county in the Forensic Border Coalition survey that properly investigated migrant deaths and tracked the final disposition of the remains so that, in the event of an identification, repatriation would be possible (Spradley et al., 2017). Willacy County had no reported unidentified deaths. During the survey, members of the coalition took the opportunity to educate justices of the peace and other county officials about laws surrounding unidentified human remains and resources that could help them fulfill their legal obligations. Due to the education and outreach of the South Texas Human Rights Center and the coalition, Brooks and Starr have changed their practices and now send remains to a medical examiner's office.

In Arizona, which has 15 counties, all unidentified human remains are taken to a centralized medical examiner's office for autopsy, and data pertaining to these deaths are recorded in a centralized depository. By comparison, Texas, which surpassed Arizona in migrant deaths in 2012, has 254 counties, of which only 13 have medical examiners. By law, each county is responsible for handling unidentified deaths found within its jurisdiction. Texas counties with a population of over one million have a medical examiner, who in turn may be called upon to examine remains recovered in nearby counties without licensed forensic pathologists. Along the Texas-Mexico border, the Webb County Medical Examiner's Office serves nine counties,[5] not all of which contract for services with this agency.

In the absence of a medical examiner, a justice of the peace holds jurisdictional authority for a county's unidentified human remains. Justice of the peace is a political office that is a holdover from frontier days and an ongoing necessity in this sparsely populated and underfunded region. Each Texas county has one or more justices of the peace, depending on the county population, with duties to reside over the justice court, act as coroner, and conduct marriage ceremonies. As an elected position, there is no requirement for a justice of the peace to be an attorney, have medical training, or have any other training for the position other than taking an 80-hour course. As in many small towns or communities, the person elected as justice of the peace is usually someone well known in the area, a resident for life and commonly of a family that has been in the county for generations. The justice of the peace is obligated to follow the Texas Criminal Code of Procedures; TCCP chapter 49 covers inquests on dead bodies, stating that if the cause of death is unknown or suspicious or if the death is unattended, an inquest can be ordered. Further, in following TCCP chapter 63 (Missing children and missing persons), a DNA sample from any unidentified person must be taken and submitted to the University of North Texas Health Science Center. Justices of the peace are also required to keep a record of the final disposition of unidentified human remains within their jurisdiction, although this practice is seldom followed in south Texas (Spradley et al., 2017). The situation is further exacerbated by the fact that no centralized database on suspected migrant deaths exists in Texas. Therefore, for many years it was impossible to obtain comprehensive data on migrant deaths throughout the state. This began to change in 2012, when migrant deaths increased dramatically in Brooks County (see figure 11.2a), resulting in widespread media coverage.

In the following years Brooks County became the focus of numerous inquiries by both migrant rights activists and academics. They found that the county failed to conduct autopsies (or refer remains to nearby medical examiners), let alone collect DNA samples or keep records of the disposition of the deceased. To make matters worse, the county had no funds to adequately investigate the deaths and did not qualify for emergency border funds because it is not located on the border itself. Armed with this information, migrant and human rights organizations lobbied the state in 2012 to increase funding for Brooks County so it could manage suspected migrant deaths adequately, including collecting DNA samples. The funding was granted, enabling Brooks County to send remains to a medical examiner's office in Webb County. Since 2013, all remains have been sent for autopsy, DNA sampling, and identification efforts

---

5. Counties: Brooks, Webb, Hidalgo, Kennedy, Cameron, Jim Hogg, Zapata, Starr, and Willacy.

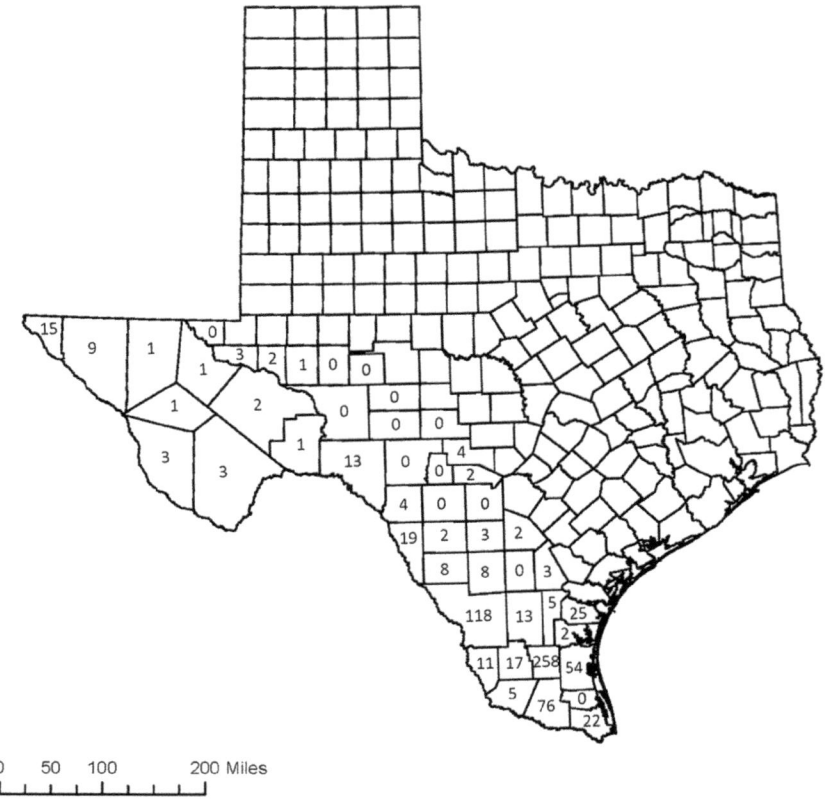

Figure 11.2a Unidentified remains submissions to the University of North Texas Center for Human Identification for DNA analysis from 2010 to 2016. At that time, Brooks County had cases on 258 unidentified migrants. Image from UNT Center for Human Identification Institute of Applied Genetics, 2017.

In 2013 forensic anthropologists Lori Baker of Baylor University and Krista Latham of University of Indianapolis led the first exhumations of more than 60 unidentified human remains buried in the Sacred Heart Cemetery in Brooks County. The remains were taken to the laboratories of Katherine Spradley, a forensic anthropologist at Texas State University and a coauthor of this chapter, launching Operation Identification, the first large-scale attempt to forensically identify the exhumed remains of missing migrants in Texas (see figure 11.2b). Further exhumations at the cemetery took place in 2014 and 2016, with more planned for the future. By early 2018, Spradley and her team of faculty and students had forensically had identified 30 individuals and enabled the repatriation of 27 remains to family members (see figure 11.3). Spradley and her team believe, based on reports provided by the Brooks County Sheriff's Office, that 50 to 100 more unidentified migrants are buried at the Sacred Heart Cemetery, without markers to locate the burial sites.

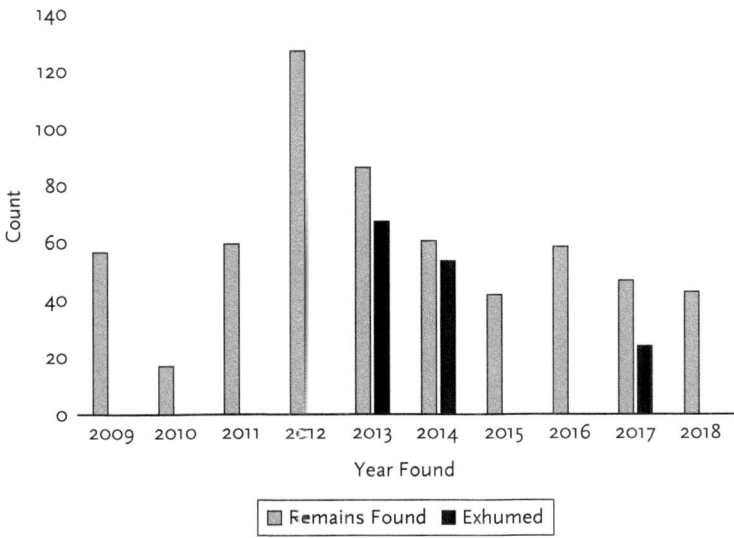

Figure 11.2b Unidentified remains recovered and exhumations by Brooks County Sheriff's Office, 2009–2018. A total of 601 sets of remains have been found in Brooks County since 2009. Exhumations occurred during 2013, 2014, and 2017 but do not represent exhumations of bodies recovered during the time period. Because there is little documentation associated with the recovered remains, most remains exhumed do not have an associated recovery year, but based on known county burial practices, all exhumed remains were found and buried prior to 2013. Graph by Kate Spradley.

NON-DNA DATA SOURCES

DNA analysis is, of course, only one method—though a very valuable one—that can aid in the identification of missing persons. Some migrants, for example, die attempting to swim across the Rio Grande; if their bodies are recovered soon enough, they may be identified based on a combination of identifiable features, including fingerprints and dental records. Moreover, human remains recovered in the desert are often found in situ, making clothing and other personal effects useful in the investigation. But DNA analysis often is vital to resolving these cases, largely because migrant remains are usually found in poor condition, and few dental or medical records are available for migrant families.

Since 2013 the Colibrí Center for Human Rights in Tucson, Arizona, has led an effort to develop a comprehensive database of missing migrants along the southwestern US border. The Colibrí team collects missing persons reports from families within the United States and Latin America and recently began to collect DNA samples from families of the missing. The Colibrí database contains information about unidentified human remains from the Pima County Office of Medical Examiner so that they can work to generate identification hypotheses for cases in Arizona. On the missing persons side (the antemortem data), the database includes biogeographical data from families, descriptions of articles of clothing, personal

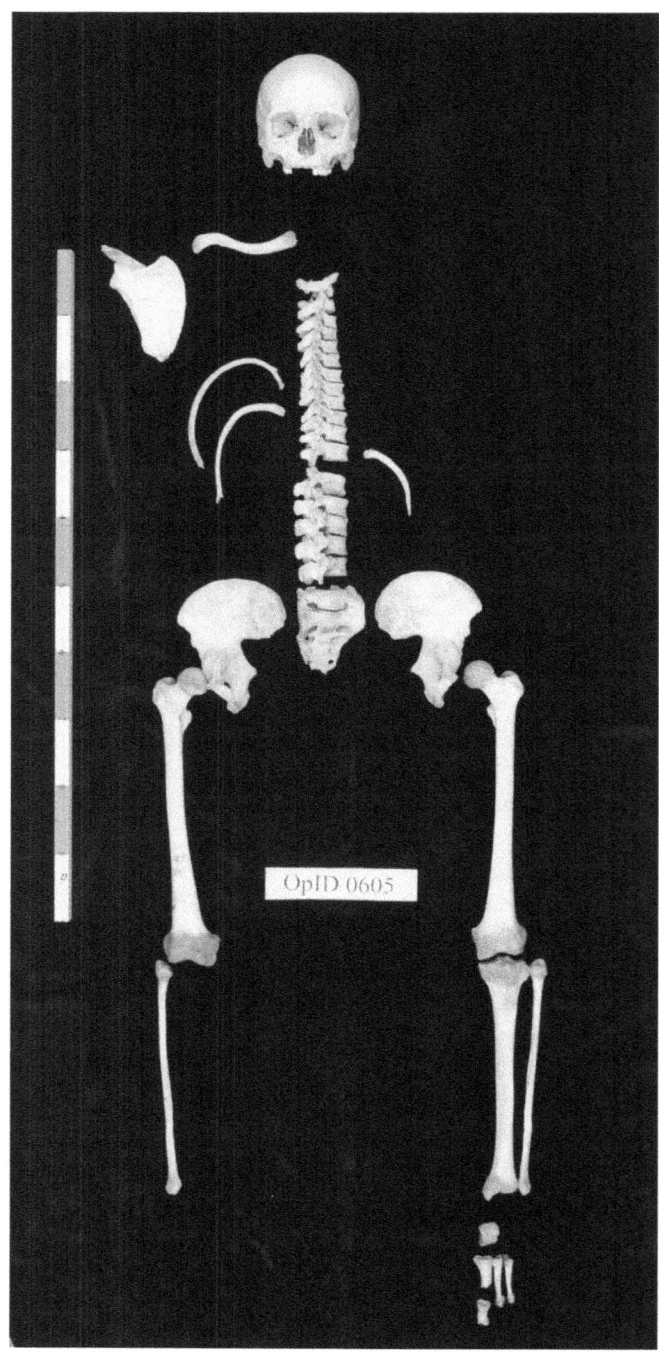

Figure 11.3 Remains of an unidentified person processed by Dr. Kate Spradley at Texas State University following an exhumation from the Sacred Heart Cemetery in Brooks County, Texas. Image courtesy of Kate Spradley.

effects, tattoos, or any other relevant identification metrics. On the remains side (the postmortem data), the database includes information from autopsy and/or anthropology reports, for example, estimated sex, age, height, geographic origin, and any trauma, pathologies, or individualizing characteristics. The Colibrí database is only accessible to Colibrí team members. This security measure protects families that might be afraid to have their personal information stored in a government database or fear extortion should any sensitive or identifying information contained in it be made public. Partner activists and academics have searching privileges only within the missing persons side in order to ascertain if multiple missing persons reports exist for the same individual or to search for a potential association to an unidentified person.

Launched in 2009, the National Missing and Unidentified Persons System, or NamUs, is an invaluable tool for missing persons' cases in the United States (US Department of Justice, National Institute of Justice, 2018). The database is searchable by the public, and anyone can make an entry for a missing persons case, which is then verified by law enforcement. One major drawback, however, is that once missing persons reports from families of missing migrants are made public, they can be perused by criminals for exploitation, even demanding ransom in exchange for information to find their loved ones (Reineke & Halstead, 2017). Suppose, for example, a mother reports her migrant adult son missing. If the report with her son's name and other personal identifiers is posted on NamUs, the smuggler who killed her son might recognize the name of the deceased and extort the mother with threats to her safety. Or perhaps her son died from dehydration, and another person extorts the mother with claims that he knows where her son is and will release him or disclose his location if she pays the extortionist. This problem stemmed from NamUs's Version 1.0, which had two layers to its database: a nonpublic layer comprised of cases entered prior to law enforcement review, and a public layer comprised of the cases that are confirmed missing persons cases. This was a purposeful design to prevent anyone from submitting cases that are not truly missing persons, such as spouses in search of their partners or parents in search of adult children who have departed voluntarily and do not wish to be found. With extra layers of protection, missing migrants cases in NamUs cannot be as easily accessed by the public. But if the cases remain in the nonpublic layer of the database, then the other NGOs that are working these cases will not be able to see information that could be relevant for identification. NamUs administration has been working toward improvements to the system for missing migrants. A proposed fix in the recently released Version 2.0 of NamUs could correct this issue; the new version also requires registration of users, making it harder for bad actors that might exploit NamUs to remain anonymous.

SO MANY STAKEHOLDERS, SO FEW RESOURCES

Governmental and nongovernmental efforts to identify remains recovered near the Mexican border and elsewhere are complicated by questions of sovereignty, privacy, and national security (see table 11.1). At present, both private and government entities

Table 11.1 ROLES OF STAKEHOLDERS IN MISSING MIGRANTS' DNA IDENTIFICATION PROCESSES

U.S. Missing Persons Index of the Combined DNA Index System (CODIS; maintained by the US Federal Bureau of Investigations, FBI, Quantico, VA) Samples from the unidentified human remains are collected by a pathologist, anthropologist, coroner, or medical examiner and sent to a CODIS laboratory. Family reference sample (FRS) DNA collection is carried out or overseen by law enforcement (or designated officials) and sent to a CODIS laboratory. For acceptance of FRS into CODIS, the law requires that (1) a law enforcement official is present for the DNA collection, (2) a consent form for FRS meets certain requirements, and (3) any claimed biological relationships are reported. The CODIS laboratory will upload DNA data from the remains into the "Unidentified Human Remains" index and DNA data from the FRS into the "Relatives of Missing Persons" index. Comparisons of the two indices for kinship associations (partial matches) might lead to further evaluation of a candidate relationship. A third "Pedigree Tree" index comprises claimed relationship ties of the FRS of the missing to the missing person, used to verify hypotheses of relationships based on kinship probabilities.

National Missing and Unidentified System (NamUs; Fort Worth, TX) The US National Institute of Justice coordinates the NamUs. NamUs includes a variety of information on missing persons and found remains, such as name, age, physical characteristics, and circumstances of a disappearance. More sensitive information is restricted only to authorized individuals involved in conducting the investigation, while the least sensitive information can be accessed or submitted by the public (Kim et al., 2016). DNA is not maintained or stored by NamUs, but whether DNA has been collected for CODIS is noted in the NamUs case report.

US Customs and Border Protection Missing Migrants Project (CBP-MMP; Tucson, AZ) In 2016, the Tucson Border Patrol initiated the Missing Migrants Project (MMP), which expanded soon after into the Rio Grande valley in south Texas. Given that the US Customs and Border Protection (CBP) is often a first contact for reports of border deaths, they have developed a network of regional contacts to respond to 911 calls for assistance and reports of missing migrants. The most valuable role of the MMP is the search and rescue efforts that can locate migrants before their demise. DNA is not collected from remains by the CBP, as they are not permitted even to handle remains on US soil (CBP agents are permitted to rescue or recover persons from the Rio Grande, which spans the border). The CBP refers family members who contact them in search of relatives to local law enforcement agencies and also to local NGOs that may assist in the search.

Argentine Forensic Anthropology Team (Equipo Argentino de Antropología Forense, EAAF; Buenos Aires, Argentina) EAAF liaisons work with families to collect FRS DNA samples for their Border Project on both sides of the border, using legal affidavits to verify the legitimacy of the collection process. They maintain memoranda of understanding relationships with multiple authorities in Mexico, El Salvador, Guatemala, and elsewhere around the world and develop best practices for communication (Doretti et al., 2017). The EAAF develops forensic data banks pertaining to missing persons and unidentified cases that could be across borders (Doretti et al., 2017). FRSs are sent to a commercial genetic testing company (Bode Technology Group, Lorton, VA) for analysis. Even though Bode is an accredited laboratory, the FRSs collected to date do not meet CODIS standards since EAAF consent language does not include information required by CODIS law (42 U.S.C. Chapter 136, Subchapter IX Part A § 14132), and since the FBI has not authorized the designated officials to collect DNA outside of the United States. FRSs collected by EAAF via officials at consulates and embassies in the United States might be eligible for CODIS, as long as the appropriate consent is collected as well. Samples from remains found outside the United States might be arranged by the EAAF to go to Bode, but they will not be eligible for CODIS. (Consular officials

Table 11.1 CONTINUED

might be allowed to collect DNA with the permission of CODIS authorities. Still, for NGO-collected FRSs to be eligible for CODIS, the consent forms would need to have CODIS requirements even if the foreign officials are accepted as legal officials for CODIS purposes.) Kinship associations made and reported by Bode are sent to the EAAF for verification with other ante- and postmortem data. The EAAF acts as a family liaison to communicate identifications. However, most US-based human remains (documented in CODIS) are not compared to the FRSs collected by the EAAF and housed at Bode.

Colibrí Center for Human Rights (Colibrí; Tucson, AZ) Because Colibrí has a long-standing relationship with the medical examiner's office of Pima County Arizona, it contributes to the identification investigations of migrant remains found in that border county. Colibrí has developed a postmortem database (Fleischman et al., 2017) of thousands of migrants' remains for these cases, as well as case information from the EAAF, STHRC, and other organizations along the border. In addition, Colibrí has developed an antemortem database of families reporting missing relatives. As part of this effort, in 2016 the organization began collecting FRSs from relatives of the missing, including undocumented persons in the United States and relatives across the border (Reineke & Halstead, 2017). Samples from remains and FRSs are sent to Bode for analysis and database comparisons. If a potential kinship association is made at Bode, then the medical examiner in AZ verifies the identification based on other ante- and postmortem data. In parallel, the AZ medical examiner will send remains samples to a CODIS laboratory for analysis.

South Texas Human Rights Center (STHRC; Falfurrias, TX) Because of the great number of migrant deaths in Brooks County, Texas, the STHRC has taken the lead in assisting the Brooks County Sheriff's Office (BCSO) in taking missing persons reports for NamUs inclusion and FRS collection from relatives of missing persons. Under a memorandum of understanding, the STHRC coordinates and assists the BCSO in acquiring an FRS to send to CODIS. If the family members are willing, the STHRC also will send a second sample under the EAAF's protocol to Bode.

Operation Identification (OpID; San Marcos, TX) Remains buried without autopsy in south Texas are exhumed and maintained at the forensic anthropology center for postmortem analysis. All unidentified human remains samples are sent to a CODIS laboratory for analysis. A second remains sample might also be sent to Bode.

Houston Migrant Rights Collective (MRC; Houston, TX) The Houston MRC helps coordinate an annual missing persons day in Harris County, Texas, to maximize access to families of missing migrants for NGOs and law enforcement (facebook.com/derechosdelosmigranteshouston).

International Criminal Police Organization (Interpol; Lyon, France) Interpol maintains a missing persons database and a DNA database similar to CODIS. However, the DNA database software is capable only of making direct matches, not kinship associations. At this time, Interpol is only involved in advising missing migrants policies for DNA comparisons and cross-border data sharing.

International Commission on Missing Persons (ICMP; The Hague, Netherlands) The general operations of the ICMP are covered in chapter 00. At this time, the ICMP takes an advisory role in developing policies and procedures among the stakeholders in managing cross-border identifications.

International Committee of the Red Cross (ICRC; Geneva, Switzerland) Like the ICMP, the ICRC has taken an advisory role in the missing migrants crisis, working with international, national, and NGO stakeholders to assess how it could be involved in the future to improve identifications.

Many more stakeholders are involved in missing migrants investigations. These stakeholders are those identified by the authors as involved in the DNA identification efforts.

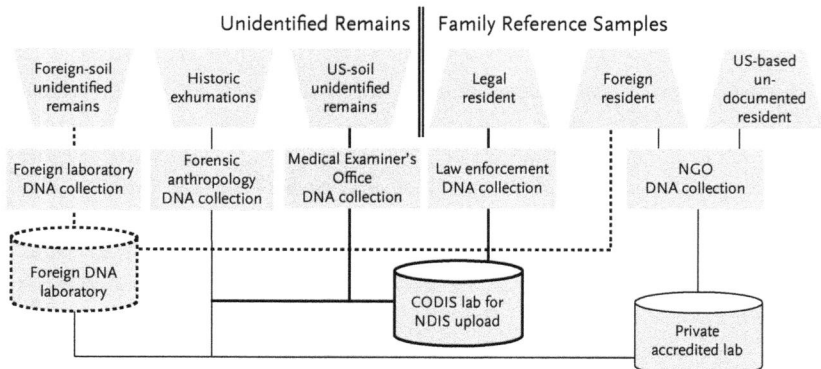

Figure 11.4 DNA collection for missing migrant identification in the United States. The routine route for identification of human remains found on US soil when compared for kinship to US residents would occur through a CODIS laboratory. In missing migrants cases, though, human remains found in another country and FRSs of non-US residents might not be entered in CODIS but rather be collected through NGOs or foreign authorities. Image courtesy of Sara H. Katsanis.

are involved in the collection of DNA from remains of presumed missing migrants and from families of the missing along the Mexican border, which has led to confusion, duplication, and inconsistent findings (Katsanis & Faith, 2016; Inter-American Commission on Human Rights, 2018). While DNA comparisons and kinship analysis can sometimes be challenging to interpret and understand, kinship associations will not occur if the DNA data of remains are in one database and the family reference sample is in another (see figure 11.4 for a graphical illustration of current identifications). At the same time, the privacy and legal challenges in missing migrants' cases mean that a centralized law enforcement database is not only impractical to share across borders but in many cases could be unethical, particularly when it comes to sharing DNA data with countries that stigmatize people based on their genetic ties.[6] (Kim et al., 2016). What is needed is a network of stakeholders working and sharing data in concert and with vigilant privacy protections for potentially sensitive data.

US-Based Law Enforcement DNA Efforts

CODIS maintains a missing persons index that, along with NamUs, is vital for assisting with identifications of missing migrants, with some caveats. CODIS was developed precisely to address the challenges of sharing DNA data across jurisdictions, at least for criminal justice purposes. But there are problems in the system when it comes to missing persons:

6. For example, some countries' legal and cultural systems might penalize a woman whose child is not related to her husband. DNA testing of family members might inadvertently reveal such misattributed parentage and result in discrimination or even violence against her.

First, the protocol that guides CODIS prescribes what DNA data can and cannot be accepted for upload into the National DNA Index System (NDIS), the national index of CODIS (US Federal Bureau of Investigation, 2017).[7] CODIS's integrity in this regard is a pinnacle in the US criminal justice system, especially as the highest level of quality of DNA evidence is absolutely essential to avoid errant data in the system for criminal investigations and convictions. However, these security regulations render CODIS inflexible. For instance, DNA data for CODIS must be collected through or on behalf of criminal justice agencies or designated government officials (US Federal Bureau of Investigation, 2017). In the context of missing persons cases, this rule applies both to the provision of family reference samples and to the entry of unidentified remains samples. CODIS requires that any voluntary DNA samples (e.g., family reference samples) be collected in the presence of law enforcement,[8] and that consent must be obtained, along with a statement of the intended uses and protections afforded to the person providing the specimen.[9] This means that if a DNA sample is collected by anyone other than a US law enforcement officer or medical examiner, the resulting DNA data may not be entered into CODIS's database, even if the laboratory testing the sample is accredited. In effect, family reference samples collected from NGO representatives or from another country are not eligible for CODIS.

Second, regulatory restrictions (61 F.R. 37495) protecting data privacy limit CODIS's ability to share such information with other federal agencies, let alone with foreign governments and NGOs searching for missing persons.[10] These privacy restrictions mean CODIS is perhaps the most secure and protected DNA database in the United States. However, these restrictions also prevent NGOs and government agencies outside of the criminal justice system from receiving information from or accessing CODIS's DNA data for use in their own investigations. While this seems good practice for criminal justice purposes, it means that unidentified remains that are genotyped and entered into CODIS are not in non-CODIS databases.

Finally, the Privacy Act of 1974 (5 U.S.C. § 552), which protects DNA data, applies only to US citizens and legal residents (Katsanis et al., 2018). This is particularly challenging for cases of missing migrants, since many of those who may be providing a family reference sample would be neither citizens nor legal residents. While DNA and identity information provided by US citizens for

---

7. CODIS requires contributing laboratories to be accredited and undergo routine FBI quality assurance audits. If the data are not held to these measures, then it is possible that data might not be to the same quality standards.

8. "Relatives of a missing person must be willing to provide a DNA sample and sign a consent form in the presence of law enforcement" (US Federal Bureau of Investigation, 2018).

9. "Law enforcement personnel must witness the voluntary collection of the Family Reference Samples and a consent and information form must be completed and signed by the person providing the DNA sample. The identity of the contributor providing the DNA sample must be verified by law enforcement (e.g., through presentation of an appropriate government-issued identification card)" (US Federal Bureau of Investigation, 2018).

10. On a case-by-case basis, DNA data from particular crimes might be granted permission to be shared. For example, DNA data might be provided to Interpol to investigate an international crime or a terrorist, under DHS authority.

missing persons cases is restricted from being shared with other arms of the government—such as immigration agencies—foreign-born family members do not have this protection. As a result, family members of missing migrants who are fearful of reporting a loved one missing to US law enforcement cannot identify their loved ones if the remains are only in the CODIS database (Reineke & Halstead, 2017).

Nonprofit Organizations

Unsurprisingly, many migrants and their relatives fear approaching the police to report a missing relative out of concern that immigration authorities will target their family or community members for deportation (Kim et al., 2016). Foreign nationals are permitted to provide DNA samples to CODIS, but like American citizens, they must do so through a law enforcement agency.[11] Many migrants come from countries where the police and government are particularly corrupt, so the idea of trusting law enforcement is a stretch for them.

Several community and nonprofit organizations in the southwestern United States have programs designed to assist families in the search for missing migrants.[12] Quite often the death of a missing migrant is accidental, by exposure to extreme temperatures or lack of water. The volunteer-run missing persons groups often take the initiative to organize and assist with identifications of migrants' remains. But some cases may be homicides[13] or otherwise require police involvement to maintain the integrity of the case for later prosecution. For this reason, it is important for remains to be processed through legal and forensic channels to investigate the death of the person and potentially as evidence to prosecute any wrongdoers in the future. Processing DNA through CODIS laboratories can take an excruciatingly long time, however. CODIS forensic laboratories are notoriously overburdened with crime scene evidence for sexual assaults (Wang & Wein, 2018) and homicides. DNA from unidentified persons cases might take longer to process than cases in which there may be a live victim or an obviously violent cause of death, which are given priority.

---

11. "Foreign nationals can be added to NDIS for the purpose of assisting with the identification of a missing family member. The DNA Identification Act of 1994 does not limit the entry of a voluntarily provided Family Reference Sample based on the nationality of the donor. However, any voluntarily provided DNA sample must be collected in the presence of law enforcement and include the appropriate consent and information documentation" (US Federal Bureau of Investigation, 2018).

12. Nonprofit organizations that assist in the search for missing migrants include Humane Borders (humaneborders.org), Coalición de Derechos Humanos (derechoshumanosaz.net), No More Deaths (forms.nomoredeaths.org/en/), and Aguilas del Desierto.

13. One forensic pathologist claims that only two to three migrant deaths are homicides (The Monitor Editorial Board, 2015).

Outside of law enforcement, private organizations and NGOs are developing processes to consolidate cross-border DNA identifications. Some of these independent efforts have emerged recently, while others have developed over time. In recent years NGOs have begun to collaborate and organize their efforts, alongside and sometimes in partnership with law enforcement efforts to improve processes (US Customs and Border Protection, 2016). Coalitions[14] have formed to develop procedures for managing human remains from the time they are found until they are returned to families. This of course includes the collection and use of DNA, both from family members and from remains found on both sides of the border.

International

On the international front, several organizations have developed programs to improve or supplement exchange of DNA data for the identification of missing persons. Given the international nature of deaths on a border of civilians, who could be from either side of that border (or beyond), engaging with international organizations is an important part of improving legal and scientific processes while maintaining a humanitarian lens on the crisis at hand.

One of the most active organization is the International Commission on Missing Persons (ICMP), which was created in 1996 after thousands of people went missing during the wars in the former Yugoslavia (see chapter 9). The ICMP is a treaty-based organization that works cooperatively with governments and local organizations to locate and identify persons missing as a result of armed conflicts, disasters, and other humanitarian disasters (International Commission on Missing Persons, 2017). The ICMP has pioneered the use of DNA-led efforts to identify the missing through locally based DNA laboratories whenever feasible. The ICMP recognizes the cross-border challenges in sharing DNA data across the US-Mexico border, and on the European home front regarding Syrian and other refugees who have died while crossing the Mediterranean Sea (International Commission on Missing Persons, 2016).

Another international organization assisting in coordinating efforts to identify the remains of missing migrants along the Mexican border is the International Committee of the Red Cross (ICRC). The ICRC is a humanitarian institution based in Geneva, Switzerland, with a mandate to protect victims of international and internal armed conflicts, including the war wounded, prisoners, refugees, civilians, and other noncombatants. While the ICRC does not maintain a DNA database on missing persons, it has a vested interest in securing identifications of the deceased and in assisting families with their searches for missing relatives (International Committee of the Red Cross, 2018). Known for its neutrality, the ICRC has encouraged discussions among stakeholders about means of improving the exchange of genetic

---

14. Examples are forensicbordercoalition.org, reunitingfamilies.org, and Operation Identification

information between the private and governmental DNA databases and NGOs in missing persons cases, and has provided expertise based on its extensive prior work in disaster victim identification (Doretti et al., 2017).[15]

A third international organization is the International Criminal Police Organization, or Interpol, based in Lyon, France. Interpol has a strong interest in developing resources and helping international missing persons efforts across borders and has assisted in identification of particular cross-border cases in Europe. It also maintains a program dedicated to sharing data on missing persons among cooperating government agencies (Interpol, 2018). However, it lacks software to enable kinship analysis among DNA samples, restricting the utility of its current DNA capabilities.

ATTEMPTS TO UNIFY STAKEHOLDERS

For two days in 2016, NGO and US government representatives met in Charleston, South Carolina, to deliberate on how to improve DNA testing and harness NamUs for missing migrant cases in the southwestern United States. Meeting participants called attention to the fragmented nature of the effort to apply DNA analysis to identification efforts along the Mexican border. On the one hand, government participants were concerned that NGOs were collecting family reference samples, which in turn was leading to the creation of too many fragmented databases (see Figure 11.4). NGOs, on the other hand, attested that family members would not provide DNA samples for CODIS to law enforcement for fear that their families might be at risk of arrest or deportation. While the chasm seemed insurmountable, the participants agreed that everyone deserves to be named in death; that every grave should have a name, no matter the cause of death; and that every family should know the fate of their loved ones.

The minutes of the Charleston meeting are not public, but the result was a better informed room of stakeholders, who spent hours inside and outside the meeting discussing options and strategies to improve the status quo (US National Institute of Justice, National Missing and Unidentified Persons System, 2016). While tensions between NGOs and the federal agencies were strong, the meeting ended with a few positive outcomes. First, meeting participants recognized the importance of Missing Persons Days, like the "Missing in Harris County Day" in Texas, for convening families for the collection of family reference samples and the signing of consent forms. Second, the group recognized the distinct role of NGOs in gathering antemortem data on the missing. Attendees agreed that the ability of the NGOs to focus solely on the border crisis and the needs of families of the missing is crucial for identification efforts. Furthermore, the sustained centralization of data from missing migrants reports by the NGOs, particularly gathering data beyond DNA, was

---

15. The ICRC has developed guidelines for disaster victim identification (International Committee of the Red Cross, 2005, 2009).

recognized as an important contribution. Finally, participants agreed that improved language was necessary for informed consent in family reference sample collection (Katsanis et al., 2018).

Attempts to reunite participants from the Charleston meeting at subsequent professional conferences in September 2016[16] and February 2017[17] were only partially successful. At each, about a dozen stakeholders updated one another on the current status of their respective efforts. Both meetings were overshadowed by one of the core issues emerging from the Charleston meeting: the retrospective comparison of DNA data from family reference samples that had already been collected by NGOs over the years and processed in a private laboratory to the DNA data from unidentified remains that were solely in CODIS, like the data collected by the EAAF.

As of 2017, the EAAF (Argentine Forensic Anthropology Team; see chapter 7) had collected almost 3,000 family reference samples from over 1,000 relatives of migrants believed to be missing along the Mexican border (Doretti et al., 2017). The DNA samples were collected by legal affidavit, mostly outside of the United States, and sent to Bode Technology, a private laboratory in Lorton, Virginia, for analysis and comparison to unidentified remains. While this approach has been successful in identifying dozens of human remains, cross-referencing the family reference samples with the unidentified remains in CODIS could potentially lead to more identifications.

That said, the Charleston meeting participants could not agree on whether or not such an arrangement would be feasible given CODIS's rather strict requirements.[18] On one side, the FBI maintains that (1) if a family reference sample was collected without consent from the donor, then it is ineligible for CODIS, and (2) if law enforcement was not present at the point of collection then CODIS cannot accept the sample DNA data. On the other side, the EAAF maintains that if the family reference sample was collected through affidavits sufficient for government officials in Mexico and other countries, then that should be sufficient for use by CODIS in the United States. So far, CODIS has turned down requests by NGOs to conduct a one-time "manual keyboard search" of its database to compare family reference samples stored at Bode Technology[19] with DNA data of human remains in CODIS's database (Doretti et al., 2017; Nelson, 2017).

In April 2018 the US Customs and Border Protection (CBP) met with some of the stakeholders in Edinburg, Texas, for an annual Missing Migrants Program Summit. Representatives from various Latin American consulates,[20] NGOs, and universities,

---

16. At the 27th International Symposium on Human Identification in Minneapolis, MN.
17. At the 69th American Academy of Forensic Sciences Annual Scientific Meeting in New Orleans, LA.
18. Journalists have reported on this challenge, for example (Nelson, 2017).
19. A manual keyboard search is "an exceptional mechanism initiated by, and with the approval of, the NDIS Custodian to compare a target DNA record against DNA records contained in NDIS without resulting in the target record being included or uploaded into the National DNA Index System" (US Federal Bureau of Investigation, 2017).
20. Present were consulate representatives from Ecuador, El Salvador, Guatemala, Honduras, and Mexico.

as well as federal, state, and local government agencies shared their resources and how their respective organizations are addressing migrant deaths. Over two days, each group described the challenges it faces in rescuing migrants and identifying migrant remains. No group was confrontational, but no group was overly transparent, either. The one tense moment—the elephant-in-the-room revelation—occurred when a representative of an NGO asked whether the US Department of Homeland Security was aware of the CBP's efforts to assist migrants in need and whether the program was at risk from any government changes, given President Donald J. Trump's strong anti-immigration stance. Tension was released by an attempt at humor and a reassurance that the leadership was aware of—and pleased with—the Missing Migrants Program. However, no one mentioned the fact that the Trump administration had just that month begun separating families crossing the border as part of its new "zero-tolerance" policy (see chapter 10), and no one mentioned the report issued that month by Derecho Humanos and No More Deaths implicating the CBP's actions as contributing to migrant deaths, accusing the agency of fueling "the disappearance of tens of thousands of migrants and refugees in the expansive wilderness north of the US–Mexico border."[21]

On the upside, a post-meeting side conversation between two representatives of the DOJ and DHS drew a crowd of a dozen or more people as the two brainstormed ways to compare data sets across the federal departments in order to expedite identifications. They toyed with ideas for harnessing NamUs 2.0 to develop an access-only system of reporting and to potentially permit access to NGOs. While these conversations were not substantive, they reinforced the notion that most stakeholders still held the common belief that all missing migrants should be identified and granted dignity in death. Unfortunately, the common sentiment has not led to a resolution in six years of conversations, negotiations, and workshops to address the technical and policy issues.

DISTRUST AMONG STAKEHOLDERS

In recent years, the increasingly tense political dialogues on immigration have intensified the strain between NGOs and government entities working toward identification of missing migrants. While their goal is apparently the same—identification of the dead—the clients are vastly different, in as far as clients can be defined. Government and law enforcement agencies seeking to identify the dead are working on behalf of US citizens in an attempt to maintain public order and deter and prevent crime. For their part, NGOs serve the families of the missing, who wish to know the

---

21. According to a recent NGO report, US government actions force migrants to cross at dangerous points, and border patrol tactics contribute to their demise (La Coalicion de Derechos Humanos and No More Deaths, 2018a). The report details how water stations have been destroyed by government workers and how the tactics of chasing migrants forces dispersement and disorientation of parties.

fate of their missing relatives and to give them a proper burial. Those who serve the government's interest usually are just as compassionate as their NGO counterparts. But this clash in missions divides the two groups, ultimately preventing them from sharing genetic information that could lead to more identifications.

Deceased migrants in the United States may have been criminals—smugglers, drug traffickers, human traffickers, rapists, or murderers—or may not. If a deceased migrant is identified as a possible criminal, that is valuable information for law enforcement and for border protection to further casework in investigating these crimes. The same holds true about family members who are criminals. The CBP and FBI certainly are interested in this information, and in a law-and-order state, they should be provided with this information to protect Americans.

The trouble is that we do not know who the deceased is until a name is supplied, whether by a DNA kinship association or otherwise. At that point, even if the deceased is not a named criminal, then the families still might be at risk if the police and/or border protection know their identities. The families might comprise a combination of US citizens and undocumented residents, which is common along the border. In possession of names and addresses of the family members of the deceased, immigration authorities could target family members for deportation.

WHO ARE JUANITA AND JUAN DOE?

Every missing migrant case has its own story, sometimes two stories. The first is that of the migrant embarking on an adventure, whether by force or by choice, and ultimately being unable to endure. The second is that of the migrant's family and their efforts to discover what happened to their loved one. The story of a migrant death never has a happy ending. At best, the story ends with a family receiving the remains of a loved one and burying them in their homeland. This was the story of the 2014 documentary *Who Is Dayani Cristal*, which chronicled the investigation into the journey of Dilcy Yohan Sandres Martinez, a migrant from Honduras who died in Arizona (Pulse Films et al., 2014). But most cases do not even get to this point. They simply end with a question mark, leaving family members in a limbo world of hope and denial.

A Successful Case of DNA-Based Identification

On June 15, 2012, María Albertina Iraheta Guardado crossed the US-Mexico border near Brownsville, Texas, only to die two days later in Brooks County. At the age of 37, Maria left Honduras for the United States shortly after her husband was killed by a stray bullet. Her plan was to join her sister in New York, find a job, and send money home to her mother and children (Rose, 2015).

María's remains were found and interred in the Sacred Heart Cemetery, next to a temporary marker that read "Unknown Female, Cantina Ranch." Several years later, a team of Baylor students, under the direction of forensic anthropologists, exhumed her remains. She was given the case number 425 and taken for analysis to the Forensic Anthropology Center at Texas State University (FACTS). When the students opened body bag 425, they found the remains of a decomposing female, clothing, and a backpack full of food. When the insoles of the shoes of 425 were searched, an identification card was discovered bearing the name María Albertina Iraheta Guardado. An ID card is a rarity in such cases, as most migrants do not carry identification or are forced to give up their identification before crossing the border.

A missing persons' report for María was found in NamUs with a Colibrí team member listed as case manager, but the NamUs entry indicated that no DNA sample was available. Because the ID card provided an identification hypothesis, and the information from the anthropology analysis was concordant with the demographic information on the ID card, the Argentine Forensic Anthropology Team was contacted, and they later collected a DNA sample from the family in Honduras. Two weeks later, the DNA association report, along with a comparison of antemortem and postmortem data, suggested that the remains of 425 belonged to María.

Once the justice of the peace in Brooks County had approved the forensic report, María's family was notified of the identification. A year passed before the family finally received María's remains, however. The reason was that the name had to be changed on the death certificate because human remains cannot be sent out of the country as "Unidentified" or "John or Jane Doe," but must have their legal name on the death certificate. Finally, María's family received her remains in April 2015.

María was buried for over a year as an "Unknown Female" in a county cemetery in rural Texas. Yet a simple search of her shoe, phone calls to appropriate agencies, and a DNA sample could have identified her sooner.

Still Unidentified and Missing

As of early 2018, FACTS had received a total of 279 unidentified sets of human remains via Operation Identification. Of these, 252 remain unidentified. Only three sets of remains had an identification card. Personal effects with information that can generate an identification hypothesis are rare. All cases are uploaded to NamUs, and all cases have a DNA sample submitted to CODIS. Of the 30 identifications at the FACTS laboratory, more than half are from working with the EAAF, two through fingerprints, and the remainder through CODIS. Of the CODIS identifications, most are through the missing persons index (largely through the South Texas Human Rights Center's involvement with the Missing in Harris County Day), while some are from the criminal offender index. If large-scale DNA cross-referencing between NGOs and the federal government were permitted, more identifications could be made.

Each year, Operation Identification receives phone calls and emails from families searching for their missing loved ones (Sacchetti, 2014). Each year there are more exclusions than identifications. Brooks County sheriff Urbino "Benny" Martinez estimates that for each set of human remains recovered in the county, there are five or ten others that are not found ("Texas' Brooks County is 'Death Valley for migrants,'" 2014). Identifying human remains depends on searching for and finding unidentified decedents on south Texas ranches. If landowners do not give permission, an unknown number of decedents will never be found, and many families will never learn the fate of their loved ones.

On August 24, 2015, Homero Roman went missing in Brooks County. He crossed with a group from Tamaulipas, Mexico, into Mission, Texas, and was later dropped off at Encino, Texas, and told to walk north to bypass the CBP checkpoint (National Missing Persons and Unidentified System, 2015). When his group was spotted by the CBP agents, some were caught, although he and others ran and scattered. Homero has not been heard from again. Search and recovery efforts by the Brooks County Sheriff's Office and the South Texas Human Rights Center still occur as new information is obtained and provided by family members. The families do not give up; they know the worst has almost certainly happened. They want to know where their loved ones are and they want to bring them home.

In testimony during the Inter-American Commission for Human Rights meeting on October 5, 2018, families of the missing testified alongside civil-society forensic experts (Inter-American Commission on Human Rights, 2018). Together, the forensic experts and families requested that the federal government allow comparison of DNA from families of the missing, collected by NGOs, to DNA from unidentified human remains in CODIS. Mothers told stories of their children who went missing when crossing the border in Texas and Arizona. One mother described the two holes she has in her heart for the two children she lost, Julio Gálvez Carrillo and Yadira Gálvez Carrillo, who crossed the border in Arizona.

THE INHUMANE STATUS QUO

If our world economy is such that we can send billions of dollars across borders with a keystroke; if we have technology for searching and targeting drone strikes on individuals through satellite surveillance; if we can de-identify coded information in bits and bytes (as in Bitcoin exchange); if we have the genetic references from millions of people from around the world; and if we have the ability to determine kinship of distant relatives with commercial genetic testing, surely we can improve genetic identifications across borders and return the remains of the deceased to their families. The technologies exist; it is simply the political will, resources, and infrastructure that are lacking.

We also must surmount the inherent bias and discrimination so often directed at migrants and other marginalized communities. Anecdotal reports suggest that a range of abuses have been directed at migrants on ranches and other tracts of land throughout south Texas. These include incidents of willful negligence shown to the remains of presumed migrants; efforts to destroy water resources for migrant travelers; and chase and scatter tactics by CBP agents designed to either capture migrants or lure them into treacherous regions where they are likely to perish (La Coalicion de Derechos Humanos and No More Deaths, 2018b; Reineke & Halstead, 2017). In 2016, No More Deaths, a ministry of the Unitarian Universalist Church of Tucson, Arizona, described many of these abuses in a 36-page report based on a survey of 58 border crossers and a review of 544 migrant cases from the Missing Migrant Crisis Line. According to No More Deaths, many of these practices were a byproduct of a 1994 Clinton-era Border Patrol strategy called "Prevention Through Deterrence," which sealed off urban entry points and funneled people to wilderness routes, risking injury, dehydration, heat stroke, exhaustion, and hypothermia. A 2010 report by the Congressional Research Service similarly characterized "Prevention Through Deterrence" as a tactic to reroute "the illegal border traffic from traditional urban routes to less populated and geographically harsher areas, providing U.S. Customs and Border Protection agents with a tactical advantage over illegal border crossers and smugglers" (Haddall, 2010). Given the Trump administration's hard-line position on migration, it is unlikely that such abuses will soon be investigated or measures taken to prevent them.

Another issue that must be confronted is the "long-term dead"—that is, the hundreds of sets of human remains buried in unmarked graves in cemeteries near the Mexican border in the United States. Most of these decedents are believed to be of migrants who perished attempting to cross the border. Some counties, like Brooks County, have begun to collaborate with university and state medico-legal experts to exhume and attempt to identify the long-term dead. But much more needs to be done, including persuading some local officials and funeral home directors, who believe that the mere burial of an unidentified corpse or set of remains bestows dignity and that nothing more needs to be done.[22] The long-term dead deserve the same respect as those who die today in counties that manage unidentified deaths in accordance with the law. Even though a death happened one year ago, two years ago, or five years ago, those individuals have the same rights as the dead today and need to be found, exhumed, and identified.

Families of missing migrants are seldom shown the respect they deserve. Our research has found that some funeral homes in Brooks County have profited from the dead, first by charging for storage and burial and then again in the pursuit of identification. In 2012, for example, a coyote contacted relatives of a migrant to let the family know that their loved one was left behind to die near the town of Falfurrias in Brooks County (Kovic, 2013). Soon thereafter, an aunt of the deceased drove to Brooks County to inquire about the whereabouts of her nephew's body. Directed to a local funeral home, she looked through pictures of dead bodies and recognized her nephew. When she requested that the remains be exhumed and DNA testing be performed, the director of the funeral home told her it would cost $250 for a DNA sample to be taken, $2,500 for DNA analysis, $2,500 for exhumation, and $100 per

---

22. The second author, as a participant in the Forensic Border Coalition Cemetery Survey Project, talked with many funeral directors and government officials in various counties in south Texas. A funeral is a way to honor the dead, and therefore many funeral directors and county officials felt that they were treating the dead respectfully through burial. While their intentions may have been good, burial without investigation robs the individual of the chance at identification and robs the individual's family of the right to know what happened to their loved one.

day for storage until the remains were identified. Although funeral homes charge for their services, DNA analysis is free through the University of North Texas, and the identification of human remains is not within the purview of funeral homes.

Repatriation of Remains after Identification

For family members, identification of a missing relative is just the start of the repatriation process (Doretti et al., 2017). Lacking any guidelines, family members must undergo endless travel to countless offices and wade through mounds of paperwork. Who has the authority to send the remains? To receive and bury the remains? What happens if remains are partial? How can we assure that the remains repatriated are the correct ones?[23] Are DNA samples returned with the body? What cultural traditions must be upheld in these processes? In some cases, the remains might have been stored disrespectfully or even cremated prior to DNA identification,[24] so there might be nothing more than a piece of paper and photographs to return to a family.

Lengthy Turnaround Times

In terms of the actual DNA testing, the length of time spent finding a kinship association to a family reference sample ideally should be limited to the ability to obtain DNA data from the remains, which are often degraded and challenging to handle. Genotyping of the family reference sample, software-based kinship associations, and review of any leads and reporting should take no more than a week. In terms of routine police casework, though, missing persons are a low priority, given the increased demand for genetic analyses for homicide and sexual assault cases. This is the case for missing persons in general, not just for missing migrant casework. The long turnaround time adds weight to the need for NGOs and family advocacy groups to assist in the caseload.

Inconsistent Family Reference Sampling

The consent process for family members of the missing to provide DNA samples is woefully inadequate (Hennessey, 2013; Willekens et al., 2016). Disaster-victim identification operations have informed development of international DNA collection

---

23. There have been rumors of remains being returned to families that might not have had the scientific evidence to support return. Without a systematic process in place over the years, remains might have been returned to families directly from funeral homes without true scientific identification having been made.
24. Some reports claim that unidentified human remains of presumed immigrants are cremated (Sanchez, 2016). Experts assert that DNA is collected from these remains prior to cremation.

standards and recommendations, including best practices for obtaining consent for collection of family reference samples (Prinz et al., 2007). Yet consent procedures differ greatly among organizations within and outside of law enforcement depending on the context of the incident (Katsanis et al., 2018). In modern medicine and research, we recognize that a person's consent to provide a DNA sample should not be the end of the engagement, that consent is a process, not just a form. In missing migrant cases, the legal information provided to families of missing persons across borders might be confusing and inaccurate, like the aforementioned example of the US Privacy Act of 1974 (5 U.S.C. § 552) applying only to US citizens and legal residents (Katsanis et al., 2018).

With so many stakeholders involved in identification efforts and so little communication among them, family reference samples may be taken from one family multiple times and not at all from another family. Families in Mexico have reported exasperation about providing DNA samples to multiple agencies or the same agency multiple times.[25]

Genetic information and the personal data included in antemortem reports can be highly sensitive. These data can be used for identity theft or even to exploit families (as discussed previously about NamUs data). Principles applied to protect privacy of genetic information ought to be a part of the consent process for family reference samples and sensitive data collected from families. By nature of the crisis of having a missing loved one, families are vulnerable to coercion by anyone offering assistance. Cohesion of efforts and communication among all stakeholders are essential to respect the privacy of the families and to minimize the intrusiveness of the process. Every time a family is approached for a family reference sample, they must recount the story and tragedy of their personal family crisis. It is vital that a humanistic approach be maintained to minimize their pain.

Ineffective Genetic Markers for Family Relationships

STRs are well suited for the identification of human remains, as they can be detected from minute specimens and specimens that have been seriously degraded, as is common with migrant remains. However, genetic technologies over the past 20 years have advanced beyond STRs to include sequencing genomic regions of interest, use of mtDNA markers, and expansion of STR assays (see chapter 1).

STR analysis of around 20 loci is well suited for detecting parent-child and perhaps sibling relationships but less reliable when dealing with family members of a missing migrant who are distantly related. Improved genomic tools like DNA sequencing and analyzing panels of thousands of variants of SNPs can detect distant relationships, however. They can also be used to infer biogeographic ancestry of

---

25. A Human Rights Watch liaison in Mexico told the authors about families being approached by police, NGOs, and other groups so many times that they refused to participate any longer.

unidentified decedents and for getting leads on skeletal decedents.[26] With so many stakeholders and organizations using DNA analysis, we need a common shared platform to cross-compare DNA data for kinship. If family reference samples are typed only for 20 STRs, then those 20 STRs must be assessed in all unidentified decedents for a kinship association to be detected. But missing persons investigations should now embrace and take advantage of the improved DNA technologies developed for medical testing and ancestry investigations.

CONCLUSION

The willful negligence of migrants on US soil that results in their deaths should be considered an international human rights issue. Yet many authorities attribute such deaths to migrants having taken irresponsible routes. Many of the deceased migrants would be alive today, however, if it weren't for the deportation policies and chase tactics used to frighten them into treacherous lands. Many would not have died if they were not prey to the exploitation and neglect of smugglers who failed the people who paid them for their services. Many would not have died if water stations were available, maintained, and secure from vandalism. These acts of omission as well as commission could be prosecutable. For this reason, a careful law enforcement approach to identification of each deceased migrant is essential as evidence of this concerted, systematic neglect.

Most important is that law enforcement, family advocacy groups, and NGOs must find a way to work together to improve information sharing, DNA-based identifications, and the repatriation of remains to families. Family advocates and human rights activists clearly have a significant role to play working with families directly and representing their emotional and privacy needs. At the same time, law enforcement has a critical role to play in providing standards for DNA identifications and serving as an umbrella for developing uniform DNA standards among organizations. Immigration officials have a role to play as well in search-and-rescue efforts to prevent migrant deaths; in the recovery of remains; and in providing a network of communication among stakeholders, including consulates, local law enforcement, and families.

How best to organize this effort may still require years of discussion and trial and error. But with the cooperation of all parties, solutions should be obtainable rather quickly. Questions remain: Should there be a cross-border DNA database accessible to multiple parties or a network of databases with limited, controlled access? Should a DNA database be maintained by law enforcement, an international organization, or an NGO? How can sensitive data be protected from secondary uses or misuse? And finally, who foots the bill for identification of deceased migrants? Should the cost of

---

26. Reports indicate that as of 2017 an increasing number of migrants across the US-Mexico border are not Latin American. Detecting a hypothesized geographic origin can aid an investigation.

DNA typing of unidentified decedents on US soil be shouldered by local agencies in the area where the person died? Or by state or federal agencies? Should the government where the deceased is a citizen be billed for expenses once a positive identification is made? Or should families of the deceased be billed for such expenses? And who should pay for the collection of family reference samples and DNA typing?

At present, NGOs, universities, and non-CODIS laboratories are covering the bulk of the costs of the exhumation and identification of the remains of missing migrants. Operation Identification conducts exhumations of unidentified migrants, develops post-ortem data reports, and coordinates DNA testing. The Argentine Forensic Anthropology Team provides DNA typing for family reference samples. Bode Technology conducts the DNA analysis of family reference samples and remains, charging only for the consumable costs, with no commercial profit. All of these services are provided essentially for free. Meanwhile, a nationwide federal missing persons grant sustains the analysis of unidentified decedents and family reference samples for CODIS (US Department of Justice, Office of Justice Programs, 2018). The Mexican government also pays for the analysis of family reference samples in their government laboratories.[27] Other entities that have raised funds to improve these processes include NamUs, ICRC, ICMP, and academic institutions.

Almost none of the efforts to identify the dead at the border are economically self-sustaining. Colibrí and the EAAF are funded through charities whose resources could run out. Academic laboratories cannot sustain a service model that outpaces its research and teaching needs. Even the US federal grant for missing persons is not sufficient to sustain the needs of the nation in this area, much less the challenges specific to the missing migrant crisis.

What is needed urgently are protocols, sustainable resources, and political will. Developing an ongoing program for identifications in the southwestern United States is different from prior discrete programs for identifying war dead. Unlike identification efforts in Argentina (chapter 7), in Bosnia (chapter 9), and after 9/11 (chapter 14), which required a finite set of resources and time, border deaths are ongoing. The missing migrant crisis at the US-Mexico border is but a microcosm of the challenges faced in other regions around the world. With the growing number of climate and economic refugees in our future, numbers of missing will only increase. The fact that the challenges outlined in this chapter have become so arduous is a testament to the neglect accorded to our fellow migrating humans. If we cannot resolve this with the tools and resources available in the United States, the situation bodes poorly for Greece, Thailand, Spain, and Italy, all struggling with similar challenges.

In the words of Mahatma Gandhi, "A nation's greatness is measured by how it treats its weakest members." With modern DNA technology and some common humanity, we can do better at the US-Mexico border, as well as other borders around the world.

---

27. Mexico decided in 2016 that they must process remains for DNA samples themselves, rather than pay for DNA to be processed by a private laboratory (Reineke & Halstead, 2017).

REFERENCES

Baker, L. E., & Baker, E. J. (2008). Reuniting families: An online database to aid in the identification of undocumented immigrant remains. *Journal of Forensic Sciences, 53*, 50–55.

Dart, T. (2017, June 16). "Shameful" raid on aid camp at US-Mexico border puts lives at risk, volunteers say. *The Guardian* https://www.theguardian.com/us-news/2017/jun/16/us-mexico-border-aid-camp-raid.

Doretti, M., Osorno Solís, C., & Daniell, R. (2017). "The Border Project: Towards a regional forensic mechanism for the identification of missing migrants. In F. Laczko, A. Singleton, & J. Black, eds., *Fatal Journeys: Improving data on missing migrants* (Vol, 3, Pt. I, Chapter 6). Geneva, Switzerland: International Organization for Migration.

Eagleman, D. (2009). Metamorphosis. In *Sum: Forty tales from the afterlives,* (pp. 23–25). New York, NY: Pantheon Books.

Fleischman, J. M., Kendell, A. E., Eggers, C. C., & Fulginiti, L. C. (2017). Undocumented border crosser deaths in Arizona: expanding intrastate collaborative efforts in identification. *Journal of Forensic Science, 62*, 840–849.

Frey, J. C. (2015, July 6). Graves of shame. *The Texas Observer* https://www.texasobserver.org/illegal-mass-graves-of-migrant-remains-found-in-south-texas/.

Haddal, C. C. (2010). *Border security: The role of the U.S. Border Patrol.* Washington, DC: Congressional Research Service.

Hennessey, M. (2013). World Trade Center DNA identifications: The administrative review process." *Promega* https://www.promega.com/-/media/files/resources/conference-proceedings/ishi-13/oral-presentations/hennesseyrev1.pdf?la=en.

Inter-American Commission on Human Rights. (2018, October 5). *Identification of the remains of migrants disappeared along the United States border.* 35th public hearing 169 Period of Sessions. University of Colorado Law School, Schaden Commons Room, Boulder, CO.

International Commission on Missing Persons. (2016). *Missing persons and Mediterranean migration.* Retrieved from https://www.icmp.int/news/missing-persons-and-mediterranean-migration/.

International Commission on Missing Persons. (2017). *Missing persons from the conflict and its aftermath: a stocktaking.* The Hague, The Netherlands: International Commission on Missing Persons Retrieved from https://www.icmp.int/?resources=missing-persons-from-the-kosovo-conflict-and-its-aftermath-a-stockt.

International Committee of the Red Cross. (2005). *Missing people, DNA analysis and identification of human remains: A guide to best practice in armed conflicts and other situations of armed violence.* (Pub. 2005 ref. 0871). Geneva, Switzerland: International Committee of the Red Cross.

International Committee of the Red Cross. (2009). *Missing people, DNA analysis and identification of human remains.* Geneva, Switzerland: International Committee of the Red Cross.

International Committee of the Red Cross. (2018). *Missing persons.* Retrieved from https://www.icrc.org/en/war-and-law/protected-persons/missing-persons.

International Organization for Migration. (2017, September 7). *UN migration agency launches new missing migrants website.* Retrieved from https://www.iom.int/news/un-migration-agency-launches-new-missing-migrants-website.

Interpol. (2018). *Missing persons.* Retrieved from https://www.interpol.int/notice/search/missing.

Katsanis, S. H., & Faith, S. (2016). Re-thinking international missing persons DNA databases. In *27th International Symposium on Human Identification.* Minneapolis, MN: Promega, Inc. https://promega.media/-/media/files/products-and-services/genetic-identity/ishi-27-oral-abstracts/12-katsanis.pdf

Katsanis, S. H., Snyder, L., Arnholt, K., & Mundorff, A. Z. (2018). Consent process for family reference DNA samples. *Forensic Science International: Genetics, 32*, 71–79.

Kim, J., Scully, J. L., & Katsanis, S. H. (2016). Ethical challenges in missing persons investigations. In S. J. Morewitz & C. Sturdy Colls, eds., *Handbook of missing persons* (Chapter 25). San Francisco, CA: Springer.

Kimmerle, E. H., Falsetti, A., & Ross, A. H. (2010). Immigrants, undocumented workers, runaways, transients and the homeless: Towards contextual identification among unidentified decedents. *Forensic Science Policy & Management: An International Journal, 1,* 178–186.

Kovic, C. (2013). *Searching for the living, the dead, and the new disappeared on the migrant trail in Texas: Preliminary report on migrant deaths in South Texas.* In, 26. Texas Civil Rights Project.

La Coalicion de Derechos Humanos and No More Deaths. (2018a). *Disappeared report: How the US border enforcement agencies are fueling a missing persons crisis.* Tucson, AZ: La Coalicion de Derechos Humanos.

La Coalicion de Derechos Humanos and No More Deaths. (2018b). *Disappeared report, part 1: The consequences of chase & scatter in the wilderness.* Tucson, AZ: La Coalicion de Derechos Humanos.

La Coalicion de Derechos Humanos and No More Deaths. (2018c). *Disappeared report, part 2: Interference with humanitarian aid: Death & disappearance on the US-Mexico border.* Tucson, AZ: La Coalicion de Derechos Humanos.

Laczko, F., Singleton, A., & Black, J. (2017). *Fatal journeys: Improving data on missing igrants* (Vol. 3, Part I). Geneva, Switzerland: International Organization for Migration.

Martinez, E. (2017, April 1). Program to identify dead and missing across US put on hold. *Reveal* https://www.revealnews.org/article/program-to-identify-dead-and-missing-across-us-put-on-hold/.

Reagan, M. (2019, July 28). Texas State program works to ID missing migrants. *The Monitor*. Retrieved from https://www.themonitor.com/2019/07/28/texas-state-program-works-id-missing-migrants/.

National Missing Persons and Unidentified System. (2015, August 31). Missing person: Homero Roman (NamUs file MP30031). . Retrieved from https://www.namus.gov/MissingPersons/Case#/30031/details.

Nelson, Aaron. (2017, July 2). Mexico: Struggle to find closure. *San Antonio Express-News* https://pulitzercenter.org/reporting/mexico-struggle-find-closure.

Prinz, M., et al. (2007). DNA Commission of the International Society for Forensic Genetics (ISFG): Recommendations regarding the role of forensic genetics for disaster victim identification (DVI). *Forensic Science International: Genetics, 1,* 3–12.

Pulse Films (production distributor), García Bernal, G. (producer), & Silver, M (director). (2014). *Who is Dayani Cristal.* . Impact Partners, Candescent Films, Ford Foundation, Diamond Docs, Silverlining Films, Canana Films, Rise Films, Mundial.

Reineke, R., & Halstead, C. (2017). Identifying dead migrants, examples from the United States-Mexico border. In F. Laczko, A. Singleton, & J. Black, eds., *Fatal journeys: Improving data on missing migrants* (Vol. 3, Part I, Chapter 5). Geneva, Switzerland: International Organization for Migration.

Rose, A. (2015, June). The mystery of case 0425. *Scientific American* https://www.scientificamerican.com/article/the-forensics-of-identifying-migrants-who-die-exhausted-after-crossing-from-mexico/.

Sacchetti, M. (2014, July 27). The unforgotten. *Boston Globe* https://www.bostonglobe.com/metro/2014/07/26/students-make-efforts-identify-immigrants-buried-unmarked-graves-near-southwest-border/4iDqnsqHzu9m8N6pPZXffI/story.html.

Sanchez, T. (2016, June 18). Remains of hundreds of unidentified immigrants are buried in Imperial County cemetery. *The Los Angeles Times* https://www.latimes.com/local/lanow/la-me-migrants-cemetery-20160619-snap-story.html.

Spradley, K. M. (2014). Toward estimating geographic origin of migrant remains along the United States—Mexico border. *Annals of Anthropological Practice, 38,* 101–110.

Spradley, K. M., et al. (2017). Searching for the unidentified in South Texas: The Forensic Border Coalition (FBC) Cemetery Survey Project." In *American Academy of Forensic*

Sciences. New Orleans, LA: American Academy of Forensic Sciences. Retrieved from https://theforensicbordercoalition.files.wordpress.com/2016/10/aafs-2017-abstract-spradley.pdf

Stern, C. (2015). *Border crossing fatalities 2013–2015* (Report, Webb County Medical Examiner).

Texas' Brooks County is "Death Valley for migrants" (2014, July 9). *NBC News* https://www.nbcnews.com/storyline/immigration-border-crisis/texas-brooks-county-death-valley-migrants-n152121.

UNT Center for Human Identification Institute of Applied Genetics. (2017). *Texas border region: Analysis and identification of unidentified human remains*. Fort Worth, TX: University of North Texas & University of North Texas Health Science Center.

US Customs and Border Protection. (2016, March 16). *Tucson Border Patrol establishes missing migrant team*. US Department of Homeland Security https://www.cbp.gov/newsroom/local-media-release/tucson-border-patrol-establishes-missing-migrant-team.

US Department of Justice, National Institute of Justice. (2018). *National Missing and Unidentified Persons System, NamUs*. Retrieved from https://namus.gov.

US Department of Justice, Office of Justice Programs. (2018). *Funding to identify missing persons*. Retrieved from https://www.nij.gov/topics/law-enforcement/investigations/missing-persons/pages/funding-program.aspx.

US Federal Bureau of Investigation. (2017, July 17). *National DNA Index System (NDIS) operational procedures manual* (Version 5). Quantico, VA: US Federal Bureau of Investigation.

US Federal Bureau of Investigation. (2018). *Frequently asked questions on CODIS and NDIS*. Retrieved from https://www.fbi.gov/services/laboratory/biometric-analysis/codis/codis-and-ndis-fact-sheet.

US National Institute of Justice, National Missing and Unidentified Persons System. (2016, March 17–18). *Missing Migrant Working Group meeting minutes*. Missing Migrant Working Group. Charleston, SC: US National Institute of Justice.

Wang, C., & Wein, L. M. (2018). Analyzing approaches to the backlog of untested sexual assault kits in the U.S.A. *Journal of Forensic Science, 63*(4): 1110–1121. 10.1111/1556-4029.13739.

Willekens, F., Massey, D., Raymer, J., & Beauchemin, C. (2016). International migration under the microscope. *Science, 352*, 897–899.

Zavis, A. (2016, December 22). The desperate trek: Haitians, Africans, Asians, the sharp rise in non-Latin American migrants trying to cross into the U.S. from Mexico. *The Los Angeles Times* https://www.latimes.com/projects/la-fg-immigration-trek-america-tijuana/.

CHAPTER 12

Taking Stock

DNA Testing and Its Complex Truths

DAWNIE STEADMAN AND SARAH WAGNER

Holding the microphone close to her lips, Blanca Luz Nava Vélez pulled no punches in her critique of the Mexican government: "The government and its institutions deceived the families.... They have told us nothing but pure lies."[1] It wasn't the first time Blanca Nava had voiced deep distrust of her government and its version of the disappearance of her son and 42 other students from the Ayotzinapa Rural Teachers College in the town of Iguala on the evening of September 26, 2014. For months she and other parents had taken to the streets, protesting against the controversial police investigation and chanting "Con vida los llevaron, con vida los queremos!" ("They took them alive, we want them back alive!"), a phrase that echoed the demands of the relatives of the "disappeared" (*los desaparecidos*) in Argentina over four decades before (Tuckman, 2014; Garibian, 2014, p. 52; Schwartz-Marin & Cruz-Santiago 2016a, p. 60; see also Rosenblatt, 2015, pp. 93–99). But on this occasion, a press conference in Mexico City, the students' families aimed their criticism directly at the science delivered by the government, and they did so with the backing of a different set of facts generated by a different set of experts: forensic scientists of their own choosing.

On that day, February 9, 2016, the Argentine Forensic Anthropology Team (Equipo Argentino de Antropología Forense, EAAF) gathered the press to unveil the results of their yearlong investigation into the disappearance of the 43 students (EAAF, 2016).[2] The case had drawn significant international attention to Mexico's

---

1. Miguel Augustin Pro Juárez, *EAAF presenta peritaje sobre Basurero de Cocula en caso Ayotzinapa* (YouTube video, February 9, 2016), https://www.youtube.com/watch?v=0q_Up-M-u0Q.
2. In addition to EAAF's investigation, the Interdisciplinary Group of Independent Experts, created in November 2014 through an agreement among the Inter-American

alarming number of missing persons (an estimated 27,000 individuals, many thought to be victims of enforced disappearance, since 2007) and the depths of state corruption (García-Deister & Smith, 2016; Schwartz-Marin & Cruz-Santiago, 2016a, 2016b; Robledo-Silvestre & Velásquez-Upegui, 2017).[3] How could more than 40 youths, last seen in police custody, disappear overnight without a trace? What happened to them, and who was responsible? Only a small bone fragment of one of the young men had been identified by DNA, though its provenance was unclear. In their version of events, Mexican officials alleged that the students had been killed by a local drug cartel, their bodies burned at a garbage dump and the ashes thrown into a nearby river. But those claims rested on contested evidence.

Decrying official stonewalling and the implausibility of the government's account, the families had early on turned to the Argentine group, pioneers in the field of forensic science and human rights investigations with operations already in-country, including their investigation of murdered and disappeared women in Ciudad Juarez and the city of Chihuahua and of missing migrants along the US-Mexico border (García-Deister & Smith, 2016; Alonso et al., 2016; see also chapter 11). The press conference provided the first glimpse of the case EAAF had built. Slide by slide, the Argentine experts set forth with prosecutorial precision what could and could not be definitively established about the disappearances and the disposal of victims' remains; their team of international scientists found that the government's explanation, relying heavily on the testimony of two alleged perpetrators, was inconsistent with the physical evidence. It was simply impossible, the Argentines charged, for those responsible to have incinerated the students' remains at the dumpsite in the manner and during the period of time in question.[4]

After the EAAF's presentation, family members joined them on the dais. Some held posters, while others wore shirts emblazoned with the faces of their children. Speaking in turn, Blanca Nava and several of the parents took aim at Mexico's attorney general: far from the "historical truth" (*verdad histórica*), the perpetrator testimony and forensic evidence proffered by the state, DNA included, were all part of a well-coordinated "historical lie" (*mentira histórica*). At a time when forensic genetics is considered the gold standard of evidence for unearthing truth and exposing guilt, the Ayotzinapa families' rejection of the government's account seemed to go against the grain. Yet the press conference, the public contestation of the Mexican

---

Commission on Human Rights, the Mexican state, and representatives of the disappeared students in Ayotzinapa, prepared its own final report, issued on April 24, 2016.

3. According to Human Rights Watch, the number of missing or disappeared persons in Mexico is now estimated to be 37,000, with 26,000 unidentified sets of remains (Wilkinson, 2019).

4. In collaboration with the Centro de Derechos Humanos Miguel Agustín Pro Juárez (Centro Prodh), EAAF commissioned the research agency Forensic Architecture to create a "cartography of violence" to document the students' disappearance; the resultant report traces "an act of violence that is no longer a singular event but a prolonged act, which persists to this day in the continued absence of the 43 students" (Forensic Architecture, *The Enforced Disappearance of the Ayotzinapa Students*; video, September 26, 2017, https://www.forensic-architecture.org/case/ayotzinapa/).

government's official account, and the families' presence on the dais with the scientific experts from an internationally recognized nongovernmental organization (NGO) had precedence. The phenomenon of relatives of missing persons demanding truth and accountability through the apparatus of forensic scientific investigation has a well-established history, particularly in Latin America.

But the Ayotzinapa cases also signaled a more recent development in the field of forensic science applied to identify missing persons and document human rights abuses—what the International Committee of the Red Cross (ICRC) has termed "Humanitarian Forensic Action" (Cordner & Tidball-Binz, 2017).[5] In particular, they exemplify an increasingly complex field of actors and institutions whose objectives of legal redress, national security, and humanitarianism do not always align.

In this chapter we examine this development and its related tensions, paying specific attention to the increasing centrality of forensic genetics in efforts to account for the missing and the violence that precipitated their disappearance. Whether deemed part of a "forensic turn" (see, e.g., Weizman & Keenan, 2012; Anstett & Dreyfus, 2015; Dziuban, 2017) or the emergence of a "forensic epistemic community" (Kovras 2017), a division has gradually emerged between (and from) smaller-scale initiatives of human rights investigations, epitomized by the work of EAAF and its model replicated in Latin American contexts, toward more large-scale DNA-led humanitarian initiatives "grounded in global law enforcement" (Smith, 2016, p. 398), typified by the International Commission on Missing Persons (ICMP), which got its start in Bosnia and Herzegovina in the late 1990s but now operates globally. If human-rights-based investigative efforts have historically presented a means to demand *legal* accountability for state-sponsored violence, identification-centered efforts increasingly are cast as *humanitarian* acts, what is owed the dead and their surviving kin. However, couched in the language of human rights and international humanitarian law (Kovras 2017, pp. 122–125), the "right to know" afforded the relatives of the missing has nevertheless become progressively distanced from criminal justice processes by which perpetrators are held accountable. Identification, then, may now serve as an end in and of itself, to varied effect.

This shift has played out in different ways and in different contexts—from war crimes trials of the former Yugoslavia to the byzantine transitional justice initiatives in "post-conflict" Colombia, the decentralized and civil-society-led exhumations of the Spanish Civil War dead, and the efforts to identify migrants who have died trying to cross the Mediterranean Sea or the US-Mexico border (see chapter 11). In this chapter we examine the role DNA occupies in this bifurcating movement. As DNA-led identification efforts, such as those promulgated by ICMP, gain traction amid heightened national security concerns, smaller-scale or more narrowly

---

5. Cordner and Tidball-Binz argue that the humanitarian forensic action of "[m]anaging the dead, including protecting their dignity but also helping to identify them to prevent and resolve the tragedy of people missing," exists as part of post-conflict or post-disaster humanitarian responses (2017, p. 65; see also Moon, 2020). For a critique of ICRC's notion of humanitarian forensic action, especially the dangers of its reductive history of forensic science and humanitarianism, see Rosenblatt, 2018.

targeted forensic investigations like those in the Ayotzinapa cases continue apace but face significant obstacles—from limited funding and other material constraints to outsized expectations about what a tool such as DNA testing can achieve (Kovras, 2017, p. 101; see also chapter 14). This is not to suggest that examples of forensic scientific responses to violent conflict and/or missing persons hew strictly to either one side or the other (i.e., institutions employing either a human-rights-centered approach or one centered on national security/global law enforcement identification); instead, responses fall along a continuum between the two models, reflecting ongoing debates within the field. In this increasingly complex terrain of forensic scientific intervention, a diverse array of partnerships has emerged, ranging from NGOs outsourcing genetic testing to private biotech firms to collaborations between Global South forensic institutions and family advocacy organizations.

In the context of competing approaches and new partnerships, forensic genetics has significantly affected the field. "DNA testing has revolutionized forensic science applications to [human rights] investigations and missing person [identification] efforts," explains forensic anthropologist and archaeologist Derek Congram, as "the standard for positive identification has increased dramatically with DNA."[6] From a forensic scientific perspective, one consequence of this redrawn standard is the tendency to privilege DNA to the exclusion of other types of forensic evidence—what forensic anthropologist Thomas Holland, former scientific director at the Defense Department's Joint POW/MIA Acccunting Command, characterizes as a kind of evidentiary "tunnel vision."[7] But the revolution in practice and standards spurred by the widespread application of DNA analysis has broader ramifications as well, including costs that often reinforce existing socioeconomic divisions and inequities, both locally and on a more global scale.

This increased reliance on forensic genetics also affects how people—families of the missing, government officials, forensic institutions—perceive the work of forensic investigations and identification efforts. As expertise and technology circulate, expectations are raised and, along with them, a mix of responses, including hope but also, in some instances, frustration and disappointment. Finally, the pivot toward widespread DNA testing exposes the parameters of forensic scientific responses, which are largely defined by the state. More often than not, states control the medico-legal structures within which DNA testing and databases for missing persons and their relatives operate. While we cannot tackle all of these issues in this chapter, we frame our "stocktaking" around three thematic areas: trust in the science and its practitioners, expectations and outsourcing, and future directions. Drawing on case studies and insights from leading figures in the field, we argue for a nuanced

---

6. Email to author, July 13, 2018. Having worked with the International Criminal Tribunal for the former Yugoslavia, the International Criminal Court, the US Departments of Defense and Justice, and the ICRC, Congram has participated in multiple exhumation and identification efforts and thus has seen firsthand the differing institutional approaches to applying forensic science in response to violent conflict and missing persons.

7. Email to author, March 19, 2016.

understanding of the costs and benefits of forensic intervention so dominated by the biotechnological tool of DNA testing.

TRUST AND THE IDENTIFICATION OF THE MISSING: ARGENTINA, BOSNIA AND HERZEGOVINA, AND THE US-MEXICO BORDER

"They have told us nothing but pure lies." Distrust in authorities' accounts of missing persons is nothing new, particularly in Latin America. In fact, it made good sense that the Ayotzinapa families turned to the Argentine forensic team when they first began to doubt the local officials' capacity to locate their missing sons. Among experts and activists in the field, Argentina is considered the birthplace of the application of forensic science to document human rights violations and identify missing persons (Joyce & Stover, 1992). Following the military dictatorship from 1976 to 1983, organizations representing the families of the more than 10,000 forcibly "disappeared" (*los desaparecidos*) and the newly appointed National Commission of Disappeared Persons requested assistance from the Science and Human Rights Program of the American Association for the Advancement of Science (AAAS) in the exhumation and identification of the disappeared (see chapter 7). Subsequent collaboration with US forensic experts, including the late forensic anthropologist Clyde Snow, gave rise to EAAF, then a fledgling NGO made up of university students turned forensic anthropologists. In the 30 years since then, EAAF has gone on to assist with human rights investigations and missing persons identification efforts across the globe, working in more than 50 countries (Steadman & Haglund, 2005).

The EAAF has profoundly shaped how forensic science became an instrument to document crimes and recover missing persons, in particular victims of state-sponsored violence. As Iosif Kovras (2017) argues, by providing training for and helping establish national forensic teams in post-conflict societies, especially those facing the phenomenon of enforced disappearance (e.g., Guatemala, Peru, Chile), EAAF quickly became the "undisputed champion in forensic investigations for human rights"; it did so, among other ways, by privileging the agency of family members and their "cry for truth" (pp. 96, 97). As Lindsay Smith explains, "the EAAF was founded on an explicit ethical relationship to family members and with an unwavering political recognition of the violence of the state" (2016, p. 402). In keeping with this ethical relationship, one of EAAF's enduring principles is to work closely with families of the missing, involving them in the identification efforts from beginning to end, and thus build trust in the scientific process and its results. According to this logic, families participate as active stakeholders, not simply beneficiaries: the "ethical relationship" on display at the February 9, 2016, press conference with the families of the missing students.

Yet even when forensic scientists stand shoulder to shoulder with families of the missing, as advocated by the Argentine team, gaps exist between expert and layperson knowledge and can lead to subtle and overt friction. This is especially true with forensic genetics. When EAAF began its work in 1984, DNA testing was not available. But as the technology developed, it was gradually incorporated into the organization's forensic investigations, in parallel with classic forensic techniques such as anthropology, archaeology, and odontology. The EAAF recognized the leap of faith the technology required, with families of the missing being asked to accept as fact that which they could not see or touch. "DNA is one of the most powerful methods we use," notes Luis Fondebrider, EAAF cofounder and current president. "And yet at the same time for the families it can be the most mysterious. The 'magic' happens inside a closed, pristine lab, and the geneticists, who produce the magic number, 99.9999%, are like the high priests. So it's different from other methods of identification, where you can see a tooth, a broken bone, a tattoo. With DNA it rests on what you can believe. But why should a relative of a missing person accept the word of someone whom they've never met before?"[8]

Trust was also critical to the initial success of ICMP's efforts to identify the tens of thousands of missing persons from the wars in the Western Balkans in the mid-1990s. The history of building that trust is complicated, bound up with the lingering effects of wartime betrayals—namely, the international community's failure to carry out its responsibility to protect the region's civilian populations. In this sense, postwar intervention, and specifically forensic scientific intervention, faced understandable skepticism. This was especially true in Bosnia and Herzegovina, where DNA was first applied as a tool of postmortem identification "on a massive scale" (Kovras, 2017, p. 100). The innovative approach depended on international backing, but that support took several years to establish. Part of the problem lay with the international community's emphasis in the immediate postwar years on the rule of law and war crimes trials. Thus, the ascendance of forensic genetics came after several years of efforts by the International Criminal Tribunal for the former Yugoslavia (ICTY) to gather evidence of genocide and crimes against humanity. The forensic response to the former Yugoslavia's missing persons (some 40,000 individuals, victims of the region's wars of secession) concentrated on documenting sites of execution and disposition and, as a parallel if secondary effort, on analyzing and identifying remains (Stover & Peress, 1998). In Bosnia and Herzegovina, and later in Kosovo, the Tribunal's Office of the Prosecutor sought to establish the ethnicity of the victims and whether they were civilians, but it did not require that its investigators "identify every victim" (Stover & Shigekane, 2002, pp. 854, 857). Indeed, as Cordner and Tidball-Binz argue, "identifying the dead was not included [in the Tribunal's exhumation program]. This was not necessary for the ICTY. Convictions for murder and genocide do not need names attached to the dead" (2017, p. 66).

When the forensic genetic-centric model emerged in the Western Balkans, it did so primarily in response to a specific set of crimes and their material consequences: the July 1995 genocide at the United Nations (UN) "safe area" of Srebrenica, in which more than 8,000 Bosniak (Bosnian Muslim) men and boys were killed, their bodies dumped into mass graves. The Bosnian Serb perpetrators returned to the graves in the weeks and months after the killings, and in an effort to hide the traces of their

---

8. Email to author, February 2016.

crimes, dug up, transported, and reburied the already decomposing remains into secondary, even tertiary gravesites. That network of mass graves presented local and international forensic personnel with extraordinary challenges: more than 8,000 victims, often multiple members of the same family; the majority of remains partial and commingled; and surviving relatives forcibly displaced, with many of them refugees resettled far beyond Bosnia and Herzegovina's borders. DNA testing was the only effective means to counter the trenches full of jumbled bones and identify the missing victims, and its implementation by ICMP, the international organization that eventually led the identification process, depended on trust gradually built within the community (see chapter 9).

Although some families of the missing had pushed early on for the more comprehensive application of DNA testing, the esoteric details of deoxyribonucleic acid defied easy translation. A chief pathologist working on the Srebrenica cases, Rifat Kešetović explained:

> Perhaps now it looks easy, right? But at the time [late 1999 and early 2000s] it wasn't easy to convince families to accept such a way. There was an enormous task—that is the collection of the relatives' blood. . . . Fundamentally they didn't understand what DNA was, how DNA is inherited, and with what probability DNA can help identifications. The majority of the people at that time knew that through blood types you could determine identity—not identity but the category of a particular group of people. And a majority of people, laymen, knew that that method didn't have a great probability, for example, if you find blood type A, you recover it, it's clear to you that around 20 to 30 percent of the population could belong to that type. They didn't know about DNA. (quoted in Wagner, 2008, p. 16)

In addition to overcoming skepticism toward authorities, given the tragic betrayal of the purported UN "safe area" of Srebrenica, families had to put their faith in a scientific process that turned on intangible proof. That ICMP collected "family reference samples" through blood stain cards rather than buccal swabs may have helped enlist the support and participation of a population that reckoned lineage along blood lines (Wagner, 2008, p. 117; see also chapter 9).

Though it has taken significant time—19 years since the program began in full force in 2001—ICMP's DNA-led identification effort has proven successful, especially in the case of the Srebrenica missing. By June 2020, ICMP had assisted in identifying over 7,000 of individual victims of the genocide using DNA-based methods (Kulukčija, 2020). More than 6,600 of them are buried at the Srebrenica-Potočari Memorial and Cemetery, where relatives gather every July 11 to commemorate their loved ones. Their tombstones mark the end of the forensic scientific effort at truth telling and social repair for individual families.

Not everyone in Bosnia and Herzegovina, however, trusts the findings of the Srebrenica investigations. Many Bosnian Serbs, especially those living in the Podrinje region (the eastern part of the country), resent the international spotlight on the Srebrenica missing, especially on the anniversary of the genocide, with some continuing to deny or diminish the crimes of July 1995. Still others go so far as to claim the coffins are empty or filled with remains from the Philippines, where ICMP helped to identify victims of the 2004 tsunami (Rohde, 2015). Though forensic science has managed to account for the vast majority of the Srebrenica missing, the enduring stalemate of exclusionary ethnonationalist politics, the frustrated, often limited, avenues of legal redress to prosecute perpetrators, and persistent efforts at genocide denial (Green, 2020) have meant that Bosnian society remains deeply divided.

The model of forensic intervention developed by ICMP nevertheless has gained traction on an international scale in recent years. "In December 2014, ICMP officially became a distinct intergovernmental/international organization, based on the understanding that there is a 'systemic global challenge that demands a coherent and effective global response'" (Kovras, 2017, p. 101), and relocated its headquarters to The Hague in the Netherlands. Having discovered it was "increasingly difficult to sustain the organization and expertise in these single-country settings, mostly in the developing world" (Smith, 2016, p. 409), ICMP touted its achievements forged from the identifications it provided in the former Yugoslavia as evidence of the success of its innovative model, primed for application on a global scale.[9] With "growing visibility and policy influence," Kovras argues, by the late 2000s, "ICMP had become the 'hegemon' in forensic investigations" (2017, p. 101), positioning itself as a broker among states (i.e., governments), judicial institutions, and other international organizations, as much as an advocate for families of the missing. ICMP's pivot toward state actors and its promulgation of the DNA-led approach have drawn fire within the field of forensic science applied in response to violent conflict and/or missing persons. For Luis Fondebrider, for example, the model of an external, state-aligned organization pledging answers through DNA represents a "new scientific imperialism" (Fondebrider, 2018).

In many ways, the Tuscon, Arizona–based Colibrí Center for Human Rights embodies the shifting terrain in the forensic field of human rights investigations and missing persons identification efforts and the tensions that have arisen with the expanding application of DNA testing. Formally established in 2013, the organization grew out of a small volunteer initiative, the Missing Migrant Project, that sought to address the crisis of missing migrants who had died crossing the US-Mexico border (on this crisis, see De León, 2015; Rubio-Goldsmith et al., 2016). Simply put, Colibrí's work turns on trust and, on a daily basis, confronts its absence—that is, family and community members' trust in Colibrí staff and in the science it undertakes, alongside their deep distrust in government officials and institutions. Given these dynamics, Colibrí serves a particularly important role for the Pima County Office of Medical Examiner (PCOME) in Tucson, which receives the remains of an average of 140 undocumented border crossers annually and maintains the remains of more than 1,000 unidentified migrants recovered since the early 2000s.

---

9. For example, ICMP explains on its website: "It has pioneered the application of state-of-the-art DNA and advanced database informatics to locate and identify large numbers of missing persons. To date, more than 70% of the estimated 40,000 persons reported missing at the end of the fighting in the Western Balkans have been accounted for" (International Commission on Missing Persons, 2018).

According to Bruce Anderson, the forensic anthropologist at PCOME, his office simply does not have the resources to locate families of missing migrants, explain the DNA process, and collect family reference samples. The core problem, says Anderson, is that "the families may not be fully engaged in modern technology or, if they don't trust their government or ours, then they are not going to step forward." For example, relatives of young men suspected of smuggling migrants are especially reluctant to report them as missing because they fear they will be held responsible for the smuggling activities. In other cases, some families do not know how to begin the process of searching for their missing relatives or face language barriers (speaking indigenous languages rather than Spanish) and thus have limited opportunities to pursue searches. Fortunately, Anderson and his staff have the assistance of Colibrí and EAAF. "Colibrí and EAAF are all over this," says Anderson, especially among indigenous communities across Central America. Both Colibrí and EAAF assist PCOME by contacting and interviewing relatives of missing persons and collecting family reference samples.[10]

Similar to EAAF, Colibrí's model of forensic response prioritizes working directly with families and building a foundation of trust, a process that may take years. The emphasis on trust is critical not only to how identifications are made but also to how they are relayed to families of the missing. When a DNA match is made, Colibrí prepares a report of all the evidence pointing to identification, which, in some cases may be DNA alone if there is an absence of clothing or other personal effects recognizable to families. Before sending the report, Colibrí staff members call the families to notify them verbally and stay on the phone to speak with as many relatives as needed to get all questions answered, especially concerning the DNA probabilities. When the report is sent in the mail, it includes a condolence card signed by the entire Colibrí team. Such seemingly simple expressions of care and solidarity with the family, remembering their names, and conducting follow-up calls are acts of individual and group remembrance (e.g., inviting them to Day of the Dead celebrations) that help the family trust Colibrí and therefore trust the identification results.[11]

Even with such efforts to build and maintain trusting relationships, relatives of the missing do not always accept the news from Colibrí. As Director Robin Reineke explains, "Families can be on either side of a spectrum: [On the one hand,] DNA is magic and gives answers quickly and unambiguously, which can raise expectations that can be difficult to manage. And, on the other hand, DNA means nothing to them." There are instances, though rare, in which families decline involvement and reject an identification altogether. Sometimes the response turns on the clash between what families remember (and thus "know" ontologically) and what they are being asked to accept on an entirely different epistemic order. Reineke explains, "When we identify skeletal remains genetically and deliver the news there [may be] nothing left of the woman they remember [clothing, hair, scars], so if they don't have any stock in DNA as a mechanism to get them to the truth then they don't believe

---

10. Telephone interview with author, July 12, 2018.
11. Telephone interview with author, July 12, 2018.

the results." She has seen families politely accept the DNA results and receive and bury the bones, but then continue their search; in the end, they do not believe the recovered remains belong to their family members. The level of erasure of the border, Reneike says, often leaves little left for families to connect to and little recourse for reckoning with their missing relatives' absence (2016, pp. 78–85). In such instances, grief has sidelined trust, whether of the state, its institutions, forensic personnel, or even the most compelling of DNA matching reports.

IMPACT ON THE FORENSIC FIELD: EXPECTATIONS, OUTSOURCING, AND CODEVELOPMENT IN MARGINALIZED REGIONS

> That was when we ran and were staying in Gulu town, when we were returning back home, we found a dead body on the road to our home. Because of drought and hunger [and] food problems, we were walking around past it to go to our garden to farm until that body decomposed and rotted to skeleton. So when people started to return back home, we realized that if we leave the skeleton there then the children will take it and begin playing with it, [which] might cause more problems, so we took our time and got a blanket, wrapped the body and buried [it].[12]

This scene, described by a survivor from Acholiland of the decades-long war (1980s–2005) between the Lord's Resistance Army (LRA) and the Ugandan government in northern Uganda, underscores how common it is to discover remains in the open or in shallow graves throughout the region (Eichstaedt, 2009). The "problems" the survivor refers to are deeply cultural; Acholi and other tribal groups in northern Uganda believe that when the dead are not buried properly, their spirits will haunt the living by spreading disease, destroying crops, and killing livestock.[13]

The survivor's story reflects not only the massive forcible displacement of two million people, many into internal displacement camps maintained by the government (Branch, 2011), but also the fact that if and when people managed to escape and to return to their villages (few were left standing), they often found human remains of murdered people whom they did not know. The LRA, in particular, was known to kidnap people from one village and kill them many kilometers away, although the Ugandan government also left trails of hastily buried bodies in its wake.

Any forensic response to these troubling dead and their improper burials must first contend with the gaps in knowledge about the violence that originally produced them. But efforts to locate and determine responsibility for the atrocities in contexts like Uganda are challenging. With individuals killed and deposited separately from

---

12. Interview with a survivor by the University of Tennessee research team, Gulu, Uganda, July 2017.
13. A proper burial is generally defined in Acholi culture as on ancestral land, in a prepared single grave, and with culturally appropriate ceremonies.

others abducted from the same village, frequently far from their point of abduction, often it is unknown when they died or which warring faction was responsible, even when there are surviving witnesses. As one informant explained, "When there is war, it becomes difficult to tell between the different soldiers/groups because the uniforms look the same and with fear you cannot know whether these are government soldiers or rebels." This problem is compounded by the fact that northern Ugandans have traditionally lacked consistent medical or dental care. As such, there is only limited antemortem information, such as X-rays, to aid in identifications. Moreover, there is a dearth of forensic infrastructure, let alone political will on the part of the government, to investigate atrocities.

All of this leads to the question: What kind of forensic response is possible in a post-conflict setting like northern Uganda? Who should decide how it is conducted, and on what terms? The postwar landscape of northern Uganda raises a critical set of ethical and practical considerations, which in turn expose the politics of forensic scientific intervention and the material and ideological implications of the forensic genetic-centric model of response. To begin with, it is important to recognize the limits of the seemingly "global" historical precedents of forensic intervention: in northern Uganda, as in many other similarly affected communities around the world, people know little, if anything, about the international courts in The Hague[14] or the existence of EAAF or ICMP, or even about the forensic efforts of the International Criminal Tribunal for Rwanda in neighboring Rwanda. Though northern Ugandans, particularly in Acholiland, have seen hundreds of NGOs come and go since 2005, these efforts have focused on reintegration of child soldiers, trauma counseling, water and food aid networks and agricultural development; none have had a forensic focus (legal, humanitarian, or security driven),[15] so the potential benefits and limitations of forensic exhumations and identifications have never been considered by these communities.

From a strictly forensic perspective, it is clear that DNA would have to be the primary modality of a massive-scale identification program in contexts like that in northern Uganda, an initiative that would require creating a DNA data bank. But such an endeavor necessitates not only careful reflection concerning the ethical, logistic, social, and political ramifications of developing and maintaining a DNA data bank, but also managing the expectations of what forensic science, and DNA specifically, can provide to communities. These are the very questions confronting a team of cultural anthropologists, forensic anthropologists, and archaeologists based

14. Though in northern Uganda there was an incident known by some local residents in which two LRA leaders were tried/prosecuted in Gulu Town by the local judiciary working with the International Criminal Court.

15. There was a forensic excavation of a mass grave site in Lukodi, Uganda, ostensibly ordered by the government shortly after a massacre, in which survivors were forced to exhume the bodies they had buried and were not informed about the purpose or results of the exhumations (Justice and Reconciliation Project, 2013). This incident has led to a lasting, negative image of forensic investigation among people in and around Lukodi (Hepner et al., 2018).

at the University of Tennessee, Knoxville, which has been conducting community-driven ethnographic work in Acholiland since 2012.

As a first step, the team is assessing community attitudes and expectations should a forensic effort to exhume and identify the missing be initiated in northern Uganda. In particular, the team seeks to determine whether survivors and community members feel forensic intervention would assist the local transitional justice efforts, broadly focused on accountability and truth telling,[16] or cause more political retribution from the government and/or psychosocial harm than good (particularly by disturbing the spirits of the deceased).

The University of Tennessee team has conducted individual interviews and convened small focus groups to explore a range of topics, including personal experiences of the war and family loss; if those interviewed know where bodies and graves are located, and how they think such improperly buried bodies affect them; how (if at all) they perceive the processes of transitional justice and forensic science; what cultural rituals should be incorporated into the exhumation and the reburial process if it did occur; and what (if anything) they know or understand about the use of DNA to find family members (Hepner et al., 2018). The team works with a number of Ugandan researchers and NGOs who are also addressing the legacy of the conflict among the living and the dead, thus creating hybrid partnerships that help reimagine the more holistic potential of contemporary forensic scientific responses.

During the interviews and focus group meetings, the team briefly describes the science of DNA and how DNA analysis can be applied to the identification of human remains. This orientation is important because most northern Ugandans get their knowledge of genetics primarily from radio advertisements for paternity testing. Interviewees can be quick to anticipate the "DNA box" as a tool to answer far more questions that also concern them, such as how and where someone died or who killed them. Thus they may see it as a "truth machine" (Lynch et al., 2008; Aronson & Cole, 2009) that can remedy their personal and social problems or as a mythical black box that leaves much room for imagination, conjecture, and disbelief. In either case, managing expectations is a constant concern.

In regions like northern Uganda, where antemortem records are scarce, DNA analysis is often the only means of identifying the remains of missing persons. Yet there are often few, if any, DNA laboratories able to perform such tests. As a result, either DNA testing has to be outsourced, which can be costly, or a DNA laboratory will need to be established, which would require major funding and long-term commitments atypical of smaller-scale, human rights-forensic investigation teams with limited resources.[17] Moreover, family members of the missing may also be reluctant to send

---

16. "Transitional justice includes a range of responses to massive human rights violations, including exposing the truth about past atrocities, holding perpetrators accountable, providing reparations for victims, and fundamentally reforming the state and social institutions that allowed—and in many cases participated in—atrocities" (International Center for Transitional Justice, 2018).

17. This includes the International Criminal Tribunal for Rwanda, which left no durable forensic footprint in the country.

their blood samples or buccal swabs across the country or the globe to places they can never dream of visiting themselves. A way around these obstacles might be to draw on the Central and South American models, particularly the Guatemalan Forensic Anthropology Foundation (FAFG), and codevelop regional indigenous forensic teams in places of need, such as East and Central Africa, which would also include resources and expertise for DNA testing.

Some observers recommend the forensic humanitarian community follow the "codevelopment" model rather than the traditional concept of "capacity building" (Cordner & Tidball-Binz, 2017; Crisp, 2010). The latter typically refers to one-off or periodic training or establishment of a particular solution or structure by a foreign agency (usually with limited local input) that is then turned over to local groups that may or may not be able to support or implement it. By contrast, the codevelopment model, based originally on the model of establishing health infrastructure in areas that lacked basic services, advocates that international service personnel coordinate with local health workers to learn from each other, develop an understanding of the local issues, and promote novel, innovative solutions that in many cases neither the foreign teams nor local health monitors could have devised alone. Codevelopment also implies long-term partnerships to effect success, much like the collaborative model developed by EAAF to assist other regional teams. A new effort to support such codevelopment-focused forensic humanitarian efforts includes is Humanitarian and Human Rights Resource Center (HHRRC) of the American Academy of Forensic Sciences.

In response to high-flow forensic DNA needs in marginalized areas that have experienced mass violence but lack the political will or material resources to establish in-country/region laboratories, efforts to privatize, or democratize, DNA access have led to new collaborations among both private biotech firms and local communities. For example, in contrast to the government-funded DNA analyses at the University of North Texas, which places DNA profiles of unidentified remains in the CODIS database, many organizations that handle the remains of undocumented migrants from the US-Mexico border, including PCOME, primarily utilize private labs such as Bode Cellmark Forensics (see chapter 11). This is due in part to expertise, as private labs such as Bode have developed the techniques to extract postmortem DNA from degraded, sun-bleached bones and teeth that were exposed in the desert for many years. But the decision to use private labs also stems from families' concerns about having their DNA maintained in a US government database. At this time, there appears to be little, if any, collaboration or information sharing between private and government labs and databases, which can lengthen the time to make antemortem/postmortem matches.

An offshoot of the codevelopment approach has been the development of citizen-led scientific interventions—from exhumations to the creation of DNA data banks—in Mexico. These citizen-science initiatives seek to provide alternative modalities of truth seeking, namely ones in which families of the missing attempt to take the matter of forensic accounting into their own hands, given the negligence—or even involvement in atrocities—of the state itself (Schwartz-Marin & Cruz-Santiago, 2016a). But such initiatives raise their own thorny set of bioethical issues, not the least of which relate to individual privacy and long-term access and authority over databases. "DNA and forensic genetics are always embedded in the institutional networks that provide them with credibility," and "dominant discourse and practice still works under the assumption that the State should be the one in charge of realizing the 'right to the truth'" (Schwartz-Marin & Cruz-Santiago, 2016a, pp. 70–71).

FUTURE DIRECTIONS

> DNA testing protocols can be expected to get faster with rapid DNA instrumentation. Improved sensitivity and technology in recent years has enabled higher amounts of data to be recovered from biological evidence. Conclusions that are stronger can be drawn in many cases with probabilistic approaches under development. (Butler, 2015, p. 5)

In an effort to prognosticate the future of DNA in forensic science, John Butler of the National Institute of Standards and Technology (NIST) identifies four phases of evolution in DNA science: exploration, stabilization and standardization, growth, and sophistication. The sophistication phase, Butler writes, includes an "expanding set of tools with capabilities for rapid DNA testing outside of laboratories, greater depth of information from allele sequencing, higher sensitive methodologies applied to casework, and probabilistic software approaches to complex evidence, [and the] need to confront privacy concerns increases as knowledge of genomic information improves" (2015, p. 3). There is no doubt that DNA technology will continue to evolve to provide faster and cheaper results, but perhaps in ways that both aid and complicate its application to humanitarian-oriented forensic responses.

It is important to remember that for a variety of reasons, DNA may not be the primary modality of identification in all contexts, just as individuated identification itself may not be the universal desired response to clandestine or improper burials and unnamed remains (Bennett, 2020). Consider northern Uganda: although it is unlikely that any other forensic method or technology of identification could successfully address the troubling presence of the unidentified dead, DNA testing, if implemented, would never entirely supplant other modes of postmortem identification. The same holds true for the missing migrants along the US-Mexico border. For example, Bruce Anderson explains that currently only about 20% of the migrants identified at PCOME over the past 20 years were identified by DNA; visual identifications, fingerprints and circumstantial evidence made up the remaining 80% of identifications. Anderson notes, however, that the rate of DNA identifications will most assuredly increase over the next several decades given that it is the only modality able to identify migrants whose remains have been scattered and degraded and who have no extant personal effects, as well as the remains of more than 1,000 unidentified migrants still stored at PCOME.

Cristina Cattaneo, director of LABANOF (Laboratorio di Antropologia e Odontologia Forense) at the University of Milan, is on the front lines of a different, though politically similar, crisis in southern Europe—namely, how to identify the more than 20,000 migrants, mostly from North Africa, who have drowned while attempting to cross the Mediterranean Sea since 1993.[18] Cattaneo has to contend with the hegemony of not only DNA but also a forensic system that privileges specific kinds of evidence, their collection, and the protection of their data. As she told *Scientific American* in 2017, "The three kings of identification are DNA, fingerprints and dental records. . . . But those are for rich people" (Nadeau, 2017). In the cases she and her team encounter, fingerprints are only viable if the individual migrant had been arrested for a major crime in a country with an online database, and DNA testing is equally problematic because of the lack of family reference samples. Further, "It is unthinkable to recover personal belongings for DNA analysis from the families in the country of origin as this in many cases would put them in danger because of the specific political scenario, or because no contact can be made at all in the country of origin. Furthermore the relatives or those claiming missing persons are spread out all over the world, particularly in Europe" (Oliviera et al., 2018, p. 122). Also hampering DNA analysis and the probative strength of potential matches is the relative paucity of studies done on allele frequencies from the missing migrants' regions or countries of origin, such as East and West Africa (Oliviera et al., 2018, p. 127).

In response to these obstacles, Cattaneo and her colleagues are building a database that includes DNA but also pushes the boundaries of traditional forensic evidence. Their approach employs a range of antemortem/postmortem comparison techniques, among them dental profiles and superimposition; facial superimposition in cases where the facial features are well preserved and/or there are facial scars or moles; and tattoos, which can be compared with pictures (not just descriptions), all of which are coupled with the existent biological profile. Like those of Anderson and Reineke, Cattaneo's efforts to identify the bodies of missing migrants, people whose presence even in death is cast as a transgression of national borders and thus security, must contend with the structural obstacles of regional (i.e., European Union) and international protocols (or lack thereof) regarding not only DNA collection but also responsibility for their bodies. In the face of such political and practical constraints, Cattaneo and her colleagues argue, "novel strategies of identification have to be sought, both anthropological, odontological and genetic—or, better yet, combined strategies" (Oliviera et al., 2018, p. 127).

The development of independent databases for humanitarian efforts such as what Cattaneo and her colleagues in Italy have pioneered, in which identification

---

18. The International Organization for Migration's Missing Migrant Project estimates that between January 2014 and June 2017, the Mediterranean Sea "account[ed] for the vast majority of deaths recorded globally," with "nearly 14,500 deaths in total . . . recorded across the region" (International Organization for Migration, 2017, p. 6). For the most current data on migrant deaths in the Mediterranean, see the Missing Migrant Project website: https://missingmigrants.iom.int/region/mediterranean.

rather than legal redress is the principal aim, is relatively new, and although they are necessary to match unknown remains to families, such databases present significant challenges of their own, especially when working with marginalized populations. The extractive colonial history of scientists and foreign states among indigenous populations (see, e.g., Reardon, 2004; Reardon & Talbear, 2012) makes a DNA database held by private or government labs in Western countries particularly problematic—from bioethical issues of informed consent to breaches of privacy at the behest of national security agendas. How, then, can the humanitarian community guarantee the people with whom they work that nonconsenting research will be permanently banned? While contracts and memoranda of understanding may evoke legal consequences for unauthorized use of DNA reference samples, explaining the layers of protection and risk in a written consent form or verbally to communities unfamiliar with the science of forensic genetics may make the process even more daunting and unappealing for families (see chapter 15). Moreover, the fear that relatives of the missing may have about giving their DNA to foreign or domestic government-controlled or -accessed databases cannot be dismissed. As Derek Congram explains, "[Take] for example, a family [that] is missing a member who joined a rebel group. The family wants to submit their DNA to the authorities to see if any of the exhumed bodies match. However, the family is wary of the implications and uses by the state—apart from comparing [them] with exhumed victims. Will the government use the DNA profile to search for other family members who might be implicated in other crimes or activity with rebel groups?"[19]

Faced with limited or nonexistent forensic infrastructure, scientists and practitioners working as part of a forensic scientific response often have little choice but to rely on foreign or government laboratories. But this too may be changing. For example, the development of Rapid DNA systems is moving from tabletop to portable, such that units can eventually be used in the field or in makeshift laboratories without requiring a clean room. What makes Rapid DNA identification promising in humanitarian settings is its ability to generate and interpret "STR profiles (colloquially termed 'DNA fingerprints') in less than 2 h[ours] in a ruggedized, field-deployable system by nontechnical users" (Grover et al., 2017, p. 1489). Portable Rapid DNA systems can now extract and sequence short tandem repeats (STRs) from blood samples and buccal swabs, as well as other tissues potentially in the future, within a single, closed system that does not require laboratory expertise because all of the chemistry is within the system (Wiley et al., 2017). In this sense, the field may soon have at its disposal a "box" that family members could see for themselves and the outputs of which could populate a database that is unaffiliated with any government agency. But no matter its sophistication, forensic scientific intervention technology does not occur in a vacuum. Rather it is applied amid the fraught conditions of violent conflict and its social, political, and economic consequences; thus, there are a number of challenges to the application of Rapid DNA to humanitarian contexts. For instance, the data still must be interpreted by and communicated to the families by

19. Email to author, July 13, 2018.

trained personnel; the types of samples used (e.g., postmortem bone and teeth) and output characteristics may be limiting; and the technical process is proprietary and thus "ownership" of the (still theoretical) database would remain an important subject of bioethical concern, especially in circumstances of state-sponsored violence.

Although DNA analysis is becoming relatively less expensive, it is far from free, and many organizations, including Colibrí and medical examiner offices working on missing migrant cases, continually struggle to find federal and private funding for their DNA identification efforts.[20] Bruce Anderson has cut thousands of bone samples from unidentified migrants, yet far too many languish at PCOME awaiting funds to complete the postmortem sequencing. Moreover, medical examiner offices and Colibrí also depend on foreign consulates to pay for family reference samples to be analyzed when a potential match is suspected from non-DNA sources, but foreign states may decline to submit their citizens' DNA to US databases like CODIS or to private labs in the United States, leaving potential matches unconfirmed. The politics of national security complicate, and at times directly thwart, not only funding but also the willingness of families to provide samples. According to Robin Reineke, recent US policies on the border have greatly reduced the number of families who will meet with Colibrí to provide reference samples and has made funding from even private sources more difficult to obtain (see chapter 11).

Finally, no matter how sophisticated, inexpensive, or portable DNA analysis becomes, families' trust in and acceptance of the results will continue to be a critical issue. Any humanitarian forensic effort must create sustainable, long-term relationships that not only build trust but also determine what additional evidence the community needs to accept identifications. Bruce Anderson calls this the "fuller identification": while the scientists can accept DNA at face value and need no further supporting evidence, this may not be the case for some families of the missing (Gowland & Thompson, 2013). "Families are more likely to grab onto a piece of clothing or a paper with a phone number on it than the scientific [findings]," says Anderson. "As anthropologists we strive to have something else to offer, [such as] the fullest possible description of the person [and] their belongings, and the place they were found."[21]

CONCLUSION

As we asked our colleagues to take stock of how DNA has been used in their field and what lies ahead with its increasingly central role in humanitarian responses to violent conflict and mass fatalities, certain trends emerged. We see that despite the technical advances and faith in its efficacy, DNA technology and its results may be meaningless to families without their trust in the science and their

---

20. Like Colibrí, LABANOF is supported strictly by private foundations, having received no funding from either the Italian government or the European Union.

21. Telephone interview with author, July 20, 2018.

participation in the very decisions about how that science should address their needs. Families of the missing, particularly those who overcome fear and risk their own security or well-being in order to advocate for their loved ones, need answers that they can understand if not touch. It is manifest that the complex truth of resolving absence through forensic scientific responses goes far beyond a DNA barcode, whether it is used for legal redress, humanitarian-centered identification efforts, or some combination of the two. More often than not, relatives of the missing want to know what happened—if their son or daughter, father or cousin, was tortured before he or she was killed; they want to know where the body was found and how long ago; they want to know if anything, any material artifact that they might recognize, was recovered with the body. As Bruce Anderson succinctly puts it: "Families must [be allowed to] participate and let them tell us what their needs are and if we can meet them we will, if not then we tell them we can't, but don't ship them two barcodes and say this is a match, we are the government, trust us. We can't do that."

REFERENCES

Alonso, A. D, Galbraith, P. D., & Nienass, B. (2016). Bringing back the dead to society: An interview with Mercedes Doretti. *Social Research*, *83*(2), 511–534.

Anstett, É., & Dreyfus, J., eds. (2015). *Human remains and identification: Mass violence, genocide, and the "forensic turn"*. Manchester, UK: Manchester University Press.

Aronson, J. D., & Cole, S.A. (2009). Science and the death penalty: DNA, innocence, and the debate over capital punishment in the United States. *Law & Social Inquiry*, *34*(3), 603–633.

Baskin, D. R., & Sommers. I. B. (2010). Crime-show-viewing habits and public attitudes toward forensic evidence: The "CSI Effect" revisited. *Justice System Journal*, *31*(1), 97–113.

Bennett, C. (2020). Is DNA always the answer? In Parra, R., Zapico, S., & Ubelaker, D., eds., *Forensic science and humanitarian action: Interacting with the dead and the Living*, (pp.521–534). Hoboken, NJ: John Wiley & Sons.

Branch, A. (2011). *Displacing human rights: War and intervention in Northern Uganda*. Oxford, UK: Oxford University Press.

Butler, J. M. (2015). The future of forensic DNA analysis. *Philosophical Transactions of the Royal Society B*, *370*(1674), 20140252.

Cordner, S., & Tidball-Binz, M. (2017). Humanitarian forensic action—its origins and future. *Forensic Science International*, *279*, 65–71.

Crisp, N. (2010). *Turning the world upside down—the search for global health in the 21st century*. Boca Raton, FL: CRC Press.

De León, J. (2015). *The land of open graves: Living and dying on the migrant trail*. Berkeley, CA: University of California Press.

Dziuban, Z., ed. (2017). *Mapping the "forensic turn"*. Vienna, Austria: New Academic Press.

EAAF. (2016). *Comunicado de prensa: Equipo Argentino de Antropología forense (EAAF) presenta peritaje sobre caso Ayotzinapa; confirma la imposibilidad científica de la "verdad histórica" oficial*.

Eichstaedt, P. (2009). *First kill your family: Child soldiers of Uganda and the Lord's Resistance Army*. Chicago, IL: Chicago Review Press.

Fondebrider, L. (2018, July 18). *Aspectos técnicos y sociales de las exhumaciones en Timor y Republica Centroafricana* [Paper presentation]. International conference, Cuerpos incómodos (Bodies out of place), Donostia/San Sebastián, Spain.

García-Deister, V., & Smith, L. (2016). Ensamblajes de la ciencia forense en América Latina. In G. Mateos & E. Suárez-Díaz, eds., *Aproximaciones a lo local y lo global: Latinoamérica en la historia de la ciencia contemporánea* (pp. 269–300). Mexico: Colección Eslabones en la Ciencia, Centro de Estudios Filosóficos, Políticos y Sociales Vicente Lombardo Toledano.

Garibian, S. (2014). Seeking the dead among the living: Embodying the disappeared of the Argentinian dictatorship through law. In J. Dreyfus & É. Anstatt, eds., *Human remains and mass violence: Methodological approaches* (pp. 44–55). Manchester, UK: Manchester University Press.

Gowland, R., & Thompson, T. (2013). *Human identity and identification*. Cambridge, UK: Cambridge University Press.

Grover, R., et al. (2017). FlexPlex27—highly multiplexed rapid DNA identification for law enforcement, kinship, and military applications. *International Journal of Legal Medicine, 131*(6), 1489–1501.

Green, M. H. (2020). Srebrenica Genocide Denial Report 2020. Srebrenica, Bosnia and Herzegovina: Srebrenica Memorial. Retrieved from https://www.srebrenicamemorial.org/en/article/230/download-srebrenica-genocide-denial-report-for-2020.

Hepner, T., Steadman, D. W., & Hanebrink, J. R. (2018). Sowing the dead: Massacres and the missing in Northern Uganda. In C. Anderson & D. Martin, eds., *Blood in the villages: Bioarchaeological and forensic evidence for massacres* (pp. 136–154). Gainesville, FL: University Press of Florida.

International Center for Transitional Justice. (2018). Retrieved from https://www.ictj.org/about

International Commission on Missing Persons. (2018). *Where we work: Western Balkans*. Retrieved from https://www.icmp.int/where-we-work/europe/western-balkans/.

International Organization for Migration. (2017). *Fatal journeys* (Vol. 3, Pt. I). Retrieved from https://publications.iom.int/system/files/pdf/fatal_journeys_volume_3_part_1.pdf.

Joyce, C., & Stover, E. (1992). *Witnesses from the grave: The stories bones tell*. New York, NY: Ballantine Books.

Justice and Reconciliation Project (JRP). (2013). *Finding community relevance in transitional justice: Drawing attention to the need for decent reburials—A case of Lukodi in Gulu District*. Gulu, Uganda: JRP.

Kovras, I. (2017). *Grassroots activism and the evolution of transitional justice: The families of the disappeared*. Cambridge, UK: Cambridge University Press.

Kulukčija, S. (2020). EU Special Representative visits ICMP project related to identifying Srebrenica genocide victims. Press Release, June 5, 2020. International Commission on Missing Persons. Retrieved from https://www.icmp.int/press-releases/eu-special-representative-visits-icmp-project-related-to-identifying-srebrenica-genocide-victims/.

Lynch, M., Cole, S. A., McNally, R., & Jordan, K. (2008). *Truth machine: The contentious history of DNA fingerprinting*. Chicago, IL: University of Chicago Press.

Moon, C. (2020). Extraordinary deathwork: New developments in, and the social significance of, forensic humanitarian action. In Parra, R., Zapico, S., & Ubelaker, D., eds., *Forensic science and humanitarian action: Interacting with the dead and the Living*, (pp.37–48). Hoboken, NJ: John Wiley & Sons.

Nadeau, B. (2017, August 9). Giving dead migrants a name. *Scientific American*. Retrieved from https://www.scientificamerican.com/article/giving-dead-migrants-a-name/.

Olivieri, L., et al. (2018). Challenges in the identification of dead migrants in the Mediterranean: The case study of the Lampedusa shipwreck of October 3rd 2013. *Forensic Science International, 285*, 121–128.

Reardon, J. (2004). *Race to the finish: Identity and governance in an age of genomics*. Princeton, NJ: Princeton University Press.

Reardon, J., & Tallbear, K. (2012). "Your DNA is *our* history": Genomics, anthropology, and the construction of whiteness as property. *Cultural Anthropology, 53*(S5), S233–S245.

Reineke, R. (2016). Missing persons and unidentified remains at the United States–Mexico border. In T. Brian & F. Laczko, eds., *Fatal journeys: Vol. 2. Identification and tracing dead and missing migrants*. Geneva, Switzerland: International Organization for Migration.

Robledo-Silvestre, C., & Velásquez-Upegui, E. P. (2017). The disappearance of 43 teacher-trainees in Mexico: An approach to critical discourse analysis in the press. *Revista Colombiana de Ciencias Sociales, 8*(2), 353–371.

Rohde, D. (2015, July 17). Denying genocide in the face of science. *The Atlantic*. Retrieved from https://www.theatlantic.com/international/archive/2015/07/srebrenica-massacre-bosnia-anniversary-denial/398846/

Rosenblatt, A. (2015). *Digging for the disappeared: Forensic science after atrocity*. Stanford, CA: Stanford University Press.

Rosenblatt, A. (2018). The Danger of a single story about forensic humanitarianism. *Journal of Forensic and Legal Medicine, 61*, 75–77.

Rubio-Goldsmith, R., Fernández, C., Finch, J. K., & Masterson-Algar, A., eds. (2016). *Migrant deaths in the Arizona desert: La vida no vale nada*. Tucson, AZ: University of Arizona Press.

Schwartz-Marin, E., & Cruz-Santiago, A. (2016a). Forensic civism: Articulating science, DNA and kinship in contemporary Mexico and Colombia. *Human Remains and Violence: An Interdisciplinary Journal, 2*(1), 58–74.

Schwartz-Marin, E., & Cruz-Santiago, A. (2016b). Pure corpses, dangerous citizens: Transgressing the boundaries between mourners and experts in the search for the disappeared in Mexico. *Social Research, 83*(2), 483–510.

Schweitzer, N. J., & Saks, M. J. (2007). The CSI effect: Popular fiction about forensic science affects public expectations about real forensic science. *Jurimetrics, 47*(3), 357–364.

Smith, L. (2016). The missing, the martyred and the disappeared: Global networks, technical intensification and the end of human rights genetics. *Social Studies of Science, 47*(3), 398–441.

Steadman, D. W., & Haglund, W. D. (2005). The scope of anthropological contributions to human rights investigations. *Journal of Forensic Sciences, 50*(1), 1–8.

Stover, E., & Peress, G. (1998). *The graves: Srebrenica and Vukovar*. Zurich, Switzerland: Scalo.

Stover, E., & Shigekane, R. (2002). The missing in the aftermath of the war: When do the needs of victims' families and international war crimes tribunals clash? *International Review of the Red Cross, 40*(848), 845–866.

Tuckman, J. (2014, November 21). Mexicans in biggest protest yet over missing students. *The Guardian*. Retrieved from https://www.theguardian.com/world/2014/nov/21/mexicans-protesting-about-missing-students-scuffle-with-police

Wagner, S. E. (2008). *To know where he lies: DNA technology and the search for Srebrenica's missing*. Berkeley, CA: University of California Press.

Weizman, E., & Keenan. T. (2012). *Mengele's skull: The advent of forensic aesthetics* Berlin, Germany: Sternberg and Portikus.

Wiley, R., Sage, K., LaRue, B., & Budowle, B. (2017). Internal validation of the RapidHIT® ID system. *Forensic Science International: Genetics, 31*, 180–188.

Wilkinson, D. (2019, January 15). *Mexico: The other disappeared*. Retrieved from Human Rights Watch website: https://www.hrw.org/news/2019/01/15/mexico-other-disappeared.

PART III
Challenges and Debates

A widow stands at the edge of a mass grave in the mountains of Iraqi-Kurdistan in 1992. A campaign by Iraqi forces resulted in the killing, disappearance, and relocation of hundreds of thousands of Kurdish men, women, and children in the late 1980s. Forensic DNA analysis has become an essential tool for identifying the remains of victims of mass violence. Once identified, the remains can be returned to family members for proper burial. The forensic findings can also be used in court to prosecute those responsible for these atrocities. (Photo: Susan Meiselas)

CHAPTER 13

Admissibility of DNA Evidence in Court

ANDREA ROTH

Forensic DNA typing has existed since the late 1980s, and has been admitted in court cases as evidence of identity since the late 1980s (Kaye, 2010, pp. 60–63). Some DNA evidence admissibility questions are now relatively uncontroversial (such as a "match" between robust single-source polymerase chain reaction-short tandem repeat (PCR-STR) profiles), and others are still contentious (such as results of "low-copy-number" testing and interpretations of complex mixtures by expert systems). This chapter offers a brief overview of the legal rules governing the admissibility of forensic DNA typing results, primarily in US court cases.

Any discussion of the admissibility of DNA in court, it should be said, is really a discussion of several different questions of admissibility. Before introducing evidence of a DNA profile "match" at trial, for example, the proponent must show not only that the DNA typing method is reliable, but also that the method of interpreting the results and calculating the statistical significance of the results is reliable. Thus, in a case involving a complex DNA mixture in which the prosecution alleges that a suspect is a likely contributor and seeks to introduce a likelihood ratio (LR) reported by a probabilistic genotyping software program like TrueAllele, the prosecution might be called upon in a pretrial reliability hearing to establish the reliability of (1) the PCR-STR method used to compare alleles among various potential contributors to a mixture; (2) the reliability of the expert system in estimating the number of contributors and whether a peak is a true allele or an artifact; and (3) the reliability of the statistical method the system uses to generate the LR, along with the reliability of the LR itself as an expression to the jury of the statistical significance of the results.

It is also worth noting that even where forensic DNA typing results are admissible as an evidentiary and constitutional matter, their meaning and probative value might be vigorously contested at trial by the opponent. The parties might disagree over whether a peak is a true allele or artifact and offer conflicting expert testimony on the matter; whether a match statistic is grossly overstated and offer conflicting expert testimony on the matter; the relevant population of potential contributors to a mixture; and the prevalence of phenomena like DNA transfer, which might offer an alternative innocent explanation for the presence of a person's DNA at a crime scene. An opponent might also question the qualifications or conclusions of a DNA expert, even if that expert succeeds in testifying. In short, the admissibility of DNA—that is, whether a judge or jury determining the facts of a case is even allowed to hear the results of forensic DNA typing—is only the first of many questions related to how the legal system treats DNA evidence in court.

THE BASIC LEGAL RULES GOVERNING ADMISSIBILITY OF DNA EVIDENCE

To be admissible in a civil or criminal trial in the United States, forensic DNA typing results must comply with the jurisdiction's rules of evidence (each state, as well as the federal system, has its own rules of evidence), as well as provisions of the US Constitution that give certain trial rights to those accused of a crime.

This section focuses on the rules of admissibility of DNA evidence applicable *at trial*. Although not all court cases involving DNA go to trial, the rules related to admissibility of evidence at trial loom large over settlement or plea negotiations, which are conducted in the shadow of a trial. And while DNA typing results might also be offered in legal proceedings beyond trial, such as sentencing proceedings, the rules governing admissibility of evidence at sentencing are generally both simple and permissive. For example, the US Sentencing Guidelines state that sentencing courts "may consider relevant information without regard to its admissibility ... at trial, provided that the information has sufficient indicia of reliability" (U.S.S.G. § 6A1.3(a)).

Reliability Requirements for Expert Testimony

The first set of legal rules governing the admissibility of DNA relate to the requirement, under statutory or common law rules of evidence, that expert testimony based on scientific methods be reliable. Nearly all DNA typing results offered to prove identity are presented through one or more expert witnesses: laboratory technicians, DNA analysts, statisticians, population geneticists, and the like. As a result, the admissibility of DNA will turn in part on the rules of evidence governing expert witness testimony. In particular, nearly every state, as well as the federal system, requires that expert testimony be based on reliable methodology.

Some courts—following the so-called *Frye* standard—delegate the question of reliability of DNA to the scientific community, allowing the admission of expert testimony based on DNA typing and interpretation methods so long as those methods are "generally accepted" within the relevant scientific community. Before the 1920s, scientific evidence was treated like most other evidence, subject only to the usual requirements of relevance, witness competence, and the like (*Spring Co. v. Edgar*, U.S., 1878). But in 1923, the D.C. Circuit Court of Appeals in *Frye v. United States* held that a criminal defendant accused of murder, James Frye, could not offer the expert testimony of Dr. William Moulton Marston—who would later create the character *Wonder Woman*—that Mr. Frye had taken and passed a polygraph examination (*Frye*; Lepore, 2015). According to the *Frye* court, a novel scientific methodology like the polygraph should not be admitted unless it is "sufficiently established" as a method "to have gained general acceptance" among the "authorities" in the field (*Frye*, p. 1014). Because the polygraph "ha[d] not yet gained such standing and scientific recognition," it was properly excluded by the trial court. The *Frye* "general acceptance" standard was highly influential and ultimately became the dominant standard in US courts for admissibility of expert testimony based on such methods.

Not until 1993, with the US Supreme Court's decision in *Daubert v. Merrell Dow Pharmaceuticals*, would the *Frye* test's dominance be challenged. In *Daubert*, the Court held that the federal rule of evidence governing admissibility of expert testimony—Rule 702—did not require that an expert's method be "generally accepted." The rule's language required only that an expert's "scientific" or other technical or specialized knowledge be helpful to the jury, which in turn required only that a method—if purportedly scientific—be scientifically valid (*Daubert*). And like all preliminary questions related to admissibility of evidence, the scientific validity of an expert scientific method must be determined by the trial judge, not by the scientific community. Thus, the Court reasoned, expert testimony is admissible so long as the expert is qualified and her method, if scientific, is deemed by the trial judge to be sufficiently reliable. In setting forth the factors to be considered by trial judges in determining scientific validity, the *Daubert* Court relied heavily on Karl Popper's view of the scientific method (*Daubert*, p. 593). Influenced by Popper's preoccupation with the concept of falsifiability, the *Daubert* Court set forth the following nonexhaustive list of factors to be considered by a judge in determining reliability of an expert method: (1) whether the method "can be (and has been) tested"; (2) whether the method "has been subjected to peer review and publication"; (3) the method's "known or potential rate of error"; (4) "the existence and maintenance of standards controlling the technique's operation"; and (5) whether the method is generally accepted in the "relevant scientific community."

In two subsequent decisions, the Supreme Court held that the *Daubert* reliability test applies not only to the method an expert uses but also to the expert's application of that method (*General Electric Co. v. Joiner*, 1997) and that *Daubert* applies not only to "scientific" methods but to all expert testimony, including nonscientific "technical" fields like tire-tread analysis (*Kumho Tire Co. v. Carmichael*, 1999). Together, *Daubert*, *Joiner*, and *Kumho Tire* are typically called the "*Daubert* trilogy" (Bernstein & Jackson, 2004). The language of Federal Rule of Evidence 702 has been amended to reflect the trilogy's holdings and now requires both that the "testimony is the product of reliable principles and methods" and that "the expert has reliably applied the principles and methods to the facts of the case" (F.R.E. 702(c),(d)). Courts applying *Daubert* to scientific methods continue to apply the nonexhaustive *Daubert* factors (testability, peer review, existence and extent of error rate, existence of standards to govern the method, and general acceptance in the scientific community) set forth in *Daubert* itself (see, e.g., Moss, 2015).

Today, most states and the federal system have shifted to the *Daubert* standard, either through court decisions or through passage of a statute or rule similar to Federal Rule of Evidence 702. Still, a significant minority of states, including New York and California, continue to adhere to *Frye* in determining admissibility of novel scientific evidence such as DNA (Jurilytics, 2017). Thus, in relying on precedent related to admissibility of a particular DNA method, litigants should be aware of which standard—*Frye* or *Daubert*—governed the precedential decision and which standard governs in the litigant's jurisdiction.

The Evidentiary Rule against Hearsay and the Constitutional Right of Confrontation

The second set of rules that might preclude admission of DNA typing results is the rule against "hearsay" and its corresponding constitutional rule, the confrontation clause of the Sixth Amendment to the US Constitution. For better or worse, the Anglo-American system prefers that the claims of human witnesses, if offered for their truth, be made live, in court, subject to the oath, physical confrontation, and cross-examination. Thus, the Federal Rules of Evidence exclude "hearsay"—an out-of-court statement offered "for the truth of the matter asserted in the statement"—as presumptively inadmissible (F.R.E. 801(c), 802). All 50 states have an analogous rule (Broun, 2013, § 244). Hearsay is thus inadmissible unless the proponent lays a foundation for admissibility under an applicable exception to the rule against hearsay, such as for "business records," "statements against penal interest," or "dying declarations" (Broun, 2013, § 245 et seq.). Even if a hearsay statement is admissible as an evidentiary matter under an exception, its admission in a criminal case against the accused might still violate the confrontation clause if it is the "testimonial" hearsay of a nontestifying declarant. Hearsay is generally "testimonial" if it is a sufficiently solemn statement that is either facially accusatory or created with the help of government officers, such as a stationhouse police confession of a defendant's alleged accomplice (*Crawford v. Washington*) or a formal affidavit of a forensic chemist about the presence of a drug in a tested substance (*Melendez-Diaz v. Massachusetts*).

In the context of DNA, these two rules arise most often when a testifying DNA expert determines a match or calculates a match statistic based in part on the hearsay report of another DNA expert who does not testify at trial.

The Fourth Amendment

The Fourth Amendment to the Constitution also gives rise to potential admissibility challenges to DNA evidence in court when offered in a criminal case against the accused. The Fourth Amendment protects against "unreasonable searches and seizures" and prohibits the issuance of a search or arrest warrant unless supported by "probable cause" (U.S. Const. amend. IV). Police may therefore apply for a search warrant to obtain a nonconsensual DNA sample from a criminal suspect only if they have probable cause to believe the DNA will show that the suspect has committed a crime. Police can also obtain DNA from a suspect by consent without violating the Fourth Amendment (*Schneckloth v. Bustamonte*; Will, 2003). To the extent forced DNA sampling in the absence of any individualized suspicion to believe the suspect is engaged in a crime is unconstitutional, any DNA test results stemming from such a Fourth Amendment violation may be inadmissible under the "exclusionary rule" as the "fruit" of a constitutional violation (*Mapp v. Ohio*; *Maryland v. King*). As discussed further later in the chapter, the primary context in which criminal defendants have argued that evidence of a DNA match violates the Fourth Amendment is in database "cold hit" cases.

THE STATUS OF RELIABILITY-BASED ADMISSIBILITY CHALLENGES TO DNA EVIDENCE

This section explores the status of reliability challenges to various forms of DNA evidence, setting forth both areas of consensus, in which the reliability of DNA typing results will not likely be disputed, and areas of controversy, in which the reliability of DNA typing results or statistical methods is more contentious.

Single-Source PCR-STR Testing Results and Random Match Probabilities (RMPs)

Some forms of DNA evidence are now universally accepted as evidence of identity in US courts as a matter of reliability. The original forms of forensic DNA testing and interpretation used in the 1980s and early 1990s were subject to much criticism during the "DNA Wars," the history of which has been ably told by others (Kaye, 2010; Lynch et al., 2008; see chapter 1). But these earlier techniques have been replaced in forensic DNA analysis by PCR-based STR discrete-allele typing. Courts now universally accept as generally reliable both the PCR process for amplification of DNA and the STR-based system of identifying and comparing alleles (Kaye, 2010, pp. 190–191).

The most common PCR-STR-based kits used in forensic analysis in criminal cases in the United States are those manufactured by Applied Biosystems (such as ProFiler/CoFiler, testing 13 core STR loci, including amelogenin, or sex; IdentiFiler, testing 16 loci, including amelogenin; and, most recently, GlobalFiler, testing 24 loci)) and Promega's PowerPlex kits (FBI, n.d.). These are the primary kits accepted by the Federal Bureau of Investigation (FBI) for uploading to the Combined DNA Index System (CODIS) for comparison purposes (FBI, n.d.). All new reference samples taken by convicted or arrested persons for upload to CODIS must contain 20 "core CODIS loci" and thus must be tested using the most recent, most highly discriminating kits (FBI, n.d.). All evidence samples from crimes or missing persons inquiries must be at least tested for the core CODIS loci, though results at all 20 loci are not necessary for comparison purposes (FBI, n.d., §§ 19, 20).

In addition, the use of the RMP to express the statistical significance of a match between two PCR-STR single-source (non-mixture) profiles has also been universally accepted by US courts.[1] In general, evidence of a DNA match is inadmissible without a corresponding match statistic expressing the statistical significance of the match to the factfinder.[2] In the words of one court, "[w]ithout the probability assessment, the jury does not know what to make of the fact that the [DNA] patterns match: the jury does not know whether the patterns are as common as pictures with two eyes, or as unique as the Mona Lisa." (*United States v. Yee*, 1991). The RMP is the product of the probabilities of a person having each of the alleles represented in a single-source PCR-STR profile (Butler, 2009, pp. 229–230). While the RMP has been the subject of some academic debate because its accuracy rests on the assumption of statistical independence among the STR loci and minimal population substructure, courts have universally accepted it as a reliable expression of the statistical significance of a match between two single-source samples under both *Frye* and *Daubert* (see, e.g., Mueller, 2008). Only where the RMP has been mistaken by a prosecutor as the chance of the defendant's innocence—the "prosecutor's fallacy" or "fallacy of the transposed conditional"—have courts commented on its potential for undue prejudice.[3] Moreover, while expert witness assertions of source attribution based

---

1. See, e.g., *State v. Roman Nose*, 667 N.W.2d 386, 398 (Minn. 2003) ("[A] random match probability statistic [product rule] is scientifically acceptable when applied to a known single source sample."); and *People v. Smith*, 132 Cal. Rptr. 2d 230, 233 (Ct. App. 2003) ("Defendant concedes, 'It is generally accepted the [polymerase chain reaction and short tandem repeats] can be completely accurate in typing genetic material from single source samples.'"). Cf. *United States v. Silva*, 889 F.3d 704, 718 (10th Cir. 2018) (expert described single-source sample analysis "as easy as you can get").

2. See, e.g., *State v. Tester*, 968 A.2d 895, 909 (Vt. 2009) ("[A]dmission of DNA match evidence, without additional evidence of the frequency with which such matches might occur by chance, is error."); *Deloney v. State*, 938 N.E.2d 724, 730 (Ind. Ct. App. 2010) (deeming DNA evidence inadmissible without "accompanying testimony explaining the statistical significance of those non-exclusion results"); *United States v. Davis*, 602 F. Supp. 2d 658, 673 (D. Md. 2009) ("DNA evidence cannot be admitted in a vacuum; the Government must also present some additional information with which a jury can accurately assess the significance of the consistency between a defendant's DNA profile and that of the evidence."); and *Commonwealth v. Mattei*, 920 N.E.2d 845, 858 (Mass. 2010) ("The challenged expert [DNA] testimony concerning the nonexclusion results should not have been admitted without accompanying statistical explanation of the meaning of nonexclusion."). But see *State v. Hummert*, 933 P.2d 1187, 1191 (Ariz. 1997) (noting that in Arizona, no numerical statistic is required as foundation for expert testimony on a DNA match).

3. The "prosecutor's fallacy," or fallacy of the transposed conditional, occurs when a lawyer (or judge or juror) mistakes the RMP (e.g., one in a million) for the probability that the defendant is not the source. Put differently, the person hearing the statistic mistakes one conditional probability (the chance the defendant would match the profile, given that he is not the source of the DNA, or the RMP) for its transposed conditional (the chance the defendant is not the source, given that he matches). See, e.g., *McDaniel v. Brown*, 558 U.S. 120 (2009) (noting that the prosecutor and the government's DNA expert both engaged in the fallacy of the transposed conditional in their statements before the jury); and (Roth, 2010), explaining the fallacy in laypersons' terms. ().

on RMPs have been the subject of some defense challenges, courts generally allow DNA experts to testify to their opinion, based on an exceedingly small RMP that the DNA profiles share a common source. For example, FBI analysts commonly testify to source attribution above any RMP threshold of 1 in 300 billion (1,000 times the US population) (Butler, 2009).

Y-STR and Mitochondrial DNA (mtDNA) Testing Results

Litigants have also made reliability challenges to Y-STR and mtDNA testing results, with little success. While forensic PCR-STR typing looks at short repeated sequences of DNA on the "autosomal" (nonsex) chromosomes (Butler, 2009), forensic Y-STR typing looks at certain short repeated sequences on the Y chromosome in male DNA samples. Men inherit the Y-STR profile of their father, and the profiles are not believed to change much over generations. Thus, the statistical significance of a Y-STR match is not calculated by an RMP; the allelic frequency tables that generate RMPs for traditional PCR-STR profiles assume statistical independence of the STR markers. In Y-STR typing, in contrast, analysts use a "counting method" to generate a match statistic. That is, they look for the Y-STR profile or *haplotype* in a database of Y-STR profiles from a relevant population, take the resulting number of "hits" (say, zero or one), and build a confidence interval around that number to express the chances of seeing additional matches in a larger population sample. Thus far, courts appear to have universally accepted Y-STR typing results as reliable when offered by the government in a criminal case, both under *Frye*[4] and *Daubert*.[5] Notably, courts have excluded Y-STR typing results in certain circumstances when offered by a criminal defendant in postconviction proceedings as evidence of innocence, even while acknowledging that Y-STR typing is generally accepted.[6] One recent court also

---

4. See, e.g., *People v. Zapata*, 8 N.E.3d 1188, 1195 (Ill. Ct. App. 2014) (holding that no new *Frye* hearing was required because Y-STR was already generally accepted); *State v. Calleia*, 997 A.2d 1051, 1065 (N.J. Ct. App. 2010), *rev'd on other grounds*, 206 N.J. 274, 20 A.3d 402 (2011) (ruling that Y-STR is generally accepted); *Commonwealth v. Jacoby*, 170 A.3d 1065, 1095 (Pa. 2017), *petition for cert pending* (ruling as a matter of first impression that Y-STR is generally accepted); *State v. Bander*, 208 P.3d 1242, 1255 (Wash. Ct. App. 2009) (no new *Frye* hearing required because Y-STR is already generally accepted); and *People v. Stevey*, 148 Cal. Rptr. 3d 1, 12 (Ct. App. 2012) (no new *Kelly-Frye* hearing required because Y-STR is already generally accepted).

5. See, e.g., *People v. Tunis*, 318 P.3d 524, 528 (Colo. Ct. App. 2013) (holding as an issue of first impression that Y-STR is reliable under *Daubert*); *People v. Wood*, 862 N.W.2d 7, 24 (Mich. Ct. App. 2014), *judgment vacated in part on other grounds*, 498 Mich. 914, 871 N.W.2d 154 (2015) (trial court did not abuse its discretion by admitting expert testimony on Y-STR after holding *Daubert* hearing); and *State v. Maestas*, 299 P.3d 892, 934 (Utah 2012) (same).

6. See, e.g., *Commonwealth v. DiCicco*, 25 N.E.3d 859, 869 (Mass. 2015) (holding that the trial judge did not abuse her discretion by excluding defendant's proffered Y-STR expert witness, because the Y-STR exclusion was based on a single potential allele, a method that is discouraged, though not prohibited, under the SWGDAM Y-STR guidelines); *People v. Stoecker*, 10 N.E.3d 843, 849 (Ill. 2014) (holding that, given the DQ alpha testing done

excluded evidence of Y-STR results in a (currently pending) Texas murder trial, but only apparently because the results did not incriminate the defendant other than the tendency to show that the DNA was male (Winkle & Goard, 2018).

Challenges to mtDNA typing results have been similarly unsuccessful. Unlike Y-STR typing, mtDNA typing isolates a long sequence of DNA in a particularly hypervariable region of DNA found in the mitochondria of one's cells (outside the nucleus). Like Y-STRs, mtDNA is not recombinant; we inherit our mtDNA sequence from our mothers, and mtDNA sequences (like Y-STRs) are not believed to change much over generations. As with Y-STR haplotypes, the statistical significance of a match between mtDNA sequences is expressed through the *counting method*, building a confidence interval around the number of matching haplotypes found in a relevant mtDNA population database. To be sure, mtDNA match statistics have been criticized for being misleading and inaccurate, given the amount of "clustering" of haplotypes based on migration patterns (Kittles et al., 2006). Nevertheless, most if not all challenges to mtDNA typing results or match statistics in US courts have been unsuccessful under *Frye*[7] and *Daubert*,[8] with very few challenges even being brought in the last decade.

Admissibility Challenges to "Low Copy Number" DNA Typing Results

The primary context in which admissibility challenges to single-source nuclear DNA comparison results are still successful is low copy number (LCN) DNA testing. Many laboratories have a different set of protocols for testing DNA samples involving an amount of input DNA lower than 1 nanogram, a level below which the identification

---

before trial, there was no "reasonable likelihood of more probative results" using Y-STR typing after conviction); and *People v. Barker*, 1-12-3238, 2015 WL 2069736, at *8 (Ill. App. Ct. Apr. 30, 2015) (same).

7. See, e.g., *People v. Stevey*, 148 Cal. Rptr. 3d 1, 11 (Ct. App. 2012) ("mtDNA evidence . . . has also gained general acceptance within the scientific community"); *People v. Klinger*, 713 N.Y.S.2d 823, 831 (N.Y. Co. Ct. 2000) (holding that mtDNA is generally accepted and reliable based on holdings in a number of other jurisdictions); *Magaletti v. State*, 847 So. 2d 523 (Fla. Ct. App. 2003) (same); *State v. Pappas*, 776 A.2d 1091, 1108 (Conn. 2001) (same); and *People v. Holtzer*, 255 Mich. App. 478 (2003) (same).

8. See, e.g., *State v. Brochu*, 949 A.2d 1035, 1049 (Vt. 2008) ("Although mtDNA evidence is relatively new, all jurisdictions to have considered the issue have uniformly found mtDNA to be reliable."); *United States v. Beverly*, 369 F.3d 516, 531 (6th Cir. 2004) (holding that trial court did not abuse its discretion in admitting mtDNA evidence because the "scientific basis for the use of such DNA is well established"); *United States v. Coleman*, 202 F. Supp. 2d 962 (E.D. Mo. 2002) (holding mtDNA admissible under *Daubert*); *Wagner v. State*, 864 A.2d 1037, 1044 (Md. Ct. App. 2005) (same); and *State v. Council*, 515 S.E.2d 508, 518 (S.C. 1999) (same). Cf. *State v. Griffin*, 384 P.3d 186, 203 (Utah 2016) (noting, while discussing probative/prejudicial balancing, that "every state that has been confronted with the question of whether mtDNA is admissible under its applicable rules of evidence has answered the question in the affirmative").

and interpretation of alleles becomes more difficult and controversial (Butler, 2015, pp. 159–160; ISHI Conference, 2017). Several trial and appellate courts have ruled LCN testing reliable, under both *Frye*[9] and *Daubert*.[10] Nonetheless, at least a handful of courts have excluded LCN testing results on reliability grounds.[11]

Admissibility Challenges to Mixture Interpretations by Human Analysts

Criminal defendants in the United States challenging the admissibility of DNA typing results have perhaps had the most traction in cases involving DNA mixtures and the presentation of a match statistic called the combined probability of inclusion, or CPI, to the factfinder.

Because DNA mixtures involve more than one contributor, it may be difficult for analysts to determine how many contributors there are to a mixture, and which alleles at each locus belong to which contributor. As a result, a simple calculation of an RMP using allelic frequencies and the product rule, as analysts do for single-source sample comparisons, is not possible in mixture statistics. And while new probabilistic genotyping software programs hold great promise for mixture deconvolution and the calculation of highly discriminating match statistics based on the consideration of thousands of permutations, analysts are more limited in their mixture interpretation abilities. Thus, analysts calculate a match statistic in mixture cases by (1) identifying all alleles at all loci, (2) determining whether the reference sample of interest (such as a criminal suspect) has alleles that are consistent with the alleles

---

9. See, e.g., *Phillips v. State*, 226 Md. App 1 (Md. Ct. Spec. App. 2015) (holding LCN DNA analysis admissible under *Frye* and that any attack on its reliability went to its weight rather than admissibility); *People v. Garcia*, 963 N.Y.S.2d 517 (N.Y. Sup. Ct., Bronx County 2013) (holding both that LCN DNA testing is not a novel science that would require a *Frye* hearing before being admissible and that LCN DNA testing conducted by the OCME in New York is generally accepted); and *People v. Megnath*, 898 N.Y.S.2d 408 (N.Y. Sup. Ct., Queens County 2010) (holding that LCN DNA testing as conducted by the OCME is admissible under *Frye*). Cf. *People v. Lazarus*, 190 Cal. Rptr. 3d 195, 239 (Ct. App. 2015) (holding that defendant was not entitled to a *Frye* hearing because there was no evidence that LCN was not generally accepted).

10. See, e.g., *United States v. Morgan*, 53 F. Supp. 3d 732 (S.D.N.Y. 2014), aff'd, 675 F. App'x 53 (2d Cir. 2017) (holding that the OCME's LCN testing methodology is reliable under *Daubert*); and *United States v. Barton*, 8:14-CR-496-T-17AEP, 2016 WL 4921036, at *6 (M.D. Fla. Sept. 14, 2016) (holding that magistrate did not abuse its discretion in denying *Daubert* challenge to LCN results). Cf. *United States v. Sleugh*, 14-CR-00168-YGR-2, 2015 WL 3866270, at *3 (N D. Cal. June 22, 2015), *appeal pending* (holding that no *Daubert* hearing was necessary to introduce LCN results).

11. See *United States v. McCluskey*, 954 F. Supp. 2d 1224 (D.N.M. 2013) (excluding LCN testing under *Daubert* because the New Mexico Department of Public Safety laboratory used different procedures and methods than the New York OCME, and there was no evidence that the NM procedures and methods would yield reliable results); and *People v. Collins*, 15 N.Y.S.3d 564 (N.Y. Sup. Ct., Kings County 2015) (holding LCN testing using the OCME's "FST" software was not admissible under *Frye*).

in the mixture, and (3) if so, calculating the chance that a randomly selected person would also be consistent with the combination of alleles present in the mixture—the CPI (sometimes referred to equivalently as the "random man not excluded" statistic) (Bieber et al., 2016; Butler, 2015).

The CPI has been the subject of considerable criticism, from both sides of the American criminal justice system. On the one hand, the CPI is a much less discriminating statistic than the RMP and tends to ignore a significant amount of relevant information about likely number of contributors and likely contributor profiles (Curran & Buckleton, 2008). At the same time, some argue that the CPI carries too great a risk of falsely inculpating innocent suspects, because it removes any loci from its statistical calculation that exhibit signs of "allelic dropout": based on a suspect's allele being absent from the mixture, rather than considering the possibility that the absence of the suspect's allele reflects that the suspect is simply not a contributor (Murphy, 2015, pp. 92–94; Butler, 2015; Curran & Buckleton, 2010).

In 2010, in response to the critiques of the CPI from the scientific community, the Scientific Working Group on DNA Analysis Methods (SWGDAM) changed its DNA mixture interpretation guidelines to require laboratories to remove certain loci from their CPI calculations. SWGDAM's concern was with peaks that were above the profiling system's "analytical threshold" (AT) (the height below which a peak carries too great a risk of being an artifact rather than genetic material) but below the system's "stochastic threshold" (ST) (the height above which stochastic effects,[12] such as allelic dropout, are unlikely to occur). The new guidelines required laboratories to remove any locus from the CPI calculation that contained any peak above the AT but below the ST.[13] In 2016, a distinguished group of scientists reiterated this call to remove any locus from the CPI calculation where any peak was within a certain range (Bieber et al., 2016; PCAST, 2016, p. 78 (citing Bieber et al. 2016)). And later in 2016, the President's Council of Advisors on Science and Technology (PCAST) concluded that the CPI method is "clearly not foundationally valid" under *Daubert* (PCAST, 2016, p. 78).

While the CPI has been introduced as a valid match statistic under both *Frye*[14] and

---

12. Stochastic effects are those due to sampling issues caused by the low number of events.

13. See Scientific Working Group on DNA Analysis Methods, "SWGDAM Interpretation Guidelines for Autosomal STR Typing by Forensic DNA Testing Laboratories" (approved January 4, 2010, § 4.6.3) ("When using CPE/CPI (with no assumptions of number of contributors) to calculate the probability that a randomly selected person would be excluded/included as a contributor to the mixture, loci with alleles below the stochastic threshold may not be used for statistical purposes to support an inclusion. In these instances, the potential for allelic dropout raises the possibility of contributors having genotypes not encompassed by the interpreted alleles.").

14. See, e.g., *State v. Bigger*, 227 Ariz. 196, 205, 254 P.3d 1142, 1151 (Ariz. Ct. App. 2011) (CPI is "generally accepted," even when applied to LCN samples); and *Phillips v. State*, 126 A.3d 739, 751 n.11 (Md. Ct. Sp. App. 2015), *aff'd on other grounds*, 152 A.3d 712 (Md. 2017) (holding in footnote that the use of a CPI statistical computation for a steering wheel DNA sample was admissible because the laboratory "analyzed the steering wheel sample in a generally accepted manner.").

Daubert[15] in numerous criminal trials in the United States, at least one recent court has rejected the CPI as unreliable,[16] and other courts have recently reversed convictions based on the presentation of mixture statistics to a jury that, viewed in retrospect, are vastly more inculpatory—sometimes by several orders of magnitude—than they would have been under the post-2010 guidelines (see, e.g., Moran, 2017; Texas Forensic Science Commission, 2015). Moreover, some courts have excluded a CPI on undue prejudice grounds in cases where the statistic is only minimally discriminating (such as excluding only 50% of the population).[17] Litigants involved in DNA mixture cases should therefore be aware of the guideline change and how it might affect match statistics presented at both past and current trials.

Admissibility Challenges to Complex Mixture Interpretations by Expert Systems

Some laboratories have begun to address the problems of the CPI by employing expert systems to interpret DNA mixtures. Unlike human analysts, expert systems, or probabilistic genotyping software (PGS) in this case, can consider much more information, including data sets estimating allelic dropout averages at various loci, than human analysts can. Instead of calculating the CPI, these expert systems calculate an LR or similar statistic that purports to compare the probability of seeing the mixture given the competing hypotheses that the suspect (or other person of interest) is or is not a contributor to the mixture. The resulting LRs tend to be much more discriminating than the CPI; a typical LR reported by the program TrueAllele would state, "A match between Mr. [Defendant] and the fingernails is 189 billion times more probable than a coincidental match to an unrelated Caucasian." (Perlin, 2010). Expert systems have also been wielded by lawyers for criminal defendants as evidence of innocence in high-profile exonerations (see, e.g., McCall, 2018). While some PGS are open source, most are proprietary. One proprietary program, the "FST" software developed by New York's Office of the County Medical Examiner (OCME), was excluded by one trial judge under *Frye*, prompting the OCME to both make the source code public and shift to using a different program (Jacobs, 2016). Now, the two main

---

15. See, e.g., *State v. Haughey*, 3 A.3d 980, 992 (Conn. Ct. App. 2010) (defendant could not demonstrate that the trial court abused its discretion in admitting CPI evidence).

16. *United States v. Williams*, 3:13-CR-00764-WHO-1, 2017 WL 3498694, at *13 (N.D. Cal. Aug. 15, 2017) (excluding DNA analysis from the SERI laboratory, which used a "suspect-centric" form of CPI analysis "not based on sound methodology."). Cf. *People v. Smith*, C062513, 2011 WL 4528254, at *22 (Cal. Ct. App. Sept. 29, 2011) (unpublished) (holding that "the CPI was an improper statistical analysis to be used in this case because [the expert] knew he was dealing with a mixture that had significant allelic dropout," but finding no prejudice).

17. See, e.g., *People v. Pike*, 53 N.E.3d 147, 170 (Ill. App. Ct. 2016), *reh'g denied* (May 2, 2016), *appeal denied*, 89 N.E.3d 761 (Ill. 2017) (holding that it was error, but not plain error, for the trial court to admit a "50% inclusion probability statistic" derived from CPI calculations "because the statistic was irrelevant").

programs used in forensic DNA testing in the United States are TrueAllele, owned by the Pittsburgh, Pennsylvania, company Cybergenetics; and STRMix, owned by the New Zealand research institute ESR (PCAST, 2016, p. 80).

The 2016 report of the President's Council of Advisors on Science and Technology (PCAST) cited expert systems as an improvement over existing human analysis of complex mixtures, concluding that such methods have been established as foundationally valid for mixtures with three or fewer contributors, where the minor contributor constitutes at least 20% of the intact DNA in the mixture and the DNA amount exceeds the minimum required by the method for analysis (PCAST, 2016, p. 82). The report suggested, however, that use of the software beyond its empirically established range could be problematic. To be sure, the PCAST report has itself been subject to criticism both for failing to more fully solicit the participation of forensic examiners and law enforcement and for placing a premium on properly designed "black box" validation studies as a prerequisite for foundational validity (National District Attorney's Association, 2016; Budowle, 2017).

So far, the LRs from both STRMix and TrueAllele have been admitted in numerous courts across the country, in both *Frye* and *Daubert* jurisdictions.[18] In fact, TrueAllele has not been excluded on reliability grounds by any court, although one California trial court—later reversed by a higher court—had deemed the failure to disclose TrueAllele's source code a barrier to its admissibility (*People v. Chubbs*, 2015). STRMix has been deemed inadmissible in two cases. In the first, the New York *Hillary* case, the trial judge excluded the evidence under *Frye* not based on a ruling that STRMix is an inherently unreliable method, but on the lack of internal validation studies by the local laboratory that conducted the testing (*Hillary* Order, 2016, pp. 9–10). Notably, the inculpatory LR generated by STRMix in *Hillary* was contradicted by TrueAllele results on the same sample, which indicated that Mr. Hillary was likely not a contributor (Roth, 2017, p. 2019). In the second case, in June 2018, a Texas trial judge excluded under *Daubert* the STRMix results from male DNA found on the thigh of a female murder victim, after human analysis "came up inconclusive" (Winkle & Goard, 2018). In another recent case, currently pending appeal, a California state judge has conditioned the admissibility of STRMix under *Frye* on the government providing the source code to the defense, which it thus far has refused to do (*People v. Dominguez*, 2018). While the research institute ESR offers limited access to STRMix's source code to defense experts before trial under a nondisclosure agreement, the company has declined to allow broader access, citing a trade secret privilege. While some legal commentators have suggested that no such trade secret privilege should exist in criminal cases (Wexler, 2018; Chessman, 2017), no appellate court has yet been

---

18. See, e.g., *People v. Bullard-Daniel*, 54 Misc. 3d 177, 191 (N.Y. Co. Ct. 2016) (holding STRMix results admissible under *Frye*). The Cybergenetics website lists numerous cases in which TrueAllele has been admitted under either *Frye* or *Daubert*. See TrueAllele Admissibility, cybgen.com, https://www.cybgen.com/information/admissibility/page.shtml. Likewise, the STRMix website lists numerous cases around the country and globe in which STRMix has been admitted. See https://strmix.esr.cri.nz/#news.

persuaded by such arguments to uphold a trial court's order requiring source code disclosure.

In sum, the results of expert systems TrueAllele and STRMix have thus far been widely held admissible as evidence of both guilt and innocence. Nonetheless, if defense requests for access to source code continue to be granted, and if disclosure of source code is deemed a condition of admissibility, then the proprietors of these programs may have to subject their source code to further scrutiny or face possible exclusion of results in certain cases. Moreover, to the extent these systems continue to be used on complex mixtures beyond the empirically established range of the software—with multiple contributors, extreme peak height differential, or low template DNA—and to the extent different systems continue to generate contradictory results, courts might be more receptive to reliability challenges in the future.

Admissibility of Evidence Related to DNA "Transfer"

Under certain circumstances, DNA can "transfer" from one individual or surface to another (*direct transfer*), or even from that second person/surface to a third person/surface (*secondary transfer*) (Butler 2009, p. 80; 2011, pp. 18–19). The likelihood of transfer occurring is a function of a number of factors, including the type of surface touched and whether the individuals involved are "shedders" or "non-shedders" (Fonneløp et al., 2017). In a given case, the likelihood of transfer might be a critical issue for the factfinder, in terms of what inference to draw from the presence of a person's DNA at a crime scene. For example, in one recent case, a state appellate court ruled that the presence of a defendant's DNA on a handgun found in his house was insufficient evidence to convict him for possession of the gun, because of the high likelihood of transfer (*Finley v. State*, 2014). And in a high-profile California case, a homeless man, Lukis Anderson, accused of killing a wealthy Silicon Valley investor, was eventually exonerated after the presence of his DNA on the victim's fingernails was explained by DNA transfer; the same EMTs who responded to the murder scene had assisted Anderson earlier in the day and could have transferred traces of Anderson's DNA to the victim (Worth, 2018). While numerous courts have allowed expert testimony as to transfer, several courts have also denied motions for postconviction relief filed by defendants claiming that a DNA transfer expert would have made a difference at trial.[19]

---

19. See, e.g., *Adams v. State*, 161 Idaho 485 (Ct. App. 2016) (holding that the addition of expert testimony on DNA transfer would not have made a difference to the trial outcome); and *Sancier v. Comm'r of Correction*, 139 Conn. App. 644 (2012) (acknowledging that DNA transfer is "theoretically possible" but ruling that transfer evidence would not have affected outcome). Cf. *State v. Freeman*, No. 28150, 2008 WL 142299, at *1 (Mo. Ct. App. Jan. 16, 2008), *rev'd en banc*, 269 S.W.3d 422 (Mo. 2008) (rejecting defendant's claim that the DNA evidence against him was insufficient because of the possibility of transfer).

THE STATUS OF CONSTITUTIONAL ADMISSIBILITY CHALLENGES TO DNA EVIDENCE

This section explores the status of constitutional challenges to DNA evidence under the confrontation clause and Fourth Amendment.

Confrontation Clause Challenges to Reliance on Hearsay DNA Reports of Nontestifying Analysts and to Proprietary Expert Systems

Because of the rule against hearsay and the confrontation clause, the proponent of a forensic DNA report cannot offer *the report itself* into evidence without calling the report's author to the witness stand. However, the proponent of the testimony can circumvent the hearsay rule by having the testifying analyst simply explain to the factfinder that her expert opinion is *based upon* the other analyst's report. Under Federal Rule of Evidence 703 and its state analogs, so long as the hearsay report itself is not offered as evidence, a testifying expert is free to "base an opinion" upon hearsay or other inadmissible evidence (*Williams v. Illinois*; F.R.E. 703).

Although the evidentiary rules allow a testifying DNA analyst to rely on another analyst's hearsay report in rendering an opinion, there is still an open question as to whether such testimony might violate the confrontation clause in a criminal case if offered against the accused. In *Williams v. Illinois* (2012), the Supreme Court heard a rape case in which the state's DNA analyst testified that the defendant's PCR-STR profile, which was tested and developed at the testifying analyst's state laboratory, "matched" the profile developed from the victim's vaginal swabs, which were tested and analyzed at a different laboratory, Cellmark. The defendant argued that the analyst's testimony violated the confrontation clause because the analyst's expert opinion was based in large part on the analysis and conclusions of Cellmark's analyst, who did not testify. A majority of justices of the Supreme Court concluded that the testimony did not violate the confrontation clause, but no one argument received a full five votes. Four justices concluded that the nontestifying expert's report did not implicate the clause because it was technically offered only to explain the basis of the testifying analyst's opinion, rather than for its "truth." An additional justice concluded that the testimony did not implicate the clause because the hearsay report of the nontestifying analyst was not sufficiently formal or solemn to count as testimonial hearsay, given that the Cellmark analysis was conducted before a suspect had been identified.

Because neither of the theories of admissibility in *Williams* garnered five votes, and because of the changing composition of the Supreme Court, the *Williams* decision leaves unresolved whether future DNA cases with slightly different facts might present a confrontation clause problem. For example, a recent decision by New York's highest state court reversed a burglary conviction on confrontation clause grounds where the testifying DNA analyst, who opined that the defendant's DNA matched the DNA from the crime scene but who had not conducted, witnessed, or supervised the DNA testing in the case, simply read to the jury the hearsay report of another, nontestifying DNA analyst colleague (*People v. Austin*, 2017).

The other potential confrontation clause challenge to DNA evidence relates to the results of proprietary expert systems offered at trial. A reported LR from TrueAllele is not considered "hearsay" under American rules of evidence because it is not an assertion by a human witness; TrueAllele cannot be placed under oath, "cross-examined," or physically confronted. But some commentators have argued that expert systems offering accusatory claims against criminal defendants should perhaps be considered "witnesses against" the defendant for purposes of the confrontation clause (see, e.g., Roth, 2017). While "confrontation" of a proprietary algorithm would not be synonymous with "cross-examination," it might involve disclosure of source code; disclosure of prior statements of the algorithm related to the same subject matter; or a right to some sort of technical transparency report that reveals relevant assumptions of the program, such as the program's estimate of allelic dropout rates, or stutter percentages, at various loci. With the exception of a dissenting California Supreme Court justice, however, no appellate court has yet been persuaded that machine-generated results might implicate the confrontation clause (*People v. Lopez*, 2012). Indeed, at least one US Supreme Court justice has intimated that "raw data" from a machine would likely not implicate the confrontation clause (*Bullcoming v. New Mexico*, 2011, Sotomayor, J., dissenting).

Fourth Amendment Challenges to DNA Database "Cold Hit" Results

Thus far, Fourth Amendment challenges to forcible DNA sampling of those convicted of or arrested for certain crimes, and the uploading of the resulting DNA profiles to CODIS for comparison purposes, have been unsuccessful. Each state, as well as the federal system, maintains a DNA database, authorized by statute, containing the PCR-STR profiles of people convicted of, or arrested for, various crimes (Roth, 2013). These official statutory databases are all interconnected through CODIS, allowing local police anywhere in the country to compare a crime scene DNA profile to the 14+ million profiles in CODIS to look for a match, or "cold hit," to an unsolved case. In 2013 the Supreme Court upheld the constitutionality of Maryland's arrestee database, on grounds that states can reasonably require arrestees to give DNA for identification purposes, just as they are forced to give fingerprints (*Maryland v. King*, 2013). Most recently, the California Supreme Court held that California's arrestee database was also constitutional under *King*, a critical decision, given that California's database differs significantly from Maryland's database in that it does not allow for automatic expungement of an arrestee's record and is more expansive in the crimes it covers (*People v. Buza*, 2018).

To the extent that Fourth Amendment challenges to database "cold hits" might be successful in the future, they will probably relate to more controversial tactics such as familial searching, searching of genealogy websites, or collection of "abandoned" DNA. *Familial searching* entails searching a DNA database not just for a perfect match, but for a partial match, indicating that a person with partially matching might be *related to* the perpetrator (Murphy, 2010; Chapter 4). While familial searching has been banned in a few states, including Maryland, it is explicitly permitted in 12 states, including California (Rainey, 2018). While proponents argue that the practice helps catch elusive criminals (Rainey, 2018), critics argue that it has a significant racially disparate impact and unfairly allows law enforcement to scrutinize people who are not in a criminal or even an arrestee database, in the absence of any suspicion of wrongdoing (Murphy, 2010).

No Fourth Amendment challenge has yet been successful against the collection of "abandoned" DNA, which people inadvertently leave on coffee cups, cigarette butts, and the like. The prevailing logic is that because the DNA has been "abandoned," it is akin to garbage, which is devoid of Fourth Amendment protections because the owner has abandoned it and thus has no "legitimate expectation of privacy" in the contents under existing constitutional doctrine (Joh, 2006). A Fourth Amendment challenge against government searches of commercial genealogy databases, onto which members have voluntarily posted their DNA profiles for comparison with other members, might be precluded on the same grounds. California police recently identified a suspect in the Golden State Killer case based on a search of the open-source genealogy website GEDMatch, conducted without a warrant, without probable cause, and in a way that violated the terms of service (by creating a fake name associated with the crime scene DNA profile they uploaded for comparison purposes) (Zhang, 2018; see chapter 15).

The Supreme Court's recent decision in *Carpenter v. United States* might breathe new life into such challenges, however. In *Carpenter*, the Court held that the government cannot subpoena historical cell phone location records without a warrant, even though the defendant had shared his location with his cell phone company (*Carpenter v. United States*, 2018). In doing so, the Court significantly limited the reach of the "third-party doctrine" that a suspect has no legitimate expectation of privacy in information he shares with others.[20] The full implications of *Carpenter* for DNA database search challenges remains to be seen.

CONCLUSION: DNA ADMISSIBILITY ISSUES ON THE HORIZON

The next decade will inevitably bring further DNA admissibility issues as the technology advances. For example, the use of Rapid DNA machines—which can develop a profile from a sample in as little as 90 minutes—on crime scene evidence samples will surely be the subject of *Daubert* and *Frye* reliability challenges. While reference samples developed with Rapid DNA are eligible for upload to CODIS, crime scene

---

20. See *Carpenter v. United States*, 585 U.S. __ (2018), slip op. at 11 ("Given the unique nature of cell phone location records, the fact that the information is held by a third party does not by itself overcome the user's claim to Fourth Amendment protection.").

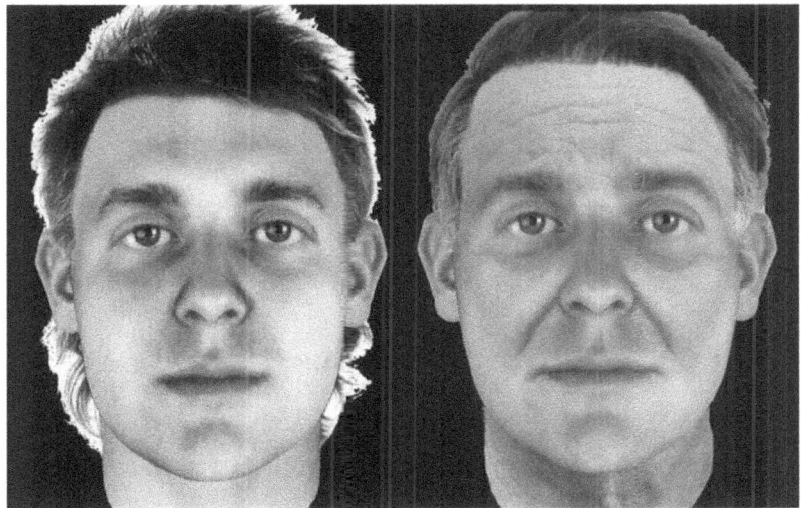

Police released images of how an attacker connected to several crimes, including at least a rape and a murder that happened more than 20 years ago in Montgomery County, Md., might have looked at 25 years old, left, and at 45 years old. (Courtesy of Montgomery County Police)

Figure 13.1 Composite sketch of a crime suspect created by Maryland police using DNA phenotyping. *Source:* "Maryland composite suspect sketch created through phenotyping. Courtesy of Montgomery County Police.

(evidence) samples are not (FBI, n.d.). Likewise, the use of DNA *phenotyping* to identify crime suspects may well face admissibility challenges. Phenotyping involves estimating the physical characteristics of a suspect based on a crime scene DNA profile (Southall, 2017). As figure 13.1 shows, the technique is already being used to develop composite sketches of suspects.

Because phenotyping is used only to initially identify a suspect, not to ultimately prove that the suspect matches the DNA from a crime scene, it is unlikely that the practice will trigger reliability challenges at trial. However, the reliability of the technique may well be relevant to Fourth Amendment challenges to searches and seizures based on a suspect's alleged similarity to an estimated phenotype.

In sum, the foundational validity of PCR-STR forensic DNA typing for single-source robust samples and the use of the RMP to express the statistical significance of two matching single-source profiles are well established as reliable. Nonetheless, several aspects of forensic DNA typing may still continue to raise significant admissibility issues.

REFERENCES

Bernstein, D., & Jackson, J. (2004). The Daubert trilogy in the States. *Jurimetrics, 44*, 351–366.

Bieber, F., et al. (2016). Evaluation of forensic DNA mixture evidence: Protocol for evaluation, interpretation, and statistical calculations using the combined probability of inclusion. *BMC Genetics, 17*, 125–140.

Broun, Kenneth S., et al., eds. (2013). *McCormick on evidence*. 7th ed. Eagan, MN: Thomson Reuters.

Budowle, B. (2017, June 17). *Open letter and affidavit re: PCAST Report*. University of North Texas Health Science Center.

Bullcoming v. New Mexico, 564 U.S. 647 (2011).

Butler, J. (2009). *Fundamentals of DNA typing*. Cambridge, MA: Academic Press.

Butler, J. (2011). *Advanced topics in forensic DNA typing: Methodology*. Cambridge, MA: Academic Press.

Butler, J. (2015). *Advanced topics in forensic DNA typing: Interpretation*. Cambridge, MA: Academic Press.

Carpenter v. United States (2018, June 22). Slip opinion. Retrieved from https://www.supremecourt.gov/opinions/17pdf/16-402_h315.pdf.

Chessman, C. (2017). A "source" of error: Computer code, criminal defendants, and the Constitution. *California Law Review, 105*, 179–228.

Crawford v. Washington, 541 U.S. 36 (2004).

Curran, J., & Buckleton, J. (2008). A discussion of the merits of random man not excluded and likelihood ratios. *FSI Genetics, 2*, 343–348.

Curran, J., & Buckleton, J. (2010). Inclusion probabilities and dropout. *Journal of Forensic Sciences, 55*, 1171–1173.

Daubert v. Merrell Dow Pharmaceuticals, Inc., 509 U.S. 579 (1993).

Federal Bureau of Investigation (FBI). (n.d.). *Rapid DNA—CODIS fact sheet*. Retrieved from https://www.fbi.gov/services/laboratory/biometric-analysis/codis/rapid-dna.

Finley v. State, 139 So. 3d 940 (Fla. Dist. Ct. App. 2014).

Fonneløp, A., et al. (2017). The implications of shedder status and background DNA on direct and secondary transfer in an attack scenario. *FSI Genetics, 29*, 48–60.

Frye v. United States, 293 F. 1013 (D.C. Cir. 1923).

General Electric Company v. Joiner, 522 U.S. 136 (1997).

International Symposium on Human Identification (ISHI) Conference. (2017, July 24). *Under the microscope—interview with Bruce Budowle*. Retried from https://www.ishinews.com/under-the-microscope-bruce-budowle-2/.

Jacobs, S. (2015, January 5). Judge tosses out two types of DNA evidence used regularly in criminal cases. *N.Y. Daily News*. Retrieved from http://www.nydailynews.com/new-york/nyc-crime/judge-tosses-types-dna-testing-article-1.2065795.

Joh, E. (2006). Reclaiming "abandoned" DNA: The Fourth Amendment and genetic privacy. *Northwestern University Law Review, 100*, 857–884.

Jurilytics (2017, March 2). Daubert *and* Frye *in the 50 states*. Retrieved from https://jurilytics.com/50-state-overview.

Kaye, D. (2010). *The double helix and the law of evidence*. Cambridge, MA: Harvard University Press.

Kittles, R., et al. (2006). Database limitations on the evidentiary value of forensic mitochondria DNA evidence. *American Criminal Law Review, 43*, 53–88.

Kumho Tire Co. v. Carmichael, 526 U.S. 137 (1999).

Lepore, J. (2015). On evidence: Proving *Frye* as a matter of law, science, and history. *Yale Law Journal, 124*, 1092–1158.

Mapp v. Ohio, 367 U.S. 643 (1961).

Maryland v. King, 133 S. Ct. 1958, 569 U.S. __ (2013).

McCall, M. (2018, April 18). Two men exonerated by DNA evidence in gang rape sue Hammond, Indiana State Police. *Chicago Tribune*. Retrieved from http://www.chicagotribune.com/suburbs/post-tribune/news/ct-met-lawsuit-against-indiana-police-st-0419-story.html.

Melendez-Diaz v. Massachusetts, 557 U.S. 305 (2009).

Moran, G. (2017, October 24). Judge reverses murder conviction, saying crucial DNA information not disclosed. *San Diego Union Tribune*. Retrieved from http://www.sandiegouniontribune.com/news/courts/sd-me-dna-evidence-20171024-story.html.

Moss, K. (2015). The admissibility of TrueAllele: A computerized DNA interpretation system. *Washington & Lee Law Review, 72*, 1033–1076.

Mueller, L. D. (2008). Can simple population genetic models reconcile partial match frequencies observed in large forensic databases? *Journal of Genetics, 87*, 101–108.

Murphy, E. (2010). Relative doubt: Familial searches of DNA databases. *Michigan Law Review, 109*, 291–347.

Murphy, E. (2015). *Inside the cell: The dark side of forensic DNA*. New York, NY: Nation Books.

National District Attorney's Association. (2016, November 16). *Letter to president regarding PCAST report*. Retrieved from https://csidds.files.wordpress.com/2016/11/ndaa-pcast-response-final.pdf.

PCAST. (2016). President's council of advisors on science and technology. *Forensic Science in Criminal Courts: Ensuring Scientific Validity of Feature- Comparison Methods*. https://obamawhitehouse.archives.gov/administration/eop/ostp/pcast/docsreports.

People v. Austin, New York Court of Appeals, slip opinion no. 07300 Oct. 19, 2017.

People v. Buza, 4 Cal. 5th 658 (2018).

People v. Dominguez, Super. Ct. of Cal., No. SCD#230596, Order Granting Motion to Compel Discovery (March 29, 2018).

People v. Hillary, Decision & Order on DNA Analysis Admissibility (Catena, J.). New York, St. Lawrence County Court No. 2015-15 (2015).

People v. Lopez, 286 P.3d 469, 494 (Cal. 2012) (Liu, J., dissenting).

People v. Superior Court ex rel. Chubbs, No. B258569, 2015 WL 139069 (Cal. Ct. App. Jan. 9, 2015).

Perlin, M. (2010, December 29). *Explaining the likelihood ratio in DNA mixture interpretation*. Promega Proceedings. Retrieved from https://www.promega.ee/~/media/files/resources/.../ishi%2021/.../perlin.pdf.

Rainey, J. (2018, April 28). Familial DNA puts elusive killers behind bars: But only 12 states use it. NBCNews. Retrieved from https://www.nbcnews.com/news/us-news/familial-dna-puts-elusive-killers-behind-bars-only-12-states-n869711.

Roth, A. (2010). Safety in numbers? Deciding when DNA is enough to convict. *New York University Law Review, 85*, 1130.

Roth, A. (2013). *Maryland v. King* and the wonderful, horrible DNA revolution in law enforcement, *Ohio State Criminal Law Journal, 11*, 295–309.

Roth, A. (2017). Machine testimony. *Yale Law Journal, 126*, 1972–2053.

Schneckloth v. Bustamonte, 412 U.S. 218 (1973).

Southall, A. (2017, October 19). Using DNA to sketch what victims look like: Some call it science fiction. *New York Times*. Retrieved from https://www.nytimes.com/2017/10/19/nyregion/dna-phenotyping-new-york-police.html.

Spring Co. v. Edgar, 99 U.S. 645 (1878).

State v. Tester, 968 A.2d 895 (Vt. 2009).

Texas Forensic Science Commission. (2015, August 21). *Letter to Texas criminal justice community*. Retrieved from https://www.tidc.texas.gov/.../Memo-Presentation-from-Texas-Forensic-Science-Commission.pdf.

United States Sentencing Commission. (2016/2018). *United States sentencing guidelines*. Retrieved from https://www.ussc.gov/.

United States v. Davis, 602 F. Supp. 2d 658 (D. Md. 2009).

United States v. Yee, 134 F.R.D. 161 (N.D. Ohio 1991).

Wexler, R. (2018). Life, liberty, and trade secrets: Intellectual property in the criminal justice system. *Stanford Law Review, 70*, 1343–1429.

Will, J. (2003). DNA as property: Implications on the constitutionality of DNA dragnets. *University of Pittsburgh Law Review*, *65*, 129–143.

Williams v. Illinois, 567 U.S. 50 (2012).

Winkle, K., et al. (2018, June 20). *Judge throws out DNA evidence in UT murder suspect's case*. KXAN. Retrieved from http://www.kxan.com/news/local/austin/judge-throws-out-dna-evidence-in-ut-murder-suspect-s-case/1250971292.

Worth, K. (2018, April 19). Framed for murder by his own DNA. *Wired Magazine*. Retrieved from https://www.wired.com/story/dna-transfer-framed-murder/.

Zhang, S. (2018, June 1). How a tiny website became the police's go-to genealogy database. *The Atlantic*. Retrieved from https://www.theatlantic.com/science/archive/2018/06/gedmatch-police-genealogy-database/561695/.

CHAPTER 14

Immediacy and Authority

Identification Efforts in Bosnia and Herzegovina and the World Trade Center

AMY MUNDORFF AND SARAH WAGNER

Time is a complicated metric when applied to the identification of victims of mass fatalities. Forensic genetic data, on one end of the spectrum of time, have helped answer questions of time writ large—the "deep time" of ancient DNA research on human evolution and migrations (Hagelberg et al., 2015). On the other end, forensic genetics addresses much narrower, microcosmic slices of time, such as estimating rather precisely the time of death (Ferreira et al., 2018). Our concern in this chapter is how time enters into deliberations and processes of postmortem identification efforts in the wake of mass fatality.

For our purposes, the salient aspect of time is the temporal relationship between death and the decisions taken in response to it in the aftermath of mass fatalities. We are particularly interested in decisions that prioritize genetic testing as the principal means to identify otherwise unrecognizable human remains and how the composition of *decision makers* may reflect the time-from-event dynamic. In DNA-led efforts, forensic genetic matches between remains and reference samples steer the postmortem identification process, while ante- and postmortem information derived from other lines of evidence, such as forensic anthropological data, dental comparison, or material evidence, serves to support the genetic proof of individual identity (see, e.g., Yazedjian & Kešetović, 2008). Urgent or immediate, gradual or evolving, the span of time between the event and the implementation of a DNA-led process can tell us much about how scientific responses emerge and about the dynamics that shape their outcomes.

Focusing on the application of DNA technology in the aftermath of mass fatalities sheds light on how identification efforts are structured for either legal or humanitarian purposes, or both, as well as how families of the missing enter into and experience these efforts and how authority over results is constituted and enacted. In exploring these issues, we contrast the forensic genetic effort to identify the victims of the terror attacks on the World Trade Center (WTC) of September 11, 2001, which was initiated within hours of the catastrophe, with the effort to identify the victims of the Srebrenica genocide of July 1995 in Bosnia and Herzegovina, which took place several years after the event. These efforts, which ran nearly concurrently even though the two events were not concurrent, represent two of the largest DNA-led human identification efforts to date. In contrasting these two events and their respective forensic responses, we address issues often overlooked in analyzing the social and political import of identification efforts—namely, how time influences, and even structures, the processes of and expectations surrounding individual postmortem identification and how its pressures of urgency are felt and received, from a gradual approach to an immediate call to action.

Seen through the lens of time and its material, political, and even emotional effects, the examples of 9/11 and Srebrenica also invite analysis of the interplay between the scientific community and family members of the missing during the identification-of-remains process and its anticipated outcomes. Following anthropologist Lindsay Smith's lead, we focus attention on these two sets of actors because (a) they exist as part of a larger "hybrid network" that also includes "legal institutions, law enforcement, and material and ideological systems," and (b) they reveal the often-competing needs, expectations, and epistemologies surrounding the care of missing persons (2017, p. 400).

The scientific interventions in postwar Bosnia and Herzegovina and the WTC serve as important milestones within the larger history of forensic science applied to human rights investigations and missing persons identification efforts. In many ways, these interventions paved the way for DNA technology's central role in the intertwining processes of legal redress and humanitarian action. As Smith argues, the field of human rights forensics has undergone a significant reorientation away from human-rights-based responses typified by smaller-scale Global South initiatives (see, e.g., chapters 7 and 8) toward a more security-centered Global North approach:

> Human rights forensics, a movement that began as an indictment of the state for its forced disappearance of its citizens, its obfuscation of their lives, their deaths and even their burials, increasingly has become allied with state and military officials in order to fund its identification work. . . . Rather than the holistic, investigatory approach espoused by [Clyde] Snow in the early years of human rights genetics, which took each disappeared person as a mystery to be painfully reconstructed through any source available, including testimonials and skeletal analysis for trauma, a DNA-led approach, pioneered by [the International Commission on Missing Persons], has taken increasing precedence, where ante-mortem data and skeletal analysis, although important, are subordinated to high-throughput typing of skeletal samples and reference samples looking for a match. (Smith, 2017, p. 412).

Given this reorientation, it is doubly important to understand how the two events and their responses unfolded and with what results. We begin with the example of Bosnia and Herzegovina, where the first DNA-led identification was made on November 16, 2001, just two months after the terrorist attacks of September 11.

SREBRENICA: DNA-LED IDENTIFICATIONS IN POSTWAR BOSNIA AND HERZEGOVINA

At the end of the Yugoslav secession wars that spanned the 1990s, some 40,000 individuals were missing, with their remains either unrecovered or recovered but unidentified. Bosnia and Herzegovina had the highest total of missing persons—an estimated 31,500—among the constituent republics (and successor states) of former Yugoslavia. Among the missing in Bosnia and Herzegovina were more than 8,000 victims of the genocide near the United Nations (UN) "safe area" of Srebrenica.

In the wake of the 1992–1995 war in Bosnia and Herzegovina, Srebrenica stood out as the worst atrocity to occur in Europe since World War II. Refugees had sought shelter there from across the eastern part of the country when the war first broke out but soon found themselves largely cut off from humanitarian assistance. They lived in dire conditions and under the constant threat of violence for the three years that followed, despite the enclave being designated a UN "safe area" and the arrival of its peacekeeping forces in the spring of 1993. In July 1995, when the Army of Republika Srpska (Vojska Republike Srpske or VRS) decided to expel the 40,000 Bosniak (Bosnian Muslim) refugees from the safe area, UN "blue helmets" proved toothless, failing to intervene as the VRS captured and killed over 8,000 Bosniak men and boys fleeing from the enclave.[1]

The majority of the victims were executed en masse and their bodies dumped into mass graves. Forensic evidence provided at several of the Srebrenica-related trials at the International Criminal Tribunal for the former Yugoslavia (ICTY) gradually sketched the network of associated graves. For example, the indictment of the assistant commander for security of the Drina Corps (one of the six geographically based corps of the VRS), Vujadin Popović, outlined the temporal span and location of the secondary burial sites associated with the Srebrenica crimes:

> From about 1 August 1995 through about 1 November 1995, VRS and MUP personnel participated in an organised and comprehensive effort to conceal the killings and executions in the Zvornik and Bratunac Brigade zones of responsibility by reburying bodies exhumed from initial mass graves at the following locations: Branjevo Military Farm. Kozluk; the "Dam" near Petkovci; Orahovac; and Glogova; and transferring them to secondary graves at: twelve sites along the Cancari Road (containing bodies from Branjevo Military Farm and Kozluk); four sites near Liplje (containing bodies from the "Dam" near Petkovci); seven sites

---

1. For detailed, eyewitness accounts of the humanitarian conditions in the "safe area," the fall of Srebrenica, and the genocidal campaign that ensued, see Suljagić (2005) and Nuhanović (2019).

near Hodzici (containing bodies from Orahovac); and seven sites near Zeleni Jadar (containing bodies from Glogova). (Prosecutor v. Popović, 2002)

Srebrenica's legacy of graves, particularly secondary mass graves—trenches full of commingled and partial bodies created when the perpetrators sought to hide traces of their crimes by digging up and scattering the contents of the numerous primary mass graves—triggered an unprecedented DNA-led identification effort (Jugo & Wagner, 2017, pp. 200–206; see also Congram et al., 2015; Yazedjian & Kešetović, 2008). Spearheaded by the International Commission on Missing Persons (ICMP), an organization initiated by President Bill Clinton in 1996 to address the phenomenon of missing persons in the aftermath of the secession wars, the identification process front-loaded DNA as the primary means of identifying the remains of the missing. Within a short period of time, forensic genetics became the engine driving the entire program. Operating out of DNA laboratories in Tuzla, scientists collected blood samples from tens of thousands of surviving relatives in Bosnia and Herzegovina and among the diaspora, largely resettled in Europe and North America. These reference samples were compared to bone samples cut from the skeletal remains exhumed from mass graves and recovered on the forest floor (see chapter 9; Jugo & Škulj, 2015, pp. 43–46; Wagner, 2008, pp. 82–122).

According to Tom Parsons, ICMP's director of forensic sciences, new DNA extraction techniques and enhanced polymerase chain reaction (PCR) amplification methods "opened the way to a massive campaign to collect blood samples from family members of the missing and establish a DNA database of reference profiles against which DNA profiles from bone samples could be compared. This was the launch of a new type of 'DNA-led' identification that was quite different from the conventional approach, in which a hypothetical identity for a set of remains is developed and then subjected to a yes-no DNA test" (International Commission on Missing Persons, 2016).

In the case of Srebrenica, the innovative application of DNA testing proved highly successful.

As of 2020, 6,955 individual victims have been identified through DNA testing, with an additional 34 individuals identified by traditional forensic methods International Commission on Missing Persons (2020, June).

This is an impressive result given the conditions of the secondary mass graves, from which the majority—upwards of 65%—of identified victims were exhumed.

The centrality and symbolism of forensic genetics were not lost on ICMP as it set its sights on operations outside of the western Balkans. In many ways, DNA has become the organization's signature brand. Soon after signing an agreement with the Dutch government in October 2015, ICMP moved its headquarters from Sarajevo to The Hague. By relocating ICMP to The Hague, Kathryne Bomberger, then its director, explained, "we intend to augment our efforts to work with other organizations, including on initiatives to address the issue of missing migrants" (quoted in Kidd, 2015). DNA would be its primary entrée into these new spheres of collaboration. It was also the centerpiece of its widely distributed "fact sheet," prepared in anticipation of the twentieth anniversary of the Srebrenica genocide. Emblems of DNA testing—a partially filled vial, a drop of (presumably) blood, a double helix—gave form to the numbers attesting to the organization's scientific achievements in the face of Srebrenica's mass graves and scattered remains.

But *how* did forensic genetics come to dominate the process in Bosnia and Herzegovina? And how did it then become the model that would later be exported to other conflict and disaster settings worldwide? The question is not so much about the conditions that gave rise to it—namely, the forensic obstacles presented by the commingled remains in the vast network of primary, secondary, and even tertiary mass graves, which led to the need for DNA testing (Jugo & Wagner, 2017, p. 205)— but rather how the forensic practitioners, families, and their advocates came to accept the central role of forensic genetics and how they responded to the complexities and, in some instances, the unexpected complications that such a DNA-led system introduced.

The process unfolded over several years and involved multiple actors. Several accounts map the various stakeholders in the immediate postwar years, from local (i.e., entity-level institutions within Bosnia and Herzegovina, such as that nation's Federal Commission on Missing Persons) to the constellation of nongovernmental organizations (NGOs; e.g., the Boltzmann Institute, the [entity-level] Federation Red Cross, and the US-based Physicians for Human Rights) and international institutions (e.g., the International Committee of the Red Cross and the ICTY). They also testify to the shifting postwar dynamics and the unstable, fragmented dispersal of authority regarding the forensic scientific response to the war's missing persons and especially the Srebrenica mass graves (see, e.g., Kešetović & Wagner, 2016, pp. 48–49; Jugo & Škulj, 2015, pp. 40–46; Rosenblatt, 2015, pp. 47–54; Wagner, 2008, pp. 85–101; Stover & Shigekane, 2002, pp. 853–856; Vollen, 2001; Stover & Peress, 1998, pp. 167–181, 194–199).

By examining the chronology of which organizations or institutions spearheaded the identification process, we can glean three important points. First, the application of DNA testing was not an immediate response. Rather, six years passed between the signing of the Dayton Peace Agreement in December 1995 and November 2001, when ICMP announced the first major result of its DNA-led process, the successful identification of the remains of a 16-year-old victim of the Srebrenica genocide. A year and a half later, in 2003, 882 victims, the overwhelming majority of them identified by DNA, were buried at the Srebrenica-Potocari Memorial and Cemetery as part of three collective funerals held that year. The results of the forensic genetic intervention were finally starting to flow. But the time that had lapsed was significant; compounding the already complex conditions of the Srebrenica graves, most of them still undetected, wartime inaction on the part of the international community had laid the groundwork for the slow, at times disjointed, forensic response. As Nettelfield and Wagner write: "In the case of Srebrenica, the lack of intervention during the war and the moral paralysis of those members of the international community charged with protecting its citizens create[d] the context for postwar intervention" (2014, p. 5).

Second, from the early stages of the recovery and identification efforts, families of the Srebrenica missing, represented by its most politically mobilized members, who formed associations first in Tuzla and later Sarajevo, demanded accountability and action, including eventually pushing for a more systematic application of DNA testing (rather than simply as a tool to confirm or exclude a presumed identification). In the immediate postwar years, their sense of urgency, however, bumped up against the structural obstacles of diffuse bureaucratic centers for both legal redress and humanitarian intervention. Laurie Vollen, who had helped establish Physicians for Human Rights' Srebrenica Identification Project in 1997, recounted families' heightening frustration: "They despaired over the slow pace of the exhumation process and demanded that the remains be recovered and collected out of respect for their loved ones and to demonstrate the nature and scale of the crimes committed" (2001, p. 339). In this sense, identified and returned remains could attest to the scale and intensity of the crimes of July 1995. "But," as Vollen points out, "the ICTY's timetable for exhuming the Srebrenica graves held the unearthed remains essentially hostage to prosecutorial priorities and The Hague's logistical capacity. Survivor voices had little, if any, effect on the pace of the investigations" (2001, p. 339). The "prosecutorial priorities" derived from the tribunal's mandate of "holding leaders accountable, bringing justice to victims and giving them a voice, establishing the facts and developing international law and strengthening the rule of law" (quoted in Nettelfield, 2010, p. 3). In the case of Srebrenica, ICTY investigations of mass graves meant that exhumations focused on documenting the crimes themselves. As Lara Nettelfield explains, "ICTY was the first institution to oversee exhumations of mass graves in Eastern Bosnia," and its experts would go on to testify about the manner and logistics of mass executions, the presence of blindfolds and ligatures, and the dispersal of remains from primary to secondary mass grave sites (2010, p. 106). Though families sought to engage directly with the recovery and identification process in these early years, they were unable to do much more than demand accountability and action from the sidelines. When they were drawn into the identification process as active participants, it was principally as providers first of antemortem data and later of DNA samples. Families of the missing also served as advocates for ICMP's early outreach campaign to convince other relatives of the need to provide DNA reference samples (Kešetović & Wagner, 2016, p. 49).

Vollen's observation regarding the families' thwarted efforts in the immediate postwar years to speed the recovery process underscores the third point: identification efforts originally took a back seat to war crimes investigations led by the ICTY. Not until four years after the war's end did the more humanitarian-oriented work of individual identifications finally receive the same level of financial and technical support. In 1999, ICMP began to implement its mandate to assist the Bosnian authorities and local institutions in recovering and identifying the remains of the war's missing (Wagner, 2008, pp. 96–97).

Rifat Kešetović, the director of ICMP's Podrinje Identification Project, noted the lack of local resources, infrastructure, and expertise in the early years to address the enormous obstacles presented by the Srebrenica cases:

> Unfortunately, the state of Bosnia and Herzegovina, even then, did not recognize the need and the timing to establish a specialized scientific institution that could

take control of the identification process. Furthermore, an attempt at massive identification through classical or traditional methods was, for me, not an alternative from a professional standpoint. (Kešetović & Wagner, 2016, p. 49)

Without available local resources and limited by the goal of the ICTY's investigative teams, which was to pursue criminal prosecutions, the work of identifying the missing was not undertaken until international funds and expertise were provided for the creation of ICMP.

As Kešetović explains, the "pioneering venture" of the DNA-led model raised its own quandaries, questions that the forensic staff had to put to the families: "Is it necessary to inform the family if even one bone is found? Should families bury incomplete, identified remains, or wait until the other missing body parts are found? What should be done if other missing parts are found after the burial? How should genetic information be protected? What is the proper course of action in cases when misattributed paternity is revealed, and so on?" (Kešetović & Wagner, 2016, p. 50). Rather than guiding the DNA-led identification efforts from the beginning, these dilemmas—both technical and moral, in Kešetović's assessment—shaped the process and therefore families' experiences. Indeed, with the DNA technology's success in identifying partial remains exhumed from the secondary mass graves and, in rarer instances, revealing past errors of re-association, the staff at the Podrinje Identification Project (subsequently placed under the aegis of the Missing Persons Institute and the Tuzla Canton) acknowledged the anxiety produced by the piecemeal recovery process. The project eventually designed a consent form with the aim of making families fully aware of the unpredictable course of recovery and identification. Among other things, the form posed the following questions to the representative of the missing person (the primary contact among the surviving relatives):

- Do you wish to be notified about new DNA test results?
- If yes, do you wish to be notified (a) of every new DNA test result, or (b) when the mortal remains are relatively complete?[2]
- Do you want the additional remains to be buried (a) along with those already buried remains, or (b) in the collective ossuary at the Srebrenica-Potočari Memorial and Cemetery?
- Do you wish to be present for the re-exhumation/re-association/burial?
- Do you want the remains that have been found to date to be buried in the originally planned burial site? (Wagner 2014, p. 502).

2. The question of what constituted a relatively complete body had been a source of debate for several years prior to the creation of this form. In November 2004, representatives from ICMP, the Federal Commission for Missing Persons (the Federation entity-level institution), and the various family associations gathered to discuss the backlog of partial bodies that had not yet been identified. Drawing on forensic scientific guidelines and theological advice from the Islamic Community (Islamska zajednica), the attendees determined that the previous threshold of 70% would no longer serve as the requisite "completeness"; rather, the responsible forensic institutions would follow the families' preference that they be notified of the identification of "even one bone" (Wagner, 2004, pp. 179–180).

The option of burial of the newly recovered remains in the collective ossuary at the Srebrenica-Potočari Memorial and Cemetery was especially important because it signaled the acknowledged limits of scientific intervention. It is known—and accepted—by the forensic specialists and Srebrenica families alike that not every bone will be re-associated and therefore identified. There will remain skeletal elements, whole and fragmented, either too small or too degraded to yield any conclusive DNA evidence of individual identity. Those unnamed fragments and small bones, often of the hands and feet, will be buried together in one central grave, a communal ossuary in the memorial center near the central covered prayer space of the *musala*. The decision to bury these remains together emerged from discussions with the family associations, the Islamic Community, and the memorial center designers. Though a rose bed marks its future location, the yet unfilled communal ossuary offers relief to the distress arising from the lack of individual recognition.

WORLD TRADE CENTER: DNA-LED IDENTIFICATIONS AFTER 9/11

In January 2004, 16 months after the WTC attack, the US government officially reported a total of 2,749 people missing (Brondolo, 2004). But actually, the recovery efforts had begun within hours of the building's collapse. Unlike the slow response to the Srebrenica massacre, there were high expectations in the United States that the DNA-led identification process would begin immediately. In fact, by September 12, 2001, the Office of Chief Medical Examiner's (OCME) DNA laboratory in New York City had already received nearly 100 postmortem DNA samples from its mortuary operations. That number would eventually reach into the tens of thousands (Shaler, 2005). While the news reports, mayor's office, victim's families, and scientists around the world were touting how DNA would lead the identification process, and postmortem samples were already being collected, the infrastructure to *make* DNA identifications on such a massive scale was not in place. DNA identification for skeletonized remains was still in its infancy and was performed manually on a case-by-case basis.

Broadcast live on network television, in the attack thousands of people appeared to be vaporized in front of our eyes. A DNA-led approach on such a massive scale was compelled by the need to gain control over a chaotic situation. Nationalism played a role; "we"—the US government, local and state officials, and the American people—would not let the terrorists win by rendering the victims unidentifiable. Trauma and emotions also played a role as family members needed to know whether their missing loved ones were dead or wandering the city with amnesia. And economics played a role: bills had to be paid and estates settled; death certificates had to be issued. As the days passed, remains began to decompose, and fragile DNA started to degrade. The fires continued to burn in three locations at the disaster site, rendering some remains completely calcined (so badly burned that nothing organic survived).

At the same time, the mayor of New York directed the OCME to identify *every fragment* of human remains (NIJ, 2006). Many of the families may have understood this statement as an assurance that every fragment of human remains would be *identifiable*. Thus began the disconnect between the performative politics at the national and local levels of responding to the crisis, the circumstances and contingencies of the scientific process, and the needs and expectations of the families of the victims.

To understand the roots of that disconnect, we need to examine the intensity and scale of violence wrought by the terror attacks, alongside the tightly compressed timeline of destruction. On September 11, 2001, terrorists hijacked four airplanes flying over the United States. One was flown into the Pentagon building in Washington, D.C.; another crashed in a field outside of Shanksville, Pennsylvania; and two were flown into the WTC buildings in downtown New York City. American Airlines (AA) Flight 11 departed Boston's Logan airport at 7:59 that morning carrying 92 individuals. At the same time, United Airlines (UA) Flight 175 also departed Boston's Logan airport. This flight was carrying 65 individuals. Unbeknownst to the world, each of those flights carried five terrorists, intent on crashing the airplane. At 8:46 a.m., AA Flight 11 crashed into the North Tower of the WTC complex between the 93rd and 99th floors. Seventeen minutes later, at 9:03 a.m., UA Flight 175, carrying 10,000 gallons of jet fuel and traveling at approximately 590 mph, 150 mph faster than recommended at such low altitudes, crashed into the South Tower of the WTC complex between the 77th and 85th floors. At 9:59 a.m., 56 minutes after it was hit, the South Tower collapsed. The collapse lasted 10 seconds and registered 2.1 on the Richter scale (Marchi & Chastain, 2002). Twenty-nine minutes later, at 10:28 a.m., the North Tower collapsed.

A total of 102 minutes separated the first airplane's impact and the second tower's collapse (NCTA, 2004). The Twin Towers were each 110 stories, or approximately 1,350 feet, tall, with 10.4 million square feet of office space. As many as 90,000 people entered and exited the buildings every day, with approximately 50,000 people working in the towers and another 40,000 passing through in the underground mall. At the time the first plane struck the North Tower, an estimated 16,000–18,000 civilians were in the WTC complex (NCTA, 2004). As a result of an adept evacuation by the Port Authority Police Department for New York and New Jersey (PAPD) (NCTA, 2004), nearly everyone below the impact zone was evacuated safely. Approximately 1,942 of the 2,346 civilians killed during the collapse were thought to have been at or above the level of impact (Hirsch, 2008). An additional 403 "members of service" (i.e., police, firefighters, EMS, FBI) were also killed responding to the attacks that morning. Of the people still in the buildings at the time of the collapses, only 18 individuals survived. The PAPD's mass evacuation success was in response to lessons learned from the 1993 bombing of the WTC buildings, which killed 6 individuals and led to changes in safety protocols for the complex. All seven buildings, constituting more than 12 million square feet of rental floor space, were destroyed as a result of the attacks (Marchi & Chastain, 2002). The WTC recovery site, entombing the remains of 2,749 victims from 27 different countries, covered 17 acres of downtown New York City (Gill, 2005).

The buildings' devastation mirrored a different decimation—that of the bodies of the victims. Most of the more than 20,000 recovered fragments would not include fingers to fingerprint, or teeth for dental comparison—they were simply body fragments, with nearly 5,000 measuring no more than one inch (Mundorff et al., 2009). The ensuing response prioritized DNA collection and analysis from the onset, but not without facing significant obstacles at the site (Mundorff, 2014; Mundorff, et al., 2014; Toom, 2016, pp. 694–696). First, victim remains had to be recovered from which a sample could be taken. Second, a person had to be reported missing and have either antemortem DNA or family DNA reference samples on file for comparison. The WTC site took months to "unbuild" (Langewiesche, 2002), with official excavations lasting through May 2002 and additional sifting of debris transported to the Staten Island landfill continuing through August 2002. While the public perception was that the identification process began immediately, several months passed before most of the remains were recovered from the debris. Once remains were recovered, often the DNA was too compromised to allow complete short tandem repeat (STR) profiles. Initial testing resulted in more than 60% of the samples yielding partial profiles or no results at all (Shaler, 2005), further delaying DNA identifications. Collecting antemortem samples or family reference comparisons was equally challenging. Director of Forensic Biology Robert Shaler later reflected on the decision to collect antemortem information at the family assistance center (FAC) in the immediate aftermath, concluding that the process had moved forward too quickly. "A huge lesson learned is that there must be a proper mechanism for collecting information from biological relatives. The forms used were not adequate, and many family samples simply have insufficient documentation with regard to the biological relationship to the missing person." (Hirsch, 2002 p. 2).

Further complicating the process, it was not until January 2004 that the official list of missing persons from the WTC disaster was finalized. In the chaos that followed the disaster, more than 20,000 people were initially reported missing. Some people were reported multiple times and by different people, resulting in duplicate reports for a single person or reports with conflicting information for the same individual. There were also instances of multiple victims having the same first, middle, and last name, which added a level of complexity to the review of potential duplicate records (Hennessey, 2002, 2014).

In addition, cases of potential fraud slowed the timeline for finalizing the list, because a confirmed missing person meant a death certificate. A WTC victim death certificate allowed access to funds established for the victims' families. A small number of individuals preyed upon the grief of a nation and fraudulently reported a missing family member for profit. Therefore, every reported missing person had to be vetted before being officially included on the missing person list. As a result, the immediacy in implementing the identification process may have contributed, in part, to the disconnect between politics, science, and the victims' families.

The chief medical examiner worked closely with legal and city personnel to put in place a process to issue a death certificate for individuals whose remains had not been recovered. Yet families desiring a death certificate before remains were identified had to first "prove" their loved one died on 9/11. The burden of proof was established with an affidavit stating that a missing person was at the WTC or on one of the airplanes and had not been seen since September 11. The information was confirmed with a second affidavit from the employer or airline, which was then reviewed by the OCME legal department before going to the Surrogate's Court for judicial approval. Creating the process to issue a death certificate without remains took only three weeks to implement, and by the one-year anniversary more than 1,600 had been issued.

Yet even with the list of missing becoming formalized, remains recovered and their DNA samples analyzed with STR markers, and family reference samples submitted for comparison, the OCME lacked the software to compare STR data on such a scale because it simply did not exist (Shaler, 2005 p. 78). On October 12, 2001, the OCME successfully confirmed the first DNA identifications, but the process was tedious and slow, and mostly done by hand. The matching pace changed significantly in mid-December, with the installation of the first iteration of the Mass Fatality Identification System, the automated DNA matching software known as M-FISys (Cash et al., 2003). Designed by Gene Codes Forensics specifically for the WTC identification project, M-FISys was the first program to integrate STR, mitochondrial DNA, and eventually, single nucleotide polymorphism (SNP) data and had the ability to combine multiple test results with partial profiles into a full virtual profile.

While DNA identification of the victims was garnering all the press, prompting the convening of brilliant scientific minds from all over the world (Biesecker et al., 2005) and costing millions in research, traditional identification methods such as fingerprint and dental matching were successfully identifying victims at a quick pace. More than 250 fragments were identified by traditional modalities in the first month, before DNA confirmed a single ID. But these methods were not groundbreaking and did not capture the media's attention, even though their success dominated victim identifications for the first 28 weeks.

The DNA-led program only took precedence with the exhaustion of traditional methods, when most remains with fingers to print or teeth to X-ray had been identified, and DNA was the only means left to identify the thousands of unidentified remains. Eventually, DNA identifications surpassed fingerprint and dental identifications. At the same time, additional fragments were commonly linked by DNA to already identified individuals as more antemortem and postmortem profiles and family reference samples were added to the database. However, with success came unforeseen problems.

In the public imagination, the idea of a DNA-led approach to the WTC identifications took hold quickly, bolstered by the mayor's directive to DNA test every fragment, the unprecedented number of fragmentary remains without any other means of identification, and the at times outsized expectations of the families. So why was it so difficult to anticipate how the identification process would unfold? Although referring only to the DNA component of the massive identification effort, John Snyder of the New York State Police Forensic Investigation Center answered this question shortly after 9/11 by comparing the immediate efforts to "building the plane while we're flying it" (Shaler, 2005, p. 172). While the decision to lead with DNA might have been immediate, building the capacity and infrastructure took time. To parse out Snyder's statement in a practical sense, it meant that postmortem samples would take months to recover, as would receiving, accessioning, and testing comparative samples. It also meant that software robust enough to handle the statistics had to be developed, and new technologies for testing compromised samples needed to be validated—all of which took time and none of which was guaranteed to be successful. Maybe the relevant question is not why was it difficult to anticipate how the process would unfold, but where were the connections and disconnections concerning the understanding of this process between the scientists and the families?[3]

Years later we glean three important points to help us appreciate how a top-down approach is understood by the families of the missing and how an open and continued dialogue between practitioners and families can help encourage realistic expectations. First, in light of the unknown number of victims and the degree of fragmentation of remains, a decision was taken to subject every piece of human remains, no matter how small, to DNA testing. This decision, coupled with the promise to continue testing remains in perpetuity and take advantage of advances in DNA technology, resulted in some families receiving notifications of additional identifications over and over again. Although the fragment-to-victim ratio should have been around $8/1$[4] (Mundorff, 2014), in actuality it was not uncommon for a single individual to have dozens of identified fragments. And these discrete identifications for the same individual may have occurred weeks, months, or even years apart. When it became apparent that repeatedly notifying grieving families of additional identified remains was causing further trauma, a release authorization form was created to present families with notification options. Similar to what the Podrinje Identification Project in Bosnia and Herzegovina eventually used, the form presented options in the event any additional remains were identified. The next of kin could select not to be notified, thereby authorizing officials to decide on the disposition of the remains; they could be notified any time new remains were identified; or they could be notified when the OCME determined it unlikely further identifications would occur. For the second and third options, the individual was also able to select himself or herself (as primary next of kin) or a third party to be notified. Family members could, and indeed did, change their selections as the time from the disaster extended. Sometimes new forms were filled out because the originals were signed while in a deep state of grief, and individuals had no memory of signing them until they verified the signatures were actually theirs. Sometimes years would pass and next of kin whose original selection was "not to be notified" would be curious whether additional remains had been identified. In those circumstances, they would fill out a new form to allow the OCME to notify them. Interestingly, the vested interest of the practitioners could be felt when they realized they were not allowed to contact families who selected "not to be notified." One of the authors of this chapter, Amy Mundorff, recalled identifying over 100 fragments as belonging to an individual who had been identified earlier from only a single bone fragment. The additional links were established shortly after the next of kin was notified, but at that time, the relative had filled out the form

---

3. See Victor Toom's examination of a similar, related set of questions through the notions of "goods" and "bads," that is, (un)desired values and possible ways of ordering reality, as well as the normative assignment of "closure" (Toom, 2017).

4. If all the bodies fragmented equally, ~22,000 fragments shared by 2,749 victims equals approximately 8 pieces per victim.

asking not to be notified again. It weighed on Mundorff—and indeed still does—whether the relative would have preferred to know how much more of his or her family member had been identified beyond that single fragment.

The second point highlights the differences in the time-from-event dynamic between the two incidents and addresses Shaler's reflection that a proper mechanism for collecting information from biological relatives must be in place *before* the process is implemented. The sense of urgency to identify victims was evidenced by Red Cross and NYPD personnel collecting questionnaires and DNA samples from grieving family members at FACs within 24 hours of the disaster. Over 1,000 of the premature collections were compromised by problems such as transcription errors that occurred between the handwritten forms and data entry, missing and erroneous data collection, or insufficient collection of DNA samples to make an identification (Shaler, 2005).

This problem became apparent when some family members, after waiting a few months for an identification notice, called the OCME to inquire about the status of their case. There were instances in which, for example, early antemortem DNA collections failed to yield a profile or yielded a mixed profile, but there was no contact information for the family. These cases were essentially in limbo. In response, the OCME established a DNA hotline in January 2002. Announcements were made through news media and other outlets to reach as many families as possible. In the first year, the hotline received over 11,500 calls, resulting in nearly 9,000 consults and close to 3,000 appointments. Next of kin could confirm family information and help with kinship worksheets; they could find out the testing status of their DNA exemplars or submit victim personal effects; and most important, they could arrange to give additional samples if what they had on file was not sufficient for an identification. Prior to the hotline, it was nearly impossible to connect the dots between some of the reasons identifications were held up. Now it was possible for families to find out whether they had not yet received an identification because of insufficient reference material or simply because no match had been made yet.

The third point underscores the deep relationships that developed between OCME personnel and families of the victims. In his memoir detailing efforts to identify the victims from 9/11, Shaler posits that "mass fatalities are about families." "Working with families," he goes on, "is critical to getting the job [of identifying the remains of victims] done properly" (Shaler, 2005, p. 55). Early on, the OCME embraced the idea of including family members in decision-making processes. This was done through weekly public meetings at which updates were provided to family members and representatives from family groups on the identification progress. Eventually the meetings were held monthly, but the monthly meetings went on for years. Led by the chief medical examiner and the director of forensic biology, the meetings often included a pathologist or anthropologist, medicolegal investigators, and DNA analysts to address every aspect of the identification process. DNA experts would explain the time-consuming process of validating new identification technologies, for example, such as SNPs. This transparency contributed to helping family members manage their expectations, particularly how long it might take for especially challenging identifications. When the number of recovered remains decreased from the previous week in the initial months, staff would explain why a slowdown had occurred, which was usually related to safety concerns at the recovery site. Even potentially disturbing and challenging territory was openly discussed, soliciting input and comments from all attendees. For example, the unclaimed and unidentified remains could not be interred in the memorial[5] in their current state because they were continuing to decompose. Cremation, formalin, and other preservation options were not considered because of religious objections and the damage such preservatives would do to DNA for future testing. Desiccation was the least destructive preservation method available.

Likewise, the decision to desiccate the remains was openly discussed at the monthly meetings. Although the act was unavoidable, it was emotionally difficult to comprehend. The notion that a victim's remains might have to be further manipulated to facilitate drying was upsetting to some family members. A representative from a family group, attending the meetings on behalf of many families, requested that if it had to be done, it should be undertaken by someone from the OCME that the relatives knew and trusted, not by temporary personnel. The OCME anthropologist accommodated their request in order to comfort them during a difficult but necessary process.

Along similar lines, the DNA lab hosted tours for family members—not only to see the laboratory, but to see the staff who were working on their behalf in the laboratory—so as to humanize the process. Through these continued interactions, OCME staff often became advocates for individual families and family groups whom they got to know well over the years, developing deep and lasting relationships. The faith and trust exchanged were critical to navigating such a complicated endeavor. At the time, a disaster victim identification operation on the scale of 9/11, with the challenges of compromised, highly fragmentary, and commingled remains, was uncharted territory. Mistakes happened. But the OCME was always forthright with the families, and the majority of those affected by misidentification understood that it was unintentional and unforeseen. The OCME policy was always honesty with families, never jeopardizing the trust built over the days, weeks, months, and now years.

CONCLUSION

Though separated by geography and by critical differences in physical circumstances, available resources and infrastructure, not to mention political environments, the identification efforts after the Srebrenica genocide and the WTC attack are nevertheless connected by their respective and, at the time, innovative applications of DNA testing. Forensic genetics proved critical to providing families and the wider public

---

5. The New York City Office of Chief Medical Examiner has jurisdiction over the remains repository, located at the WTC site. The secure, environmentally controlled space was designed for continued access as identifications continue to be made (911memorial.org).

in both instances with answers and, when possible, to returning remains, no matter how partial or scant in number, to surviving kin.

In this sense, the identification efforts at Srebrenica and the WTC Center mark an important moment in the field of forensic science applied to post-conflict and post-disaster settings. DNA offered an extraordinarily powerful tool to address the destruction that war or a terror attack could visit upon the bodies of its victims. Yet as we have tried to demonstrate in juxtaposing the two examples and emphasizing the very different temporal pressures each effort faced, the DNA-led identification process also raised thorny questions about authority and about the relationship between the scientists and the families awaiting their results.

There are some striking overlaps Both identification efforts had to reconcile the piecemeal way in which remains were recovered and identified by DNA testing, and both projects eventually devised bureaucratic measures to help families manage expectations and anticipate the possibility of multiple identifications. In doing so, these measures returned to families themselves some modicum of control over how they might manage their grief and care for their loved ones' remains. That it took time in both instances to develop such a means to redress the multiple additional identifications, often spread out over months and years, to a single individual sheds light on our ability to anticipate or predict problems while implementing unprecedented technologies. Moreover, that it took time to grant relatives a say over this aspect of the identification process also tells us something about how the drive among forensic scientists to resolve absence can overlook, or at least not fully take into account, the experiences and needs of the very individuals for whom the forensic personnel labor so intently. In the Srebrenica investigations, the form recording families' wishes in the event additional remains were recovered and identified came about as the forensic process gradually moved from the more straightforward (more complete remains) cases to the commingled and partial bodies. For both the forensic scientists and the families, it took time to comprehend the extent of the secondary mass graves and its implications for the degree of commingling and disarticulation of remains. In the WTC situation, like the attacks themselves, the response was much more tightly compressed; therefore, the learning curve for forensic personnel and relatives of the missing stretched over weeks and months rather than years, as it had in Bosnia and Herzegovina. In fact, the impetus to create the release authorization form came at the request of a family member whose grief was still so new that she asked the OCME to figure out a way to stop calling her about new links and give her time to heal.

The lapse in time between the event of death and the establishment of a comprehensive, DNA-led approach also played a role in expectations. Here we see a marked difference between the Srebrenica and the WTC examples. In the case of the former, it took years for DNA testing to become the tool to drive identifications—six full years before the identification process yielded tangible results. Though families had long pushed for an expedited means to identify the remains of their loved ones, they had to wait; while waiting, they had to continually readjust their expectations of what care might be possible for their loved ones' remains. For the 9/11 families, the directive to the OCME in the immediate aftermath of the attack that every bit of human remains was to be identified not only set an accelerated pace but calibrated families' expectations toward immediate and complete answers. In the end, the daily interactions between families and the forensic staff at OCME worked to mediate the gaps in expectations and reorient families' sensibilities regarding the potential conditions (both in terms of time and substance) of identification.

For most families of the missing, the urgency surrounding resolution rarely fades, despite the passage of time. Though the DNA-led efforts to identify the victims of the Srebrenica genocide and the WTC attack differ significantly in the time required to develop infrastructure and deliver results, they share that pressing commonality of fleeting time. For in the end, there can never be a truly timely response to such devastating loss of life. Recognizing this inherent dissatisfaction is one of the most important lessons we can glean from these two investigations. Forensic genetics presents a powerful tool in the work of identifying human remains in the aftermath of violence, but its success depends on myriad contingencies, including the pressures and effects of time.

REFERENCES

Biesecker, L. G., et al. (2005). Epidemiology—DNA identifications after the 9/11 World Trade Center attack. *Science, 310*, 1122–1123.

Brondolo, T. J. (2004). Resource requirements for medical examiner response to mass fatality incidents. *Medico-Legal Journal of Ireland, 10*(2), 91–102.

Cash, H. D., Hoyle, J. W., & Sutton, A. J. (2003). Development under extreme conditions: Forensic bioinformatics in the wake of the World Trade Center disaster. *Pacific Symposium on Biocomputing* 638–653.

Congram, D., Green, A. G., & Tuller, H. (2015). Finding the graves of the missing: A study of geo-anthropological techniques in Bosnia-Herzegovina. In *Proceedings of the American Academy of Forensic Sciences 67th Annual Scientific Meeting*, vol. 21, p. 213. Colorado Springs, CO: American Academy of Forensic Sciences.

Ferreira, P. G., et al. (2018). The effects of death and post-mortem cold ischemia on human tissue transcriptomes. *Nature Communications, 9*(490), 1–15. https://doi.org/10.1038/s41467-017-02772-x

Gill, J. (2005). 9/11 and the New York City Office of Chief Medical Examiner. *Forensic Science, Medicine, and Pathology, 2*, 29–32.

Hagelberg, E., Hofreiter, M., & Keyser, C. (2015). Ancient DNA: The first three decades. *Philosophical Transactions of the Royal Society B: Biological Sciences, 370*, 20130371.

Hennessey, M. (2002, October 7–10). *World Trade Center DNA identifications: The administrative review process* [Paper presentation]. 13th International Symposium on Human Identification, Phoenix, AZ.

Hennessey, M. (2014). Data management and commingled remains. In B. J. Adams & J. E. Byrd. eds., *Commingled human remains: Methods in recovery, analysis, and identification*, pp. 425–445. Oxford, UK: Elsevier Science.

Hirsch, C. S., & Shaler, R. (2002). 9/11 through the eyes of a medical examiner. *Journal of Investigative Medicine, 50*, 1–3.

Hirsch, C. (2008). Personal communication to author.

International Commission on Missing Persons (2020, June). *Statistics of missing persons per municipality of disappearance: Srebrenica 1995*. Bosnia and Herzegovina. Retrieved from icpm.int.

Jugo, A., & Škulj, S. (2015). Ghosts of the past: The competing agencies of forensic work in identifying the missing across Bosnia and Herzegovina. *Human Remains and Violence, 1*, 39–56.

Jugo, A., & Wagner, S. E. (2017). Memory politics and forensic practice: Exhuming Bosnia and Herzegovina's missing persons. In Z. Dziuban, ed., *Mapping the "forensic turn": Engagements with materialities of mass death in Holocaust studies and beyond*, pp. 195–214. Vienna, Austria: New Academic Press.

Kidd, G. (2015, October 5). Int'l missing persons HQ headed to The Hague. *Nltimes.nl*. https://nltimes.nl/2015/10/05/intl-missing-persons-hq-headed-hague

Langewiesche, W. (2002). *American ground: Unbuilding the World Trade Center*. New York, NY: North Point Press.

Marchi, E. & Chastain, T. (2002). The sequence of structural events that challenged the forensic effort of the world trade center disaster. *American Laboratory, 34*, 13–17.

Mundorff, A. Z. (2014). Anthropologist-directed triage: Three distinct mass fatality events involving fragmentation of human remains. In B. J. Adams & J. E. Byrd, eds., *Commingled human remains: Methods in recovery, analysis, and identification*, pp. 365–388. Oxford, UK: Elsevier Science.

Mundorff, A. Z., Bartelink, E. J., & Mar-Cash, E. (2009). DNA preservation in skeletal elements from the World Trade Center Disaster: recommendations for mass fatality management. *Journal of Forensic Sciences, 54*(4), 739–745.

Mundorff, A. Z., Shaler, R. C., Bieschke, E., & Mar-Cash, E. (2014). Marrying anthropology and DNA: Essential for solving complex commingling problems in cases of extreme fragmentation In B. J. Adams & J. E. Byrd, eds., *Commingled human remains: Methods in recovery, analysis, and identification* pp. 257–273. Oxford, UK: Elsevier Science.

National Commission on Terrorist Attacks Upon the United States (NCTA). (2004). *The 9/11 Commission report: Final report of the National Commission on Terrorist Attacks Upon the United States, authorized edition*. New York, NY: W. W. Norton & Company.

National Institute of Justice (NIJ). (2006). *Lessons learned from 9/11: DNA identification in mass fatality incidents*. Washington, DC: US Department of Justice.

Nettelfield, L. J. (2010). *Courting democracy in Bosnia and Herzegovina: The Hague Tribunal's impact on a postwar state*. New York, NY: Cambridge University Press.

Nettelfield, L. J., & Wagner, S. E. (2014). *Srebrenica in the aftermath of genocide*. New York, NY: Cambridge University Press.

9/11 Memorial and Museum. (2018). *Remains repository at the World Trade Center site*. Retrieved from https://www.911memorial.org/remains-repository-world-trade-center-site.

Nuhanović, H. (2019). *The last refuge: A journey to Srebrenica*. London, UK: Peter Owen Publishers.

Prosecutor v. Popović. (2002, March 26). IT-02-57-I, Indictment. Retrieved from http://www.icty.org/x/cases/popovic_old/i nd/en/pop-ii020326e.htm.

Rosenblatt, A. (2015). *Digging for the disappeared: Forensic science after atrocity*. Stanford, CA: Stanford University Press.

Shaler, R. C. (2005). *Who they were: Inside the World Trade Center DNA story: The unprecedented effort to identify the missing*. New York, NY: Simon and Schuster.

Smith, L. (2017). The missing, the martyred and the disappeared: Global networks, technical intensification and the end of human rights genetics. *Social Studies of Science, 47*, 398–441.

Stover, E., & Peress, G. (1998). *The graves: Srebrenica and Vukovar*. Zurich, Switzerland: Scalo.

Stover, E., & Shigekane, R. (2002). The missing in the aftermath of the war: When do the needs of victims' families and international war crimes tribunals clash? *International Review of the Red Cross, 40*, 845–866.

Suljagić, E. (2005). *Postcards from the grave*. London, UK: Saqi.

Toom, V. (2016). Whose body is it? Technolegal materialization of victims' bodies and remains after the World Trade Center terrorist attacks. *Science, Technology, & Human Values, 41*, 686–708.

Toom, V. (2017). Finding closure, continuing bonds, and codentification after the 9/11 attacks. *Medical Anthropology, 37*, 267–279.

Vollen, L. (2001). All that remains: Identifying the victims of the Srebrenica massacre, *Cambridge Quarterly of Healthcare Ethics, 10*, 336–340.

Wagner, S. (2014). The social complexities of commingled remains. In B. J. Adams & J. E. Byrd, eds., *Commingled human remains: Methods in recovery, analysis, and identification*, pp. 491–506. Oxford, UK: Academic Press.

Wagner, S., & Kešetović, R. (2016). Absent bodies, absent knowledge: The forensic work of identifying Srebrenica's missing and the social experiences of families. In D. Congram, ed., *Missing persons: Multidisciplinary perspectives on the disappeared* (pp. 42–59). Toronto, ON: Canadian Scholars' Press.

Wagner, S. E. (2008). *To know where he lies: DNA technology and the search for Srebrenica's missing.* Berkeley, CA: University of California Press.

Yazedjian, L., & Kešetović, R. (2008). The application of traditional anthropological methods in a DNA-led process. In B. J. Adams & J. E. Byrd, eds., *Recovery, analysis, and identification of commingled human remains*, pp. 271–284. Totowa, NJ: Humana Press.

CHAPTER 15

Forensic Genetics, Ethics, Privacy, and Public Policy

THOMAS J. WHITE AND STEVEN B. LEE

On the morning of April 24, 2018, police in Sacramento, California, arrested Joseph DeAngelo, 72, as the suspected Golden State Killer. A retired policeman, DeAngelo was charged with committing a series of horrific crimes, including 12 murders and more than 50 rapes that terrorized communities in Sacramento and throughout the state from 1976 to 1986 (McNamara, 2018). The investigators found that DeAngelo's DNA profile could not be excluded from matching multiple forensic samples stored as evidence for decades as the case went cold. The big break in the case came when the genetic profile of the assailant, based on crime scene evidence, was tested against an online genealogy database, GEDmatch, and a partial match was observed through the pattern of alleles shared with about two dozen of his relatives. This allowed investigators and genealogists to construct family trees and focus on relatives within the assailant's approximate age range who were living in the Sacramento area at the time of the crimes. When DeAngelo's profile was subsequently obtained from "abandoned" samples, the profile matched the crime scene evidence at all loci tested (Fuller, 2018).

In addition to numerous fundamental human rights and legal issues, such as the Fourth Amendment's protection against warrantless government searches raised by the search for the Golden State Killer (Ram, 2018), this case touches on some ethical and privacy issues discussed in chapter 4 and in two previous comprehensive reviews (Charo, 2004; Clayton, 2019). We also consider how the case and emerging DNA technologies raise ethical questions involving (1) informed consent; (2) storage of forensic evidence and genetic profiles; (3) government misuse of potentially sensitive DNA data; and (4) familial sensitivity to unknown parental, sibling, or criminal relationships. Individuals whose genetic profiles are in a forensic database could expose their family members to scrutiny without their knowledge and without their consent, and information on unknown family ties (adoption, adultery, illegitimate children) might be revealed (Santurtún, 2017). An important ethical question then arises: Does the victim's and public interest in solving serious crimes justify government efforts to match an alleged assailant's profile to genealogical and other nonforensic databases? These could include clinical research databases from gene-disease associations studies, databases of deceased individuals (Callaway, 2017), missing persons, and remains from humanitarian disasters.

INFORMED CONSENT

Informed consent for clinical or research studies is ethically essential and respects persons' rights to decide whether participation in the research, often on a new diagnostic, drug, or medical procedure, is compatible with their interests, including protection from exploitation and harm (Grady, 2017). Potential participants are asked to read and sign a detailed written description about the purpose of a study as well as about its risks and benefits in a process that involves disclosure, discussion, understanding, agreement to participate, and authorization. There are various degrees of consent, which range from: consent as merely the absence of coercion, such as when human biological materials are obtained through a voluntary transfer; to presumed consent based on the failure to opt out; to actual affirmative consent (Rao, 2016).

For clinical research studies, after a participant in a study has been given information on the potential benefits and risks, informed consent may take the form of something like the questionnaire in Table 15.1, in which one is given a choice

Table 15.1 WHAT ARE THE RISKS OF GENETIC RESEARCH? THERE ARE SOME RISKS TO RECEIVING GENETIC RESULTS. PARTICIPANTS COULD EXPERIENCE RISKS SUCH AS PSYCHOLOGICAL OR EMOTIONAL DISTRESS, LOSS OF INSURANCE, LOSS OF EMPLOYMENT, DISCOVERY OF PREVIOUSLY UNKNOWN HEALTH CONDITIONS, THAT THEY ARE NOT THE BIOLOGICAL PARENTS OF A CHILD(REN), OR THAT THEY COULD CARRY A GENE FOR A CERTAIN DISEASE. THEREFORE, WE OFFER GENETIC COUNSELING BEFORE PARTICIPATION IN THE STUDY AS PART OF THE INFORMED CONSENT PROCESS.

Please feel free to ask questions and discuss your preferences with your study team members. They will help you complete the table. If you do nothing, you will be told. However, if you wish not to be told, please initial where indicated below.

What choices do I have for receiving these other results that do not have direct impact on care of my current cancer?	If you do NOT wish to be told of these results, please initial the boxes below.
1) Results that may have significance for biological family members.	
2) Results that are not related to your cancer, but may have potential medical impact for you.	

of how much information to receive beyond results that have a direct impact on one's care.

The International Commission on Missing Persons (ICMP), a nongovernmental organization (NGO) that investigates abuses of human rights, uses an affirmative consent form as shown here (ICMP, 2018).

I, the undersigned, have received a copy and understand the information contained in the *Information Sheet on DNA Genetic Testing and Processing* (ICMP.ST.LS.299.doc). I agree that data I herewith provide to ICMP, including my genetic data, will be used for identifying missing persons. Therefore: I, the undersigned,

☐ **Consent** ☐ **Do not consent**

That data I provide to ICMP, including my genetic data, may be used for purposes of establishing the identity of a deceased missing person.

I, the undersigned,

☐ **Consent** ☐ **Do not consent**

That data I provide to ICMP, including my genetic data, may be used in evidence in international or national criminal proceedings against persons charged with an offense such as genocide, crimes against humanity, war crimes or acts of terrorism and is subject to the protection set forth in the *Information Sheet on DNA Genetic Testing and Processing* (ICMP.ST.LS.299.doc). I understand that I will be given an opportunity to re-confirm my consent in advance of my data being used for such purposes.

Standard informed consent procedures are changing, and some clinical investigators have argued that studies on clinical records from which names and identifying information have been removed or anonymized specimens should not require informed consent. They also argue that broad consent forms should cover both the current research and unspecified future studies. In the United States, legal requirements based on statutory and case law may determine whether the treating physician must personally provide all the necessary disclosures or if the consent process, like other aspects of modern medicine, can take advantage of specialization and division of labor (Fernandez, 2018).

Unlike those who take part in research studies, convicted criminals in most countries and jurisdictions have, it is generally held, forfeited the right to refuse to consent to provide a blood, saliva, or buccal swab sample for storage and genetic profiling. In the United States, all 50 states have laws establishing genetic databases for individuals convicted of serious crimes and for evidence from crime scenes. Several states also have databases for arrested individuals or even potential suspects, although the costs and resources for testing samples from all these sources vary widely. At the national level, the FBI's Combined DNA Index System (CODIS), as of September 2019 contained nearly 14 million DNA profiles from convicted felons, over 3 million arrestee profiles, and over 900,000 forensic profiles resulting from the DNA sampling requirement of several states (FBI, 2019). Moreover, in the United States and some other countries, members of the armed and reserve military forces are required to provide blood samples to be used for the individual identification of remains recovered from conflict or accidental death. The US Armed Forces DNA Identification Laboratory (AFDIL) provides consultation, research, and education services in forensic DNA analysis and is the only human remains DNA testing laboratory in the Department of Defense helping to identify the remains of service members from both current and past conflicts. In 2017 the AFDIL maintained a reference specimen collection, accession, and storage biobank and database of over one million DNA samples (Lee, 2013; AFDIL, 2017).

Similarly, in Europe the Council of Europe, Resolution of June 9, 1997, on the exchange of DNA analysis results (97/C 193/02) invites member states "to consider establishing national DNA databases . . . [built up] in accordance with the same standards and in a compatible manner." The resolution allows member states "to decide on the conditions under which, and the offenses regarding which, the DNA analysis results may be stored in a national database" (Santurtún, 2017). Later, the Prüm Convention of 2005 established infrastructure and procedures to make it possible for European states to share particular forms of intelligence, including DNA profiles, in an effort to counter cross-border criminality and terrorist threats (Williams, 2017).

Go Yoshizawa, professor of medicine at Osaka University in Japan, analyzed informed consent forms for genomic research that had been formulated by four research programs and institutes in Japan, Malaysia, and Taiwan. The comparative analysis highlighted East Asian contexts for consent as distinct from those of other regions. In addition to identifying communicative functions, such as providing recontact options, offering interactive support for research participants, and providing opportunities for family or community engagement in the consent process, East Asian contexts also address the social function of consent forms. This consists of informing participants of possible social risks, including genetic discrimination and sharing samples and data with other agencies, and highlighting the role ethics committees can play. This resource provides a model for a more inclusive, interactive, consent process. It implies that informed consent cannot be validated solely with the completion of a consent form at the initial stage of the research (Yoshizawa, 2017) and can help to address the societal challenges of developing a one-time form intended to cover consent for all future unpredictable DNA advances and applications.

Investigations of the missing children of some of the thousands of people disappeared during Argentina's military dictatorship from 1976 to 1983 have raised complex issues for informed consent. This is due primarily to the legal and ethical difficulties involved in defining the rights of biological grandparents and "adoptive parents," or appropriators, who illegally adopted children of the disappeared (see chapter 7). The biological grandparents and other relatives and the young people suspected of being the children of the disappeared are both secondary victims and evidence that crimes against humanity were committed (Herrera, 2018). Hence, their DNA profiles are required both to repair the relationships of birth and for judicial investigation of the alleged crimes.

The vast majority of the families in the Banco Nacional de Datos Genéticos (BNDG) database who were contacted provided blood samples for DNA analysis and comparison against the pool of young people that might include one or more of their relatives. They signed affirmative consents developed by the National Commission for the Right to Identity (CONADI; Comisión Nacional por el Derecho de Identidad de las Personas, 1992), which works in conjunction with the BNDG to locate missing children, who are now young adults. The BNDG's informed consent document for individuals, called the Bilingual Unified Record in English, states: "I, of my own free will and being of sound mind, consent to having a biological sample taken from me so that a genetic profile may be obtained from it and BNDG may carry out such comparisons as it may deem necessary."[1]

In cases where children, or the alleged appropriators, refuse to voluntarily provide samples for genetic profiling, the federal courts intervene under Law 26.549, adopted in November 2009. The law provides for extraction of "DNA by means other than body inspection, such as the requisitioning of personal objects containing cells already detached from the body, by means of a house search or personal searches."[2] In these instances, the court's edict essentially overrules the need for a consent form to be signed.

DNA STORAGE AND DATA SECURITY

Because of its potential impact on privacy and human rights, one of the most contentious issues in DNA archiving is how long biological samples, DNA profiles, and other police records can be retained. To address these concerns, the Forensic Genetics Policy Initiative in the United States has provided an extensive analysis and recommendations, described later in the chapter (FGPI, 2014; FGPI, 2017).

Some countries, such as Germany, destroy each individual's DNA sample as soon as the DNA profile that is needed for identification purposes has been obtained. This protects privacy to some extent by preventing the sample from being reanalyzed to obtain personal health information. However, other countries retain samples for longer periods, in some cases indefinitely.

How long should individuals' DNA profiles and other personal information be stored on computer databases? Most countries with DNA databases keep the DNA profiles of people who have committed serious crimes such as rape and murder in the database indefinitely, but there is a wide variety of rules for entering and removing data about people who are convicted of lesser crimes (FGPI, 2014, 2017).

Concerns about the storage of forensic DNA profiles and samples extend beyond the government to anyone who can hack the given system and obtain access to a person's DNA profile. This might include organized criminal or terrorist groups or anyone seeking to track down a specific individual. For example, individuals in witness protection programs could have their identities revealed if a discarded sample

---

1. Banco Nacional de Datos Genéticos https://www.argentina.gob.ar/ciencia/
2. See Congreso Argentino, Codigo Proceso Penal, Ley 26.549, November 26, 2009. Retrieved at: http://servicios.infoleg.gob.ar/infolegInternet/anexos/160000-164999/160779/norma.htm.

were collected, matched to a DNA profile stored in a database, and linked to their old identities (FGPI, 2014, 2017).

Significant security risks exist for government, private, and commercial databases of genetic and health information. Threats to information security plague many industries, but the threats against healthcare information systems in particular are growing. Data breaches, generally described as an impermissible use or disclosure of protected information, are particularly prevalent (Gordon, 2017). Recent attacks, such as that against England's National Health Service in 2018, involved ransomware, in which confidential information is encrypted in such a way that only the attacker has the "key" to unlock the data. More worrisome is the possibility that genetic information linked to names and medical records can be used for various fraudulent and criminal activities. These include falsifying medical claims; assessing someone's suitability for a job or visa; tracking down the relatives of political dissenters or pursuing enemies, and identifying paternity and nonpaternity for personal, commercial, or criminal reasons (Samuel, 2018).

The ICMP (see chapter 9) database is on secure servers, protected physically by locked facility access and informatically by various IT security measures and a prohibition on use of the Cloud or email for sensitive information. The ICMP database is also protected legally by a treaty-level agreement that affords ICMP privileges and immunities intended to protect the data from any government or other seizure. Its data management system is carefully designed so that only select staff members involved in various parts of the DNA and matching process have access, and then only to select parts of the data. A few key staff have access to all information needed for issuing a match, and others to anonymized genetic data as needed for their work, but most have no access (Parsons, 2018).

At Argentina's BNDG, blood samples are stored and codified on FTA cards (Whatman Inc., Clifton, NJ), a paper specially treated to bind and protect nucleic acids from degradation. The blood samples and the extracted DNA are stored in separate secured areas with access restricted by electronic recognition. The BNDG cannot discard blood samples, because in cases where there is a restitution, the entire analysis is repeated using the original blood samples. All samples from potential missing grandchildren that initially yield no match against the familial database are also stored on the chance that a match will be found when more families or potential relatives have been profiled.

23andMe, a popular direct-to-consumer DNA company, gives its customers the option to have their saliva sample and/or the DNA extracted from it discarded or stored after it has been analyzed. According to the company's privacy policy, "individuals' information is secured by de-identification/pseudonymization, encryption, and data segmentation, as well as limiting access to essential personnel via multi-factor authentication, single sign-on, and strict least-privileged authorization policy, and by detecting hacking threats and managing vulnerabilities. Names and contact information are stripped from sensitive information, including genetic and phenotypic data. These data are then assigned a randomly generated ID so the data cannot reasonably be identified. 23andMe uses industry standard security measures to encrypt sensitive personal data both at rest and in transit. Additionally, data are segmented across logical database systems to further prevent re-identifiability" (23andMe, 2018).

Another direct-to-consumer company, Ancestry.com, states that it "maintains a comprehensive information security program designed to protect our customers Personal Information using administrative, physical, and technical safeguards. The specific security measures used are based on the sensitivity of the Personal Information collected. We have measures in place to protect against inappropriate access, loss, misuse, or alteration of Personal Information . . . under our control. Ancestry's Security Team regularly reviews our security and privacy practices and enhances them as necessary to help ensure the integrity of our systems and your Personal Information. . . . To request the destruction of your biological samples, you must contact Member Services" (ancestry.com, 2018).

PRIVACY

For forensic purposes, the rights to physical integrity, physical privacy, and genetic privacy with respect to protection of health information and family privacy are complex. Privacy concerns were initially addressed by using genetic markers that did not code for genes and thus conferred no information on risk for disease. Such panels of markers—for example, the short tandem repeats (STRs) described in chapter 1—are the basis of DNA profiles stored in the CODIS database. Variants in regions that code for genes as well as whole genomes have been used for individual identifications, including forensic investigations, which raises ethical concerns about revealing medical information and unknown familial relationships. Often referred to in clinical and research settings as "secondary" or "incidental" findings, such results are becoming a more frequent issue of concern in the era of genomic forensics and medicine. With each human genome expected to contain around 100 harmful variants, secondary findings of potential interest/importance are not exceptional, but inevitable, and there has been extensive review and debate in the medical literature on the ethical issues surrounding disclosure (Parker, 2013; Mackley, 2017; May, 2018). While the focus of this chapter is on the intentional or inadvertent disclosure of health or kinship information from forensic databases, much can be learned about the risks to individuals from the perspective of medical ethics. The risks of disclosure include causing anxiety or psychological harm, discrimination, stigmatization, and potential extortion of the individual or family members (Joly, 2017).

There are several scenarios for disclosing genetic information obtained through a physician-prescribed diagnostic test or as part of a clinical research study: (1) disclose only the results immediately relevant to the person's current clinical needs, (2) disclose both the results that are immediately relevant and other results that are "actionable," or (3) disclose all the results. The American College of Medical Genetics (ACMG) recommends reporting to physicians the results of incidental findings on variants in 59 genes for disorders that are medically highly actionable (May, 2015). The ACMG's initial recommendation has been modified by the addition of an "opt out" to the informed consent document, allowing patients to prohibit investigation of secondary results.

DNA analysis is increasingly used to identify badly damaged and fragmented remains of victims of conflicts and disasters. Incidental findings, such as genetics-related information for which investigators were not looking, may result from these identification efforts. Because of the critical role played by family members in providing personal samples for DNA analysis, the genetic identification of remains is particularly prone to reveal unexpected familial relationships, such as misattributed paternity or false beliefs about sibling relationships. Parker and coauthors present a cogent argument in support of a general policy of nondisclosure of incidental findings, balancing the social benefits of identification efforts with the concern for minimizing and fairly distributing the risks of participation by family members (Parker et al., 2013). The implementation of next generation sequencing (NGS; also known as massively parallel sequencing, MPS) on large numbers of new genetic loci and their flanking regions in forensic DNA laboratories may also result in additional unexpected findings, as disease state probabilities may be linked to certain polymorphisms previously deemed benign. Future expansion of forensic DNA testing into potentially clinically significant or linked markers may require considering how to approach parallel disclosure scenarios. Will forensic DNA analysts, for example, be obliged to disclose the information generated on markers that may predict predisposition to disease? This ethical issue may be further complicated when the disease predisposition is actionable and involves a young person.

Fears of privacy intrusion must also be considered in the design of noncriminal DNA databases; they are particularly important for human rights applications, where a database may be comprised of DNA and other personal information of citizens, noncitizens, undocumented residents, and high-risk populations that may already be stigmatized, victimized, and vilified (KIm & Katsanis, 2013; Katsanis & KIm, 2014)). In October 2019, US attorney general William Barr issued a rule with the expectation that federal authorities will gather DNA information on about 748,000 immigrants annually, including asylum seekers presenting themselves at legal ports of entry, and enter the genetic profiles into the CODIS database. In essence, this rule would conflate immigration with criminal activity and violate the privacy rights of innocent people (Allyn, 2019).

Two groups have surveyed scores of direct-to-consumer companies operating in the United States and United Kingdom that offer genetic testing services, sorted into four broad categories: health-related, ancestry and genealogy, family relationship (including surreptitious testing for paternity and "infidelity"), and lifestyle and wellness (including athletic ability and fitness, nutrition, child talent, diet and weight management, cosmetics, beauty and anti-aging) (Phillips, 2016; Hazel & Slobogin, 2018). They analyzed whether and to what extent the companies' privacy policies informed consumers about how their genetic information would be used and secured, with whom it would be shared, and a host of other issues. The majority of companies surveyed "failed to live up to the basic privacy principles embodied in the four Fair Information Practice Principles (Notice, Choice, Access, and Security) or Privacy Framework (Privacy by Design, Simplified Consumer Choice, and Transparency) endorsed by the Federal Trade Commission" (Hazel & Slobogin, 2018). A comprehensive review of the law of genetic privacy and its applications, implications, and limitations concluded that "few, if any, applicable legal doctrines or enactments provide adequate protection or meaningful control to individuals over disclosures that may affect them" (Clayton, 2019). A general consensus resulting from the numerous public surveys conducted and sometimes contentious discussion about the privacy of genetic information is that ensuring personal anonymity and informed consent are of paramount importance, and that more regulatory oversight is needed (Fernandez, 2015; Gornick, 2017; Fivetti, 2017; May, 2018; Samuel, 2013). With the explosive growth of direct-to-consumer genetic testing services (17 million current customers and growing) and huge, genome-wide sequencing studies ongoing in China, the United States, and the United Kingdom, crowdsourced genealogies and genome databases will enable genetic research to advance forensic investigations as well as disease risk prediction, diagnosis, and treatment (Lussier, 2018).

IDENTIFICATION OF INDIVIDUALS USING FAMILIAL DNA SEARCHING

Familial DNA searching (FDS) of forensic databases of convicted (or arrested) persons raises a range of logistical, social, ethical, and legal considerations (Kim, 2011; chapter 4). FDS involves searching a forensic database not only for an exact match to evidentiary DNA, but also for a partial match, indicating that the person in the database might be related to the perpetrator and thus provide a new lead to investigators. In 2017 the National Criminal Justice Reference Service released a study on key findings, history, policies, practices, technology, and perceptions of familial DNA in four states (Field, 2017). One state highlighted in the report, California, implemented a DNA partial match policy (Crime Scene DNA Profile to Offender[3]) in 2008 that includes the requirement to complete a "Y-STR typing of the same crime scene evidence that yielded the submitted forensic (autosomal STR) unknown profile . . . by the submitting agency and confirm it is concordant with the offender's Y-STR type obtained by [the U.S. Department of Justice (DOJ)]," thus reducing the chance of false positives (Choi, 2012; Myers, 2010). Additional policy options and recommendations have been suggested to "balance effective utilization of familial searching while minimizing harm to and affording maximum protection of U.S. citizens." The Netherlands, United Kingdom, France, United States, and Australia allow kinship analysis under certain conditions using profiles included in their police databases. Germany has also considered adopting a similar policy (*Nature*, 2018). In 2019 the Justice Policy Program of the Social and Economic Well-Being Division of the Research and Development Corporation (RAND) published a

---

3. Gima (2008); and California Department of Justice Division of Law Enforcement Information Bulletin: DNA Partial Match (Crime Scene DNA Profile to Offender) Policy http://dnaresource.com/documents/CAfamilialpolicy.pdf http://dnaresource.com/documents/CAfamilialpolicy.pdf

comprehensive study of practices on familial DNA and moderate stringency DNA testing and discussed their effects (Piquado, 2019). This study contains a literature review on familial and moderate stringency DNA searching; a survey of varying familial and moderate stringency DNA policies and data from state and local forensic laboratories; interviews with representatives of California, Texas, English, and Welsh stakeholders; and discussions on law enforcement use of genealogy and private DNA databases, concerns with racial bias, privacy concerns, policy implications, accountability and impacts of familial and moderate stringency searches on black and Latinx people (Piquado, 2019).

Notorious cold cases in the United States and Europe have been solved through familial searches of criminal databases (some of which have been described in previous chapters), and the number of reports of such developments is rising as profiles from old evidence are searched for matches on public databases (Bieber,

Table 15.2 RECENT SUCCESSES USING FAMILIAL DNA SEARCHING TO SOLVE VIOLENT CRIMES AND COLD CASES THE TABLE LISTS MANY OF THE SUCCESSES IN SOLVING VIOLENT CRIMES AND COLD CASES WITH THE USE OF FAMILIAL DNA SEARCHING. THREE OF THESE CASES RESULTED IN THE EXONERATION OF WRONGFULLY CONVICTED INDIVIDUALS WHO HAD BEEN IMPRISONED FOR NUMEROUS YEARS FOR THE CRIMES. CREDIT LINE: FAMILIAL DNA SEARCHING, FINAL REPORT: LITERATURE REVIEW, P 20 (JANUARY, 2015). FORENSIC TECHNOLOGY CENTER OF EXCELLENCE, NATIONAL INSTITUTE OF JUSTICE, USA

Year	Jurisdiction	Case/Defendant	Offense/Date
2002	U.K.	"Saturday Night Strangler" (Joseph Kappen)	Serial rape/homicide (3 victims) (1973)
2003	U.K.	Jason Thomas Ward	Rape/homicide (2002)
2003	U.K.	Jeffrey Gafoor	Homicide (1988)
2004	U.K.	Daniel Anderson	Rape (1992–1997)
2004	U.K.	Craig Harman	Manslaughter (2003)
2004	North Carolina	Willard Brown*	Rape/homicide (1984)
2005	Kansas	"BTK Killer" (Dennis Rader)	Serial homicide (10 victims) (1974–1991)
2006	U.K.	"The Shoe Rapist" (James Lloyd)	Serial rape (1980s)
2006	U.K.	Christopher Downes	Rape (1984–1985)
2006	U.K.	Graham Darbyshire	Rape (2 victims) (1993–1995)
2006	U.K.	Tahir Mahmood	Rape (1993)
2006	U.K.	Ian O'Callaghan	Rape/homicide (1994)
2007	U.K.	Ronald Castree**	Rape/homicide (1975)
2007	U.K.	Geoffrey Godfrey	Rape (1993)
2008	U.K.	Russell Bradbury	Rape (1986)

Table 15.1 CONTINUED

Year	Jurisdiction	Case/Defendant	Offense/Date
2008	U.K.	Dale Burrows	Rape (1989)
2008	New Zealand	Wayne Jarden	Rape (2 victims) (1988–1996)
2008	U.K.	Derek Young	Serial rape (3 victims) (1990–1994)
2008	U.K.	James Ben Davies	Serial rape (3 victims) (1998–2000)
2008	U.K.	David Newton	Serial rape (3 victims) (1997–2006)
2009	U.K.	David Lace***	Homicide (1979)
2009	U.K.	Robert Morley	Homicide (1985)
2009	U.K.	Harry Musson	Rape (1990)
2009	New Zealand	Joseph Reekers	Homicide (2001)
2009	Denver, Colorado	Luis Jaimes-Tinajero	Automobile thefts
2010	U.K.	Paul Stewart Hutchinson	Homicide (1983)
2010	U.K.	Phil Collins	Rape (1990)
2010	U.K.	"Isle of Wight Rapist" (Keith Davison)	Rape (1990)
2010	California	"The Grim Sleeper" (Lonnie David Franklin, Jr.)	Serial homicide (10 victims) (1985–2010)
2011	U.K.	Robert Saint	Rape (1989)
2011	California	Elvis Lorenzo Garcia	Rape (2008)
2011	U.K.	Kevin Holmes	Rape (2010)
2012	California	James Brown	Rape/homicide (1978)
2012	U.K.	"Pot-bellied Rapist" Michael Acey	Rape (1984)
2012	U.K.	David Bryant	Kidnapping/rape (4 victims) (1982–1995)
2012	Texas	Jack Wesley Melton	Rape (1994)
2012	U.K.	Jon Molt	Rape (1997)
2012	U.K.	Keith Henderson	Rape (2001)
2012	California	"Roaming Rapist of Sacramento" (Derek Sanders)	Serial rape (10 victims) (1998–2003)
2013	U.K.	Barry Howell	Rape (1989)
2013	U.K.	Savador Orozco	Rape (1990)
2013	U.K.	Ian Phipps	Rape (2 victims) (1986–1991)
2013	U.K.	Hilland Matthews	Rape (1992)
2014	Virginia	Tyrone Lamont Halloway	Rape (2001)
2014	Wisconsin	Michael L. Dixon	Serial rape (2002–2012)
2014	Wisconsin	Antoine Devon Pettis	Rape (2014)

*Led to the exoneration of Darryl Hunt, who was wrongfully convicted and spent 18 years in prison for the crime.
**Led to the exoneration of Stefan Kiszko, who was wrongfully convicted and sentenced to 16 years in prison for the crime.
***Led to the exoneration of Sean Hodgson, who was wrongfully convicted and spent 27 years in prison for the crime.

2006; FTCoE, 2015). The earliest successes occurred in 2002 in the United Kingdom. In the United States in 2004 scientists compared evidence from the Deborah Sykes homicide case that revealed a close, but not perfect, match to a North Carolina DNA offender, which led investigators to his brother, Willard Brown (Willing, 2005). In 2010 the Grim Sleeper case was solved and reported as the first successful use of 'familial' matching in a high-profile US case (Dolan, 2010). In 2017 the Los Angeles County Sheriff's Department announced that police had used a familial DNA search to solve the 1976 murder of Karen Klaas, ex-wife of the Righteous Brothers singer Bill Medley (Wang, 2017). In 2018, in a case revived through DNA evidence, French police arrested a couple 31 years after a girl, who turned out to be their daughter, was found dead. The parents were traced after the DNA of their son, tested in an unrelated criminal case, was matched with that of the unknown girl (BBC News, 2018).

IDENTIFICATION USING AGGREGATE DATA FROM GENEALOGICAL WEBSITES OR CLINICAL TRIALS

When a search of a secured criminal DNA database did not identify suspects, the investigators in the Golden State Killer case accessed a genealogical database that the public uses to search for relatives and ancestors, raising concerns about inappropriate expansion of the intended use of public databases. The case also publicized the recent expansion of DNA profiles in genealogical databases. In 2018, the largest commercial database at AncestryDNA held about nine million profiles, while 23andMe held about five million, and Family Tree DNA, MyHeritage, and GEDmatch each had about a million. For perspective, this is only about 10 percent of what could be profiled, because the largest publicly available genealogical databases, currently without genetic data, have 130 million users. The power of using long-range familial searches of genealogical databases to identify suspects via distant familial relatives is vast. Using genomic data of 1.28 million individuals, Erlich and coworkers (Erlich et al., 2018) estimated that about 60% of the searches for North American descendants of Europeans would result in a third cousin or closer match, which could allow their identification using demographic information such as geography, age, and gender. The authors further predicted "that with a database size of ~3 million US individuals of European descent . . . over 99% of the people of this ethnicity would have at least a single 3rd cousin match and over 65% [would be] expected to have at least one 2nd cousin match." Even without genetic information, crowdsourced data from 86 million profiles have been analyzed to provide information on human longevity and the geographical dispersion of families (Kaplanis, 2018).

In response to the use of a genealogical database to identify the suspect in the Golden State Killer crimes in 2018, members of the US Congress requested details directly from four of the genealogical companies, including information about how their customers' data are stored, used, and deleted upon request. Specifically, the lawmakers wanted to know what personal information is collected from customers, which employees of the companies can see that information, and which third parties can buy or access the data. They also had questions about the systems in place to make sure that the information is secure (Thielking, 2018).

GEDmatch policy, prior to and during the Golden State Killer investigation and up until May 2019, had explicitly permitted law enforcement to use its database to investigate DNA obtained and authorized by law enforcement to "identify a perpetrator of a violent crime such as homicides and sexual assault against another individual or to identify remains of a deceased individual" (Ram, 2018). GEDmatch's policy had stated: "[W]e may disclose your Raw Data, personal information, and/or Genealogy Data if it is necessary to comply with a legal obligation such as a subpoena or warrant. We will attempt to alert you to this disclosure of your Raw Data, personal information, and/or Genealogy Data, unless notification is prohibited under law" (GEDmatch, 2018). In May 2019, the GEDmatch informed consent changed to require an affirmative consent by subscribers for use by law enforcement for searching of potential relatives using profiles of forensic evidence of violent crimes (Augustein, 2019).

23andMe's policy states that it "will not provide information to law enforcement or regulatory authorities unless required by law to comply with a valid court order, subpoena, or search warrant for genetic or Personal Information" (23andMe, 2018), and AncestryDNA's policy states: "[W]e may share your Personal Information if we believe it is reasonably necessary to comply with valid legal process (e.g., subpoenas, warrants). If we are compelled to disclose your Personal Information to law enforcement, we will do our best to provide you with advance notice, unless we are prohibited under the law from doing so."

Several factors mitigate the threat to privacy here, Greytak and colleagues (2018) maintain in a compelling argument: (1) only data voluntarily uploaded and explicitly made public are searched; (2) no one is legally required to contribute to a genetic genealogy profile, and samples are not in the possession of government agencies; (3) raw genetic data are not disclosed to law enforcement; and (4) genetic genealogy is for lead generation, not conviction. However, without more firm regulation, genealogical searches could also be used for less serious crimes (Aldhous, 2019).

In September 2019, the US Department of Justice issued an interim policy on forensic genetic genealogical analysis and searches (FGGS) that laid out purpose and scope, applications, limitations, case criteria, investigative cautions, and sample and data control and dispositions (US Department of Justice, 2019). Among other restrictions, this guidance document stated that "information and data derived from FGGS is not, and cannot be, uploaded, searched, or retained in any CODIS DNA Index." This interim policy also requires that 'an investigative agency must seek informed consent from third parties before collecting reference samples that will be used for FGGS, unless it concludes that case-specific circumstances provide reasonable grounds to believe that this request would compromise the integrity of the investigation."

Genealogical firms could alter their informed consent forms to be more affirmative, as outlined in the DOJ policy (see followingbox), and GEDMatch has recently done something similar (Augenstein, 2019). One consequence of imposing a change to an opt-in consent is a slowdown of solving cases due to the reduction in searchable profiles. For example, now less than 20% of GEDMatch users have opted in (181,000 of over 1 million members). While this presently limits subsequent law enforcement FGGS, it may only be temporary, as it permits existing and new clients the opportunity to review and comprehend the company's new consent policy and to be fully informed on how their profiles might be used in law enforcement investigations (Schuppe, 2019). Further, it now appears that a court-ordered search warrant has forced the company to permit such searches (Hill, 2019).

> I, the undersigned, have received a copy and understand the information contained in the Privacy policy of Genealogy.com.

Therefore: I, the undersigned,

☐ **Consent**   ☐ **Do not consent**

that data I provide to Genealogy.com, including my genetic data, may be used by law enforcement to search for matches between profiles of my potential relatives in the genealogical database and the DNA profile of a sample obtained from forensic evidence of a violent crime.

I, the undersigned,

☐ **Consent**   ☐ **Do not consent**

that matches between profiles of my potential relatives in the genealogical database and the DNA profile of a sample obtained from forensic evidence of a violent crime may be used to create more extensive genealogies of people who match and that the information could also be used to identify potential leading suspects who might be my relatives, who might subsequently be investigated and charged in a criminal prosecution.

Further, I understand that I may reverse my consent at any time in the future.

Surveys of the general public and genealogists have reported that the majority supported police searches of genetic websites that identify genetic relatives and disclosure of such genetic-testing customer information to police, as well as police creation of fake profiles of individuals on genealogy websites. Respondents were significantly more supportive of these activities when the purpose was to identify perpetrators of violent crimes, perpetrators of crimes against children, or missing persons, than when the purpose was to identify perpetrators of nonviolent crimes (Guerrini et al., 2018; Gleeson, 2018). Since the arrest of the suspect in the Golden State Killer case, the rate of solving cases using forensic genetic genealogy has grown to about one case per week using long-range familial searches (Erlich, 2018; Jacobo, 2019). In Lancaster, Pennsylvania, for example, police used genealogical information from a close relative to identify Raymond Rowe, a popular DJ, as the suspect in the rape and strangulation of an elementary schoolteacher years before, in 1992 (Scolforo, 2018). More than three decades after a 12-year-old girl was found raped and murdered near a park in Washington State, investigators cracked the case after submitting DNA evidence collected at the scene to a public genealogical database. The accused was identified after investigators recovered a discarded restaurant napkin containing his DNA and matched it to a genetic profile from crime scene evidence (Van Dyke, 2018).

Research databases of DNA profiles and sample biobanks have also occasionally been screened as part of a forensic investigation, a practice that may result in diminished participation in research studies and diminished willingness to donate biological materials. To ward off such an occurrence, Dranseika and coworkers proposed that an "ethically preferable strategy is to make relevant types of forensic access [to research databases] legally impossible. If this cannot be achieved, the next best option is to limit relevant forensic access as much as possible. To the extent that relevant types of forensic access cannot be completely excluded by legal means, information on the potential of such forensic access should be communicated to the donors" (Dranseika et al., 2016).

In addition to research databases, clinical and genealogical databases could be linked to forensic databases, making it possible to identify an individual if personal records were not anonymized. Edge and co-workers showed that records could be matched across genotype data sets that have no shared markers based on "linkage disequilibrium" (LD) between loci appearing in different data sets (Edge et al., 2018). *Linkage disequilibrium* is where alleles (DNA markers) occur together more often than can be accounted for by chance because of their physical proximity on a chromosome Using two data sets for the same 872 people, one with 642,563 genome-wide single nucleotide polymorphisms (SNPs) and the other with 13 STRs used in forensic applications, Edge et al. found that 90–98% of forensic STR records could be connected to corresponding SNP records and vice versa. As illustrated in Figure 15.1, this use of linkage disequilibrium explains how investigators may compare evidence profiles and criminal databases that are based on STR markers to the genealogical databases that are based on the SNP panels of markers.

GOVERNMENT MISUSE OF GENETIC PROFILES

Several countries have approved the use of variants in regions that code for genes to be used in attempts to predict physical characteristics or biogeographical origin, which may ostensibly be of use in investigations (Vogel, 2018). The use of these predicted characteristics has been highly controversial and challenged (Reardon, 2017b; Sperry, 2017). However, artificial intelligence and machine learning methods for combining large databases of genomic information with phenotypic information from electronic health records may improve the predictive value (Hendersen, 2018). Similarly, collaboration between humans and machines offers tangible benefits to face-identification accuracy in forensic applications (Matheson, 2016; Phillips, 2018).

In 2015, Kuwait adopted Act 78/2015, stating that all citizens, residents, and visitors to the country must provide DNA samples to the authorities; the law establishes a penalty of one year of imprisonment and a fine of around $45,000 for any

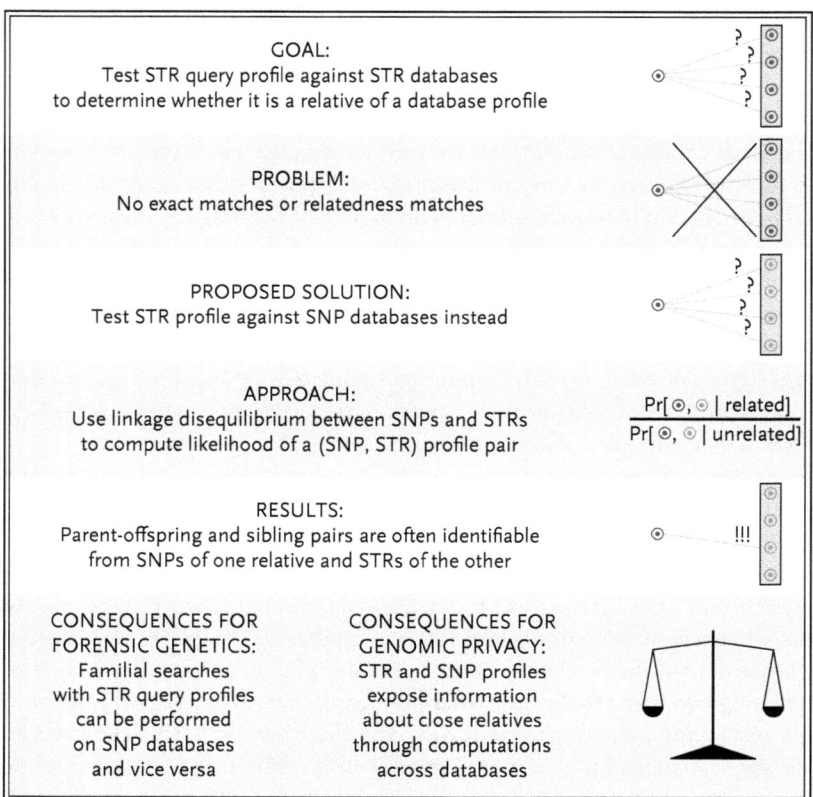

Figure 15.1 Statistical detection of relatives typed with disjoint forensic and biomedical loci. *Source:* Kim et al. (2018), *Cell* 175, 848–858, October 18, 2018 © Elsevier Inc., https://doi.org/10.1016/j.cell.2018.09.00.--

person who deliberately and without justification refuses to cooperate (Santurtún, 2017). This means that DNA databases could be used to track individuals who have not committed a crime, or whose "crime" is an act of peaceful protest or dissent. For example, in a country where freedom of speech or political rights are restricted, the police or secret services could attempt to take DNA samples from the scene of a political meeting to establish whether or not particular individuals had been present (FGPI, 2014, 2017). Familial searching could lead to significant abuses by allowing investigators or anyone who infiltrates the database to track down the relatives of political dissenters. Or it could allow groups to pursue enemies or identify paternity and nonpaternity for personal, political, or criminal reasons (FGPI, 2014, 2017). Following international objections to Kuwait's law, it was repealed in 2017. When genetic ancestry results are made public voluntarily (Warren, 2018), as Senator Elizabeth Warren did to counter malicious comments by President Donald Trump deriding her Native American ancestry, they risk being perniciously misinterpreted by politicians (Zimmer, 2018).

Police in the northwestern region of China have gathered blood samples from the region's large population of Muslim Uighur people (Cyranoski, 2017; Wee, 2019). Human Rights Watch reported that citizens in Xinjiang, China, were required to give a blood sample to get a passport and that 17.5 million people, predominantly Uighurs, were given health examinations that included providing blood samples. China also has a nationwide genetic database that has accrued entries from 40 million individuals, including 1.5 million from samples found at crime scenes.

CULTURAL PERSPECTIVES

The H3Africa Working Group on Ethics has developed guidelines to foster genomic research expertise on the African continent, with the goal of using genomic methods to address health inequities in both communicable and noncommunicable diseases (H3Africa Working Group, 2014). The guidelines emphasize the informed consent process and community engagement as equally critical for the ethical conduct of research. The guidelines do not address the use of genetic information specifically for forensic investigations, but the examples of community engagement from five previous genomic studies in Africa provide a comprehensive introduction to the ethics involving informed consent; sample collection and storage; and the development of strategies for mitigating harmful consequences, such as stigma, that might arise from incidental findings of genomic research. Thus, the guidelines would apply equally to human rights and humanitarian investigations.

In 2017, the San people of Africa drafted a code of ethics for researchers wishing to study their culture, genes, or heritage (South African San Institute, 2017). This was in response to past studies that had included insulting language such as the term "Bushmen," used scientific jargon when communicating with the San, failed to consult study communities about findings before publication, and approached individuals before asking community leaders for permission. The Aboriginal Australians and Canada's First Nations and Inuit have drawn up similar codes, which standardize the process for consultation, the system for explaining benefits for participating communities, and the protocol for data storage and access.

Currently, little legal structure or ethical guidance is available to help researchers determine ethical best practices for "paleogenomic" studies (Bardill, 2018). *Paleogenomic* refers to research involving the remains of ancestors of indigenous peoples (where *ancestors* refers to all pre-European-contact individuals in the Americas as well as post-European-contact deceased indigenous individuals from infants to elders). Results of such research can occasionally undermine or complicate community claims in treaty, repatriation, territorial, or other legal cases. Similar to the H3Africa Working Group's guidelines, the Summer internship for INdigenous peoples in Genomics (SING) Consortium emphasized that community engagement is of paramount importance in decisions about genomic research and that ancestral remains should not be regarded as "artifacts" but as human relatives who deserve respect in research. Since the deceased cannot give consent, so this reasoning goes, present-day communities should be consulted (Bardill, 2018; Reardon, 2017b).

FORENSIC SCIENCE AND DNA: SCIENTIFIC REVIEWS AND PUBLIC POLICY

Public policies, reviews, and scientific committee reviews and reports published over the last decade have helped to inform and guide directions for improving and developing rigorous standards in forensic science disciplines (see appendix 15.1).

Questioned Reliability of Pattern Evidence vs. DNA

In 2009 the National Academy of Sciences released a comprehensive review of the state of forensic science, *Strengthening Forensic Science in the United States: A Path Forward* (National Research Council, 2009). The NAS found many potential problems in several forensic disciplines, including pattern evidence, such as bitemark analyses, stating: "The committee received no evidence of an existing scientific basis for identifying an individual to the exclusion of all others." That same finding was stated in a 2001 review stating that such analysis "revealed a lack of valid evidence to support many of the assumptions made by forensic dentists during bite mark comparisons." The NAS also underscored the lack of any known error rates in toolmark and bitemark comparisons and poor performance on proficiency testing by examiners in these fields (Pretty, 2001). Several responses to the reports from the forensic science community have been published (see PCAST section).

In contrast, in its review of the state of forensic DNA, the NAS report stated: "Thus, DNA analysis—originally developed in research laboratories in the context of life sciences research—has received heightened scrutiny and funding support. That, combined with its well-defined precision and accuracy, has set the bar higher for other forensic science methodologies, because it has provided a tool with a higher degree of reliability and relevance than any other forensic technique". The NAS statement highlights the expansive history of and ongoing research, validation, and technical and judicial review that have led to the solid scientific foundation of forensic DNA.

In 2016 the President's Council of Advisors on Science and Technology (PCAST) published *Forensic Science in Criminal Courts: Ensuring Scientific Validity of Feature-Comparison Methods* (PCAST, 2016; Fields, 2016) and "recommended actions to strengthen forensic science and promote its more rigorous use in the courtroom." In the summary, PCAST concluded that there are two important gaps: "(1) the need for clarity about the scientific standards for the validity and reliability of forensic methods and (2) the need to evaluate specific forensic methods to determine whether they have been scientifically established to be valid and reliable . . . aiming to help close these gaps for . . . 'feature-comparison' methods . . . based on the presence of similar patterns, impressions, or other features in the sample and the source." PCAST cited gaps in the analysis of DNA mixtures, hair, latent fingerprints, firearms and spent ammunition, toolmarks and bitemarks, shoeprints and tire tracks, and handwriting.

The report elicited a series of rebuttal responses throughout the forensic science community and the federal government. The attorney general of the United States at the time, Loretta Lynch, released a statement advising that "the U.S. DOJ would not adopt the report's recommendations" (Fields, 2016), and the FBI noted the report's "subjectively derived" criteria and disregard of numerous published studies that would meet the report's criteria for "foundational validity" (FBI, 2016). Soon thereafter, the American Society of Crime Lab Directors issued a response drawing attention to the flaws in the report's methodology (ASCLD, 2016), and the Bureau of Alcohol, Tobacco, Firearms, and Explosives (ATF) noted PCAST's "failure to address firearms and toolmark studies that had been submitted for consideration" (ATF, 2016). The Association of Firearm and Tool Mark Examiners response pointed out that the report's insistence upon a single report being the benchmark for foundational validity suggested a "fundamental lack of understanding" of the extent of research in the field (AFTE, 2016). Included in the firearms and toolmark community response, AFTE provided a 94-page compendium, dated June 14, 2011, because in AFTE's opinion, "the 2009 National Academy of Sciences (NAS) report only provided a very limited, cursory review of the scientific literature supporting firearm and toolmark identification."

PCAST DNA MIXTURE RESPONSE

Samples with multiple contributors present unique challenges for DNA analysis, as described in chapter 3. In the section "Evaluation of Scientific Validity for Seven Feature-Comparison Methods" in the final PCAST report, the council stated: "DNA analysis of single-source and simple mixture samples includes excellent examples of objective methods whose foundational validity has been properly established" (PCAST, 2016). On the other hand, it pointed out: "Concerning validity as applied, DNA analysis, like all forensic analyses, is not infallible in practice. Errors can and do occur. Although the probability that two samples from different sources have the same DNA profile is tiny, the chance of human error is much higher." Several studies have been published that also discuss the importance of contextual and confirmation bias in all forensic science disciplines, and research is continuing to evaluate methods to understand and mitigate the impact of bias (Jeanguenat, 2017; Dror & Langenburg, 2018; Dror 2018).

PCAST also evaluated the forensic DNA complex mixture procedures and concluded "that subjective analysis of complex DNA mixtures has not been established to be foundationally valid and is not a reliable methodology." In support of this position, the council cited results from an interlaboratory study conducted in 2013 by NIST (National Institute of Standards & Technology), labeled MIX13, in which 70% of laboratories failed to reach the correct conclusion (i.e., had incorrect or inconclusive results) from a complex mixture, highlighting the need to improve complex mixture analysis (Butler, 2018; Augenstein, 2018). The study authors stated at that time, however, that "new probabilistic genotyping software will improve the way DNA mixtures are handled," and Buckleton (2018) recently demonstrated that probabilistic genotyping is better at analyzing mixtures than methods that were available in 2013. PCAST also addressed the validity and limitations of probabilistic genotyping methods:

> Given the problems with subjective interpretation of complex DNA mixtures, a number of groups launched efforts to develop computer programs that apply various algorithms to interpret complex mixtures in an objective manner. The programs clearly represent a major improvement over purely subjective interpretation. They still require scientific scrutiny, however, to determine (1) whether the methods are scientifically valid, including defining the limitations on their reliability (that is, the circumstances in which they may yield unreliable results) and (2) whether the software correctly implements the methods. PCAST finds that, at present, studies have established the foundational validity of some objective methods under limited circumstances (specifically, a three-person mixture in which the minor contributor constitutes at least 20 percent of the intact DNA in the mixture) but that substantially more evidence is needed to establish foundational validity across broader settings. (PCAST, 2016)

The Texas District Attorney's Association (TDAA) summary response stated that the PCAST report "took significant issue with the interpretation of complex DNA mixtures (mixtures with more than two contributors) and . . . the report noted that the laboratory processing of complex mixtures was the same as for single source and simple mixtures, it found that complex mixture interpretation was unreliable due to the lack of standards or guidelines for that approach" (Kaminar, 2018). As a result, the report held that the entire combined probability of inclusion (CPI) statistic used in complex mixtures lacked validity. The report also addressed the use of probabilistic genotyping software, which it called "promising," but it also claimed that such software still lacked sufficient testing to be considered "foundationally valid."

Dr. Bruce Budowle of the University of North Texas's Center for Human Identification stated that

> the report conflates two issues regarding complex mixtures and CPI. According to Budowle, PCAST begins by properly addressing the lack of detailed guidelines relating to interpretation of mixtures. However, PCAST then holds that because there are insufficient guidelines concerning interpretation of mixtures, the CPI statistic used to calculate the likelihood of an individual being a contributor to the interpreted mixture is invalid. Budowle observes that this is an error because the mathematical principles from which that likelihood is derived are the same ones used in single-source random match probability (RMP), which the report had determined to be valid only pages earlier. Regarding PCAST's rejection of probabilistic genotyping as insufficiently studied to be valid, Budowle writes that PCAST failed to contact any of the laboratories that had conducted internal validation studies before implementing probabilistic genotyping software to determine whether their research was consistent with the published articles available. In fact, "There is no

indication that the PCAST Committee made any effort to become informed to opine on the reliability and validity of probabilistic genotyping." (Kaminar, 2018)

ADVANCING AND IMPROVING FORENSIC SCIENCE THROUGH RESEARCH

In the research and advances section in the preceding summary, it was noted that "while a number of the PCAST Report's 'scientific findings' are methodologically flawed, we should not discount recommendations or efforts to improve forensic disciplines" (Kaminar, 2018). Although there may be flaws in some of the PCAST report, the respective forensic science disciplines have been responding and continue to improve their disciplines. For example, the field of latent print analysis is undergoing efforts to move from subjective matching to objective probability reports. The Defense Forensic Science Center (DFSC) has developed, validated, and implemented a software application called FRStat, which facilitates the evaluation and reporting of the statistical strength of friction ridge skin comparisons (Swofford, 2018). This software expresses results as "an estimate of the relative probability of a given amount of correspondence when impressions are made by the same source rather than different sources" (Swofford, 2018). This moves latent print analysis in the direction of DNA analysis as an objective comparison method. Additional publications and ongoing research have also enhanced our understanding of the extent of the empirical research conducted in the firearms and toolmark field, including a study of confirmation bias These views have been codified by Ron Nichols in *Firearm and Toolmark Identification: The Scientific Reliability of the Forensic Science Discipline* (Nichols, 2018). Additional research in the firearms and tool mark fields and statistical analyses in this area are also underway (Nichols, 2018; NIST, 2010). As discussed previously, DNA analysis of complex mixtures is moving toward adoption of probabilistic genotyping, which can help to reduce the subjectivity in interpretation of profiles (Butler, 2018).

Research, testing, validation, and verification on real-world samples, including forensic complex mixtures using probabilistic genotyping, new technologies, and applications of massively parallel sequencing (MPS) in phenotype and ancestry estimations, continue to expand and improve forensic DNA science and broaden our capabilities (McCord & Lee, 2018).

Next Generation Sequencing of Phenotype-Informative Genetic Markers

Forensic DNA laboratories now have the ability to type hundreds of genetic markers including both STRs and SNPs that provide information about individual identity, phenotype, and ancestry information on multiple samples simultaneously from less than 1 ng of input DNA (Børsting, 2015). The new capabilities are revolutionary and permit the detection of sequence differences in STR alleles of the same size, improving mixture resolution capabilities (see chapter 5). Further, the ability to predict phenotype from ancient (Fortes, 2013) and extant samples (Walsh, 2013) provides potential investigative leads for cases in which no full or partial matches are observed from any databases. In addition, new applications using "SNP microhaplotypes"—genetic markers with strings of two or more bi-allelic markers—can improve mixture resolution capabilities (Kidd, 2017, 2018). Forensic applications to the prediction of ancestry, eye, hair, and skin color along with facial features resulting from years of research (Rajeevan, 2012; Maroñas, 2015; Walsh, 2017; Bulbul, 2018; Kidd, 2011; Kayser, 2018; Claes, 2018; Matheson, 2016) have in part been developmentally validated and are commercially available (Lee, 2016; Jager, 2017; Chaitanya, 2018).

One such commercial entity, Parabon Nanolabs, offers law enforcement a commercial service called SnapShot to predict phenotype and "a variety of tools for solving hard cases including Genetic Genealogy, DNA Phenotyping and Kinship Inference," (Parabon, 2019). According to a recent report, Parabon Nanolabs has analyzed hundreds of cases for law enforcement to predict physical traits like hair, eye, and skin color from DNA. Parabon has reported that it identified people related to suspects in 20 of the 100 cases it has analyzed. Although age and environmental variables such as smoking are not addressed with its technology, research on epigenetic patterns for age prediction (Snir, 2018; Freire-Aradas, 2017; Florath, 2014) and other phenotype characteristics such as body height/stature, hair loss/baldness, hair structure, and facial shape (Kayser, 2015) and smoking-associated DNA methylation patterns (Alghanim, 2018) is underway and may provide useful additional tools for investigative leads. A recently formed group, the VISible Attributes Through Genomics, or VISAGE, Consortium brings together leading forensic genetics experts as 13 partners from EU universities and police or justice organizations with the goal of "providing reliable intelligence information on appearance, age, and ancestry of unknown trace donors" (VISAGE, 2019). The VISAGE Consortium has provided important visible trait genetic research results for policymakers as they develop frameworks for forensic DNA phenotyping (Samuel & Prainsack, 2018).

Policy Issues Regarding Searches of Nonforensic Databases

Searching by law enforcement of nonforensic DNA databases has provided resolution to dozens of cases. Some have suggested that in addition to achieving the successes reported, searching genealogy databases "helps to remedy the racial and ethnic disparities that plague traditional forensic searches . . . due to the fact that disparities in the criminal justice system are . . . reproduced in the imbalanced racial and ethnic makeup of these forensic databases. . . . . Genealogical databases, by contrast, are biased toward different demographics" (Ram, 2018). A similar logic has been used to support a proposal for a "population-wide DNA database" (Kaye, 2014). The searching of nonforensic databases raises additional concerns regarding further expansion of suspicion-less searching of ordinary citizens and their privacy, unlike forensic databases, regarding which the US Supreme Court has approved suspicion-less genetic searches for individuals with diminished expectation of privacy (*Maryland v. King*, 2013). An article by Erlich (2018) proposed a measure to mitigate some of the risks and to restore control to consumer genetics providers and consumers. In this proposal, direct-to-consumer providers should cryptographically sign the text file containing the raw data available to customers, which would offer better protection in cases where laws cannot deter misuse.

In the report "Neuroforensics: Exploring the Legal Implications of Emerging Neurotechnologies: Proceedings of a Workshop," a workshop panel of participants reviewed different aspects of the emerging use of neurotechnologies in the legal system (Bain, 2018that the idea of "predicting future behavior such as violence risk—for example, [in which] genetic studies have shown links between certain genotypes and violent behavior reported by Tiihonen et al 2015, is particularly fraught, with a high potential for misuse of science to determine treatments and responses in parole and civil commitment hearings" (quoted in Bain, 2018, p. 4).

The research, validation, and implementation of such new genetic markers and technologies may provide crime laboratories additional tools to solve more crimes, but they also raise further concerns about privacy. As new phenotype and ancestry markers are added and stored in databases, the likelihood of additional potential linkage to secondary/incidental information increases. With the new data come heightened privacy, disclosure, and retention concerns, as there is currently a lack of protection or policy regulating law enforcement access to nonforensic databases (Murphy, 2018; Ram, 2018).

Finally, the proper execution of the initial investigation of a crime or disaster by first responders is pivotal to an investigation. Kruse has shown that the initial crime scene examination is as crucial to the creation of forensic evidence as are the laboratory analyses; statements by plaintiffs, witnesses, and suspects that are elicited by police investigators; and interrogations that prosecutors and defense attorneys bring to the courtroom (Kruse, 2016). All of these factors, in part or whole, have played a significant role in wrongful convictions (US DOJ, 1996), which may be resolved with sound, validated, standardized forensic science research, methods, and interpretation. The Innocence Project has reported 350 DNA exonerations to date over its 25-year history (Innocence Project, 2018).

CONCLUSION

The future of the criminal justice system depends on cutting-edge advances in scientific research, careful and thorough validation, proficiency testing, external review, standardization, laboratory accreditation, quality assurance, and continual review and improvement. At the same time, the implementation of these advances needs to be balanced with protection of individual rights.

There is currently very little oversight on how law enforcement utilizes nonforensic DNA databases such as those for genealogy, mass disasters, missing persons, and clinical research. Protections and policies need to be enacted to ensure that appropriate constraints and enforcements are in place to control the use of these nonforensic databases. The potential of artificial intelligence and machine learning to combine extensive electronic health records of phenotype data with genomic data to identify individuals and predict phenotypes is an area that warrants particular attention (Hendersen, 2017). Evaluation of informed consent; disclosure; sample and data retention; and cultural, ethical, and societal concerns also needs to be addressed and periodically revisited as research and technology continue to change the boundaries of our technical capabilities.

Scrutiny in the form of external scientific boards; technical, legislative, and policy reviews; and the development and enforcement of standards continue to play important roles in informing our decisions when weighing the benefits of implementing scientific technological advances with the need to protect individual rights and public privacy. In "A response to the Forensic Genetics Policy Initiative's Report: Establishing Best Practice for Forensic DNA Databases," Samuel and colleagues state: "[W]e urge that any discussion of human rights safeguards in the context of forensic genetics is opened up to the debate about the governance of forensic genetics and its associated social, ethical and regulatory concerns in this wider sense" (Samuel et al., 2018, p. e21). The areas that the authors list as requiring increased efforts are (1) systematically considering the ethical and social dimensions of NGS and capabilities to conduct forensic DNA phenotype profiling; (2) the growing possibility of integrating data from both forensic and nonforensic databases; (3) the ethical challenges of decentralized data and the new accountability deficits created with data generated by laboratories that may lack quality control and validation requirements; (4) the impact of commercial interests and the need to make them transparent and accountable to the public; and (5) the need to stimulate regulation, awareness, debate, and education on the potential uses of DNA-based information as the applications of forensic genetic technologies expand and diversify, to assess their potential benefits, limitations, and uncertainties (Samuel, 2018).

As research and advances in forensic genetics and genomics continue to raise our forensic scientific capabilities, resulting in new and exciting technological advances being implemented, we must "balance the benefit of solving crime with the public interest in avoiding unwarranted government scrutiny" (Ram, 2018).

REFERENCES

AFTE, Board of Directors1 and Editorial Committee 2. (2015). Comments on NCFS views document: "Scientific literature in support of forensic science and practice". *AFTE Journal, 47*(2). Retrieved from https://afte.org/uploads/documents/position-ncfs-2015.pdf.

Albright, T. (2017). Why eyewitnesses fail. *Proceedings of the National Academy of Sciences of the USA, 114*, 7758–7764.

Aldhous, P. (2019, May 14). Arrest of a teen on an assault charge has sparked new privacy fears about DNA sleuthing. *BuzzFeedNews* https://www.buzzfeednews.com/article/peteraldhous/genetic-genealogy-parabon-gedmatch-assault

Alghanim, H., et al. (2017). Detection and evaluation of DNA methylation markers found at SCGN and KLF14 loci to estimate human age. *Forensic Science International: Genetics, 31*, 81–88.

Allyn, B., & Rose, J. (2019, October 21). Justice Department announces plan to collect DNA from migrants crossing the border. *NPR 24 Hour Program*. https://www.npr.org/2019/10/21/772035602/justice-department-announces-plan-to-collect-dna-from-migrants-crossing-the-bord

American Society of Crime laboratory Directors, Inc. (ASCLD). (2016, September 29). *Board statement on PCAST report on forensic science*, https://pceinc.org/wp-content/uploads/2016/10/20160930-Statement-on-PCAST-Report-ASCLD.pdf

Augenstein, S. (2019, May 20). GEDMatch changes are "blow" to law enforcement—and forensic genealogy. *Forensic Magazine*. Retrieved from https://www.forensicmag.com/news/2019/05/gedmatch-changes-blow-law-enforcement-and-forensic-genealogy?fbclid=IwAR04m8HwG8U05LwKoQWzMvRki8NDRIXP7aRbpy6Ra3WbKs2miRc7IpRCa2I.

Ancestry.com. (2019). *Your privacy: Legal or regulatory process*. Retrieved from https://www.ancestry.com/cs/legal/privacystatement. https://www.ancestry.com/cs/legal/privacystatement

Aviv, R. (2017, June 29). Remembering the murder you didn't commit. *The New Yorker*. https://www.newyorker.com/magazine/2017/06/19/remembering-the-murder-you-didnt-commit

Bain, L., et al. (2018). *Neuroforensics: Exploring the legal implications of emerging neurotechnologies: Proceedings of a workshop*. Washington, DC: National Academies Press. https://doi.org/10.17226/25150

Bardill, J., et al., and the Summer internship for INdigenous peoples in Genomics (SING) Consortium. (2018). Advancing the ethics of paleogenomics. *Science, 360*, 384–385.

BBC News. (2018, June 14). Martyr of the A10/DNA leads to France arrests over 1987 murder. https://www.bbc.com/news/world-europe-44479378

Bieber, F. H., Brenner, C. H., & Lazer, D. (2006). Finding criminals through DNA of their relatives. *Science, 312*, 1312–1315.

Bell, S., et al. (2018). A call for more science in forensic science. *Proceedings of the National Academy of Sciences of the USA, 115*, 4541–4544.

Børsting, C., & Marling, N. (2015). Next generation sequencing and its applications in forensic genetics. *Forensic Science International: Genetics, 18*, 78–89.

Buckleton, J. S., et al. (2018). NIST interlaboratory studies involving DNA mixtures (MIX13): A modern analysis. *Forensic Science International: Genetics, 37*, 172–179.

Bulbul, O. (2018). Improving ancestry distinctions among Southwest Asian populations. *Forensic Science International: Genetics, 35*, 14–20. https://doi org/10.1016/j.fsigen.2018.03.010.

Butler, J. M., et al. (2018). NIST interlaboratory studies involving DNA mixtures (MIX05 and MIX13): Variation observed and lessons learned *Forensic Science International: Genetics, 37*, 81–94.

Caliebe, A., Krawczak, M., & Kayser, M. (2018). Predictive values in forensic DNA phenotyping are not necessarily prevalence-dependent. *Forensic Science International: Genetics, 33*, e7–e8. https://doi.org/10.1016/j.fsigen.2017.11.006.

Callaway, E. (2017). Gene study raises thorny ethical issues. *Nature, 550*, 169–170.

Chaitanya, L., et al. (2018). The HIrisPlex-S system for eye, hair and skin colour prediction from DNA: Introduction and forensic developmental validation. *Forensic Science International: Genetics, 35*, 123–135.

Charo, R. A. (2004). Ethical and policy guidance. In D. Lazar, ed., *DNA and the criminal justice system: The technology of justice* (Chapter 8). Cambridge, MA: MIT Press.

Choi, J. (2012). California and the future of partial match DNA investigations. *Hastings Constitutional Law Quarterly, 39*, 713–737.

Clayton, E. W., et al. (2019). The law of genetic privacy: Applications, implications and limitations. *Journal of Law & Biosciences, 96*, 1–36.

Claes, P., et al. (2018). Genome-wide mapping of global-to-local genetic effects on human facial shape. *Nature Genetics, 50*, 414–423.

Cyranoski, D. (2017). China set to expand DNA database. *Nature, 545*, 395–396.

Dolan, M. (2010, July 10). In Grim Sleeper case, a new tack in DNA searching. *Los Angeles Times*. Retrieved from http://articles.latimes.com/2010/jul/10/local/la-me-0710-grim-sleeper-dna-20100710.

Dranseika, V., Piasecki, J., & Waligora, M. (2016). Forensic uses of research biobanks: Should donors be informed? *Medicine, Health Care & Philosophy, 19*, 141–146.

Edge, D., et al. (2017). Linkage disequilibrium matches forensic genetic records to disjoint genomic marker sets. *Proceedings of the National Academy of Sciences of the USA, 114*, 5671–5676,

Erlich, Y., et al. (2018). Identity inference of genomic data using long-range familial searches. *Science, 362*, 690–694.

Fernandez, C. V., O'Rourke, P. P., & Beskow, L. M. (2015). Canadian Research Ethics Board leadership attitudes to the return of genetic research results to individuals and their families. *Journal of Law, Medicine & Ethics, 43*, 514–522.

Fernandez, H., Joffe, S., & Feldman, E. A. (2018). Informed consent and the role of the treating physician. *New England Journal of Medicine, 358*, 2433–2438.

Field, M. S., et al. (2017). *Study of familial DNA searching policies and practices: case study brief series*. National Criminal Justice Reference Service, Document # 251081. https://www.ncjrs.gov/pdffiles1/nij/grants/251081.pdf

Fields, J. (2016, September 20). White House Advisory Council report is critical of forensics used in criminal trials. *Wall Street Journal*. Retrieved from www.wsj.com/articles/white-house-advisory-council-releases-report-critical-of-forensics-used-in-criminal-trials-1474394743.

FLN-TWG. (2018). Forensic Laboratory Needs Technology Working Group—Opening a New Channel to Improve Forensics https://nij.ojp.gov/topics/articles/forensic-laboratory-needs-technology-working-group-opening-new-channel-improve

Forensic Genetics Policy Initiative. (2014). *DNA databases and human rights*. Retrieved from http://dnapolicyinitiative.org/resources/dna-databases-and-human-rights/

Forensic Genetics Policy Initiative. (2017). *Establishing best practice for forensic DNA database*. Retrieved from http://dnapolicyinitiative.org/wp-content/uploads/2017/08/BestPractice-Report-plus-cover-final.pdf.

Fortes, G. G. (2013). Phenotypes from ancient DNA: Approaches, insights and prospects. *Bioessays, 35*, 690–695.

Freire-Aradas, A., Phillips, C., & Lareu. M. V. (2017). Forensic individual age estimation with DNA: From initial approaches to methylations tests. *Forensic Science Review, 29*, 121–144.

FTCoE (Forensic Technology Center of Excellence, RTI). (2015). *Final report: Familial DNA searching: current approaches*. Retrieved from https://rti.connectsolutions.com/p49iz1rzbpi/.

Fuller, T. (2018, April 26). How a genealogy website led to the front door of the Golden State Killer suspect. *The New York Times*. https://www.nytimes.com/2018/04/26/us/golden-state-killer.html

Garrison, N. A., et al. (2013). Forensic familial searching: Scientific and social implications. *Nature Reviews Genetics, 14*, 445.

GEDMatch.com. (2018). *Terms of service and privacy policy*. Retrieved from https://www.gedmatch.com/tos.htm.

Gima, L. (2008). *DNA partial match (crime scene DNA profile to offender) policy*. California Department of Justice. (Information Bulletin 2008-BFS-01). https://oag.ca.gov/system/files/attachments/press_releases/n1548_08-bfs-01.pdf

Gleeson, M. (2018, 14 November). *DNA and family tree research*. Retrieved from http://dnaandfamilytreeresearch.blogspot.com.

Gordon, W. J., Fairhall, A., & Landman, A. (2017). Threats to information security—public health implications. *New England Journal of Medicine, 377*, 707–709.

Greytak, E. M., et al. (2018). Privacy and genetic genealogy data. *Science, 361*, 857.

Gornick, M. C., et al. (2017). Effect of public deliberation on attitudes toward return of secondary results in genomic sequencing. *Journal of Genetic Counseling, 26,* 122–132.

Grady, C. (2017). Informed consent. *New England Journal of Medicine, 376,* 856–859.

Guerrini, C. J., et al. (2018). Should police have access to genetic genealogy databases? Capturing the Golden State Killer and other criminals using a controversial new forensic technique. *PLoS Biology, 16,* e2006906.

Gupta, S. (2017). Written in blood. *Nature, 549,* S24–S25.

H3Africa Working Group on Ethics and Regulatory Issues for the Human Heredity and Health (H3Africa) Consortium. (2014, February). https://h3africa.org/index.php/consortium/working-groups/ethics-and-regulatory-issues/

Hendersen, J., et al. (2017). *Granite: diversified, sparse tensor factorization for electronic health record-based phenotyping.* [Paper presentation]. 2017 IEEE International Conference on Healthcare Informatics. Park City, UT https://joyceho.github.io/assets/pdf/paper/henderson-ichi2017.pdf doi:10.1109/ICHI.2017.61

Hill, K., & Murphy, H. (2019, November 5). "Game-changer" warrant let detective search genetic database. *The New York Times.* Retrieved from https://www.nytimes.com/2019/11/05/business/dna-database-search-warrant.html.

Innocence Project. (2018). *Innocence Project Anniversary 25* https://25years.innocenceproject.org/.

International Commission on Missing Persons (ICMP). (2018). *Information sheet on DNA genetic testing and processing: ICMP policy on personal data processing and protection.* https://www.icmp.int/wp-content/uploads/2016/08/icmp-pol-dg-04-12-W-doc-icmp-policy-on-personal-data-processing-and-protection.pdf

Jacobo, J. (2019). *Investigators use genetic genealogy to solve 1983 cold case sexual battery on 12-year-old girl.* https://abcnews.go.com/US/investigators-genetic-genealogy-solve-1983-cold-case-sexual/story?id=60293542.

Jäger, A. C., et al. (2017). Developmental validation of the MiSeq FGx forensic genomics system for targeted next generation sequencing in forensic DNA casework and database laboratories *Forensic Science International: Genetics, 28,* 52–70.

Johnston, J., et al. (2017). Supporting women's autonomy in prenatal testing. *New England Journal of Medicine, 375,* 505–507.

Joly, Y., et al., (2017) Comparative approaches to genetic discrimination: chasing shadows? *Trends in Genetics, 33,* 299–302

Kaminar, B. I. (2018, January–February). Responding to PCAST-based attacks on forensic science. *The Texas Prosecutor.* https://www.tdcaa.com/journal/responding-to-pcast-based-attacks-on-forensic-science/

Kaplanis, J., et al. (2018). Quantitative analysis of population-scale family trees with millions of relatives. *Science, 360,* 171–175.

Katsanis, S. H., & Kim, J. (2014). DNA in immigration and human trafficking. In D. Primorac & M. Schanfield, eds., *Forensic DNA Applications: An Interdisciplinary Perspective,* Chapter 22. Boca Raton, FL: CRC Press.

Kayser, M. (2015). Forensic DNA phenotyping: Predicting human appearance from crime scene material for investigative purposes. *Forensic Science International: Genetics, 18,* 33–48.

Kidd, J. R., et al. (2011). Analyses of a set of 128 ancestry informative single-nucleotide polymorphisms in a global set of 119 population samples. *Investigative Genetics, 2,* 1–13.

Kidd, K. K., et al. (2017). Evaluating 130 microhaplotypes across a global set of 83 populations. *Forensic Science International: Genetics, 29,* 29–37.

Kidd, K. K., et al. (2018, June 21). Selecting microhaplotypes optimized for different purposes. *Electrophoresis, 39,* 2815–2823. doi:10.1002/elps.201800092.

Kim, J., et al. (2011). Policy implications for familial searching. *Investigative Genetics, 2,* 22–30.

Kim, J., & Katsanis, S. H. (2013). Brave new world of human-rights DNA collection. *Trends in Genetics, 29*, 329–332. doi: 10.1016/j.tig.2013.04.002 PMID: 23706944

Kruse, C. (2016). *The social life of forensic evidence*. Berkeley, CA: University of California Press.

Kukla-Bartoszek, M., et al. (2018). Investigating the impact of age-depended hair colour darkening during childhood on DNA-based hair colour prediction with the HIrisPlex system. *Forensic Science International: Genetics, 36*, 26–33.

Lazar, D., ed., (2004). *DNA and the criminal justice system: The technology of justice*. Cambridge, MA: MIT Press

Lee, S. B., Crouse, C. A., & Kline, M. C. (2013). An assessment and preparation of biological specimens for DNA analysis: Optimizing storage and handling of DNA extracts. In G. S Jaiprakash & H. L. Ray, eds., *Forensic DNA analysis: Current practices and emerging technologies* (pp. 19–39). Boca Raton, FL: CRC Press.

Lee, S. B., Varlaro, J., & Holt, C. (2016, June). Stepping into the future of forensic genomics: Developmental validation of a next-generation sequencing forensic DNA sample-to-answer system. *Forensic Magazine*. Retrieved from https://www.forensicmag.com/article/2016/07/future-forensic-genomics-developmental-validation-ngs.

Lussier, A. A., & Keinan, A. (2018). Crowdsourced genealogies and genomes. *Science, 360*, 153–154.

Mackley, M. P., & Capps, B. (2017). Expect the unexpected: screening for secondary findings in clinical genomics research. *British Medical Bulletin, 122*, 109–122.

Maraoñas, O., et al. (2015). The genetics of skin, hair, and eye color variation and its relevance to forensic pigmentation predictive tests. *Forensic Science Review, 27*, 13–40.

Matheson, S. (2016). DNA phenotyping: Snapshot of a criminal. *Cell, 166*, 1061–1064.

May T. (2015). On the justifiability of ACMG recommendations for reporting of incidental findings in clinical exome and genome sequencing. *Journal of Law, Medicine & Ethics, 43*, 134–142.

May, T. (2018, June 20). Sociogenetic risks—Ancestry DNA testing, third-party identity, and protection of privacy. *New England Journal of Medicine, 379*, 410–412. doi: 10.1056/NEJMp1805870

McCord B., & Lee, S. B. (2018). Novel applications of massively parallel sequencing (MPS) in forensic analysis. *Electrophoresis, 39*, 2639–2641.

McNamara, M., & Flynn, G. (2018). *I'll be gone in the dark: One woman's obsessive search for the Golden State Killer*. New York, NY: HarperCollins.

Murphy, E. (2018). Forensic DNA typing. *Annual Review of Criminology, 1*, 497–515.

Myers, S. P., et al. (2011). Searching for first-degree familial relationships in California's offender DNA database: Validation of a likelihood ratio-based approach. *Forensic Science International: Genetics, 5*, 493–500.

National Academies of Sciences, Engineering, and Medicine, Committee on Strengthening Forensic Science at the National Institute of Justice & Committee on Law and Justice, Division of Behavioral and Social Sciences and Education. (2015). *Support for forensic science research: Improving the scientific role of the National Institute of Justice*. Washington, DC: National Academies Press.

National Research Council. (1996). *The evaluation of forensic DNA evidence*. Washington, DC: National Academies Press.

National Research Council. (2009). *Strengthening forensic science in the United States: A path forward*. Washington, DC: National Academies Press. Retrieved from https://www.ncjrs.gov/pdffiles1/nij/grants/228091.pdf.

Nature. (2018, May 3). Family connections. *557*, 5.

Nature. (2017, March 20). DNA justice. *543*, 589–590.

Nichols, R. (2018). *Firearm and toolmark identification: The scientific reliability of the forensic science discipline*. San Diego, CA: Academic Press.

NIST. (2010). *Statistics for ballistics identification*. Retrieved from https://www.nist.gov/programs-projects/statistics-ballistics-identification.

Obama Archives. (2014). *Strengthening forensic sciences: A progress report*. Retrieved from https://obamawhitehouse.archives.gov/sites/default/files/microsites/ostp/forensic_science_progress_2-14-14.pdf.

Parabon. (2019). Parabon® Customers Net 55 Solved Cases in First Year of Snapshot® Genetic Genealogy Service New DNA technology helps crack cases that otherwise may have never been solved. https://parabon-nanolabs.com/news-events/2019/05/parabon-customers-net-55-solved-cases-in-first-year-of-snapshot-genetic-genealogy-service.html.

Parker, L. S., London, A. J., & Aronson, J. D. (2013). Incidental findings in the use of DNA to identify human remains: An ethical assessment. *Forensic Science International: Genetics*, 7, 221–229.

Parsons, T., & International Commission on Missing Persons. (2018). *Policy on personal data processing and protection*. https://www.icmp.int/?resources=icmp-policy-on-personal-data-processing-and-protection

PCAST. (2016). President's council of advisors on science and technology. *Forensic Science in Criminal Courts: Ensuring Scientific Validity of Feature- Comparison Methods*. https://obamawhitehouse.archives.gov/administration/eop/ostp/pcast/docsreports.

Phillips, J. P., et al. (2018). Face recognition accuracy of forensic examiners, super-recognizers, and face recognition algorithms. *Proceedings of the National Academy of Sciences of the USA*, 115, 6171–6176.

Piquado, T. et al. (2019). *Forensic familial and moderate stringency DNA searches: Policies and practices in the United States, England, and Wales*. Santa Monica, CA: RAND Corporation. Retrieved from https://www.rand.org/pubs/research_reports/RR3209.html.

Pivetti, M., et al. (2017). Support for the Forensic DNA Database and public safety concerns: An exploratory study. *Open Psychology Journal*, 10, 104–117.

Pretty, I. A., & Sweet, D. (2001). The scientific basis for human bitemark analyses—A critical review. *Science and Justice*, 41, 85–92.

Rajeevan, H., et al. (2012). Introducing the Forensic Research/Reference on Genetics knowledge base, FROG-kb. *Investigative Genetics*, 3, 18. doi: 10.1186/2041-2223-3-18.

Ram, N., Guerrini, C. J., & McGuire, A. L. (2018). Genealogy databases and the future of criminal investigation. *Science*, 360, 1078–1079.

Rao, R. (2016). Informed consent, body property and self-sovereignty. *Journal of Law, Medicine & Ethics*, 44, 437–444.

Reardon, S. (2017a, September 14), Experts pan study claiming DNA can predict facial traits. *Nature*, 549, 139–140.

Reardon, S. (2017b, October 12). Navajo Nation reconsiders ban on genetic research. *Nature*, 550, 165–166.

The Royal Society. (2017). *Forensic DNA analysis: A primer for courts*. Retrieved from royalsociety.org/science-and-law.

Samuel, G., et al. (2018). A response to the forensic genetics policy initiative's report "establishing best practice for forensic DNA databases". *FSI Genetics*, 36, e19–e21. doi: 10.1016/j.fsigen.2018.07.002.

Samuel, G., & Prainsack, B. (2018). Forensic DNA phenotyping in Europe: Views "on the ground" from those who have a professional stake in the technology. *New Genetics & Society*, 38(2), 119–141. doi: 10.1080/14636778.2018.1549984

Santurtún, A., Lema, C., & Zarrabeitia, M. T. (2017). Fundamental rights regarding forensic databases: Review and analysis of Kuwait's law 78/2015. *Revista Española de Medicina Legal*, 43, 79–86.

Schuppe, J. (2019, October 23). Police were cracking cold cases with a DNA website: Then the fine print changed. *NBC News*. https://www.nbcnews.com/news/us-news/police-were-cracking-cold-cases-dna-website-then-fine-print-n1070901

Scolforo, M. (2018, June 25). DNA evidence leads to DJ's arrest in teacher's 1992 killing. Associated Press. https://apnews.com/e8f310c806754ff2aea9de868e0f3e69/DNA-evidence-leads-to-DJ's-arrest-in-teacher's-1992-killing

Shermer, M. (2015, September 1). Can we trust crime forensics? *Scientific American*. Retrieved from https://www.scientificamerican.com/article/can-we-trust-crime-forensics/.

South African San Institute. (2017). *San code of research ethics*. Retrieved from https://link.springer.com/chapter/10.1007/978-3-030-15745-6_7#:~:text=In%20 2017%2C%20the%20South%20African,simple%20process%20of%20 community%20approval.

Sperry, B. P., Allyse, M., & Sharp, R. P. (2017). Genetic fingerprints and national security. *American Journal of Bioethics, 17*, 1–3.

Swofford, H. J., et al. (2018). A method for the statistical interpretation of friction ridge skin impression evidence: Method development and validation. *Forensic Science International, 287*, 113–126.

Thielking, M. (2018, June 21). *Lawmakers press genetic testing companies for details on their privacy policies*. STAT. Retrieved from https://www.statnews.com/2018/06/21/congress-genetic-testing-companies-privacy-policies/.

Tiihonen, J., et al. (2015). Genetic background of extreme violent behavior, *Molecular Psychiatry, 20*, 786–792.

23andMe. (2018). *Privacy statement*. Retrieved from https://www.23andme.com/about/privacy/.

US Department of Justice. (2019, September 2). *Interim policy: Forensic genetic genealogical DNA analysis and searching*. https://www.justice.gov/olp/page/file/1204386/download

Van Dyke, M. B. (2018, June 24). A discarded napkin and a DNA website just solved a 32-year-old murder case. *BuzzFeedNews*. https://www.buzzfeednews.com/article/mbvd/gary-hartman-arrest-murder-12-year-old-michella-welch

Vilares, I., et al. (2017). Predicting the knowledge–recklessness distinction in the human brain. *Proceedings of the National Academy of Sciences of the USA, 114*, 3222–3227.

VISAGE. (2019). About the VISAGE Consortium http://www.visage-h2020.eu/#about.

Vogel, G. (2018). German law allows use of DNA to predict suspects' looks. *Science, 360*, 841–842.

Walsh, S., et al. (2013). The HIrisPlex system for simultaneous prediction of hair and eye colour from DNA. *Forensic Science International: Genetics, 7*, 98–115.

Wang, A. B. (2017, January 30). Police use DNA to solve 1976 murder of Karen Klaas, ex-wife of Righteous Brothers singer. *The Washington Post*. https://nationalpost.com/news/world/police-use-dna-to-solve-1976-murder-of-karen-klaas-ex-wife-of-righteous-brothers-singer

Warren, E. (2018). *My family's heritage*. Retrieved from www.elizabethwarren.com/heritage.

Wee, S.-L. (2019, February 21). China uses DNA to track its people, with American expertise. *The New York Times*. https://www.nytimes.com/2019/02/21/business/china-xinjiang-uighur-dna-thermo-fisher.html

Williams, R., & Wienroth, M. (2017). Social and ethical aspects of forensics genetics: A critical review. *Forensic Science Review, 29*, 145–169.

Willing, R. (2005, June 7). Suspects get snared by a relative's DNA. *USA Today*. Retrieved from https://usatoday30.usatoday.com/news/nation/2005-06-07-dna-cover_x.htm.

Yoshizawa, G., et al. (2017). Social and communicative functions of informed consent forms in East Asia and beyond. *Frontiers in Genetics, 8*, 99.

Zimmer, C. (2018, October 21). Before arguing about DNA tests, learn the science behind them. *The New York Times*. https://www.nytimes.com/2018/10/18/opinion/sunday/dna-elizabeth-warren.html

APPENDIX 15.1

In 2009, the National Research Council (NRC) released a comprehensive review of the state of forensic science titled *Strengthening Forensic Science in the United States: A Path Forward*. The report concluded that "the forensic science system, encompassing both research and practice, has serious problems that can only be addressed by a national commitment to overhaul the current structure that supports the forensic science community in this country. Among the areas determined to be flawed and in need of more research are: accuracy and error rates of forensic analyses, sources of potential bias and human error in interpretation by forensic experts, fingerprints, firearms examination, tool marks, bite marks, impressions (tires, footwear), bloodstain-pattern analysis, handwriting, hair, coatings (for example, paint), chemicals (including drugs), materials (including fibers), fluids, serology, and fire and explosive analysis" (Shermer, 2015).

The NRC report covered multiple topics, including "the fundamentals of the scientific method as applied to forensic practice—hypothesis generation and testing, falsifiability and replication, and peer review of scientific publications; the assessment of forensic methods and technologies—the collection and analysis of forensic data; accuracy and error rates of forensic analyses; sources of potential bias and human error in interpretation by forensic experts; proficiency testing of forensic experts; and infrastructure and needs for basic research and technology assessment in forensic science."

The report also made several recommendations:

1. Create a National Institute of Forensic Sciences[4] (NIFS).
2. Standardize terminology and reporting practices.
3. Expand research on the accuracy, reliability, and validity of the forensic sciences.
4. Remove forensic science services from the administrative control of law enforcement agencies and prosecutors' offices.
5. Support forensic science research on human observer bias and sources of error.
6. Develop tools for advancing measurement, validation, reliability, information sharing, and proficiency testing and to establish protocols for examinations, methods, and practices.
7. Require the mandatory accreditation of all forensic laboratories and certification for all forensic science practitioners.
8. Have laboratories establish routine quality assurance procedures.
9. Establish a national code of ethics with a mechanism for enforcement.
10. Support higher education in the form of forensic science graduate programs, to include scholarships and fellowships.
11. Improve the medico-legal death investigation system.

---

4. National Research Council. (2009) https://www.ncbi.nlm.nih.gov/pmc/articles/PMC4573542/

12. Support AFIS interoperability through the development of standards.
13. Support the use of forensic science in homeland security.

There has been significant progress since 2009. In 2014 the National Academy of Sciences (NAS) published a progress report highlighting three broadly defined areas, with a clear focus on increasing forensic science research (Obama Archives, 2014). Significant improvements were described in sponsoring forensic science research and strengthening the federal research agenda; scientific capacity; new technology and tools; international collaborations in development of standards, guidelines, and best practices; selected workshops and symposia; and education and training.

In 2015 the NAS published a second report, *Support for Forensic Science Research: Improving the Scientific Role of the National Institute of Justice*. In this report the NAS lauded the progress made by NIJ since 2009: "The committee assessed the agency's current research operations and its progress toward that goal since 2009. The committee believes that NIJ has made some very useful changes to its process for soliciting and awarding research grants, thereby improving the agency's scientific capability."

These improvements included

1) making its processes to identify the needs of forensic science practitioners more transparent;
2) increasing the level of autonomy and independence for its scientific peer-review process;
3) obtaining final sign-off authority for its research awards;
4) expanding the size of its research and development portfolio across forensic science disciplines;
5) expanding outreach and dissemination to the practice and research communities;
6) attracting new investigators to forensic science research;
7) increasing the number of graduate student fellowships; and
8) formalizing partnerships with other federal agencies involved in forensic science research, including NIST, the FBI Laboratory, DFSC, and the Bureau of Alcohol, Tobacco, Firearms and Explosives.

In the same report, the NAS set out recommendations aimed at improving the NIJ's forensic science research infrastructure:

Currently, the priority issues emphasized in the agency's solicitations appear to be reactive, and it is not clear how the priorities announced by NIJ relate to an overall long-term research agenda for forensic science. For this reason, the committee believes that the development of a strategic plan for forensic science research and development with short-, mid-, and long-term goals and priorities will help NIJ build a portfolio of cumulative knowledge and provide stability for researchers. Such a strategic plan should guide all internal decision-making, from the development of solicitations to funding decisions. At a minimum, this plan will need to include a research agenda with foundational research outcomes, technology transfer outcomes, efficiency outcomes, and justice system outcomes. The perspectives of both researchers and practitioners should be integrated into the process of identifying and prioritizing the research needs to be used in developing such a strategic plan for NIJ's forensic science research and development program. (Obama Archives, 2014)

Since this report was issued, significant additional progress has been made, including (1) creation of the National Commission on Forensic Science (although it recently was discontinued under the Trump administration), the development of the Organization of Scientific Area Committees for Forensic Science[5] (OSAC; comprised of both researchers and practitioners), and the establishment of the American Academy of Forensic Science Academy Standards Board and Consensus Bodies[5] (NAS, 2009); (2) development of an OSAC lexicon[5] (addressing in part recommendation 2 on terminology); and (3) conducting research in the need areas defined by each OSAC.

In March 2018 the NIJ formed the Forensic Laboratory Needs Technical Working Group (FLN-TWG), a committee of approximately 50 experienced forensic science practitioners from local, state, and federal agencies and laboratories. Through the FLN-TWG, NIJ "reaches out to the forensic science practitioner community to identify, discuss, and prioritize operational needs and requirements. These needs and requirements help inform NIJ's planned and ongoing research and development activities, and ensure that future research and development investments meet practitioner-driven needs." The most recent list of needs and requirements was developed based on a meeting of the FLN-TWG held in February 2018 and includes research needs in forensic biology/DNA, controlled substances and forensic toxicology, impression and pattern and trace evidence, bloodstain pattern analysis, fire and arson investigation, footwear and tire tread, forensic document examination, toolmarks/controlled substances, toolmarks/forensic pathology/forensic anthropology, crime scene examination, forensic pathology, forensic anthropology, and medico-legal death investigations.

---

5. https://www.nist.gov/topics/forensic-science/organization-scientific-area-committees-osac;andhttps://www.nist.gov/topics/forensic-science/organization-scientific-area-committees-osac/osac-lexicon.

Conclusion

The Future of Forensic DNA Analysis

The first applications of DNA technology in criminal cases took place in the United States and United Kingdom more than 30 years ago. What have we learned over the past three decades from the use of forensic DNA analysis in criminal and human rights investigations and humanitarian disasters? And what challenges, opportunities, and potential pitfalls lie ahead?

DNA IN THE COURTROOM

In the first few years of forensic DNA analysis, two different technologies were applied to criminal investigations: restriction fragment length polymorphism (RFLP) analysis of variable number tandem repeat (VNTR) loci and polymerase chain reaction (PCR) amplification of DNA sequence polymorphisms. Subsequently, all forensic tests have been based on PCR, whether analyzing length polymorphism, as in the STR loci, or sequence polymorphism, as in mitochondrial DNA. Since the initial work applying PCR to the analysis of a few autopsy samples in 1986, millions of forensic samples have been analyzed, and many thousands of cases have been resolved, using essentially the same DNA procedures: DNA extraction, PCR amplification, and analysis of the DNA sequence by a variety of techniques. The analytic technology, however, has continued to evolve, making possible, with each new refinement, the analysis of ever more challenging samples, including those with trace DNA and degraded DNA, as well as those with complex mixtures reflecting multiple contributors.

Since forensic DNA typing methods were introduced in US courts, the basic evidentiary and constitutional rules governing admissibility have been well established. These include the *Frye* and *Daubert* tests for reliability of expert methods, confrontation clause and hearsay limitations, and Fourth Amendment case law related to DNA sampling from a criminal suspect. Polymerase chain reaction-short tandem repeat

(PCR-STR) forensic DNA typing of single-source robust samples and the use of the random match probability (RMP) to express the statistical significance of a match have been well established as reliable. However, recent advances in forensic DNA typing, such as next generation sequencing (NGS), may still raise important admissibility issues.

One of the most significant impacts of DNA analysis on criminal justice systems worldwide has been the exoneration of the wrongfully convicted. The work of the Innocence Project and other nongovernmental organizations (NGOs) has resulted in new laws setting evidentiary preservation standards that enable DNA testing long after a trial has been concluded, reforms in the way eyewitness identifications and confessions are obtained, and changes in the manner in which law students and law enforcement personnel are trained. In addition, expert witness testimony based on unvalidated forensic science techniques, such as hair morphology, tooth bite analysis, and blood splatter analysis, is now being held to a higher standard than in the past and increasingly challenged or debunked.

Identifying a perpetrator in crimes without a suspect by checking an evidentiary genetic profile against a database of genetic profiles of convicts (or even arrestees and suspects) has already become standard practice, although exactly how this should be done remains controversial. If a match is found, a "cold hit," an investigation can be initiated and the individual whose DNA matched the evidence sample can be prosecuted if still alive. Typically, a cold hit means that the profile of the evidence matched the individual in the database at all of the STR genetic markers used to genotype the contributors to the database. Searching felon databases has proved to be a highly effective strategy. However, contentious issues remain concerning the appropriate composition of the database (those convicted of violent crimes, all convicted felons, all arrestees, etc.) as well as the trade-offs between civil liberties, privacy, equity, and effective law enforcement.

When a search of a criminal database reveals no cold hits (i.e., no complete matches at all genetic markers), broadening the search criteria to include partial matches expands the search to relatives of individuals in the database (familial searching). Scanning the database with specialized software to detect partial matches can identify relatives of a suspect who may be a potential source of the evidence. The DNA data can then be used as an investigative tool for pursuing multiple persons of interest (POIs) revealed by the search. The familial search strategy is controversial and has been prohibited by statute in Maryland due to concerns about identifying individuals as potential suspects, most of whom are likely to be innocent and not in the felon databases.

However, searching databases of criminal offenders has proven useful in identifying and prosecuting suspects in some high-profile cases, such as the Grim Sleeper, the nickname for serial killer Lonnie David Franklin Jr., who was convicted in 2016 for 10 murders based, in part, on familial DNA analysis. To balance public safety and victims' interests against invasion of privacy (unwanted government intrusion), California and some other states have recommended that familial searching should be performed only when all other avenues and options for identifying potential suspects in high-profile cases have been exhausted.

Initially, familial searching cases were performed on felon databases, but in more recent years genealogical databases have been used to arrest and prosecute notorious serial killers like the Golden State Killer, a former police officer arrested in 2018 for a string of rapes and murders conducted in California from 1976 to 1986. Despite these advances, guidelines for searching criminal DNA databases (convicted offenders, arrestees) and noncriminal databases (genealogical, armed services, migrants, victims of war, and humanitarian disasters) clearly need to be established.

The analysis of forensic samples with multiple contributors continues to be challenging. Estimating the number of contributors in these mixtures and accounting for shared alleles, STR artifactual "stutter peaks," and missing data (allele "drop-outs") pose both technical and statistical challenges. Until recently, the interpretation of mixed samples was expressed as a combined probability of inclusion (CPI). However, recent developments in software that performs probabilistic genotyping (PG) on the standard STR profiles have greatly increased confidence in the interpretation of mixtures. Different software programs incorporate different assumptions and statistical models, but they all yield a likelihood ratio (LR), rather than a CPI. The development and large-scale implementation of PG is transforming the ability of the forensic community to successfully analyze these challenging specimens. Since a large proportion of forensic samples are mixed, the impact on the criminal justice system will be huge.

Another important technical development that promises to significantly help the interpretation of mixtures is NGS, also known as massively parallel sequencing or single molecule sequencing. Although several different commercial NGS platforms exist, they all share the common feature of millions of parallel clonal sequencing reactions: reactions initiated with a single DNA molecule. This property allows the sequencing of the individual components of a mixture; counting the sequence profiles that correspond to the different components of a mixture can yield a robust quantitative analysis of the mixture. Analysis of haploid lineage genetic markers, like the Y chromosome polymorphism or mitochondrial DNA polymorphisms, in which each individual in the mixture contributes only one sequence instead of the two autosomal chromosomal alleles, can further help the deconvolution of forensic mixtures by NGS. Regardless of the method used to generate an LR for inclusion in a mixture, the result must be presented to the jury, and exactly how best to do this has been the subject of significant discussion and considerable controversy.

Another development in forensic DNA technology that promises to have a major impact on the criminal justice system is the deployment of Rapid DNA instruments, which are roughly the size of a microwave oven. These devices can automate the complete STR genotyping analysis of a sample from DNA extraction to interpretation of the electopherogram within 90 minutes. The prospect of obtaining an STR profile of a suspect and creating a database of arrestee profiles raises technical, legal, and ethical issues that need to be addressed.

Some recent advances in forensic DNA technology are more likely to contribute to the investigative aspect of law enforcement rather than the presentation of data for individual identification in the courtroom. Genetic markers of physical appearance as well as of biogeographic ancestry can be valuable for investigations, as can recent analyses of microbiome (bacterial and viral) sequences. In some cases, these microbiome sequences can have true probative value.

The history of forensic DNA evidence has been one of dramatic technical innovation and, given the adversarial nature of the criminal justice system, significant contention. The initial controversies about the use of DNA in forensics have been largely settled, and the admissibility of genetic profiles from single-source samples is well established. The use of objective DNA data in the courtroom has made the conviction of the guilty more likely, protected the innocent, and helped exonerate the wrongfully convicted.

However, as DNA technology evolves and DNA results from complex mixtures, trace DNA, and the search of noncriminal databases are introduced into the courtroom, issues of reliability and general acceptance (the *Daubert* and *Frye* admissibility standards), as well as of privacy, will and should be discussed and contested. The FBI's plan to develop a Rapid DNA network to allow a quick check of an arrestee's DNA profile against the Combined DNA Index System (CODIS) database will likely be the subject of *Daubert* and *Frye* reliability challenges. In addition, the use of DNA profiles based on crime scene samples for phenotyping to predict the likely physical characteristics of a suspect, and microbiome analyses to determine the source of the evidence or identify who committed an offense, may face admissibility challenges. The reliability of these techniques may also be the subject of Fourth Amendment challenges to searches and seizures based on a suspect's alleged similarity to an estimated phenotype or human microbiome. Finally, the comparison of evidentiary DNA profiles to genealogical databases may be challenged on both Fourth Amendment grounds and the reliability of methods of constructing a genealogy used to identify leads based on partial matches in the profile(s) of a suspect's relatives.

DNA IN THE SERVICE OF HUMAN RIGHTS

If forensic DNA analysis has revolutionized criminal justice systems worldwide, it has also provided families and forensic scientists with an indispensable tool for identifying victims of war crimes and human rights abuses and for using that evidence to bring those responsible to justice. One of the first forensic investigations into the whereabouts of the disappeared took place in Argentina in 1983, led by forensic anthropologist Clyde Snow and one of the editors of this book, Eric Stover. Snow and Stover later helped assemble a team of Argentine medical and archaeology students who set out to document the fate of more than 15,000 persons who had disappeared during the previous seven years of military rule. Known as the Argentine Forensic Anthropology Team, the young scientists began exhuming the individual and mass graves believed to contain the remains of the disappeared. The team's objectives in doing so were threefold: to collect evidence for trials of the nine members of the military junta and other accused, to support the right of the families of the disappeared to know the fate of their loved ones and give them a proper burial, and to establish a historical record of the manner and cause of death of dissidents and others who had disappeared during the military dictatorship (Joyce & Stover, 1991).

Snow and his Argentine colleagues went on to train forensic teams in Chile and Guatemala in the early 1990s. By the mid-1990s, the search for the missing had spread to other parts of the world, an effort led largely by the Argentine and Guatemalan teams and two American organizations, the American Association for the Advancement of Science (AAAS) and Physicians for Human Rights (PHR). By 1999, 97 forensic scientists from around the world had traveled to more than 30 countries to investigate the whereabouts of the missing and to train local forensic scientists in the procedures for unearthing mass graves. They brought with them ever-improving new technologies designed to bring greater accuracy to their investigations. Satellite imagery made it possible to generate maps to be used in pinpointing graves hidden in remote locations. Electronic mapping systems had replaced the standard archaeological technique of a baseline and string grid, which meant teams now could save time and produce more accurate data. Most important, advances in DNA analysis enabled geneticists to identify remains of the disappeared that had confounded more traditional anthropological methods and identify more of the infants and young children who had been taken from their parents (who were often killed) and given up for illegal adoption (Stover & Shigekane, 2004).

The first application of DNA analysis to reunite disappeared children with their biological relatives took place in Argentina in the early 1990s. What has been particularly salient in Argentina is the extent to which government agencies, family associations, and the judiciary, while correctly applying checks and balances, have cooperated with one another to resolve these cases. A significant breakthrough happened in 1987 when the Argentine Congress passed a law creating the Banco Nacional de Datos Geneticos (National Genetic Data Bank, or BNDG). Through its biobank of genetic profiles of grandparents and other family members of the disappeared children, BNDG has been able to identify more than 20 children, now young adults. Still, there have been complications. Many of the recovered children—now young adults—have gone through the complex and painful process of coping with finding out who their birth parents in fact were and how they came to be illegally adopted. For most—though not all—of the children, discovering the truth of their biological identity has helped liberate them from the violence that surrounded their births and the lies that defined their lives as children.

Unlike what has happened in Argentina, the effort to locate children who disappeared during El Salvador's civil war has received no support—financial or otherwise—from the Salvadorian government. Many of these children were taken from their parents by soldiers and placed in orphanages or given up for adoption overseas. Since the mid-1990s, Pro-Búsqueda de Niñas y Niños Desaparecidos (Association in Search of Disappeared Children), a family association based in San Salvador, has worked with geneticists and human rights organizations, including the UC-Berkeley Human Rights Center, to locate and reunite these children with their families. At the time of writing, Pro-Búsqueda had registered close to a thousand documented cases of disappeared children in El Salvador and had been able to reunite hundreds of children with their biological families.

The BNDG and Pro-Búsqueda stand out as exemplars of the ways in which family members can work with geneticists and human rights activists—and, in the case of

Argentina, court investigators—to locate their disappeared children. Today, BNDG is running training workshops in DNA analysis for geneticists and activists in Peru, Colombia, and El Salvador. Pro-Búsqueda's familial DNA database could help with the identification of the remains of Salvadorian migrants who died while attempting to enter the United States.

Since the mid-1990s, a growing tension has emerged between the humanitarian needs of families of the missing and the evidentiary needs and limitations of international war crimes tribunals in the aftermath of mass killings. On the one side are families who wish to know the fate of their missing relatives and, if they have died, to receive their remains. On the other side are international war crimes tribunals, which are charged to investigate large-scale killings but may lack the resources or political will to undertake forensic investigations aimed at identifying *all* of the dead. This conundrum arose with the establishment in 1993 of the International Criminal Tribunal for the former Yugoslavia (ICTY), which undertook forensic investigations of mass graves primarily for criminal and not necessarily for humanitarian purposes (Stover & Shigekane, 2004; Wagner, 1998). Fortunately, the creation of the International Commission of Missing Persons (ICMP) three years later helped to resolve this tension. By working with Croat, Bosnian Muslim, and Serb family associations of the missing, ICMP was able to launch a DNA-led effort that has resulted in a large number of identifications of the missing throughout the former Yugoslavia. At the same time, ICMP's adherence to rigorous investigatory methods and procedures produced court-admissible evidence, which later played a key role in several prosecutions at the ICTY.

While many American forensic scientists have lent their expertise, usually through nonprofit organizations like PHR, AAAS, and ICMP, to assist in the investigation of missing persons in Latin America and elsewhere, US federal and state authorities have failed to launch a coordinated forensic program to identify the thousands of migrants who have perished after crossing into the United States from Mexico. This lack of attention can be attributed to several factors. First, it is clear that the powers that be simply don't care or believe such an effort is worthwhile, given that the vast majority of those who have died were illegal migrants. Second, with no formal reporting system in place, family members of migrants are unable to report a missing persons case or fear doing so will risk their own deportation. And finally, even if a family member reports a missing migrant and provides a DNA sample, the process of analysis may take years or, in some cases, never happen because laboratories prioritize typing of sexual assault and homicide cases over typing of unidentified migrant remains. Fortunately, this situation is beginning to change. Several universities and local NGOs in the southwestern United States have joined forces to improve the preservation of migrant remains and promote outreach efforts to family members to obtain DNA samples.

No large-scale, cross-cultural study has ever been conducted to understand the long-term impact of disappearance on relatives of the missing. However, anecdotal evidence and country- and population-specific studies suggest that the absence of a missing loved one can cause significant trauma-related stress to both family members and their communities. Without the remains, some family members may fall into a limbo world of "ambiguous loss," torn between hope and grief, unable to return to the past or plan for the future (Boss, 1999; Wagner, 2008; Stover & Shigekane, 2004; International Committee of the Red Cross, 2002).

Lacking bodies and funerals, relatives of the missing are often unable to visualize the deaths of their loved ones and accept them as real. Nor can they fulfill their communal and religious obligations to the dead, whether of respect or grief. In some cultures and religious groups, funerary rituals are explicitly carried out for the deceased but are also rites of passage for the community, a mechanism for restoring the rent in the social fabric that death has caused. Moreover, psychologists dealing with survivors of trauma have long postulated that personal efficacy and control is a major determinant in recovery. In dealing with trauma, Judith Hermann argues that "no intervention that takes power away from the survivor can possibly foster her recovery, no matter how much it appears to be in her immediate self interest" (Herman, 1992, p. 133). Another way of looking at this phenomenon is through the "control over one's destiny" hypothesis developed by public health researchers. This theory postulates that gaining a sense of control in one's life often leads to better health outcomes (Syme, 1998; Lindheim & Symes, 1983). As such, families of the missing should have a voice in the decision-making processes about when and how investigators convey forensic findings to them.

Experience in several countries suggests that involving family members throughout the forensic investigation of the missing, not merely as observers but as active participants, can have positive results, especially as they are able to observe the technical skills required and the care that is taken to honor the dead. Including family members in the investigative process must go beyond the routine collection of DNA samples and antemortem data, such as medical records and dental X-rays, to help identify the deceased. Families should be actively involved in the consultative and decision-making processes of locating, exhuming, reburying, and memorializing the dead (Wagner, 2008; Stover & Shigekane, 2004; Kleinman et al., 1997; Brown, 1985). It is also paramount that all investigators—from forensic archaeologists and anthropologists working at exhumation sites to geneticists ensconced in their laboratories—are aware of the cultural, social, and political aspects of those they are trying to help. And finally, geneticists and other forensic investigators must work closely with families to help them manage their expectations. This is especially important when years or decades have passed between the disappearance of an individual and the actual initiation of a forensic investigation. Furthermore, investigators who work with commingled remains, as was the case in the aftermath of the 9/11 attacks in New York, must recognize that some family members may find it terribly distressing when they are informed repeatedly of "positive hits" based on DNA analysis. In such situations, investigators must be prepared at the onset of an investigation to present family members with a release authorization form that provides a range of notification options.

Genetic information can also serve as a useful biometric for preventing human trafficking, exposing illegal immigration and fraud, and protecting national security. For example, human traffickers posing as married couples and armed with false identification documents have been known to pass through US immigration with teenagers whom they falsely claim as their biological children. DNA analysis could easily prevent such criminal behavior. Yet using genetic information as a biometric without proper regulatory safeguards could also threaten social constructs of nationality, race, gender, and family identity based on adoption, culture, and transgender considerations. What is clear is that DNA collected from migrants should never be used to deny their human rights. These include denying immigrants legal entry into a country or stigmatizing or excluding certain populations from earning citizenship or residency based on their DNA profiles. In October 2019, Attorney General William Barr issued a proposed rule seeking to allow the federal government to collect DNA samples from more than 740,000 immigrants every year, including asylum seekers presenting themselves at legal ports of entries, and to enter the genetic profiles into the CODIS database. In essence, this rule would conflate immigration with criminal activity and violate the privacy rights of innocent people.

Similarly, human rights organizations have expressed grave concerns that "[t]he police in China are collecting blood samples from men and boys from across the country to build a genetic map of roughly 700 million males, giving the authorities a powerful tool for their emerging high-tech surveillance state" (Wee, 2020). Chinese officials say the DNA database is being used to catch criminals, the vast majority of whom are men and boys. But activists fear the DNA samples are being collected "without consent because citizens living in an authoritarian state have virtually no right to refuse" and that this genetic information could be used to track down and punish the relatives of dissidents (Wee, 2020). They also fear the DNA database could become yet another tool in a growing, countrywide surveillance network that includes street cameras, facial recognition systems, and artificial intelligence.

CHALLENGES AND DEBATES

DNA analysis holds enormous potential for identifying missing persons, establishing the truth about criminal acts, and determining who is responsible—or has been wrongly accused. Indeed, the future efficacy of the criminal justice system depends on advances in genetic research, careful and thorough validation, proficiency testing, external review, standardization of procedures, laboratory accreditation, quality assurance, and continual review and improvement. At the same time, these advances need to be implemented in a manner that safeguards individual rights, upholds ethical standards, and protects privacy. "Law enforcement agencies and cooperating genetic genealogy websites are operating in in a world of few limits," writes Elizabeth Joh, a law professor at University of California, Davis. "There are not only few rules about which crimes to investigate, but also unclear remedies in the case of mistakes, the discovering of embarrassing or intrusive information, or misuse of the information" (Joh, 2019). In this regard, research geneticists must be vigilant about how law enforcement agencies—and now direct-to-consumer genetic testing companies—use the fruits of their labors. As Clayton and coworkers have recommended, "It may be time to shift attention from attempting to control access to genetic information to considering the more challenging question of how these data can be used and under what conditions, explicitly addressing trade-offs between individual and social goods in numerous applications" (Clayton et al., 2019, p. 32). The US Department of Justice's interim policy on forensic genetic genealogical analysis and searches requires that "an investigative agency must seek informed consent from third parties before collecting reference samples that will be used for searches, unless it concludes that case-specific circumstances provide reasonable grounds to believe that this request would compromise the integrity of the investigation." Although GEDmatch has adopted an affirmative "opt-in" informed consent to allow law enforcement to use its customers' genetic data for forensic searches, it now appears that a court-ordered search warrant has forced the company to permit such searches.

At present, there is little oversight on how law enforcement—and potentially immigration agencies—utilize information on genealogy, mass disasters, missing persons, and health records contained in nonforensic databases. Protections and policies need to be enacted to ensure that appropriate constraints and enforcements are in place to control access and use of these databases. The potential of artificial intelligence and machine learning to combine extensive electronic health records of phenotype data with genomic data to identify individuals and predict phenotypes is an area that warrants particular attention. Evaluation of informed consent forms, disclosure, and sample and data retention, as well as cultural, ethical, and societal concerns, will also need to be addressed and periodically revisited as research and technology continue to change the boundaries of our capabilities. As research and advances in forensic genetics and genomics raise our scientific and technological capabilities, we must "balance the benefit of solving crime with the public interest in avoiding unwarranted government scrutiny" (Ram et al., 2018).

One of our aims in assembling this book has been to help the general public better understand forensic DNA concepts and methods. This knowledge is particularly important for jurors who may have to adjudicate evidence based on DNA typing. An excellent resource for potential jurors is a report entitled *Forensic DNA Analysis: A Primer for Courts*, which was prepared by the Royal Society in conjunction with the Royal Society of Edinburgh, Judicial College, Judicial Institute, and Judicial Studies Board of Northern Ireland (Royal Society, 2017). We hope that government agencies and professional organizations worldwide will also publish similar guidelines and standards to assist jurors as they evaluate the complex evidence generated by DNA technologies.

REFERENCES

Boss, P. (1999). *Ambiguous loss: Learning to live with unresolved grief*. Cambridge, MA: Harvard University Press.

Brown, D. L. (1985). People-centered development and participatory research. *Harvard Educational Review*, 55, 69–75.

Clayton, E. W., et al. (2019). The law of genetic privacy: applications, implications and limitations. *Journal of Law & Biosciences*, 96, 1–36.

Cuoto, R. A. (1987). Participatory research: methodology and critique. *Clinical Sociology Review*, 1987, 83–90.

de Jong, J, ed. (2002). *Trauma, war, and violence*. New York, NY: Plenum Publishers.

Hermann, J. (1992). *Trauma and recovery: The aftermath of violence—from domestic abuse to political terror*. New York, NY: Basic Books.

International Committee of the Red Cross. (2002, June). *Proceedings of a conference on the missing: Action to resolve the problem of people unaccounted for as a result of armed conflict or internal violence and to assist their families*. Geneva, Switzerland: ICRC Press.

Joh, E. (2019, June 11). Want to see my genes? Get a Warrant. *New York Times*. Retrieved from https://www.nytimes.com/2019/06/11/opinion/police-dna-warrant.html.

Joyce, C., & Stover, E. (1991). *Witnesses from the grave: The stories bones tell*. Boston, MA: Little, Brown.

Kleinman, A., Das, V., & Lock, M., eds. (1997). *Social suffering*. Berkeley, CA: University of California Press.

Lazer, D., ed. (2004). *DNA and the criminal justice system: The technology of justice*. Cambridge, MA: MIT Press.

Lindheim, R., & Syme, S. L. (1983). Environments, people, and health. *Annual Review of Public Health, 4*, 335–339.

Ram, N., Guerrini, C. J., & McGuire, A. L. (2018). Genealogy databases and the future of criminal investigation. *Science*, 360, 1078–1079.

Royal Society. (2017, November). *Forensic DNA analysis: A primer for courts*. Retrieved from http://royalsociety.org/science-and-law.

Stover, E., & Shigekane, R. (2004). Exhumation of mass graves: Balancing legal and humanitarian concerns. In E. Stover & H. Weinstein, *My neighbor, my enemy: Justice and community in the aftermath of mass atrocity*, pp. 85–103. Cambridge, UK: Cambridge University Press.

Syme, S. L. (1998). Social and economic disparities in health: Thoughts about intervention. *The Millbrank Quarterly, 76*, 493–505.

Wagner, S. (1998). *To know where he lies: DNA technology and the search for Srebrenica's missing*. Berkeley, CA: University of California Press.

Wee, S.L. (2020, June 17). China is collecting DNA from tens of millions of men and boys, using U.S. equipment. *New York Times*. Retrieved from https://www.nytimes.com/2020/06/17/world/asia/China-DNA-surveillance.html?action=click&module=News&pgtype=Homepage.

INDEX

Tables, figures and boxes are indicated by t, f and b following the page number

For the benefit of digital users, indexed terms that span two pages (e.g., 52–53) may, on occasion, appear on only one of those pages.

AAAS. *See* American Association for the Advancement of Science
AABB. *See* American Association of Blood Banks
abandoned/discarded DNA, 74, 83, 90, 228, 305–6, 329
Aboriginal Australians, 345
Abuelas de Plaza de Mayo, 8–9, 150–51n.8, 152, 156, 158, 165, 169, 181–82
 as citizen detectives, 153–54
 contributions of, 168–69
 formation of, 152
 Logares case, 156–58
 overcoming obstacles, 159–60
 Poblete Hlaczik case, 160–61
 Pro-Búsqueda compared with, 179
 search for answers, 154–55
 Vázquez case, 162, 163
Achekzai, Ali, 79
Acholiland, Uganda, 277, 278
ACHPR (African Charter on Human and Peoples' Rights), 195
ACHR (American Convention on Human Rights), 195
ACMG (American College of Medical Genetics), 335–36
Act 78/2015 (Kuwait), 343–44
Adams v. State, 303n.19
adapters, 108
admissibility of DNA evidence, 6–7, 10, 291–307, 366
 basic rules governing, 292–95
 Castro case and, 21–22
 Daubert standard and (see *Daubert v. Merrell Dow Pharmaceuticals*)
 DNA Wars, 22–24, 295
 exonerations and, 36–37

Frye standard and (see *Frye v. United States*)
 issues on the horizon, 306–7
 MPS/NGS and, 113–14
 People v. Bailey and, 36–37
 People v. Wesley and, 36–37
 Pestinikas case and, 20
 Rapid DNA and, 116
 status of constitutional challenges to, 304–6
 status of reliability-based challenges to, 295–303
adoptions, international, 218, 219
AFDIL. *See* Armed Forces DNA Identification Laboratory
Affidavit of Relationships form, 229–30
affirmative consent, 330–31, 341–42
Affymetrix, 118
AFIP (Armed Forces Institute of Pathology), 75
Africa, 238, 282, 345
African Americans, 44. *See also* race
African Charter on Human and Peoples' Rights (ACHPR), 195
AFTE (Association of Firearm and Tool Mark Examiners), 347
Agreement on the Status and Functions of the International Commission on Missing Persons, 195
Aguilas del Desierto, 252n.12
AIMS. *See* ancestry informative markers
Albania, 204
Alexander, Michelle, 44
Alfonsín, Raúl Ricardo, 155, 158–59, 181–82
allele drop-in, 55, 56, 62

allele drop-out, 52–55, 57–58, 59–61, 60f, 62, 365
  admissibility challenges and, 300, 301–2
  defined, 50
alleles, 17–18
  contaminating, 56
  iso-, 106
  mixtures and, 25–26, 49–52, 51f, 54–55, 57–58, 59–61
allele stacking, 55–56
Allen, Fred, 154–55
All of Us Research Program, 77
Alvarenga, Ester, 187
amelogenin, 30–31, 30–31n.8, 111–12, 114–15, 203, 295–96
American Academy of Forensic Science
  Academy Standards Board and Consensus Bodies, 361
  Humanitarian and Human Rights Resource Center, 280
American Airlines (AA) Flight 11, 319
American Association for the Advancement of Science (AAAS), 272, 367, 368
American Association of Blood Banks (AABB), 218–19, 222–23
American Civil War, 196
American College of Medical Genetics (ACMG), 335–36
American Convention on Human Rights (ACHR), 195
American Red Cross Society, 196
American Society of Crime Lab Directors, 347
America's Most Wanted (television program), 78–79
Amgen, 76–77
AmpFLPs, 29
analytical threshold (AT), 57f, 57, 300
Ancestry.com, 3–4, 118–19, 201, 335
AncestryDNA, 340, 341
ancestry estimation
  microbial forensics and, 139–40
  migrant tracking and, 216, 228–29
  MPS/NGS and, 106, 110–11, 112
ancestry informative markers (AIMS), 110–11n.6, 120
ANDE, 114–15, 116
Anderson, Bruce, 276, 281, 282, 284
Anderson, Lukis, 23, 303
Andrews, Tommy Lee, 20–21
anthrax attacks, 129, 130–32, 136–37
Applied BioSystems (ABI), 30–31, 295–96
Arditti, Rita, 168–69
Argentina, 181–82, 197, 264, 268, 272–73, 366. *See also* children (Argentinian)
  innocence organizations in, 38
  number of disappearances in, 150–51, 150–51n.8
  political upheaval in, 149–50
  response to crimes of military dictatorship, 158–59
  return to democracy, 155–56
Argentine Anticommunist Alliance (Triple A), 150
Argentine Forensic Anthropology Team. *See* Equipo Argentino de Antropología Forense
Arias de Franicevich, Juana Elena, 168
Arizona, 5, 37, 240–41, 242–43, 245–47, 257
Armed Forces DNA Identification Laboratory, US (AFDIL), 75–76, 331–32
Armed Forces Institute of Pathology (AFIP), 75
Armenia, 38
Army of Republika Srpska. *See* Vojska Republike Srpske
arrests
  DNA collected upon, 73–74
  Rapid DNA use and, 115, 116–18
  success related to DNA data, 78–79
artificial intelligence and machine learning, 343, 351–52, 371
Asiatic Barred Zone Act of 1917, 212t
Asociación Pro-Búsqueda de Niñas y Niños Desaparecidos de El Salvador, La. *See* Pro-Búsqueda
Association in Defense of the Wrongly Convicted, 37
Association of Firearm and Tool Mark Examiners (AFTE), 347
asylum seekers, 214, 369–70
ATF (Bureau of Alcohol, Tobacco, Firearms, and Explosives), 347
Atlacatl Battalion, 175
attorney incompetence, 43–44
attributed identification, 211
Aum Shinrikyo cult, 130–32
Australia, 38, 337–38, 345
Automated Biometric IDENTification System (IDENT), 215–16, 215t, 217
autopsies, 241–43
autosomal chromosomes, 30–31, 50–51, 297–98, 337–38, 365
Ayala, Fermín Recinos, 173, 188, 189
Ayotzinapa Rural Teachers College (missing persons from), 268–71, 272

Baker, Lori, 244
Balding-Nichols formula, 18

Bali bombing, 204–5
Ban, Jeff, 25
Banco Nacional de Datos Genéticos (BNDG), 159–60, 161, 162–69, 332–33, 367–68
　creation of, 159–60
　DNA storage and data security, 334
　evolution of techniques, 164f, 164–65
　number of recovered children identified by, 164
　processing DNA samples, 165f, 165–66
Bangladesh, 238
Barcode of Life Data System (BOLD), 134
barcoding, 108, 134
Barnert, Elizabeth, 181, 182, 187
Barr, William, 336, 369–70
Barton, Clara, 196
Battelle, 111–12
Bauer, Santiago, 162–63
Baylor University, 244, 257–58
Baylor University Reuniting Families Project, 240n.3
Berg, Paul, 107–8n.2
Bergés, Jorge, 168
Bilingual Unified Record, 332–33
biocrime, 128–30, 136
biogeographic ancestry testing, 216. *See also* ancestry estimation
biographical identification, 211
bioinformatics, 164–65
biometrics, 208, 211, 214–16, 220, 221, 369–70
　commonly used, 214
　databases using, 215t
　genetic information as, 216
　landmark laws and policies for, 212t
bioterrorism, 128–30, 132, 136–37
bite mark comparisons, 78, 346, 364
Blake, Edward, x, 19–20, 24
blood group typing/analysis, 15–16, 40, 50, 218
Blooding, The (Wambaugh), 18–19n.4
blood samples, 68, 75–76, 331–32
blood spatter analysis, 364
Bloodsworth, Kirk, 38
BNDG. *See* Banco Nacional de Datos Genéticos
Bode Cellmark Forensics, 280
Bode Technology, 255, 264
BOLD (Barcode of Life Data System), 134
Bolivia, 38
Boltzmann Institute, 315
Bomberger, Kathryne, 314–15

Bosnia, xi–xii, 10–11, 198, 199, 204, 264, 270, 311–18, 322–23. *See also* Srebrenica
　DNA-led identifications in, 313–18
　number of missing persons in, 313
　trust issue in, 273–75
Bosnian Muslims (Bosniaks), xi–xii, 273–74, 313, 368
Bosnian Serbs, 194, 273–75, 368
Brazil, 204
Brenner, Charles, 80, 181
Breyer, Stephen, 117
British Heart Foundation, 77
Britos, Laura, 153–54
Britos, Tatiana, 153–54
Brody, Reed, 178
"Broken System, A: Error Rates in Capital Cases, 1973-1995" (report), 38
Brooks County, Texas, 238–39, 241–42, 243–44, 257, 258–59, 260–61
Brooks County Sheriff's Office, 244, 245f
Brown, Anthony, 79
Brown, Jerry, 81, 91
Brown, Willard, 79, 338–40
Bruton v. United States, 42
BTK killer, 82
Buckholtz, Joshua, 351
Buckland, Richard, 18–19
Budowle, Bruce, 22, 348–49
Bullcoming v. New Mexico, 305
Bureau of Alcohol, Tobacco, Firearms, and Explosives (ATF), 347
Bureau of Records of Missing Men of the Armies of the United States, 196
Burke, Edmund, 78
Bush, George H. W., 175
Bush, George W., 72–73, 75–76, 210–11
Butler, John, 2, 281
Butte County, California camp fire, 116
Buza, Mark, 73–74

Cáceres, Justina, 162–63
Cáceres Monié, Julio César, 160–61
Cacha, La, 152
California, 5, 73–74, 93, 294, 305–6, 337–38
California Consumer Privacy Act (CCPA), 91
California Innocence Project, 37, 43–44
CAM (Central American Minors Refugee/Parole program), 223
Cambridge Analytica, 91–92
Cameron County, Texas, 241
Cameroon, 204
Campo de Mayo Hospital, 150–51
Cámpora, Héctor, 149

Canada, 37, 38, 58–59, 345
   DNA data banks/databases of, 70, 71t, 73, 78–79, 80–81, 85, 94
   familial searching disallowed in, 82
   restorative justice in, 96
Cancer Research UK, 77
capillary electrophoresis, 6b, 30, 49, 50–51, 52–54, 105, 107–8
Cardozo Law School, 7, 37
Carlotto, Estela de, 152, 153, 154, 155, 169
Carpenter v. United States, 89–90, 306, 306n.20
Carter, Jimmy, 175
Carter, Rubin "Hurricane," 37
case resolution, 94–95
case-to-case hits, 93
case-to-offender hits, 93
Casteliero, Paul, 35
Castro, Jose (Joseph), 21–22, 37. See also *People v. Castro*
Cattaneo, Cristina, 282–83
cause-of-death investigations, 139
Cavallo, Gabriel, 161
CBP. *See* Customs and Border Protection, US
CDC (Centers for Disease Control and Prevention), 136–37
Cellmark, 26–27, 304
cell phone record searches, 89–90, 306
Center of Legal and Social Studies (Buenos Aires), 154
Centers for Disease Control and Prevention (CDC), 136–37
Central America, 221, 238
Central American Minors Refugee/Parole program (CAM), 223
Centurion Ministries, 34, 35, 37
Cetus Corporation, 17, 19–20, 24
Chakraborty, Ranajit, 22
chao khao, 226–27
Chicago Tribune, The, 230
Chihuahua, Mexico, 269
children. *See also* children (Argentinian); children (Salvadoran)
   human trafficking in, 218, 222, 223–24
   international adoptions of, 218, 219
   migrant, 211, 221–24, 225–26
children (Argentinian), 4–5, 8, 149–69, 332–33, 367–68
   Abuelas' search efforts (*see* Abuelas de Plaza de Mayo)
   data bank searches for (*see* Banco Nacional de Datos Genéticos)
   data on appropriating families, 167f, 167–68
   early search efforts, 151–52
   fate of, 166f, 166
   legal victories in identification efforts, 162–63
   post-recovery life, 169
children (Salvadoran), 8–9, 173–90, 367–68
   complexities of family reunification, 186–88
   DNA typing and search for, 178–79
   finding of first, 177–78
   number of abducted, 185
   number of located, 185
   Pro-Búsqueda's efforts on behalf of (*see* Pro-Búsqueda)
Children's Hospital Oakland Institute (CHORI), 189–90
Chile, 38, 204, 367
China, 225, 336–37, 345
Chinese Exclusion Act of 1882, 212t
chip identification numbers, 215–16
Chorobik de Mariani, María Isabel "Chicha," 152, 153–54, 155
Citizenship and Immigration Services, US (USCIS), 214, 218–19, 220
Ciudad Juarez, Mexico, 269
CLAMOR (Human Rights Defense Committee for the Southern Hemisphere) (Brazil), 156–57
Clayton, E. W., 370–71
clinical trials, 340–43
Clinton, Bill, 193–94, 259–60, 314
Coalición de Derechos Humanos, 252n.12
codevelopment model, 280–81
CODIS. *See* Combined DNA Index System
coerced-compliant confessions, 42
coerced-internalized confessions, 42
cold cases, 4, 338–40, 338t
cold hits, 7, 92, 93, 189, 294–95, 305–6, 364
Colibrí Center for Human Rights, 245–47, 248t, 258, 264, 275–77, 283–84
Colin Pitchfork case, 15, 16–17, 18–19
Colombia, 4–5, 38, 168, 204, 270–71, 367–68
Columbia University Law School, 38
Combined DNA Index System (CODIS), 5, 30, 36, 73–74, 78, 81, 82, 93, 97–98, 295–96, 305, 341, 366
   expansion of, 72–73
   human remains identification and, 240–41, 248t, 250–52, 254, 255, 258, 259, 264
   introduction of, 70

migrant tracking and, 216–17, 225, 228, 336, 369–70
MPS/NGS and, 105–6, 111–12
number of profiles in, 331–32
privacy issues and, 335
Rapid DNA and, 114, 115–17, 306–7
type and number of DNA samples in, 70*t*
US Missing Persons Index of, 248*t*
combined probability of inclusion (CPI), 49, 50, 52–55, 57–58, 62, 63, 365
admissibility challenges and, 299–302
validity of questioned, 348–49
Comey, James, 5
Comisión Nacional de Búsqueda de Personas Desaparecidas en El Salvador (CONABÚSQUEDA), 189
Comisión Nacional por el Derecho de Identidad de las Personas (CONADI), 160, 163, 165, 166, 168–69, 332–33
common cold, 135–36
Commonwealth v. DiCicco, 297–98n.6
Commonwealth v. Jacoby, 297–98n.4
Commonwealth v. Mattei, 295–97n.2
complete matches, 4–5, 6*b*, 364
CONABÚSQUEDA (Comisión Nacional de Búsqueda de Personas Desaparecidas en El Salvador), 189
Conacastada, La, 188
CONADEP. *See* National Commission on the Disappearance of Persons
CONADI. *See* Comisión Nacional por el Derecho de Identidad de las Personas
confessions, 25, 41–43, 44–45
confrontation
  challenges to proprietary expert systems and, 305
  challenges to reliance on hearsay DNA reports and, 304–5
  constitutional right of, 294
Congram, Derek, 282–83
Congressional Research Service, 259–60
contaminating alleles, 56
contamination, 23
Contreras, Maria Maura, 147, 173–74, 176, 179–80n.3, 188–89
Contreras et al v. El Salvador, 188n.5
Contreras Recinos, Serapio Cristian, 147, 173, 189
Convicting the Innocent (Garrett), 42, 43
convictions
  success related to DNA data, 78–79, 94–95
  wrongful (*see* wrongful convictions)
Cook, Jay, 84

Cortina, Jon de (Padre Jon), 8–9, 173–74, 175–77, 178–79, 182, 183, 188
Costa Rica, 38
Cotton, Ronald, 41
Council of Europe Resolution of 6/9/1997, 332
counting method, 31–32, 297–98
courtroom, DNA in, 363–66. *See also* admissibility of DNA evidence
CPI. *See* combined probability of inclusion
Crawford v. Washington, 294
cremation, 261, 261n.24
Crimean War, 196–97
crimes against humanity, 194, 197, 199
criminal investigations, 6–8, 363–66
  admissibility of DNA evidence (*see* admissibility of DNA evidence)
  DNA data banks/databases used in, 7, 67–74, 78–90
  exonerations and (*see* exonerations)
  into fabricated crimes, 39–40
  familial searching in (*see* familial DNA searching)
  genealogics in (*see* genealogics)
  historical overview of DNA technology in, 6–7, 15–32
  microbial forensics in (*see* microbial forensics)
  mixed samples used in (*see* mixed samples)
  recent developments in DNA technology, 7–8, 105–21
Criminal Justice Information Services Division Advisory Policy Board, 97–98
Cristiani, Alfredo, 175n.2
Crowell, Kathleen, 39–40
CSI (television program), 2
Culberson amendment (National Defense Authorization Act of 2003), 75–76
cultural perspectives, 345
Customs and Border Protection, US (CBP), 214, 238–39, 255–56, 257, 259–60
Customs and Border Protection Missing Migrants Project, US (CBP-MMP), 248*t*
Cuylenborg, Tanya Van, 84
Cybergenetics, 301–2
Cyprus, 197–98, 204
cytochrome oxidase I gene, 134
Czech Republic, 38

data banks/databases. *See* DNA data banks/databases
Daubert trilogy, 293

Daubert v. Merrell Dow Pharmaceuticals,
 10, 37, 113–14, 116, 296–97, 306–7,
 363–64, 366
  CPI results under, 300–1
  explained, 293–94
  low copy number DNA results
   under, 298–99
  mixture interpretations under,
   300–1, 302–3
  mtDNA results under, 298
  Y-STR results under, 297–98
Davis, Jim, 116
Dayton Peace Agreement, 315
DeAngelo, Joseph James (Golden State
 Killer), 4, 84–85, 86, 118–19, 306,
 329–30, 340–41, 342–43, 365
death penalty cases, ix–xi, 24–25,
 34–36, 38
death squads
  Argentina, 150, 153
  El Salvador, 174
deCODE Genetics, 76–77
deconvolution, 49–50, 51–52, 107
Dedrickson, Kirsten, 225
Defense Department (DOD), US, 75,
 97–98, 331–32
Defense Forensic Science Center, 349
degraded DNA, 27, 105
  human remains identification and,
   200–1, 202
  in mixtures, 57, 107
  MPS/NGS and, 109–10, 111
  problem of, 31–32
  Rapid DNA not suitable for
   analyzing, 116–17
Deloney v. State, 296–97n.2
De Los Santos, Jorge, 35
De Los Santos v. O'Lone, 35
democracy, 196
dental records, 204–5, 245, 282, 321
Dentons, ix–xi
Derecho Humanos, 256
Diabetes UK, 77
diatom test, 139
digital signatures, 215–16
Di Lonardo, Ana Maria, 156, 157
diploid somatic cells, 26, 50–51
direct DNA transfer, 303
direct-to-consumer (DTC) DNA test kits,
 3–4, 68, 77, 83, 99, 336–37
Disaster Victim Identification Guide
 (Interpol), 204–5
discarded DNA. *See* abandoned/
 discarded DNA
District of Columbia, 72–73, 82

DNA analysis
 defined, 6*b*
 future of, 363–71
DNA Analysis Backlog Elimination Act of
 2000, 70
DNA data banks/databases, 3, 7, 22–23,
 22–23n.5, 67–100, 364–65
 of Argentina (*see* Banco Nacional de
  Datos Genéticos)
 case resolution and, 94–95
 challenges to, 73–74
 of El Salvador, 179–82
 expansion of, 72–73
 familial searching (*see* familial DNA
  searching)
 genealogical (*see* genealogics)
 general background, 69–72
 goals and results to date, 77–78
 immigration detainee samples in,
  75, 216–17
 incremental *vs.* universal, 224–25
 measuring desirable outcomes, 94–97
 military personnel samples in, 68, 75–76
 nonforensic, policy issues on searches
  of, 350–51
 privacy and, 87–88, 89–92
 recommendations for
  improvement, 97–99
 reduction/prevention of future crimes
  and, 95–96
 research- and health- related, 76–77, 91
 societal interests and, 96–97
 tallying output, 92–94
 third-party doctrine and, 89–90
DNA data mining. *See* DNA data banks/
 databases
DNA fingerprinting
 coining of term, 16–17
 defined, 6*b*
 evolution of technique, 26–29
 "DNA Fingerprinting Dispute Laid to Rest"
  (Lander & Budowle), 22
 "DNA Fingerprinting on Trial" (Lander), 22
DNA Identification Act of 1994, 70, 116–
 17, 212*t*, 252n.11
DNA Identification Act of 1998
 (Canada), 70
DNA-led identifications, 9, 193–95
 advancing access to, 200–3
 key technical limitations, 200–1
 of Srebrenica victims, 313–18
 of WTC attack victims, 318–24
DNA polymerase, 17, 17n.3, 19–20
DNA sequencing technologies, 134–35
DNA storage and data security, 333–35

DNA Technology in Forensic Science (report), 22
DNA transfer, 303
DNA-View, 181
DNA Wars, 22–24, 295
DOD DNA Registry, 75
Dotson, Gary, 7, 24, 27, 34–35, 39–40
double-blind lineups, 45
Dove, Edward S., 221
Dover Air Force Base, 75
Drina Corps, 313
Due Obedience law (Argentina), 161
due process, 35–36
Dunant, Henry, 196–97
Durán, Rebeca, 186–87

EAAF. *See* Equipo Argentino de Antropología Forense
Ebola hemorrhagic fever, 135–36
ECHR. *See* European Convention on Human Rights and Fundamental Freedoms
Ecuador, 38
EDNAP (European DNA Profiling Group), 70–71
Ejército Revolucionario del Pueblo (ERP), 149
electropherograms, 49, 50–51, 52, 53f, 54, 58, 59–61
Ellis Island, 209–10, 211
El Mozote, 174–75
El Salvador, 4–5, 6, 168, 367–68. *See also* children (Salvadoran)
　duration of war in, 173–74
　repression in, 174–75
　unaccompanied minor migrants from, 222f
　US intervention in, 175
Emergency Immigration Act of 1921, 212t
e-passports, 215–16
Equipo Argentino de Antropología Forense (EAAF), 152, 270, 272–73, 278, 366
　Ayotzinapa missing persons case and, 268–69
　trust issue and, 276
　US-Mexico border cases and, 240–41, 248t, 255, 258, 264
Erlich, Henry, 181
ERP (Ejército Revolucionario del Pueblo), 149
Escuela de Mecánica de la Armada (ESMA), 150–51
estimation of postmortem interval (PMI), 138–39
ethics, 4, 329–35
　cultural perspectives, 345
　DNA storage and data security, 333–35
　of familial searching, 337–40
　of genealogics, 340–43
　government misuse of genetic profiles, 11, 343–45
　informed consent (*see* informed consent)
　of migrant tracking with genetic information, 227–30
　MPS/NGS and, 112–14
　Rapid DNA and, 116–18
ethnic cleansing, 194
eugenics, 227, 228
EurofinsScientific, 113–14
Europe, 30–31, 58–59, 70–71, 112–13, 332, 338–40
European Convention on Human Rights and Fundamental Freedoms (ECHR), 195, 197–98
European Court of Human Rights, 197–98
European DNA Profiling Group (EDNAP), 70–71
European Union, 91
Evaluation of DNA Evidence, The (report), 23–24
evidence
　defined, 6b
　fabricated, 40–41
exclusionary rule, 88, 294–95
exclusions, 6–7, 17–18, 21, 24, 25
exhumations, 244, 245f
exonerations, ix–x, 7, 24, 34–46, 78, 79, 92, 351, 364, 366. *See also* wrongful convictions
　admissibility of DNA evidence and, 36–37
　challenges facing the innocence movement, 44–45
　early innocence cases, 34–35
　expansion of innocence movement, 37–38
　lessons from the innocence movement, 39–44
　percentage related to DNA analysis, 34
expert systems
　admissibility challenges to mixture interpretations by, 301–3
　proprietary, confrontation clause challenges to, 305
expert testimony, reliability requirements for, 292–94
eyewitness identification, 38, 41, 45
Eyewitness Identification Reform Act, 45

Index [379]

Facebook, 91–92
face recognition, 214, 215–16, 217
facial superimposition, 282
FACTS. *See* Forensic Anthropology Center at Texas State University
FAFG (Guatemalan Forensic Anthropology Foundation), 279–80
Fair Information Practice Principles, 336–37
Falkland Islands occupation, 155
fallacy of the transposed conditional, 296–97, 296–97n.3
falsifiability, 293
familial DNA searching (FDS), 3–4, 7, 67–68, 79–82, 99–100, 305–6, 364. *See also* genealogics; kinship analysis
 ethical issus, 337–40
 potential abuse of, 343–44
 recent successes in crime solving, 338t
 SNP arrays in, 80, 118–19
familial matches. *See* partial matches
family, as a social construct, 229–30
family reference samples (FRSs)
 in ICMP investigations, 199, 203–4, 274
 inconsistent, 261–62
 in US-Mexico border searches, 240–41, 247–50, 250f, 251–52, 254, 255, 261–62
FamilyTreeDNA, 3–4, 7, 84–85, 87, 340
Farabundo Martí, Agustín, 174
Farabundo Martí National Liberation Front (FMLN), 173–76, 179, 182
FBI. *See* Federal Bureau of Investigation
FDS. *See* familial DNA searching
Federal Bureau of Investigation (FBI), 3, 5, 22, 39–40, 61, 255, 257, 296–97, 366
 DNA data banks/databases and, 70, 73–74, 97–98
 migrant tracking and, 215–17
 PCR-STR kits accepted by, 295–96
 Rapid DNA and, 115–17
 Rapid DNA Program Office, 97–98
 SWGDAM (*see* Scientific Working Group on DNA Analysis Methods)
Federal Commission on Missing Persons (Herzegovina), 315
Federal Rule of Evidence 702, 37, 293, 294
Federal Rule of Evidence 703, 304
Federation Red Cross, 315
Feitlowitz, Marguerite, 151
Ferra, Ana María, 162–63
Fillingim, Angela, 187–88
fingerprints, 17–18, 204–5, 211, 216, 245, 282, 321, 349
Finley v. State, 303

Firearm and Toolmark Identification (Nichols), 349
First Nations of Canada, 345
Flaim, Amanda, 226–27
FlexPlex, 114–15
FLN-TWG (Forensic Laboratory Needs Technical Working Group), 361
Florida, 5
FMLN. *See* Farabundo Martí National Liberation Front
"foggy mirror test," 43
Fondebrider, Luis, 272–73
food-borne illnesses, 132, 134
foot-and-mouth disease, 132
ForenSeq DNA Signature Prep Kit, 111
Forensic Anthropology Center at Texas State University (FACTS), 257–58
Forensic Border Coalition Cemetery Survey Project, 241–42, 260n.22
forensicbordercoalition.org, 253n.14
Forensic DNA Analysis: A Primer for Courts (Royal Society), 371
forensic genetic genealogy. *See* genealogics
Forensic Genetics Policy Initiative, 333
Forensic Laboratory Needs Technical Working Group (FLN-TWG), 361
Forensic Science in Criminal Courts (PCAST), 346–47
Forensic Science International: Genetics, 112
Forensic Science Regulator, UK, 62
Forensic Sciences Associates, 19–20, 24
Forensic Sciences Service, UK, 19
Forgive Me (Crowell), 40
former Yugoslavia, 4–5, 180, 193–94, 253, 270–71, 273, 368. *See also* International Criminal Tribunal for the former Yugoslavia; individual countries
Fourteenth Amendment, 82
Fourth Amendment, 73, 74, 82, 117, 329–30, 363–64, 366
 admissibility challenges arising from, 294–95
 challenges to database cold hit results under, 305–6
 genealogics and, 87–88
 third-party doctrine and, 89–90
France, 196–97, 337–40
Franco, Francisco, 151
Frank, typhoon, 204
Franklin, Benjamin, 34–35
Franklin, Lonnie David, Jr. (Grim Sleeper), 3, 81, 338–40, 364
Franks, Harold, 39

From Madness to Hope: The 12-Year War in El Salvador (report), 176
FRSs. *See* family reference samples
FRStat, 349
fruit of the poisonous tree doctrine, 83
Frye, James, 292–93
Frye v. United States, 10, 21 36–37, 114, 116, 294, 296–97, 306–7, 363–64, 366
  CPI results under, 300–1
  explained, 292–93
  low copy number DNA results under, 298–99
  mixture interpretations under, 300–3
  mtDNA results under, 298
  Y-STR results under, 297–98
FSTG software, 301–2
Full Stop (Argentina), 158, 161
Funes, Mauricio, 179–80n.3
Furman v. Georgia, 35

Gafoor, Jeffrey, 79–80
Gálvez Carillo, Julio, 259
Gálvez Carillo, Yadira, 259
Gandhi, Mahatma, 264
García, Eduardo, 184–85
Garcia, Elvis Lorenzo, 81–82
Garrett, Brandon L., 42, 43
GEDmatch, 3–4, 7, 84–85, 87–88, 89–90, 118–19, 306, 329, 340, 341–42, 370–71
genealogics, 3–4, 68, 365. *See also* familial DNA searching
  ethical issues, 340–43
  explained, 83–87
  Fourth Amendment and, 87–88
Genealogy.com, 342
Gene Codes Forensics, 321
General Amnesty Law for the Consolidation of Peace (El Salvador), 176n.2
General Electric Co. v. Joiner, 293
General Protection Regulation (GDPR), 91
GeneReader MPS system, 111–12
genetic markers, 2
  applications of investigational value, 119–20
  defined, 16
  emerging technology, 118–19
  haploid lineage, 63
  ineffective for human remains identification, 262–63
  phenotyping and, 349–50
  statistical independence of (*see* statistical independence)

genetic polymorphism. *See* polymorphism
Geneva Convention on the Status of Refugees and its Protocol, 195
Geneva Conventions, 195
genocide. *See* Bosnia; Herzegovina; Srebrenica
genotype recycling, 219, 221
genotypes, 25, 26, 49–50, 51–52 *See also* heterozygotes; homozygotes
Germany, 38, 333, 337–38
Gideon v. Wainwright, 35
Gilbert, Walter, 107–8n.2, 134–35
Gilmore, Jim, 26
Gima, Lance, 180
Ginsburg, Ruth Bader, 74
GlaxoSmithKline, 77
GlobalFiler, 30–31, 111, 295–96
GlobalFiler Express, 114–15
Godsey, Mark, 39
Golden State Killer. *See* DeAngelo, Joseph James
Gonzalez, Angel, 41–42
Gorsuch, Neal, 90
government misuse of genetic profiles, 11, 343–45
Grandmothers of the Plaza de Mayo. *See* Abuelas de Plaza de Mayo
Greece, 264
Green, Kevin, 92
Gregg v. Georgia, 35–36
Grim Sleeper. *See* Franklin, Lonnie David, Jr.
Grinspon, Mónica, 156
Guardado, María Albertina Iraheta, 257–58
Guatemala, 4–5, 272 367
  adoptees from, 218, 219
  unaccompanied minor migrants from, 222f, 222
Guatemalan Forensic Anthropology Foundation (FAFG), 279–80
guilt supported by DNA testing, x–xi
Guinda de Mayo, 176n.1, 176–77, 178–79, 180, 183, 188, 190
Guy, Martin, 78

Hague, The, 275, 278, 314–16
hair analysis, 24, 39–40, 109, 364
Haiti, 238
hand geometry, 214
haploid lineage genetic markers, 63
haplotypes, 30–32, 119, 297–98
Harmon, Brian, 180, 181
Harris, Joseph, 36–37
HART. *See* Homeland Advanced Recognition Technology

Index [381]

Hartl, Dan, 22
Hatch, Orrin, 5
Hawthorne, Andre Wayne, 78
H.B. 30, 98
HCV. *See* hepatitis C virus
Health and Human Services Department, US, 211
health-related data banks, 76–77, 91
hearsay
  evidentiary rule against, 294
  of nontestifying analysts, 304–5
hepatitis C virus (HCV), 132, 135–36, 137–38
Hermann, Judith, 369
Herrera Piñero, Mariana, 163–64, 169
Herzegovina, 10–11, 198, 199, 204, 270, 311–18, 322–23. *See also* Srebrenica
  DNA-led identifications in, 313–18
  number of missing persons in, 313
  trust issue in, 273–75
heterozygotes, 51, 54–55, 59–61
  defined, 50–51
  isometric, 106
Hidalgo County, Texas, 241–42
HID Ion AmpliSeq Identity Panel, 111
Highway 281, 238–39
Hillary case, 302–3
HIPAA, 91
History of the Peloponnesian War (Thucydides), 196
hits, 68–69, 73, 92–94
  case-to-case, 93
  case-to-offender, 93
  cold, 7, 92, 93, 189, 294–95, 305–6, 364
HIV. *See* human immunodeficiency virus
HLA. *See* human leukocyte antigen
Hlaczik, Gertrudis María, 160–61
HLA-DQA1 gene (HLA-DQ alpha), 17, 19–20, 23–26, 27–29, 28*f*, 30
Holland, Tom, 271
Homeland Advanced Recognition Technology (HART), 215*t*, 217, 225
Homeland Security Department (DHS), US, 75, 210–11, 214, 217, 224, 255–56
homozygotes, 51, 54–55, 59–61
  defined, 50–51
  example, 53*f*
Honduras, 222*f*, 257
Houston Migrant Rights Collective, 248*t*
Howe, Brian, 39
H3 Working Group on Ethics, 345
Humane Borders, 252n.12
Human Genome Project (HGP), 128
human immunodeficiency virus (HIV), 132, 137

Humanitarian Forensic Action, 270
human leukocyte antigen (HLA), 155, 156, 156n.4, 157, 164–65
Human Microbiome Project, 128, 129, 130
human remains identification, 10, 70, 238–64, 311–26, 336, 368–69. *See also* Bosnia; DNA-led identifications; former Yugoslavia; Herzegovina; Srebrenica; US-Mexico border; western Balkans; World Trade Center attacks (2001)
  attempts to unify stakeholders, 254–56
  codevelopment model and, 280–81
  distrust among stakeholders, 256–57
  expectations of community, 279
  forensic investigations, 197–99
  future directions, 281–84
  genetic marker ineffectiveness, 262–63
  impact on forensics in marginalized regions, 277–81
  international organizations and, 253–54
  large scale, 193–205
  lengthy turnaround times in, 261
  non-DNA data sources, 245–47
  nonprofit organizations and, 252–53
  notification options, 317–18, 322–23
  outsourcing of DNA testing, 279–80
  photo of processed remains, 246*f*
  Rapid DNA and, 116, 283–84
  repatriation after, 261
  resources for, 247–54
  roles of stakeholders, 247–54, 248*t*
  as a state responsibility, 195–97
  strengthening investigative capacities, 199–200
  trust issue, 272–77
human rights and humanitarian disasters, 4–5, 6, 8–10, 366–70. *See also* children (Argentinian); children (Salvadoran); human remains identification
Human Rights Defense Committee for the Southern Hemisphere (CLAMOR) (Brazil), 156–57
Human Rights Watch, 268–69n.3, 345
human trafficking, 9, 208, 214, 218, 222, 223–24, 369–70
Hunt, Darryl, 79
Hurban, Ignacio, 169
hurricane Katrina, 204
hybrid networks, 312

IAFIS/NGI. *See* Integrated Automated Fingerprint Identification System
ICC. *See* International Criminal Court

[382]   *Index*

ICCPR (International Covenant on Civil and Political Rights), 195
ICE. *See* Immigration and Customs Enforcement
Iceland, 76–77
ICMP. *See* International Commission on Missing Persons
ICRC. *See* International Committee of the Red Cross
ICTR. *See* International Criminal Tribunal for Rwanda
ICTY. *See* International Criminal Tribunal for the former Yugoslavia
IDENT. *See* Automated Biometric IDENTification System
Identical (Turow), xi
identical twins, xi, 80–81, 90–91, 113–14
Identification Data Management System (iDMS), 204
IdentiFiler, 30–31, 295–96
iDMS (Identification Data Management System), 204
Illegal Immigration Reform and Immigrant Responsibility Act of 1996, 212t
Illinois, 37
Illinois v. Dotson, 24, 27. *See also* Dotson, Gary
Illumina, 106, 109–10, 111–12, 118
immigrants. *See* migrants/immigrants
Immigration Act of 1891, 212t
Immigration Act of 1924, 212t
Immigration and Customs Enforcement (ICE), US, 211, 214, 224
Immigration and Nationality Act of 1952, 212t
Immigration Reform Act of 1986, 212t
inclusions, 6–7, 17–18, 21, 25
India, 238
Indiana, 112
individual identification, 6–8. *See also* criminal investigations
influenza, 135–36
Information Sheet on DNA Genetic Testing and Processing (ICMP), 331
informed consent, 4, 11, 330–33, 345, 370–71
  affirmative, 330–31, 341–42
  genealogical firms and, 341–42
innocence movement
  challenges facing, 44–45
  early cases, 34–35
  expansion of, 37–38
  lessons from, 39–44
Innocence Project, 7, 34, 37, 351, 364
Innocence Protection Act, 38

Innocent (Turow), xi
inquests, 243
Integrated Automated Fingerprint Identification System (IAFIS/NGI), 215–16, 215t, 217
Inter-American Commission on Human Rights, 154, 259
Interdisciplinary Group of Independent Experts, 268–69n.2
Interim Policy on Forensic Genetic Genealogical DNA Analysis and Searching, 86–87
International Commission on Missing Persons (ICMP), 9, 10–11, 111–12, 180, 193–94, 195–96, 198–99, 201–2, 270–71, 278, 314, 315–16, 317, 368
  apprehension over success of, 198
  creation of, 193–94
  cross-border identifications and, 248t, 253, 264
  DNA storage and data security, 334
  experience of, 203–4
  family participation and, 199, 203–4
  headquarters moved to The Hague, 314–15
  impact of forensic model of, 275
  informed consent form of, 331
  legal status of, 195
  number of family references in database, 204
  trust issue and, 273, 274–75
International Committee of the Red Cross (ICRC), 248t, 253–54, 264, 270, 315
International Covenant on Civil and Political Rights (ICCPR), 195
International Criminal Court (ICC), xi–xii, 195
International Criminal Police Organization. *See* Interpol
International Criminal Tribunal for Rwanda (ICTR), 195, 278
International Criminal Tribunal for the former Yugoslavia (ICTY), 9, 194, 195, 198, 199, 273, 313, 315–16, 317, 368
international humanitarian law, 195
International Organization for Migration, 282n.18
International Society for Forensic Genetics (ISFG), 54, 62
Interpol, 204–5
  DNA databases and, 70–71, 78–79
  human remains identification and, 248t, 254
Interpol Global DNA Profiling Survey, 71
Inuit, 345

Index [383]

investigations aided, 73, 92–94
Ion Torrent, 106, 109–10, 111
Iran, 225
Iraq, 204, 225
Iraqi-Kurdistan, 289
Ireland, 38
iris imaging, 214
ISFG. *See* International Society for Forensic Genetics
ISIS, 224
isoalleles, 106
isometric heterozygotes, 106
Israel, 38
Italy, 38, 264

Jackson, Ricky, 39
jailhouse informants, 45
Jan Bashinski DNA Lab. *See* Richmond DNA lab
Japan, 38, 130–32, 332
Jeffreys, Alec, 16–17, 18–19, 24
Jim Hogg County, Texas, 241–42
Joh, Elizabeth, 370–71
Justice Department, California, 86. *See also* Richmond DNA lab
Justice Department, US (DOJ), 4, 70, 75, 80–81, 86–87, 256, 341–42, 370–71
justices of the peace, 243

Kagan, Elena, 74
Kaine, Tim, 26
Karadžić, Radovan, 194–95, 198, 199
Katrina, hurricane, 204
Katz v. United States, 90
Kelly, Ian, 19
Kešetović, Rifat, 274, 316–17
Kidd, Kenneth, 22, 201–2n.3
King, Alonzo Jay, Jr., 74, 117–18. See also *Maryland v. King*
King, Mary-Claire, 155, 156, 157
King Ranch, 239
kinship analysis, 200–3. *See also* familial DNA searching
Kirschner, Robert, 178–79
Klaas, Karen, 338–40
Kly, Joseph, 19, 20
Kosovo, 204, 273
Kovras, Iosif, 272, 275
Kumho Tire Co. v. Carmichael, 293
Kumra, Raveesh, 23
Kuwait, 224, 225, 343–44

Laboratorio di Antropologia e Odontologia Forense (LABANOF), 282
Landa, Ceferino, 160–61

Lander, Eric, 21, 22
Latham, Krista, 244
Lavallén, Rubén, 156–57
Law 23.492 (Argentina). *See* Full Stop
Law 26.549 (Argentina), 163, 333
Law and Order (television program), 2
lawyer incompetence, 43–44
Lazar, David, 80
LDIS, 70
Leahy, Patrick, 38
Leicester University, 16–17
Leiro, Raquel Teresa, 156
length polymorphism, 6–7
Lewontin, Richard, 22
Lexicon of Terror (Feitlowitz), 151
liberation theology, 174
libraries (MPS), 108–9, 111
Libya, 204, 225
lies, 39
Lifecodes, Inc., 20–21, 26–27, 37
ligation, 108
likelihood ratio (LR), 49, 52–54, 55, 61, 62–63, 291, 302–3, 365
  defined, 18
  familial searching and, 80
  MPS/NGS and, 110–11
Lincoln, Abraham, 196
lineups, double-blind, 45
linkage disequilibrium, 30n.7, 343
Linkoping University, 201–2
local DNA index system, 70
Lockyer, Bill, 180
Logares, Claudio, 156
Logares, Paula Eva, 156–58
"Long Journey Home" (study), 186
Lopez Rega, José, 150
Lord's Resistance Army (LRA) (Uganda), 277
Louisiana, 5, 73–74
low copy number (LCN) DNA, 22–23
  admissibility challenges to, 291, 298–99
  in mixtures, 50
LR. *See* likelihood ratio
LRA. *See* Lord's Resistance Army
Lukodi, Uganda, 278n.15
Lynch, Loretta, 347

MacCrate, Robert, 37–38
MacCrate Report, 37–38
machine learning. *See* artificial intelligence and machine learning
Madres de Plaza de Mayo, 150–51n.8, 152
Madrigal, Rafael, 43–44
Magaletti v. State, 298n.7
Magnacco, Jorge Luis, 162n.6

Malaysia, 332
Malaysian Airlines flight MH 17, 199, 204
manual keyboard search, 255, 255n.19
Mapp v. Ohio, 35, 294–95
Mariani, Maria Isabel. *See* Chorobik de Mariani, María Isabel "Chicha"
Marion, William Jackson, 34–35
Marshall, Thurgood, 43
Marston, William Moulton, 292–93
Martinez, Dilcy Yohan Sandres, 257
Martinez, Urbino "Benny," 258–59
Martínez de Perón, María Estela "Isabel," 149–50
Martínez Gómez, Montserrat, 185–86
Maryland, 82, 305–6, 364
Maryland DNA Collection Act (MDCA), 74, 117–18
Maryland v. King, 74, 117–18, 294–95, 305, 350–51
Mass Fatality Identification System (M-FISys), 321
massively parallel/next generation sequencing (MPS/NGS), 7–8, 31, 105–7, 363–64
  admissibility and, 113–14
  advantages in forensic DNA, 107–8
  commercial forensic kits, 111–12
  ethics, 112–14
  human remains identification and, 201–2
  introduction to the technology, 108–11
  microbial forensics and, 135–36
  mixture interpretation and, 63–64, 106, 107–8, 365
  phenotyping and, 106, 110–11, 112, 349–50, 352
  privacy and, 112, 114, 335
  properties of, 106
Matanza, La, 174
matches, 291. *See also* inclusions; random match probability
  complete, 4–5, 6*b*, 364
  defined, 6*b*
  partial, 3, 4–5, 6*b*, 337–38, 364
  requirements for, 2–3
Mayfield, John C., III, 76
McCloskey, Jim, 35, 37
McCowen, Christopher M., 96–97
McCray, Thomas, 78–79
McDaniel v. Brown, 296–97n.3
McNair, Dwayne, 113–14
measles, 135–36
medical examiners, 241–43, 251
medical malpractice, 137–38
Medical Research Council (UK), 77

Mediterranean region, 199, 200, 253–54, 270–71, 282, 282n.18
Medley, Bill, 338–40
Melendez-Diaz v. Massachusetts, 294
Melley, Brian, 230
Mendel, Gregor, 27n.6
Menem, Carlos Saúl, 158–59
Mexico, 238. *See also* US-Mexico border
  citizen-science initiatives in, 280–81
  innocence organizations in, 38
  missing persons of, 268–70
M-FISys (Mass Fatality Identification System), 321
Michigan State University, 226–27
microbial cultures, 132–33
microbial forensics, 8, 128–40
  answers sought in, 130
  applications and cases, 136–40
  defined, 129–30
  interest areas in, 130–32
  methodological evolution, 132–36
microbiome, 8, 128–29, 365–66
  defined, 129
  Human Microbiome Project, 128, 129, 130
  interest areas in research, 130–32
  relative abundance in different areas of body, 131*f*
microbiota, 129
microsatellites, 163–64, 165
Mignone, Emilio, 154
Mignone, Isabel, 154
migrants/immigrants. *See also* migrant tracking; US-Mexico border
  drop of in US, 209
  family-based programs, 218–21
  family separation, 211, 221–24
  ratio of to US population, 1925-2015, 210*f*
  by region, 1881-2016, 210*f*
  statelessness and, 225–27
  total US, 1925-2015, 209*f*
  unaccompanied minors, 221–24, 222*f*
migrant tracking, 9, 208–31, 369–70
  biometrics in (*see* biometrics)
  databases for, 215*t*
  DNA samples from detainees, 75, 216–17
  ethics of using genetic information in, 227–30
  projected extent of information gathered, 336
  proof of identity in (*see* proof of identity)
  Rapid DNA use and, 115, 217, 224
  verifying claimed relationships, 218–27

military personnel, blood collection from, 68, 75–76, 331–32
MinION, 110
Minnesota, 4
Miranda v. Arizona, 35
MiSeq, 111–12
Missing Migrant Crisis Line, 259–60
Missing Migrant Project, 275, 282n.18
Missing Migrants Program Summit, 255–56
missing persons, 268–70, 366–68. *See also* children (Argentinian); children (Salvadoran); human remains identification
mitochondrial DNA (mtDNA), 6–7, 71, 83–84
  admissibility of testing results, 10, 114, 297–98
  BNDG analysis of, 163–64, 165
  degraded DNA analysis and, 31–32, 105
  human remains identification and, 203, 240, 262
  in mixtures, 63–64, 107–8, 365
  MPS/NGS and, 106, 107–8
  WTC victim identification and, 321
MIX05 study, 58–59
MIX13 study, 58–59, 58–59n.2, 62, 347–48
mixtures, 7, 22–23, 31, 49–64, 105–6, 349, 365
  admissibility challenges to, 291, 299–303
  allele drop-in and, 55, 56, 62
  allele drop-out and (*see* allele drop-out)
  allele stacking and, 55–56
  capillary electrophoresis and, 49, 50–51, 52–54, 107–8
  challenges in interpreting, 55–59
  defined, 24–25, 107–8
  determining number of contributors in, 55
  misinterpretation of, 57–59
  MPS/NGS analysis of, 63–64, 106, 107–8, 365
  PCAST on, 347–49
  stochastic threshold and (*see* stochastic threshold)
  STR analysis of (*see under* short tandem repeats)
  stutter peaks and, 55, 56, 62
  technical innovations aiding in analysis of, 59–64
  two-person, 49–50, 51–52, 59–61
  Washington case and, 24–26
Mladic, Ratko, 194

molecular phylogenies, 137–38, 137*t*
Montoneros, 149, 150, 152
Moreira, Mercedes Beatriz, 160–61
Mothers of the Disappeared, 183
MPS/NGS. *See* massively parallel/next generation sequencing
mtDNA. *See* mitochondrial DNA
Mullis, Kary, 17
Mundorff, Amy, 322–23
Muslim Uighur people, 225, 345
MyHeritage, 3–4, 7, 340

NamUs. *See* National Missing and Unidentified Persons System
Napoleon III, 196–97
National Academy of Sciences (NAS), 39–40, 346, 347, 360–61
National Commission on Forensic Science, 361
National Commission on the Disappearance of Persons (CONADEP) (Argentina), 150–51n.8, 155, 158, 167–68, 272
National Criminal Justice Reference Service, 337–38
National Defense Authorization Act of 2003 (Culberson amendment to), 75–76
National DNA Data Bank of Canada, 70, 71*t*, 73, 78–79, 80–81, 94
National DNA Database (NDNAD) (UK), 70, 71*t*, 79–80
National DNA Index System (NDIS), 70, 116–17, 216–17, 251
National Genetic Data Bank (Argentina). *See* Banco Nacional de Datos Genéticos
National Health Service (UK), 334
National Institute of Justice (NIJ), 97–98, 111–12, 360–61
National Institute of Standards and Technology (NIST), 58–59, 97–98, 281, 347–48
National Institutes of Health (NIH), 77
nationality, as a social construct, 228–29
National Missing and Unidentified Persons System (NamUs), 247, 248*t*, 250, 254, 256, 258, 262, 264
National Registry of Exonerations, 39, 44
National Research Council (NRC), 22, 23–24, 359–60
Naturalization Act of 1790, 212*t*
Nature, 22, 229
Nava Vélez, Blanca Luz, 268, 269–70
NDIS. *See* National DNA Index System
NDNAD. *See* National DNA Database (UK)

Netherlands/Holland, 38, 199, 337–38
Neufeld, Peter, 37, 39, 45
"Neuroforensics: Exploring the Legal Implications of Emerging Neurotechnologies" (report), 351
New Jersey, 45
New Jim Crow, The (Alexander), 44
New Mexico, 112
Newshour (television program), 187
New York, 94, 294
New York Office of Chief Medical Examiner (OCME), 10–11, 301–2, 318–19, 320–21, 322–24, 325–26
New York State Police Forensic Investigation Center, 321–22
New Zealand, 38
next generation identification (NGI). *See* Integrated Automated Fingerprint Identification System
next generation sequencing (NGS). *See* massively parallel/next generation sequencing
NGOs. *See* nongovernmental organizations
Nichols, Ron, 349
Nielsen, Kirstjen, 211n.1
Nightingale, Florence, 196–97
NIJ. *See* National Institute of Justice
NIST. *See* National Institute of Standards and Technology
Nixon, Richard, 35–36
No More Deaths, 252n.12, 256, 259–60
nongovernmental organizations (NGOs)
  Ayotzinapa missing persons case and, 269–70
  exoneration cases and, 364
  migrant identification and, 239, 251, 253–55, 256
  Srebrenica identifications and, 315
  Ugandan missing persons and, 279
North Carolina, 4
North Carolina Actual Innocence Commission, 45
NRC. *See* National Research Council
nucleic-acid-level detection, 133–36
"Nunca Mas" (report), 158

Obama, Barack, 223
OCME. *See* New York Office of Chief Medical Examiner
Office of Refugee Resettlement (ORR), 211, 223–24
Ohio Innocence Project, 39
oligonucleotide probes, 118
Olimpo, El, 160–61
Olson, John, 84

1,000 Genomes Project, 201–2
Operation Identification, 240–41, 244, 248t, 253n.14, 258–59
Organization of Scientific Area Committees for Forensic Science (OSAC), 361
ORR. *See* Office of Refugee Resettlement
Orrego Benavente, Cristián, 155, 178, 180, 181, 182, 183, 184, 189–90
OSAC (Organization of Scientific Area Committees for Forensic Science), 361
Oxford Nanopore Technologies, 110, 135, 202

Pacific BioSciences (PacBio), 110, 135
Padre Jon. *See* Cortina, Jon de
Page Act of 1875, 212t
paleogenomic studies, 345
palm prints, 214
Panama, 38
Parabon NanoLabs, 84–85, 119–20, 350
ParaDNA, 114–15
Parker, Gerald, 92
Parsons, Tom, 314
partial matches, 3, 4–5, 6b, 337–38, 364
passports, 214
paternity misattribution, 230, 247–50n.6, 317, 336
pattern evidence *vs.* DNA, 346–47
Patton, Robert N., Jr , 97
Pavón de Agular, Elsa, 156–58
Payne, Christopher, 78
PCAST. *See* President's Council of Advisors on Science and Technology
PCOME. *See* Pima County Office of Medical Examiner
PCR. *See* polymerase chain reaction
PCR-STR method. *See* polymerase chain reaction-short tandem repeat method
Pegoraro, Susana, 162–63
Penchaszadeh, Victor, 154, 155
Pennsylvania, 4
Pennsylvania v. Pestinikas, 15, 16–17, 19–21, 27, 32
People's Revolutionary Army (Argentina), 149
People v. Austin, 304–5
People v. Bailey, 36–37
People v. Barker, 297–98n.6
People v. Bullard-Daniel, 302–3n.18
People v. Buza, 73–74, 305
People v. Castro, 2–3, 21–22, 37
People v. Chubbs, 302–3
People v. Collins, 298–99n.11
People v. Dominguez, 302–3

People v. Garcia, 298–99n.9
People v. Holtzer, 298n.7
People v. Klinger, 298n.7
People v. Lazarus, 298–99n.9
People v. Lopez, 305
People v. Megnath, 298–99n.9
People v. Pike, 300–1n.17
People v. Smith, 296–97n.1, 300–1n.16
People v. Stevey, 297–98n.4, 298n.7
People v. Stoecker, 297–98n.6
People v. Tunis, 297–98n.5
People v. Wesley, 36–37
People v. Wood, 297–98n.5
People v. Zapata, 297–98n.4
Pericles, 196
perjury, 39
Perón, Isabel. *See* Martínez de Perón, María Estela "Isabel"
Perón, Juan Domingo, 149, 150
persons of interest (POI), 58, 62–63, 364
Peru, 4–5, 38, 168, 272, 367–68
Pestinikas, Walter and Helen, 19–21. See also *Pennsylvania v. Pestinikas*
PG. *See* probabilistic genotyping
PGS. *See* probabilistic genotyping software
phenotyping, 83, 371
  composite sketch created using, 307f
  MPS/NGS and, 106, 110–11, 112, 349–50, 352
  potential admissibility challenges, 10, 306–7, 366
Phillips, Chris, 201–2n.3
Phillips v. State, 298–99n.9, 300–1n.14
photo arrays, 45
Physicians for Human Rights (PHR), 178, 315–16, 367, 368
Pima County Office of Medical Examiner (PCOME), 245–47, 275–76, 280, 281, 284
Pitchfork, Colin, 19. *See also Colin Pitchfork* case
Plan Sistemático, 162n.6
Plaza Libertad, 182
PMI. *See* postmortem interval, estimation of
Poblete, José Liborio, 160–61
Poblete Hlaczik, Claudia, 160–61, 162
Podrinje Identification Project, 316, 317, 322–23
POI. *See* persons of interest; proof of identity
Poland, 38
polio, 135–36
polygraph tests, 292–93
PolyMarker test, 27–29, 28f

polymerase. *See* DNA polymerase
polymerase chain reaction (PCR), x, 3, 6b, 6–7, 105
  advantages of, 22–23, 27
  all forensic DNA analysis based on, 17, 363
  defined, 2
  description of technique, 17n.3
  Dotson case and, 24
  evolution of, 27–29
  human remains identification and, 314
  introduction of technique, 2, 17
  microbial forensics and, 133
  mixtures analyzed with, 24–25
  Pestinikas case and, 19–21
  RFLP compared with, 17, 27–29
  RFLP replaced by, 23
  in targeted library preparation, 108–9
  VNTRs and, 29f, 29–31
polymerase chain reaction-short tandem repeat (PCR-STR) method, 291, 295–97, 363–64
polymorphism, 16n.1, 16–17. *See also* single nucleotide polymorphisms
Popović, Vujadin, 313–14
Popper, Karl, 293
Population Registry of Forensic DNA of El Salvador, 189–90
postmortem forensic microbiology, 138–40
postmortem interval (PMI), estimation of, 138–39
power of discrimination (Pd), 27–29
PowerPlex, 30–31, 114–15, 203, 295–96
PowerSeq Auto/Mito/Y System, 111–12
PowerSeq 46GY System, 111–12
Precision ID GlobalFiler MPS STR Panel, 111
President's Council of Advisors on Science and Technology (PCAST), 302, 349
  on forensic reliability, 346–47
  on mixtures, 347–49
Presumed Innocent (Turow), xi
Prevention Through Deterrence, 259–60
primary identifiers, 204–5
primers, PCR, 17n.3, 29, 30, 108
Priority 1 refugees, 220n.3
Priority 2 refugees, 220n.3
Priority 3 refugees, 220–21, 220n.3, 223, 229–30
privacy, 4, 9, 11, 364, 366
  DNA data banks/databases and, 87–88, 89–92
  human remains identification and, 247–50, 251–52, 336
  major issues, 335–37

migrant tracking and, 223–24, 225, 227–28
MPS/NGS and, 112, 114, 336
Privacy Act of 1974, 227–28, 251–52, 261–62
Privacy Framework, 336–37
probabilistic genotyping (PG), 50, 61–63, 105–6, 349, 365
probabilistic genotyping software (PGS), 301–2, 347–48
probability-based statistical analysis, 7
probability of inclusion (PI), 50. *See also* combined probability of inclusion
probe capture strategy, 109, 111
Pro-Búsqueda, 8–9, 168, 178, 179–80, 181–82, 186, 187, 188, 367–68
  current activities of, 183–86
  establishment of, 174
  first forensic geneticist of (*see* Vásquez, Patricia)
  government's failure to assist, 181–82
  living legacy of, 189–90
  Serrano-Cruz case and, 179–80n.3
product rule, 18, 26–27
ProFiler/CoFiler, 30–31, 295–96
Promega Corporation, 30–31, 52, 111–12, 119–20, 203, 295–96
proof of identity (POI), 211–16, 225–26, 228–29
Proposition 69 (California), 73–74
"prosecutor's fallacy," 296–97, 296–97n.3
Prosecutor v. Radovan Karadžić, 198
proxy searching. *See* familial searching; genealogics
Prüm Convention, 332
Public Law No. 109-162, 72–73
public policy, 11, 346–47
Puerto Rico, 38, 72–73, 93
Puga, Teodoro, 157

Qiagen, 52, 111–12, 201–2
QiaSeq, 201–2
Quality Assurance Standards for Forensic DNA Testing Laboratories, 61

rabies, 135–36
race
  the death penalty and, 35
  as a social construct, 228–29
  wrongful convictions based on, 44
Race and Wrongful Convictions in the United States (study), 44
Rachell, Ricardo, 78
Rader, Dennis (BTK killer), 82
Radical Party (Argentina), 155

Radio Venceremos, 175
Rae-Venter, Barbara, 84
Ramos, Maria Magdalena, 178
RAND (Research and Development Corporation), 337–38
random match probability (RMP), 17–18, 26–27, 30–31, 363–64
  admissibility of results, 296–97
  Castro case and, 21
  CPI compared with, 300
  defined, 17–18
  DNA Wars and, 22–23
  mixtures and, 49–50, 52–54
  MPS/NGS and, 110–11
rape, microbial forensics and, 137–38
Rapid DNA, 5, 7–8, 105–6, 306–7, 365, 366
  described, 114–18
  human remains identification and, 116, 283–84
  migrant tracking and, 115, 217, 224
Rapid DNA Act of 2017, 5, 97–98, 116–17, 212*t*, 217
Rapid DNA Program Office, 97–98
RapidHit 200, 114–15, 116
RapidHIT 200 (upgrade), 114–15
rapidly mutating Y STRs, 119
RDP (Ribosomal Database Project), 134
Reagan, Ronald, 175
recidivism reduction, 95–96
Red Cross, 173–74, 177, 196, 248*t*, 253–54, 270, 315, 323
Red Inocente, 38
Regeneron, 77
Reid technique, 44–45
Reineke, Robin, 276–77, 282, 284
relationship testing, 216
relative fluorescent units (RFUs), 49, 53*f*, 57*f*, 57, 59–61, 60*f*
research
  advancing forensic science through, 349–51
  data banks for, 76–77, 91
Research and Development Corporation (RAND), 337–38
Resolution 8 of 67th General Assembly, 70–71
"Response to the Forensic Genetics Policy Initiative's Report, A" (Samuel), 352
restriction endonuclease, 16–17n 2
restriction fragment length polymorphism (RFLP), 3, 17, 363
  Andrews case and, 20–21
  Castro case and, 21–22
  challenges to, 2–3, 21–22

restriction fragment length polymorphism (RFLP) (*Cont.*)
  defined, 2
  description of technique, 16–17, 16–17n.4
  DNA Wars and, 22–23
  Dotson case and, 24
  introduction of technique in courts, 2
  PCR compared with, 17, 27–29
  PCR replacement of, 23
  Pestinikas case and, 19–20
  Pitchfork case and, 18–19
  VNTRs and, 29, 105
reunitingfamilies.org, 253n.14
RFLP. *See* restriction fragment length polymorphism
RFUs. *See* relative fluorescent units
Rhode Island, 112
Ribosomal Database Project (RDP), 134
Richmond DNA lab (California DOJ), 180–81, 183
Rio Grande valley, 238–39
Ríos, Alcira, 162
RMP. *See* random match probability
Roa, Buscarita, 161
Roberts, Richard, 21
Rogers, Curtis, 84
Roman, Homero, 259
Romas, xi–xii
Romero, Oscar, 174–75, 182
Rome Statute, 197
Rowe, Raymond, 342–43
Royal Society, 371
Rwanda, 195, 278
Ryan, George, x–xi

Sábato, Ernesto, 150–51n.8, 155–56, 158
Sacred Heart Cemetery, 241–42, 242f, 244, 257–58
Saint-Jean, Ibérico Manuel, 150
Samuel, G., 352
Sancier v. Comm'r of Correction, 303n.19
Sanger, Fred, 107–8n.2, 134–35
Sanger sequencing method, 107–8, 107–8n.2, 134–35
San people, 345
Sarajevo, 314–16
sarin, 130–32
SARS, 135–36
Scalia, Antonin, 74, 117, 118
Scheck, Barry, 37
Schneckloth v. Bustamonte, 294–95
School of the Americas, 175
Science, 16–17
Scientific American, 282

scientific reviews, 346–47
Scientific Working Group on DNA Analysis Methods (SWGDAM), 3, 54, 61, 62
  establishment of, 22
  mixture interpretation guidelines, 300
  Rapid DNA and, 116–17
SDIS, 70, 82
search warrants, 89–90, 294–95, 341–42
secondary DNA transfer, 23, 303
Sensabaugh, George, 181
Sentencing Guidelines, US, 292
September 11 attacks, 136–37, 209, 210–11, 264, 369. *See also* World Trade Center attacks (2001)
sequential photo arrays, 45
Serbia, 204
Serrano Cruz, Erlinda, 179–80n.3
Serrano Cruz, Ernestina, 179–80n.3
Servini de Cubría, María, 162
Sfiligoy, Carlos, 153
Sfiligoy, Inés, 153
Shaler, Robert, 319–20, 323–24
Sheindlin, Gerald, 2–3
short tandem repeats (STRs), 6b, 30–31, 71–72, 335
  BNDG analysis of, 163–64, 165
  commercial development of, 30
  degraded DNA and, 31
  familial searching and, 81–82
  genealogics and, 343
  human remains identification and, 194, 201, 203, 240, 262–63, 283–84
  mixtures and, 24–25, 31, 49, 51, 52–54, 58, 107
  MPS/NGS and, 106, 109–10
  PCR and (*see* polymerase chain reaction-short tandem repeat method)
  Rapid DNA and (*see* Rapid DNA)
  WTC victim identification and, 319–20, 321
  X-chromosome, 30–31, 52, 106
  Y-chromosome (*see* Y chromosome short tandem repeats)
shotgun libraries, 108, 109
sickle cell anemia, 17
Simón, Julio Héctor "El Turco Julián," 161
SING (Summer internship for INdigenous peoples in Genomics) Consortium, 345
single-locus probes, 26–27
Single Molecule Real Time (SMRT) system, 110
single nucleotide polymorphisms (SNPs)
  array profiles, 118–19
  degraded DNA and, 31, 105

familial searching and, 80, 118–19
genealogics and, 83–84, 86–87, 343
human remains identification and, 201–2
mixtures and, 31
MPS/NGS and, 106, 107–8, 110–12
WTC victim identification and, 321
single-source DNA samples, 18, 49–51, 52–54, 53f, 55, 56
six-pack photo arrays, 45
16S rRNA gene, 134
Sixth Amendment, 82, 294
60 Minutes (television program), 41
Smith, Gordon, 38
Smith, Lindsay, 272, 312
Smith v. Maryland, 89
SMRT (Single Molecule Real Time) system, 110
SnapShot6, 350
Snow, Clyde, 152, 152n.3, 156, 272, 312, 366, 367
SNPs. *See* single nucleotide polymorphisms
Snyder, John, 321–22
Somalia, 225, 228–29
SOS Children's Village, 177, 178
Sotomayor, Sonia, 74, 305
South Africa, 38, 204
South America, 238
Southeast Asian tsunami, 204
Southern blotting, 16–17n.2
South Texas Human Rights Center, 241–42, 248t, 258, 259
Spain, 264
Spanish Civil War, 270–71
Spradley, Kate, 246f
Sprenkels, Ralph, 174, 176–78, 179, 183
Spring Co. v. Edgar, 292–93
Srebrenica, 199, 203, 311–18
  accountability demanded for massacre, 315–16
  DNA-led identifications in, 313–18
  ID efforts secondary to war crime investigations in, 316
  lag in identification efforts, 315, 318, 325–26
  notification options for relatives, 317–18
  number of victims identified, 314
  number of victims in, 313
  trust issue in, 273–75
  as a UN safe area, 194, 198, 273–74, 313
  WTC identification process compared with, 311–13, 318, 324–26
Srebrenica Identification Project, 315–16
Srebrenica-Potočari Memorial and Cemetery, 274, 315, 317, 318

Starr County, Texas, 241–42
State Department (DOS), US, 182, 220–21, 225, 226
statelessness, 225–27
state level DNA system. *See* SDIS
State v. Bander, 297–98n.4
State v. Bigger, 300–1n.14
State v. Brochu, 298n.3
State v. Calleia, 297–98n.4
State v. Council, 298n.8
State v. Freeman, 303n.19
State v. Griffin, 298n.8
State v. Haughey, 300–1n.15
State v. Hummert, 296–97n.2
State v. Maestas, 297–98n.5
State v. Pappas, 298n.7
State v. Roman Nose, 296–97n.1
State v. Tester, 296–97n.2
statistical independence (of genetic markers), 22, 27, 27n.6, 30–31
stochastic threshold, 54–55, 57f, 57–58, 59–61, 60f, 62, 63, 300
stochastic variation, 52
Stone, Errol, x–xi
storage of DNA. *See* DNA storage and data security
Stover, Eric, xi–xii, 155, 156, 178–79, 180, 182, 187
Straub, Peter, 76
Strengthening Forensic Science in the United States: A Path Forward (report), 39–40, 346, 359–60
Strickland v. Washington, 43
STRmix, 62–63, 301–3, 302–3n.18
STRs. *See* short tandem repeats
stutter, 31, 52–54, 53f
stutter bands, 30
stutter peaks, 55, 56, 62
Sudan, 225
Summer internship for INdigenous peoples in Genomics (SING) Consortium, 345
Support for Forensic Science Research (report), 360–61
Supreme Court, California, 73–74, 305
Supreme Court, New York, 2–3
Supreme Court, US
  on admissibility of scientific evidence, 37, 293, 304–5
  on attorney competence, 43
  on the death penalty, 35–36
  on DNA collection upon arrest, 74, 117–18, 305
  on suspicion-less genetic searches, 350–51
  on third-party doctrine, 89–90, 306

Swedish National Board of Forensic
    Medicine, 201–2
SWGDAM. *See* Scientific Working Group on
    DNA Analysis Methods
Sydnor, Charles E., III, 98
Sykes, Deborah, 79, 338–40
Syria, 225, 253

Taiwan, 38, 332
Talbott, William Earl, II, 84, 86
targeted libraries, 108–9
terrorism, 209. *See also* bioterrorism
testimonial hearsay, 294
Testimony (Turow), xi–xii
Texas, 5, 112, 221, 238–39, 240–44,
    254–55, 337–38
  Brooks County (*see* Brooks
    County, Texas)
  Cameron County, 241
  Hidalgo County, 241–42
  Jim Hogg County, 241–42
  Starr County, 241–42
  Webb County, 241–43
  Willacy County, 241–42
Texas Criminal Code of Procedures
    (TCCP), 243
Texas District Attorney's Association
    (TDAA), 348
Texas State University, 241, 244, 257–58
Thailand, 226–27, 264
ThermoFisher Scientific, 30, 52, 106, 109–
    10, 111, 119–20
third-party doctrine, 89–90, 306
Thomas, Anwar, 113
Thompson, Jennifer, 41
Thucydides, 196
Tillmar, Andreas, 201–2n.3
Tinsley, Kenneth, 26
toolmark comparisons, 346, 347
trace DNA. *See* low copy number DNA
transfer of DNA. *See* DNA transfer
Transportation Security Administration
    (TSA), 214
Triple A (Argentine Anticommunist
    Alliance), 150
TrueAllele, 62–63, 291, 301–3,
    302–3n.18, 305
Trump, Donald, 97–98, 210–11, 221, 223,
    225, 255–56, 260, 343–44, 361
TruSeq, 111–12
TSA (Transportation Security
    Administration), 214
tsunami, Southeast Asian, 204
Turkey, 197–98
Turow, Scott, ix–xii

Tuscon, Arizona, 245–47
Tuzla, 315–16
23andMe, 3–4, 77, 90, 118–19, 201, 334–
    35, 340, 341
twins. *See* identical twins
typhoon Frank, 204
Tzanck, Arnault, 154

Uganda, 10, 277–80, 281
Uighur people, 225, 345
UK Biobank, 77
Uniform Code of Military Justice, 76
Unitarian Universalist Church, 259–60
United Airlines (UA) flight 175, 319
United Kingdom, 2, 3, 6–7, 38, 62, 334,
    336–40, 363
  DNA data banks/databases of, 70, 71*t*,
    72, 77, 79–80
  first criminal case using DNA
    technology, 15
  migration in, 227, 228–29
United Nations (UN), 197–98
  El Salvador and, 176–77, 189
  safe areas of, 194, 198, 273–74, 313
United Nations High Commissioner
    for Refugees (UNHCR), 195,
    225–26, 227–28
United Nations Security Council
    Resolution 827, 198
United Nations Truth Commission,
    176–77, 189
United States v. Barton, 298–99n.10
United States v. Beverly, 298n.8
United States v. Coleman, 298n.8
United States v. Davis, 296–97n.2
United States v. Kincade, 73
United States v. McCluskey, 298–99n.11
United States v. Miller, 89
United States v. Morgan, 298–99n.10
United States v. Silva, 296–97n.1
United States v. Sleugh, 298–99n.10
United States v. Williams, 300–1n.16
United States v. Yee, 296–97
*Universal Declaration of Human
    Rights,* 195
University of California Berkeley Human
    Rights Center, 180, 181–82, 184,
    185, 367
University of Edinburgh, 221
University of Indianapolis, 244
University of Milan, 282
University of North Texas, 260–61, 280
  Center for Human Identification,
    244*f*, 348
  Health Science Center, 243

University of Santiago de Compostela, 201–2
University of Tennessee, Knoxville, 278–79
USCIS. *See* Citizenship and Immigration Services, US
US-Mexico border, 6, 9–10, 200, 238–64, 270–71, 280, 281, 368
  attempts to unify stakeholders, 254–56
  disposition of unidentified remains, 241–44
  distrust among stakeholders, 256–57
  family separations at, 221–24
  homicide cases at, 252, 252n.13
  inhumane status quo at, 259–63
  international organizations assisting in identification, 253–54
  landowner permission required for searches, 239, 239nn.1–2
  non-DNA data sources for remains identification, 245–47
  nonprofit organizations assisting in identification, 252–53
  resources for remains identification, 247–54
  roles of stakeholders, 247–54, 248t
  still unidentified and missing, 258–59
  successful case of DNA-based identification, 257–58
  unaccompanied minors at, 221–24, 222f
  US-based law enforcement DNA efforts, 250–52
US Missing Persons Index of Combined DNA Index System (CODIS), 248t
US Visitor and Immigration Status Indicator Technology program (US-VISIT), 214–16, 215t, 217

Vallejo-Nágera, Antonio, 151
variable number tandem repeats (VNTRs), 26–27, 105, 363
  defined, 6b, 16–17
  PCR and, 29f, 29–31
Vasquez, David, 42–43
Vásquez, Patricia, 182–83, 189
Vázquez, Evelyn, 162–63
Vázquez, Policarpo, 162–63
Vernon, Edward, 39
Verogen, Inc., 83, 84–85, 87–88, 89–90, 109–10, 111, 119–20
Videla, Jorge Rafael, 150, 167–68
Virginia, 94
Virginia v. Washington, 24–26
VISible Attributes Through Genomics (VISAGE), 350

Vlacovsky, Joseph, 76
VNTRs. *See* variable number tandem repeats
voice recognition, 214
Vojska Republike Srpske (VRS), 313
Vollen, Laurie, 315–16
VRS. *See* Vojska Republike Srpske

Wagner v. State, 298n.8
Wambaugh, Joseph, 18–19n.4
war crimes, xi–xii, 194, 195, 196, 199, 270–71, 316, 368
Warren, Elizabeth, 343–44
Washington, Earl, Jr., 24–26, 64
Washington State, 4, 37
Webb County, Texas 241–43
Webb County Medical Examiner's Office, 242–43
Weinstein, Barry, 25
Weir, Bruce, 22
Wellcome Trust, 77
Wesley-Bailey hearings, 36–37
western Balkans, 6, 9, 193–94, 203–4, 273–74, 275n.9. *See also individual countries*
West Nile fever, 135–36
WGS. *See* whole genome sequence
White, Byron, 42
White, Lynette, 79–80
Who Is Dayani Cristal (documentary), 257
whole genome sequence (WGS), 108, 113–14
Wilder, Douglas, 25
Willacy County, Texas, 241–42
Williams, Rebecca, 25, 26
Williams v. Illinois, 304–5
Wisconsin, 37
witness protection program, 333–34
World Trade Center (WTC) attacks (2001), 10–11, 136–37
  condition of bodies, 319–20, 322–23, 322–23n.4
  described, 319
  DNA-led identification of victims, 318–24
  notification options for relatives, 322–23
  number of casualties, 319
  official list of missing persons, 320
  Srebrenica identifications compared with, 311–13, 318, 324–26
  traditional identification methods, 321
Worthington, Christa, 96–97

Wright, Anthony, 13
wrongful convictions, 24, 351. *See also* exonerations
   compensation for, 38
   estimated percentage of, 35–36
   race-based, 44
WTC attacks. *See* World Trade Center attacks (2001)
Wyoming, 112

X chromosome short tandem repeats (X-STRs), 30–31, 52, 106
Xinjiang, China, 225, 345

Yale University, 201–2

Y chromosome short tandem repeats (Y-STRs), 30–32, 52, 71, 81, 86, 105, 337–38
   admissibility of test results, 297–98
   MPS/NGS and, 106, 111–12
   rapidly mutating, 119
Yemen, 225
Yoshizawa, Go, 332
Yrigoyen, Hipólito, 149
Yugoslavia, former. *See* former Yugoslavia

Zain, Fred, 40–41
Zamora, Margarita, 183, 190
Zepa, 194, 198
zeromode wave guide (ZMV), 110